SCOPE 22

Effects of Pollutants at the Ecosystem Level

Executive Committee of SCOPE

President: Professor R. O. Slatyer, Department of Environmental Biology, Australian National University, PO Box 475, Canberra ACT 2601, Australia.
Past-President: Professor G. F. White, Institute of Behavioral Science, University of Colorado, Boulder, Colorado 80309, USA.
Vice-President: Professor G. A. Zavarzin, Institute of Microbiology, USSR Academy of Sciences, Profsojusnaja 7a, 117312 Moscow, USSR.
Secretary-General: Professor J.W.M.La Rivière, International Institute for Hydraulic and Environmental Engineering, Oude Delft 95, PO Box 3015, 2601 DA Delft, The Netherlands.
Treasurer: Sir Frederick Warner, FRS, 11 Spring Chase, Brightlingsea, Essex CO7 0JR, UK.

Members

Dr M. A. Ayyad, Botany Department, Faculty of Sciences, University of Alexandria, Moharran Bay, Alexandria, Egypt.
Professor M. J. Kostrowicki, Institute of Geography, Polish Academy of Sciences, Warsaw, Poland.
Professor S. Krishnaswamy, School of Biological Sciences, Madurai Kamaraj University, Madurai 625 021, India.
Professor T. Rosswall, Department of Microbiology, Swedish University of Agricultural Sciences, S-75007 Uppsala, Sweden.
Professor Yasuo Shimazu, Department of Earth Science, Faculty of Science, Nagoya University, Furo-cho, Chikusa, Nagoya, Aichi 464, Japan.

Editor-in-Chief

Professor R. E. Munn, Institute for Environmental Studies, University of Toronto, Toronto, Canada, M5S 1A4.

SCOPE 22

Effects of Pollutants at the Ecosystem Level

Edited by
Patrick J. Sheehan
Donald R. Miller
Gordon C. Butler
Division of Biological Sciences
National Research Council of Canada
Ottawa, Canada

and

Philippe Bourdeau
Directorate General for Science, Research and Development
Commission of the European Communities
Brussels, Belgium

With the editorial assistance of
Joanne M. Ridgeway
Division of Biological Sciences
National Research Council of Canada
Ottawa, Canada

Published on behalf of the
Scientific Committee on Problems of the Environment (SCOPE)
of the
International Council of Scientific Unions (ICSU))
by
JOHN WILEY AND SONS
Chichester · New York · Brisbane · Toronto · Singapore

Copyright © 1984 by the
Scientific Committee on Problems of the Environment (SCOPE)

All rights reserved.

No part of this book may be reproduced by any means, nor transmitted, nor translated into a machine language without the written permission of the copyright holder.

Library of Congress Cataloging in Publication Data:

Main entry under title:

Effects of pollutants at the ecosystem level (SCOPE 22)
 Includes index.
 1. Pollution—Environmental aspects. I. Sheehan, Patrick J.
QH545.A1E33   1984   574.5′222   83-7021

ISBN 0 471 90204 7

British Library Cataloguing in Publication Data:

Effects of pollutants at the ecosystem level.—
 (SCOPE; 22)
 1. Ecology 2. Pollution
 I. Sheehan, Patrick J.   II. International Council
 of Scientific Unions. *Scientific Committee on
 Problems of the Environment.*   III. Series
 574.5′222   QH545.A1

ISBN 0 471 90207 7

Typesetting by Thomson Press (India) Limited, New Delhi
and printed by Page Bros. (Norwich) Limited

International Council of Scientific Unions (ICSU)
Scientific Committee on Problems of the Environment (SCOPE)

SCOPE is one of a number of committees established by a non-governmental group of scientific organizations, the International Council of Scientific Unions (ICSU). The membership of ICSU includes representatives from 68 National Academies of Science, 18 International Unions and 12 other bodies called Scientific Associates. To cover multidisciplinary activities which include the interests of several unions, ICSU has established 10 scientific committees, of which SCOPE, founded in 1969, is one. Currently, representatives of 34 member countries and 15 Unions and Scientific Committees participate in the work of SCOPE, which directs particular attention to the needs of developing countries.

The mandate of SCOPE is to assemble, review and assess the information available on man-made environmental changes and the effects of these changes on man; to assess and evaluate the methodologies of measurement of environmental parameters; to provide an intelligence service on current research; and by the recruitment of the best available scientific information and constructive thinking to establish itself as a corpus of informed advice for the benefit of centres of fundamental research and of organizations and agencies operationally engaged in studies of the environment.

SCOPE is governed by a General Assembly, which meets every three years. Between such meetings its activities are directed by the Executive Committee.

R. E. Munn
Editor-in-Chief
SCOPE Publications

Executive Secretary: Ms V. Plocq

Secretariat: 51 Bld de Montmorency
75016 PARIS

SCOPE 1:   Global Environmental Monitoring 1971, 68pp (out of print)
SCOPE 2:   Man-Made Lakes as Modified Ecosystems, 1972, 76pp
SCOPE 3:   Global Environmental Monitoring System (GEMS): Action Plan for Phase 1, 1973, 132pp
SCOPE 4:   Environmental Sciences in Developing Countries, 1974, 72pp

Environment and Development, proceedings of SCOPE/UNEP Symposium on Environmental Sciences in Developing Countries, Nairobi, February 11–23, 1974, 418pp

SCOPE 5:   Environmental Impact Assessment: Principles and Procedures, Second Edition, 1979, 208pp
SCOPE 6:   Environmental Pollutants: Selected Analytical Methods, 1975, 277pp (available from Butterworth & Co. (Publishers) Ltd., Sevenoaks, Kent, England)
SCOPE 7:   Nitrogen, Phosphorus, and Sulphur: Global Cycles, 1975, 192pp (available from Dr Thomas Rosswall, Swedish Natural Science Research Council, Stockholm, Sweden)
SCOPE 8:   Risk Assessment of Environmental Hazard, 1978, 132pp
SCOPE 9:   Simulation Modelling of Environmental Problems, 1978, 128pp
SCOPE 10:   Environmental Issues, 1977, 242pp
SCOPE 11:   Shelter Provision in Developing Countries, 1978, 112pp
SCOPE 12:   Principles of Ecotoxicology, 1978, 372pp
SCOPE 13:   The Global Carbon Cycle, 1979, 491pp
SCOPE 14:   Saharan Dust: Mobilization, Transport, Deposition, 1979, 320pp
SCOPE 15:   Environmental Risk Assessment, 1980, 176pp
SCOPE 16:   Carbon Cycle Modelling, 1981, 404pp
SCOPE 17:   Some Perspectives of the Major Biogeochemical Cycles, 1981, 175pp
SCOPE 18:   The Role of Fire in Northern Circumpolar Ecosystems, 1983, 344pp
SCOPE 19:   The Global Biogeochemical Sulphur Cycle, 1983, 495pp
SCOPE 20:   Methods for Assessing the Effects of Chemicals on Reproductive Functions, 1983, 568pp

SCOPE 21: The Major Biogeochemical Cycles and Their Interactions, 1983, 554pp

SCOPE 22: Effects of Pollutants at the Ecosystem Level, 1984

SCOPE 23: The Role of Terrestrial Vegetation in the Global Carbon Cycle: Measurement by Remote Sensing, 1984

Funds to meet SCOPE expenses are provided by contributions from SCOPE National Committees, an annual subvention from ICSU (and through ICSU, from UNESCO), an annual subvention from the French Ministere de l'Environnement et du Cadre de Vie, contracts with U. N. Bodies, particularly UNEP, and grants from Foundations and industrial enterprises.

Contents

Preface xi

List of Contributors xiii

Chapter 1   Introduction 1
             G. C. Butler

PART I   THEORY AND SURVEY

Chapter 2   Chemicals in the Environment 7
             D. R. Miller

Chapter 3   Distinguishing Ecotoxic Effects 15
             D. R. Miller

Chapter 4   Effects on Individuals and Populations 23
             P. J. Sheehan

Chapter 5   Effects on Community and Ecosystem Structure and Dynamics 51
             P. J. Sheehan

Chapter 6   Functional Changes in the Ecosystem 101
             P. J. Sheehan

             Reference List for Chapters 2 to 6 147
             J. M. Ridgeway

PART II   CASE STUDIES

Chapter 7   Introduction to Case Studies 193
Case 7.1   Thames Estuary: Pollution and Recovery 195
             M. J. Andrews

Case 7.2	Clearwater Lake: Study of an Acidified Lake Ecosystem P. Stokes	229
Case 7.3	Comparison of Gradient Studies in Heavy-Metal-Polluted Streams P. J. Sheehan and R. W. Winner	255
Case 7.4	Ecological Effects of Hydroelectric Development in Northern Manitoba, Canada: The Churchill-Nelson River Diversion R. A. Bodaly, D. M. Rosenberg, M. N. Gaboury, R. E. Hecky, R. W. Newbury and K. Patalas	273
Case 7.5	Accidental Oil Spills: Biological and Ecological Consequences of Accidents in French Waters on Commercially Exploitable Living Marine Resources C. Maurin	311
Case 7.6	Impact of Airborne Metal Contamination on a Deciduous Woodland System M. Hutton	365
Case 7.7	A Case Study of the Use of Fenitrothion in New Brunswick: The Evolution of an Ordered Approach to Ecological Monitoring M. F. Mitchell and J. R. Roberts	377
Case 7.8	Rehabilitation of Mine Tailings: A Case of Complete Ecosystem Reconstruction and Revegetation of Industrially Stressed Lands in the Sudbury Area, Ontario, Canada T. H. Peters	403

PART III CONCLUSIONS

Chapter 8	Conclusions and Recommendations P. J. Sheehan, D. R. Miller, G. C. Butler and Ph. Bourdeau	425
Subject Index		429

Preface

In its concern with major environmental problems, SCOPE was bound to consider with particular attention the impact of pollution, and especially chemical pollution, on human health as well as on nonhuman targets. Man's activities have resulted in an increasing circulation of naturally occurring substances (heavy metals, sulphur, nitrogen, etc.) and the massive introduction of synthetic chemicals (xenobiotics) in the environment as well as of radionuclides. The fate and transformation of pollutants, from source to target, and their effects on the environment and on the living things in it, constitute the subject matter of ecotoxicology, which SCOPE selected as one of its main project areas.

The first activity in the Ecotoxicology Project resulted in the publication of SCOPE 12, *Principles of Ecotoxicology* (1978), prepared by a small scientific committee and edited by Gordon Butler. An attempt was made to bring together basic concepts and methods of toxicology, the essentials of pollutant transfer and transformation in the environment, and the quantitative assessment of effects of pollutants on species, populations, communities and ecosystems in the terrestrial and aquatic environment.

The response of whole ecosystems was found to be least understood and a need for further exploration and elaboration of this problem was identified by the committee. Another subject deemed as requiring additional work was the relevance of tests to predict the environmental behaviour of chemicals. Both suggestions were accepted by the SCOPE Executive. A scientific advisory committee was appointed in November 1979 to organize the two follow-up studies.

SCOPE Scientific Advisory Committee on Ecotoxicology
- Ph. Bourdeau, Brussels, Belgium (Chairman)
- G. C. Butler, Ottawa, Canada
- F. Korte, Neuherberg-Munich, West Germany
- C. R. Krishna-Murti, Lucknow, India
- D. R. Miller, Ottawa, Canada
- R. Truhaut, Paris, France

A working group was constituted for each study to prepare the reports. The present report, entitled *Effects of Pollutants at the Ecosystem Level*, has been prepared by the first group. The second report is in preparation.

As was done for SCOPE 12, experts were invited to draft chapters of an outline approved by the advisory committee. It was felt by the editors that a presentation of case studies concerning particular ecosystems which had been severely damaged or were on their way to recovery, should be included in the volume. An invitation was issued to SCOPE national committees to contribute such case studies from their respective countries, and individuals were approached for the same purpose.

The manuscripts were examined and harmonized at meetings of the writers and an editorial group. The report is very much the result of a collective effort. It focuses on the response of whole ecosystems to man-imposed stress, in terms of ecosystem structure and function, rather than primarily on damage to individuals or populations resulting from chemical insult. Ecosystem recovery is also considered inasmuch as it can throw light on the deterioration process, since it may progress through similar steps, although in reverse order.

Thanks are due primarily to the authors who enthusiastically volunteered chapters and case studies. Financial support for the project was provided by the National Research Council of Canada, the European Economic Community, the Andrew Mellon Foundation and the Mobil Foundation, Inc., and is gratefully acknowledged. The hospitality extended by NRC of Canada for a meeting of writers and editors on 30 March 1981, and by the Monitoring and Assessment Research Centre, Chelsea College, University of London for a meeting on September 2–4, 1981, is deeply appreciated.

List of Contributors

M. J. ANDREWS	Thames Water Authority Metropolitan Pollution Control Northumberland House Modgen S. T. Works Isleworth, Middlesex TW7 7LP England
R. A. BODALY	Canada Department of Fisheries and Oceans Freshwater Institute 501 University Crescent Winnipeg, Manitoba, Canada R3T 2N6
Ph. BOURDEAU	Directorate General for Science, Research and Development Commission of the European Communities B-1049, Brussels, Belgium
G. C. BUTLER	Division of Biological Sciences National Research Council of Canada Ottawa, Ontario, Canada K1A 0R6
M. N. GABOURY	Manitoba Department of Natural Resources Box 40, 1495 St James Street Winnipeg, Manitoba, Canada R3H 0W9
R. E. HECKY	Canada Department of Fisheries and Oceans Freshwater Institute 501 University Crescent Winnipeg, Manitoba, Canada R3T 2N6

LIST OF CONTRIBUTORS

M. HUTTON   Monitoring and Assessment Research Centre
The Octagon Building
459A Fulham Road
London SW10 0QX
England

C. MAURIN   French Fisheries Research Institute
(I.S.T.P.M.)
Nantes, France

D. R. MILLER   Division of Biological Sciences
National Research Council of Canada
Ottawa, Ontario, Canada
K1A 0R6

M. F. MITCHELL   Environmental Secretariat
Division of Biological Sciences
National Research Council of Canada
Ottawa, Ontario, Canada
K1A 0R6

R. W. NEWBURY   Canada Department of Fisheries and Oceans
Freshwater Institute
501 University Crescent
Winnipeg, Manitoba, Canada
R3T 2N6

K. PATALAS   Canada Department of Fisheries and Oceans
Freshwater Institute
501 University Crescent
Winnipeg, Manitoba, Canada
R3T 2N6

T. H. PETERS   Agriculture Department
INCO Metals Company
Copper Cliff, Ontario, Canada.

J. M. RIDGEWAY   Division of Biological Sciences
National Research Council of Canada
Ottawa, Ontario, Canada
K1A 0R6

J. R. ROBERTS   Environmental Secretariat
Division of Biological Sciences
National Research Council of Canada
Ottawa, Ontario, Canada
K1A 0R6

LIST OF CONTRIBUTORS

D. M. Rosenberg   Canada Department of Fisheries and Oceans
                   Freshwater Institute
                   501 University Crescent
                   Winnipeg, Manitoba, Canada
                   R3T 2N6

P. J. Sheehan   Division of Biological Sciences
                   National Research Council of Canada
                   Ottawa, Ontario, Canada
                   K1A 0R6

P. Stokes   Institute for Environmental Studies
                   University of Toronto
                   Toronto, Ontario, Canada
                   M5S 1A4

R. W. Winner   Department of Zoology
                   Miami University
                   Oxford, Ohio 45056
                   USA

Effects of Pollutants at the Ecosystem Level
Edited by P. J. Sheehan, D. R. Miller, G. C. Butler and Ph. Bourdeau
© 1984 SCOPE. Published by John Wiley & Sons Ltd

CHAPTER 1
Introduction

GORDON C. BUTLER

Division of Biological Sciences
National Research Council of Canada
Ottawa, Ontario, Canada K1A 0R6

This report is concerned with changes in ecosystems. It is normal for them to be in a constant state of change but the complex interactions among their many components endow them with a measure of short-term stability so that they can withstand brief disturbances. This homeostasis does not, however, prevent long-term changes of an evolutionary type or in response to a sustained cause of harm. Because of this homeostasis the process of damage may begin slowly and subtly and may not be detected until it is well advanced. Such early changes are difficult to detect among the natural variations displayed by all ecosystems. Thus, in a programme of continuous surveillance there is, in addition to the problem of detection, the difficulty of assessment. Are the observed changes temporary and reversible, as are the 'spontaneous' ones, or are they the early stages of more permanent and important damage? The problem of assessment is rendered more difficult by the fact that not all changes, even permanent ones, are for the worse. Ecosystems, like species (or along with species), are subject to slow evolutionary changes by which they may become equally viable and better adapted to changed conditions.

Opinions on whether a change is for the better or worse will vary with the assessor. For example, chemical and thermal pollution of a body of water often results in an increase in productivity and biomass but not everyone regards the resulting eutrophication as an improvement.

Several properties of ecosystems have been considered as indices of welfare or damage in the hope that they might simplify the problem of assessment by providing quick answers. Some of these properties are listed below.

1. The total biomass of a population, a species, a compartment, or an ecosystem.
2. The number of different species present. Indices of diversity have been devised and recommended for assessment.
3. Trends in reproductive success and populations of various species, especially the most sensitive ('sentinel') species.

4. The metabolic activities of an ecosystem such as photosynthesis or the throughput of energy or chemicals.

All these attributes of ecosystems are useful to know but much effort has been wasted (at least by this writer) in trying to decide which are the most definitive indices of ecosystem welfare. Probably the answer is that no change in these or other attributes can be ignored in making a proper assessment. The most useful diagnostic information should include:

1. The history of changes in a number of similar ecosystems including the one under diagnosis.
2. As many as possible numerical descriptors and their 'normal' variation.
3. Any external or new influences, present and predicted, either injurious or restorative in effect.

At different times and with different ecosystems some signs will assume greater importance but in all cases they must be evaluated as a group. With practice the diagnostician will know what is a normal or self-restorable range of variation and what are the danger signs. The appraisal should result in a summary (quantitative if possible) of the present condition and a prediction of future developments. These are needed to guide the administrator on possible interventions and the kind of quality standards to promulgate. The analysis will also provide guidance on how to monitor the ecosystem.

Many of these aspects of the subject of assessing environments were discussed in SCOPE Publication 5 (1975) which dealt largely with the methodologies of environmental impact assessment including modelling and monitoring.

SCOPE Publication 12, *Principles of Ecotoxicology*, which may be considered the progenitor of the present report, devoted 13 chapters to the quantitative assessment of the effects of pollutants on species, populations and communities. Chapter 15 considered 'Ecosystem response to pollution'. It divided the problem into three parts:

1. How polluted is the ecosystem? This can be measured by monitoring important compartments and, with the aid of models, doses to important receptors may be calculated.
2. What are the effects on ecosystems? These are illustrated by ten examples where a variety of pollutants have been observed to act on different ecosystems.
3. How do you assess these effects in terms of ecosystem characteristics?

SCOPE 12 was prepared and published under the guidance of a preparatory committee and in its final report to the SCOPE Executive this committee concluded that the weakest part of the publication was that dealing with the responses of whole ecosystems and that this subject required further study. The Executive decided subsequently to embark on such a study and the present report is the result.

INTRODUCTION

The objective of the report is to provide information to scientific advisors of decision-makers and to scientists studying effects of environmental pollution in ecosystems.

Since an ecosystem consists of 'communities of living organisms together with their habitat (or abiotic environment) and including the interactions among these components', it is described not only by specifying the living and nonliving things it contains but also by giving an account of what goes on in it, for example, energy and material flows as well as productivity. Characteristics and classifications of ecosystems can be, and have been, discussed at length. They provided an indication of the broad range of systems which it is possible to consider. A more limited range will be included in this report, by reference to concrete examples.

The chapters that follow begin with a survey of established results in the published literature of the subject, looking at, first, the distribution of chemicals in the environment, then the difficulty of detecting the effects of pollutants in the presence of large background variations and, finally, the nature of these effects on populations, communities and whole ecosystems. This survey is followed by a collection of case studies, selected as illustrations of the problems encountered in the diagnosis, assessment and subsequent handling of problems at the ecosystem level. The final chapter presents some of the lessons to be learned from these experiences and attempts to identify some priorities for handling such problems in the future.

Part I
Theory and Survey

Effects of Pollutants at the Ecosystem Level
Edited by P. J. Sheehan, D. R. Miller, G. C. Butler and Ph. Bourdeau
© 1984 SCOPE. Published by John Wiley & Sons Ltd

CHAPTER 2
Chemicals in the Environment

DONALD R. MILLER
Division of Biological Sciences
National Research Council of Canada
Ottawa, Ontario, Canada K1A OR6

2.1 Types and Amounts Present.. 7
2.2 Distribution, Transport and Transformation...................... 10
2.3 Uptake by Biota.. 11
2.4 Fate in the Environment.. 12
2.5 Effect on the Physical Environment............................... 13

2.1 TYPES AND AMOUNTS PRESENT

Many figures are quoted for the number of chemical substances in common use today, and for the number of new ones introduced every year. Probably reliable are the values given by Maugh (1978), who arrives at the figure of 60 000 chemicals in use with several thousand additional ones being added annually.

Initially, it might seem that such numbers render impossible the task of identifying priority chemicals as potential environmental hazards, at least on anything but an intuitive basis. However, such attempts have been made (for example, MITRE, 1976) and, although the earlier studies were largely overwhelmed by the data required and ended up using very simple sets of attributes as the basis for priority identification, the experience gained has increased our understanding of what is necessary (Goodman, 1974; Harriss, 1976). Today, large data banks are being organized for precisely this purpose, most notably by the International Registry for Potentially Toxic Chemicals, an activity of UNEP (IRPTC, 1978).

As a rule, it is necessary to know certain quite specific attributes about a chemical substance to predict whether it will be a danger to the environment in general. It must be released in large quantities; it must persist in the ecosystem it enters (or be transformed into an equally or more seriously toxic material); and it must find its way to target organisms by having some kind of affinity for biological materials. Also, of course, it must be in some sense toxic. This chapter surveys briefly what is known about these various steps, and looks at how we can arrive at a useful analysis of which chemicals ought to be of primary concern.

Entries into the environment which have been seen to cause considerable damage can be classified into three categories, depending on their geographical extent (Harriss, 1976): point spills, chronic local releases and widespread releases.

By *point spills* we mean occasions when a significant amount of a chemical has entered an ecosystem at a point (in both space and time) and effects of contamination are expected in a well-defined more or less local area. The assumption is that the substance does not rapidly diffuse away, but remains in the immediate vicinity at a noticeably high concentration (or perhaps moves, but in such a way that levels remain high as it moves).

Such cases would normally occur when large quantities of a substance were being stored, transported or otherwise handled in concentrated form. This kind of situation would probably involve the commercial handling of an industrial chemical or fuel, and would include oil spills, leaks or spills of chlorine or other gases, PCBs, acids and so forth.

The fact is that we have a considerable amount of experience in dealing with such situation; authorities such as Transport Ministries and the Coast Guard or related services have recognized and planned for certain dangers for many years (Transport Canada, 1979). Lists have been drawn up in most countries of those regarded as most hazardous (Jones, 1978). For substances in this category, assessment of toxic effects is a matter of record, and policy is typically directed towards emergency procedures to be followed in case of spills.

Chronic local releases are cases in which discharges have taken place over such periods of time, and in such quantities, that a larger region (for example, a river system, a catchment basin or the landscape downwind of a source) has been contaminated. This type of situation usually results from, or is associated with, a large industrial or municipal source (Holdgate, 1979). The effluent may be a general loading of organic material or a noxious gas, in which case the problem is simply that the ecosystem cannot incorporate such quantities as it gets, or it may be the release of much smaller quantities of much more toxic trace elements. Thus, a single industrial source might produce both types of problem, by releasing organic waste material and at the same time discharging trace amounts of heavy metals. Similarly, an agricultural system might load an ecosystem with organic materials and at the same time release significant quantities of pesticides. Either one, however, would still constitute a case of chronic local release of a damaging substance.

Such cases are the most numerous today, and are rather difficult to further classify. Attempts have been made to identify the kinds of releases most likely to be found in this category, and typically result in the following as the most important:

1. General nutrient discharge, leading to eutrophication.
2. The class of chemical generally referred to (somewhat inaccurately) as heavy metals, and their organic derivatives in particular.

CHEMICALS IN THE ENVIRONMENT

3. Manufactured organics known (and intended) to be toxic, specifically the pesticides.
4. The acidifying gases, oxides of sulphur and nitrogen (MITRE, 1976; SRI, 1977; IRPTC, 1978).

We will consider each of these in the chapters to follow.

The most serious type of entry is a *widespread release*, by which we mean release of a substance in sufficient quantity, and over a wide enough area, that there could result a noticeable pollution of a significant part of the entire earth's surface. If this should happen at a level at which the ecosystem itself was affected, the problem would be grave indeed, particularly so if the substance originated with an activity which was very widespread (so that the sources of pollution are many) and was associated with an activity on which many other activities depend (such as basic energy generation). And this is not to be assumed impossible; there are something like a thousand substances manufactured in such quantities as to be capable of polluting the entire globe (Butler, 1978), and there are many others which are released inadvertently as by-products of widespread operations.

It is important to distinguish those cases in which the ecosystems involved may be affected, as opposed to cases in which all that can be said is that the substance is detectable (sometimes only by extremely delicate instrumentation) in farflung locations. The latter, obviously, may attract considerable attention but are not necessarily a problem. The former are so potentially serious that, if they should happen, we may already be too late to save some valued part of the environment.

It is instructive to examine those cases known so far in which truly global pollution has been recognized. There are at least six: radioactive fallout (UNSCEAR, 1972); DDT (Edwards, 1973); PCBs (Nesbit and Sarofim, 1972); freons and similar fluorinated hydrocarbons (NASA, 1977); and, more recently, production of carbon dioxide (Rotty, 1979); and, of course, the oxides of sulphur and nitrogen and their end-product, acid rain (Whelpdale, 1978; NRCC, 1981a).

The first three of these have not affected the structure and function of the ecosystem itself, only of certain targets in it (albeit rather important ones). The others, however, are very much of a threat to the more general system: freons and CO_2 by changing the physical parameters to which all organisms must adjust (radiation level and temperature, respectively); while acid rain has been clearly seen to be overtly toxic to ecosystems, particular aquatic systems, over a wide geographical area. It is in this area that potential dangers are greatest and advance warning would seem to be most urgent.

It should be explicitly mentioned here that there is considerable dispute about some of these cases. While each one certainly represents a situation in which an unnatural chemical substance has been discharged into the environment, it does not at all follow that the environment itself has in some sense ended up with deleterious damage. In the case of radioactivity, for example, only in very local areas (detonation points) or experimental setups (the gamma forest) has there been a marked effect on the ecosystem generally. Ecologists have had to search with great care and tenacity to discover whether DDT residues were affecting ecosystems, in spite of the obvious fact that concentrations in the fat of certain mammals were enormous. The long-term effects of PCBs, at very low concentrations, remain a matter of debate.

The cases of freons and CO_2 are interesting in quite a different sense. These substances have clearly not had an adverse effect on the ecosystems exposed to them. However, on the basis of certain modelling studies and theoretical exercises, it became clear to some—and not to others—that a danger was present and would become real if action were not taken straight away. This is another profoundly important issue. If we wait until deleterious effects can be clearly demonstrated, remedial measures will be impossible. This means that, with a threat such as CO_2 production, we must consider the risk of a situation that has never occurred, and for which historical fact, the only convincing argument for many people, can not even in theory be available. Nonetheless, as we become increasingly capable of poisoning our own environment, we must be prepared to depend on such theoretical arguments and models as are available to assess the consequences of particular policies. It is not a comfortable situation.

A further category might be added to our three kinds of spills or discharges, but as yet little is known about the possible dynamics and there are few examples to examine. This is the situation in which a substance may be quite dilute in the environment, but be concentrated geographically as well as in particular organisms (as opposed to simple biomagnification in a particular food chain, a phenomenon which we reasonably well understand). A possible case could arise from the dumping in the deep ocean of radioactive wastes. Escaped radioactivity might concentrate in fish which then congregate near the shore or even migrate to particular areas, thus constituting a source of radioactivity for a target, perhaps man (Grimwood and Webb, 1977). However, in such cases it is not usually assumed that a threat to the ecosystem generally exists, only that an unsuspected pathway to man might exist. Since only the former is the thrust of the present work, such cases are not discussed further.

2.2 DISTRIBUTION, TRANSPORT AND TRANSFORMATION

A good deal of information is now available about how chemicals distribute themselves, move (or get moved) and change their chemical form in the environment (Butler, 1978; Beijer and Jernelov, 1979). Furthermore, we know how to predict at least some of this behaviour using knowledge of physical and chemical properties (SRI, 1977) and, recently, automated computer models of such behaviour have been developed and are gaining acceptance (Roberts *et al.*, 1981).

Concerning distribution, procedures are generally available for measuring partitioning in aquatic systems (Roberts *et al.*, 1981), vapour-particulate distribution in the atmosphere (Natusch *et al.*, 1974), and binding to particles of various types in soils (Schnitzer and Khan, 1972). Such information, together with information about relative binding strengths, allows us in many cases to make quantitative predictions about release rates and subsequent transport in aquatic systems (Ottawa River Project Group, 1979) and atmospheric environments (Whelpdale, 1978; Miller and Buchanan, 1979). Biological involvement is also beginning to be described in quantitative terms, such as uptake from water by aquatic biota (Roberts et al., 1981) and uptake from soils by plants (Schnitzer and Khan, 1972). Influence of biota on physical transport has also been studied (Boddington et al., 1979).

Quantitative prediction of transport has advanced greatly in recent years (Bignoli and Bertozzi, 1979), although some areas remain less well understood than others. Probably best understood are aquatic transport (Hakanson, 1980) and surface runoff of pesticides and agricultural fertilizers (Holdgate, 1979). Short-range atmospheric transport has been well studied (Pasquill, 1974), while long-range transport is qualitatively understood but so far beyond the power of existing computers (Munn and Bolin, 1971). Groundwater, that is, subsurface transport, will require much more study (CEC, 1979a).

Transformation, for the present regarded as transformation to a chemical form which is still of concern in terms of toxicity, has been widely studied and continues to be an active area of research, particularly for metals (Summers and Silver, 1978; Butler, 1978; Beijer and Jernelov, 1979; Chau et al., 1980). Even such a familiar question as the methylation of mercury is not resolved; originally it was thought that biological transformation took place at a more-or-less constant rate (Jensen and Jernelov, 1969) and many subsequent studies have regarded this as axiomatic (Fagerstrom and Jernelov, 1971; Spangler et al., 1973; Bisogni and Lawrence, 1975; Olson and Cooper, 1976, for example). It is now realized (Miller, 1977; Beijer and Jernelov, 1979) that various processes of demethylation as well as methylation are at work, and the net result is probably an equilibrium level of methyl versus inorganic forms. Similar analyses for other heavy metals and other elements are at a much less advanced stage (Summers and Silver, 1978).

Analytic procedures for the various organometallic compounds are not trivial, and transformation from one to another may be spontaneous and fairly rapid (Beijer and Jernelov, 1979). Few standardized procedures are available at present (Oladimeji et al., 1979). Knowledge of the relative amounts of each species present is critical because of their great differences in toxicity (Friberg et al., 1979). Much work remains to be done in this area.

2.3 UPTAKE BY BIOTA

We can borrow from the extensive body of literature in the field of pharmacology to identify the factors most important in understanding and predicting the dynamics of absorption and retention of chemicals by living organisms. The

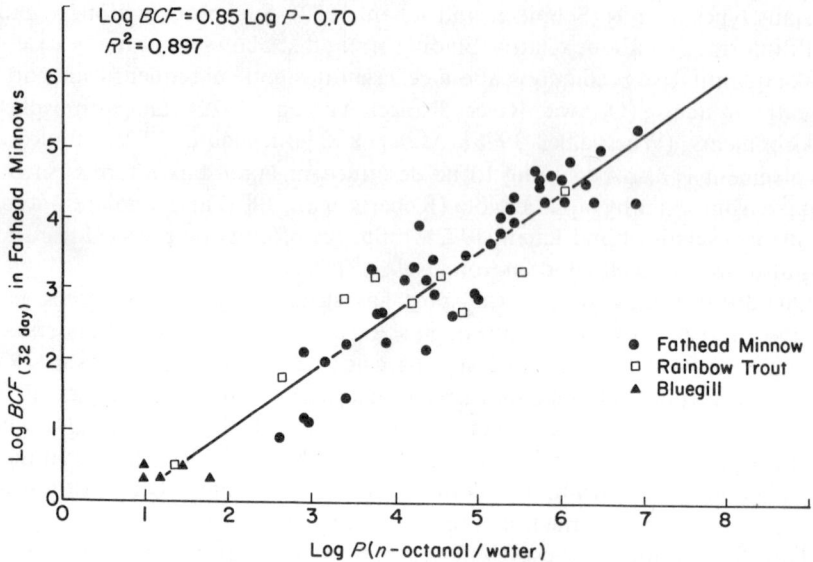

Figure 2.1 Biological concentration factor. (From Veith *et al.*, 1979. Reproduced by permission of the *Journal of the Fisheries Research Board of Canada*)

most important are lipid solubility, molecular size and degree of ionization (Stein, 1967; La Du *et al.*, 1971). Other factors are of lesser importance, such as number of hydrogen bond-forming groups in the molecule (Stein, 1967).

As an effective predictor of bioconcentration in environmental situations, lipid solubility seems to be of greatest importance, and, furthermore, it can be reliably predicted by several simple measurements such as water–olive oil partition coefficient or, more generally in recent years, the water–octanol partition (Neely *et al.*, 1974; Chiou *et al.*, 1977; Veith *et al.*, 1979; see Figure 2.1). Other important environmental observations, such as the overwhelming concentration of organic mercury in fish but not in their environment, can be explained using arguments based on preferential uptake of the organic form (Holdgate and White, 1977, Appendix D).

In general, environmental transfers of pollutants are so sufficiently well understood that specific pathways can be predicted in many cases (SRI, 1977); tabulated collections of data for use in quantitative analyses are available (Verschueren, 1977), and some progress has been made towards interactive computer programmes to process this data almost automatically (Roberts *et al.*, 1981).

2.4 FATE IN THE ENVIRONMENT

As used here, 'fate' refers to the ultimate disposition of the chemical in the ecosystem, either by chemical or biological transformation to a new form which is nontoxic (degradation) or, in the case of an ultimately persistent substance such as heavy metal, by sequestering in a marine sediment or other location which is expected to remain undisturbed.

For substances which are effectively degraded, whether by hydrolysis, photolysis, microbial degradation or whatever, it would seem necessary to collect and tabulate parameters which could serve as predictors for the rates at which degradation would occur. This has been attempted in some large studies (IRPTC, 1978; Callahan et al., 1979) but work is impeded simply because most such parameter values do not exist in the literature. On the other hand, work is progressing in establishing standard protocols for their determination (SRI, 1977; Sheehan et al., in press).

Releases into the environment of persistent chemicals lead to an exposure level which ultimately depends on the length of time the chemical remains in circulation, and how many times it is recirculated in some sense, before ultimate removal. This question must be carefully analysed through a study of global dynamics, as has been done in relatively few cases (Nriagu, 1979). A particular question which needs to be addressed more often for metals is what fraction of the general global circulation is to be attributed to man's activities, an examination which can be made either on theoretical grounds (Garrels et al., 1975) or experimentally (Friberg et al., 1979). Results are not always in agreement; for mercury, for example, estimates of the fractional global circulation which originates with man range from 5 per cent (Weiss et al., 1971) to as much as 30 per cent (Miller and Buchanan, 1979).

2.5 EFFECT ON THE PHYSICAL ENVIRONMENT

There is some literature on the general ways in which pollution can affect the physical and chemical nature of the environment (Goodman, 1974; Munn, 1978), but the best approach is probably through examples of the various possibilities.

The generation of *heat*, usually by use of river or lake water for cooling of a power generating plant, has been well studied in terms of the total amount of heat generated and the temperature rise in the environment that will result (Talmage and Coutant, 1980). However, the general and ultimate alterations that will be produced in the environment are hard to predict (Krenkel, 1979). Certainly, it is not clear that all effects are necessarily adverse; increase in temperature may lead to much higher growth rates of commercially important species, for example (Malouf and Breeze, 1978; Mann, 1979). The general attitude is that any change from the original state is to be avoided, although the possibility of making a change to a somehow more desirable ecosystem does come through in thoughtful reviews (Holdgate, 1979).

Changes in the *acidity* of a system, particularly an aquatic system, resulting from acid rainfall are generally understood and the ecological consequences have been described at length (for review, see, for example, NRCC, 1981a). Similarly, changes in the acidity of terrestrial ecosystems have been widely studied in terms of the effect of a given change in pH, although the extent to which such a change is due to rainfall as opposed to natural acidification processes in the soil is not known (Hutchinson and Havas, 1980). What is known, of course, is that pH changes can drastically affect the structure and function of the ecosystem, both directly and indirectly by, for example, increasing the concentration of heavy metals in the water through increased leaching from sediments (Miller and Akagi, 1979). This general theme is further developed in the following chapters and in the case studies.

There are, of course, many other categories of physical and chemical effects, but detailed examination would be beyond the scope of this work. The reader is referred to the case studies for specific situations in which major changes in the ecosystem have resulted from alterations in the physical environment.

Effects of Pollutants at the Ecosystem Level
Edited by P. J. Sheehan, D. R. Miller, G. C. Butler and Ph. Bourdeau
© 1984 SCOPE. Published by John Wiley & Sons Ltd

CHAPTER 3
Distinguishing Ecotoxic Effects

DONALD R. MILLER
Division of Biological Sciences
National Research Council of Canada
Ottawa, Ontario, Canada K1A 0R6

3.1 Nature of the Problem 15
3.2 Identifying Parameters to Monitor 16
3.3 Mathematical Background 17
3.4 Estimating Baseline Levels 18
3.5 Detecting Changes 20

3.1 NATURE OF THE PROBLEM

There is a fundamental difficulty involved in the quantitative examination of ecosystems, with which workers in the field are only now beginning to deal, and that is the feature called environmental fluctuation, biological variability or one of several other names.

The concept is simple enough: because of differences in physical conditions (temperature, rainfall, etc.), nutrient availability (varying runoff, for example), or other reasons, ecosystems do not behave in constant and repeatable ways. If the same system is observed under what appear to be uniform conditions over several seasons, for example, very substantial fluctuations are found in such variables as population levels of particular species, or indices of species distribution.

The problem, then, is how to decide whether an observed change in some parameter represents a deviation caused by presence of a pollutant, or whether such changes as are seen are part of the 'natural' fluctuations inherent in the system.

The question is by no means trivial and, indeed, may be the most important question facing us as we try to refine our techniques for detecting, as a prelude to dealing with, ecotoxic effects. In most cases we do not have the kind of baseline measurements that allow us to state with confidence just how large inherent fluctuations are, as would be required in order to apply classical statistical procedures to decide when the deviation should be regarded as significant in the mathematical sense. If enough information existed about the variability of the so-called undisturbed system, we could apply mathematical procedures which are well understood. Unfortunately, since such complete baseline data are lacking in many cases, we are faced with the difficult problem of deciding when a change indicates toxic effects at the ecosystem level without knowing how the system behaves in the absence of pollutants.

The problem may profitably be compared to that faced by the physician in the early stages of examination of a patient. The first thing that must be determined is whether the patient is sick at all, and physicians would largely agree that a knowledge of the patient is the most important piece of information contributing to that decision. Most of the observations that can be made, such as pulse rate, respiration rate, flushing of skin, fluid balance and even core temperature are subject to variations as the individual responds to his chemical, physical and psychological environment. The knowledge of, or ability confidently to predict, the normal ranges of such variations lies at the base of the diagnosis.

Fortunately, physicians have had many centuries of experience in observing which variations are natural and which are not, and, furthermore, identifying which parameters are generally indicative of 'poor condition', whatever that means.

Practitioners of ecotoxicology are relative beginners. So far, only limited information is available about normal ranges of variation in those parameters we might like to use as diagnostic aids; in fact, many scientists do not yet fully appreciate how many observations are necessary or how long observation must continue, to estimate with confidence a static quantity, let alone monitor a widely fluctuating variable, the base value of which may or may not be changing.

3.2 IDENTIFYING PARAMETERS TO MONITOR

Not only is there the problem of not knowing the level of fluctuation that may be regarded as natural but, more seriously, there is no broad agreement about what specific quantities to examine. We do not yet know how to take the 'pulse' of an ecosystem. Fortunately, this larger problem seems on its way to a solution, and this will be explained later in the book (cf. Chapter 6).

The state of our understanding of how to recognize and interpret environmental change seems to be as follows. If we can identify one or a small number of clearly defined variables (as the atmospheric scientists have done with temperature, carbon dioxide levels and ozone concentration) and record it or them over a long period, there is little difficulty in deciding whether a shift has occurred. (We should rather say that there is little difficulty in deciding how large a shift has to occur in order to be detectable; we do not wish to imply that any shift at all could be demonstrated.) Why, then, can not the same be done at the ecosystem level?

What is needed is the specification of quantities that have certain desirable properties, including the following:

1. indicative of overall condition of the ecosystem;
2. comparable for a variety of ecosystems;
3. easily and reliably measured;
4. related to variables used in quantitative (modelling) studies.

It is stated elsewhere in this book that we ought to be concerned not only with the toxic effects some chemical might exert on certain individuals or even certain species, but also with the effects on overall structure and function (species diversity, material flow and nutrient cycling, etc.). Thus, our measurements should relate to a level of complexity or organization above that of organisms, say, at least at the community level (Jacobs, 1975; Whittaker, 1975). This sort of measurement is clearly necessary if an overall assessment of ecosystem performance is required. We might add that this sort of quantity is needed to calibrate models of ecosystem dynamics. The whole question of measurements of ecosystem function will be fully explored in Chapter 6; here it is only necessary to say that our ability to identify appropriate variables is developing quickly.

3.3 MATHEMATICAL BACKGROUND

In purely mathematical terms, the problem is quite well understood, and comprehensive treatises on quantitative approaches are available (for example, Box and Jenkins, 1970; Poole, 1978). Nonetheless, it is surprising how rarely works appear which address in quantitative terms the questions of just how much data must be gathered to estimate parameters or to prove that apparent changes are, in fact, real (good examples are Platt *et al.*, 1970; Platt, 1975; see also Green, 1979; Cairns *et al.*, 1979).

In summary, the problem may generally be handled in ways familiar to those acquainted with standard statistical procedures. If one wishes to estimate an environmental parameter which is assumed to be static, or stationary, one first must specify the accuracy required in terms, for example, of a 95 per cent confidence interval. Then, after a certain amount of observation devoted to estimating the nature of the underlying distribution, it is possible to specify how many additional observations are required to achieve the specified accuracy. More generally, guides may be constructed describing the trade-off that applies between sampling economy and estimation accuracy.

If we wish to determine whether a particular quantity is changing in time, we proceed to make a series of observations and fit a model of some degree of complexity. The accuracy with which the various parameters in the model can be estimated, in particular judging whether long-term trends are real (parameter changes being significantly different from zero), as well as the confidence that the particular model is appropriate, increases with the number and duration of the observations in a quantifiable manner. Since good references are available in the statistical literature (Box and Jenkins, 1970; Poole, 1978), further details are not given here. It is appropriate to mention, however, that much of this information is couched in jargon, and notation, that make it quite inaccessible to practising field biologists and ecotoxicologists; there is a need for some quite practical guidebooks in this area. Specifically, there must be more emphasis on the quite frightening data requirements if the kind of accuracy desired is really to be achieved (see following section). To this writer's knowledge, the best reviews of the problem include those by Elliott (1977), Eberhardt (1978) and Green (1979), each of which contains quite practical advice on the estimation of population levels. It is to be hoped that the work of the International Statistical Ecology Program (ISEP) will provide assistance in this area; some publications are currently available (for example, Cairns *et al.*, 1979; Patil and Rosenzweig, 1979), and several others are in preparation at the time of writing.

In connection with this, we are led to think of theoretical studies, the so-called ecosystem model studies, for in no other approach to such systems have the problems of state variable identification and numerical prediction been so clearly given priority (Pielou, 1969).

Many modelling studies have been carried out. Unfortunately, upon examining the available literature, the first observation indicates that by far the larger part of such work has involved only deterministic models, in which the most likely values are predicted, without an analysis of the associated uncertainty (CEC, 1979b). From the present perspective, such modelling approaches are simply not appropriate.

On the other hand, since the early 1970s there has been a gradual increase in interest in uncertainty analysis. Omitting various writers who have simply pointed out that uncertainties make analysis difficult, without presenting studies of how these difficulties might be resolved, we would identify Reichle *et al.* (1973c), Miller (1974) and Burns (1975) as establishing a technique whereby uncertainties may be deliberately introduced into the parameters governing model behaviour so that output uncertainty may be calculated or experimentally measured. The approach may be used to compare predicted behaviour with permissible errors for validation purposes (Miller *et al.*, 1976), and has the advantage for the present discussion that fluctuations in the whole system may be predicted on the basis of fluctuations in individual parameter values. Since these represent climatic and related variables, the values of which are often well known from other sources, the approach is not always subject to the need for full-scale and expensive baseline studies. Although techniques of modelling of ecosystems need to be further developed, the additional approaches of uncertainty analysis are easy to put into place as developments are made (Goodall, 1972; Gowdy *et al.*, 1975; Majkowski *et al.*, 1981).

3.4 ESTIMATING BASELINE LEVELS

If we are attempting to estimate the average or mean value of some numerical parameter which is assumed to be sensibly constant, at least over the duration of the measuring period, simple formulae are available for estimating the standard error of the mean for a simple collection of measurements. In order for this standard error to be less than some preassigned value, it is only necessary to make the number of repetitions of the measurement large enough. The same is true for quantities known to vary from place to place, if it is agreed that what we are after is the overall average value (even if that precise value is true for only a few specific locations). If the number of observations is large (say, 50 or so) it is legitimate to assume that the value of the standard deviation so calculated will span an interval containing about two-thirds of the observations. For any number of samples, the use of the t-distribution makes it possible to calculate a confidence interval of whatever precision we like.

The surprising thing is how often field work is not examined in accordance with these considerations, and when it is, how poor the actual estimates turn out to be. Hales (1962) illustrated the former point by giving several examples of quite nonmathematical rules of thumb that had been published in various places.

One of the earliest works that points out the magnitude of the problem is that of Needham and Usinger (1956), in which it is stated that for a 95 per cent level of significance, working with organisms on a riffle in Prosser Creek (California), 194 samples would be required to give significant figures for total weights, and 73 samples for total numbers. In a similar study in the Logan River, Utah, Hales (1962) showed that the required numbers of samples depended also on genus. To have a 90 per cent chance of estimating total numbers with confidence limits of $\pm$ 25 per cent, for example, would take 18 samples for Diptera and 34 samples for Ephemeroptera in a typical location.

It is worth pointing out that the situation is not always bad, but depends on the parameter being examined. In the study quoted above. Needham and Usinger (1956) pointed out that only two or three samples would give, with a 95 per cent probability, at least one representative of each of the common genera of insects being examined.

Thoughtful analyses of these problems are increasing. Treshow and Allan (1979) studied the dynamics of a pinyon pine–Utah juniper woodland community to determine baseline conditions and annual variations, and concluded that 4 years of study were required to establish a reliable baseline.

Other examples could be quoted; however, the point is not that large quantities of data are needed in some cases while not in others, but rather that a great deal of biological research is carried out in which the question is not even addressed. The difficulty is not in the statistical analysis, but in the practical aspects of research design. Cases where sufficient sampling has been carried out to enable coefficients of variation to be determined were surveyed by Eberhardt (1978), who then described how necessary sample size may be calculated. Other examples are described by Green (1979), and Resh (1979) has given a detailed account of the sources of variability in aquatic insect studies, and sample sizes necessary to achieve given levels of reliability of data.

We may digress for a moment to mention that the problem is being addressed in some important ways through the international programme of Biosphere Reserves. Some years ago it was generally recognized that for comparison purposes it was necessary to have a certain collection of ecosystems, representative of as many classes of ecosystem as possible, preserved in states as close to the natural systems as possible (MAB, 1974). These will serve, and in many cases are serving, to allow precisely what we identify as missing, namely the carefully planned study of how systems behave when they are left as much on their own as possible. Only with this information will we be equipped to state with certainty whether a change recognized in an environmental situation really represents an 'effect' or not. There are, at the time of writing, 177 Biosphere Reserves in 46 countries, and more are being identified regularly (Anon, *Nature and Resources*, 1980).

There is a problem, the resolution of which has not been found. The problem is that some pollution exists and is measurable on a truly global basis. This means, of course, that even the most isolated Biosphere Reserves exhibit low but detectable amounts of toxic substances, so the question arises as to whether their condition really represents baseline or undisturbed behaviour (Brown, 1981). It is to be hoped, however, that such low amounts would not cause disturbances at the level we could observe.

3.5 DETECTING CHANGES

In the previous section it was emphasized that, in general, quite inadequate attention is paid to the question of how much data must be gathered in order to say that an ecosystem has changed at all, much less to blame the change on a particular pollutant. However, there have been cases where such changes have been documented, and it is instructive to mention some of them.

The Continuous Plankton Recorder Programme of the North Sea and the North Atlantic surveys some 300 species of plants and animals on commercial shipping routes and has been producing data since 1948. Glover (1979) examined these data and detected a long-term trend downward in both copepod abundance and zooplankton biomass for both the North Sea and the North Atlantic, but with the North Sea showing a remarkable reversal in the early 1970s. Interestingly, levels of fluctuation, which decreased with biomass, have not correspondingly increased. This data bank is a very extensive one, and provides a fertile ground for various investigations (Colebrook, 1978).

Two quite different examples of timewise variation were discussed by Gilboy *et al.* (1979). One consisted of individual tree rings analysed for metals, the other a moving filter method for particulates in air giving a resolution time of 2 hours. Their analysis confirmed the need for a resolution in the latter case at least this fine or even better (perhaps 1 hour) if atmospheric fluctuations were to be followed. It seems clear that many ecosystems would not require such precision, although this would ultimately depend on the magnitude of the fluctuations.

An example of research in which the problems of monitoring spatial and temporal variation have been considered is the study of long-term exposure of a forest to air pollutants described by Legge and coauthors (Legge, 1980; Legge *et al.*, 1981; Legge, 1982). This 8-year programme combined remote sensing, controlled laboratory fumigation experiments, and detailed field studies to determine the effects of sulphur gas on foliar accumulation of sulphur and essential nutrients, soil changes, pine tree physiology, and forest productivity. S-gas emissions were continuously monitored at both the incinerator and the flare stacks at the West Whitecourt Gas Plant in Alberta. This monitoring demonstrated that the incinerator contributed most of the sulphur emissions, but, on individual days, during gas plant operating upset, the flare stacks contributed substantial levels of sluphur pollutants. Intensive on-site air quality monitoring was undertaken at 2, 16 and 28 m above ground to measure the varying concentrations of SO_2 reaching the lodgepole × jack pine forest at chosen experimental sites. Analogous sampling locations were chosen based on ecological variables such as slope, aspect, soil type, species density and diversity, and environmental variables other than pollutants, including temperature, wind, solar radiation and precipitation. Pollutant variables such as the effluent composition and the concentration (including the factor of distance from source) were used to locate sites along a pollutant gradient. Concentrated biological surveillance of foliar ATP and photosynthetic rates established a positive relationship with both parameters, inversely correlated with sulphate–sulphur accumulation. Statistical analysis of basal area increment data revealed that distance from the S-gas source, time in years, and their interaction had significant effects on the woody production of pine trees.

Examples of studies in aquatic systems are those of Cushing (1979) and Myers *et al.* (1980). Cushing discussed the case of an exploited economic fish population, for which data are abundant. Myers *et al.* studied the fish of Bantry Bay, Ireland, and concluded that observed declines of fish stocks were not in fact related to the explosion of the tanker *Betelgeuse*, as had been assumed (Cross *et al.*, 1979). Tont and Platt (1979) used spectral analysis to study phytoplankton diversity of the California coast, and found considerable cyclic activity associated with time periods ranging from a small number of weeks to several years. These they attributed to upwelling events caused by wind changes.

Longhurst *et al.* (1972) pointed out that populations of zooplankton and anchovy eggs fluctuate no more in the polluted Los Angeles Bight than elsewhere off the California coast, and cited other examples on the basis of which they concluded that there is a danger of incorrectly ascribing natural fluctuations in animal populations to the effects of pollutants.

A study specifically directed to the question of adequacy of monitoring is that of Naiman and Sibert (1977). By taking intensive (3-hour) samples of various quantities, they determined that the data collected at 2-week intervals, the current monitoring practice, were adequate for temperature, salinity, nutrients and chlorophyll *a*, but not for DOC, ATP and heterotrophic activity.

A quantitative analysis of sampling strategies for trace element levels and benthic invertebrate populations in the New York Bight, involving an explicit optimization procedure applied to stratified sampling, was described by Saila *et al.* (1976). It was found that a fully optimized sampling plan required only three replicates within a station and a small number of stations (seven) for significant results.

In his discussion of the design of monitoring systems, Holdgate (1979) pointed out that surveys designed to describe average or integrated exposure of a target may not be relevant, for more damage may be done by short-lived peaks, that is, periods when exposure exceeds certain limits, and that a sampling system should be able to measure the variation in contamination levels so as to allow for the estimation of possible 'worst case' exposures. Examples of advance baseline studies include such cases as that of the Surrey, Virginia power plant (and many others) where water quality parameters were monitored for some time before the plant went into operation, and also afterwards (Bolus *et al.*, 1973). Such studies generally have been limited to physical parameter measurements.

Some attempts to review the problem have appeared; the reader is referred to papers by Cushing (1979) and Cairns and van der Schalie (1980) and to later chapters of this book, especially Chapter 6.

Effects of Pollutants at the Ecosystem Level
Edited by P. J. Sheehan, D. R. Miller, G. C. Butler and Ph. Bourdeau
© 1984 SCOPE. Published by John Wiley & Sons Ltd

CHAPTER 4
Effects on Individuals and Populations

PATRICK J. SHEEHAN

Division of Biological Sciences
National Research Council of Canada
Ottawa, Ontario, Canada K1A 0R6

4.1	Types of Response	24
	4.1.1 Introduction	24
	4.1.2 Behavioural	29
	4.1.3 Biochemical	31
	4.1.4 Morphological	32
	4.1.5 Physiological	34
	4.1.6 Altered Performance	36
4.2	Interaction with Other Environmental Stresses	40
4.3	Life History	43
4.4	Population Interactions	44
4.5	Impact on Populations and Its Relevance to Ecosystem Response	47
	4.5.1 Relevance of Population Response Extrapolations	49

The adverse effects of toxic pollutants on organisms have most often been identified with their lethal impact. As an endpoint, mortality is easily recognized and can be quantified under both field and laboratory conditions. The value of recognizing what concentrations of a toxic substance can cause a lethal response in the population is obvious, since such an effect can have great impact on the community and ecosystem. Dramatic organism kills associated with the introduction of toxic materials are definitive evidence of environmental conditions unsatisfactory for the support of life. As mortality is the cessation of all biological activity, leaving us nothing further to examine, there is a great deal of interest in establishing which sublethal responses occur at lower pollutant concentrations.

Recently, the description and quantification of sublethal responses to pollutants in individuals and populations in contaminated ecosystems have constituted a major thrust of field research in ecotoxicology. Interest in monitoring pollutant effects has stimulated a number of general reviews of material relevant to the topic of species response to toxic stress (see Butler, 1978; NAS, 1981). Vernberg (1978) presented an approach to understanding the organisms' ability to survive in a multistress, although not necessarily chemically polluted, environment. Warren's text on biology and water pollution examined such stresses as pH changes, cation concentrations, and sewage, and their effects on individuals and populations, as well as communities, in freshwater ecosystems (Warren, 1971). Lockwood (1976) presented a further survey of the biochemical and physiological effects of pollutants on aquatic species. Impacts with specific reference to marine organisms have been the subject of considerable study (for example Anderson *et al.*, 1974; Vernberg and Vernberg, 1974; Cole, 1979; McIntyre and Pearce, 1980; Vernberg *et al.*, 1979, 1981). Terrestrial animal studies have been primarily concerned with pesticide effects (Brown, 1978; Moriarty, 1978). Air pollutant effects on plants and plant–soil relationships have recently been the subject of a number of reviews (Mudd and Kozlowski, 1975; Mansfield, 1976; Drablφs and Tollan, 1980; Hutchinson and Havas, 1980).

Because of the existence of such an extensive literature base on the topic of organism response to pollutants, a comprehensive general review is not required here; rather the purpose of this chapter is to examine how pollution responses of individual populations can be related to their ultimate impact on the dynamics of the polluted ecosystem.

4.1 TYPES OF RESPONSE

4.1.1 Introduction

Response to a toxic substance can be categorized according to the dose rate, and according the severity of damage:

1. Acute toxicity causing mortality.
2. Chronically accumulating damage ultimately causing death.
3. Sublethal impairment of various aspects of physiology and morphology.
4. Sublethal behavioural effects.
5. Measureable biochemical changes.

Of particular interest is the progression through which injury to individuals may affect the success of the population which in turn may cause impacts at community and ecosystem levels.

The general concentration-response model (Figure 4.1) for pollutant effects on organisms describes a curve crossing the three zones of response: non-measureable, measurable sublethal and lethal, as concentration of the toxic substance increases. Defining the threshold for a particular response is difficult; with some of these substances or mixtures there may be a real but exceedingly low threshold, one that is beyond our ability to detect. Just as the concentration-response model is the cornerstone of laboratory toxicity testing, it is also the concept which relates population response to the environmental level of the pollutant. Evaluating the effect of single species toxicity in terms of its impact on the polluted ecosystem, requires an understanding of the thresholds of acute-

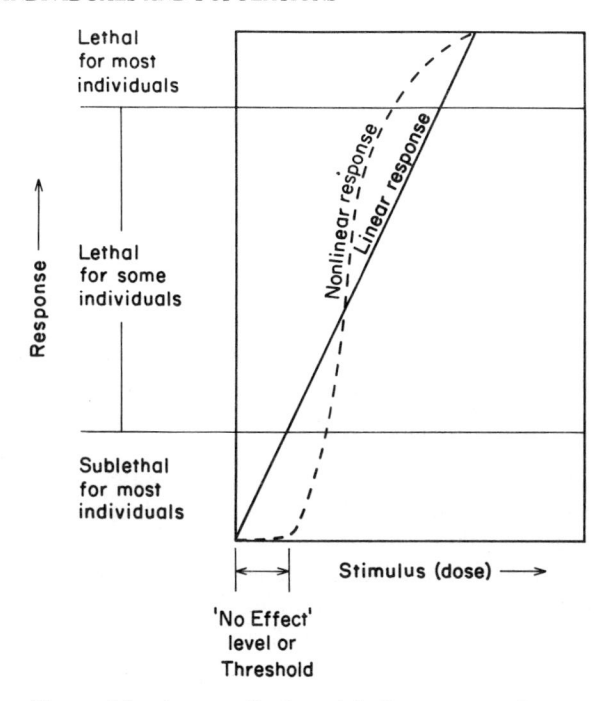

Figure 4.1 A generalized model of response to dose or concentration

lethal and critical chronic effects, and the relationships through which these responses affect the propagation and survival of the population and influence the success of other species, through ecological interactions.

The quantification of acute pollution effects on exposed populations can be estimated either from direct counts of dead organisms or from life table statistics on individual populations, provided baseline data is available. Laboratory bioassays with sensitive species, representative of the contaminated community, can be used to confirm that the suspected toxic substances are indeed lethal at environmental concentrations. However, such acute and gross destruction of life rarely provides much insight into the processes responsible for ecosystem breakdown.

A less dramatic but perhaps more important example of pollution-induced population decline results from continuous, gradual pollution input, leading to accumulation of individual injury and general deterioration of the environment. Evidence tying population declines to pollutant toxicity in these cases is not always obvious due to the extended time frame over which adverse changes have occurred. An example of chronic environmental deterioration is the influence of acid precipitation and runoff into some unbuffered lakes in eastern Europe, Scandinavia and North America, during the past several decades, which has reduced populations, particularly of fishes, and dramatically changed the structure and possibly the functioning of the aquatic systems (Drabløs and Tollan, 1980; NRCC, 1981a).

It is perhaps more important to the 'well-being' of the ecosystem to detect chronic stress quickly and accurately than it is to assess the lethal endpoint. Early detection allows corrective action to be applied before irreparable damage has occurred.

The recognition of chronic pollutant effects at various levels of biological organization, from molecular, through whole organism, to the ecosystem, is partly dependent upon the time since exposure. In general, the period before which induced changes become observable is longer for each increasing level of biological complexity. A conceptual chronology of induced effects following a population's exposure to toxic pollutants, developed as part of the present work, is depicted in Figure 4.2.

The bioaccumulation of pollutants can occur from water, air (gases and small particulates), and through the food chain. The rate at which accumulation occurs depends upon the availability of the pollutant, environmental conditions, and the organism's ability to assimilate it (see general reviews by Edwards, 1973; Livingston, 1977; Bryan, 1979). The ultimate level of accumulation depends on the internal processes of excretion, detoxification and storage (see Anderson *et al.*, 1974; Moriarty, 1978). Certain organisms have been shown to have some ability to regulate accumulation of specific pollutants. For example, fish can regulate levels of copper and zinc in muscle (Saward *et al.*, 1975), but methylmercury is not regulated (McKim *et al.* 1976). There is little evidence for metal regulation by seaweeds (Byran, 1979), and external factors appear to control accumulation in terrestrial plants (Hughes *et al.*, 1980b). Bivalve molluscs do not metabolize aromatic hydrocarbons as readily as do fish and crustaceans and therefore their tissue levels of these compounds are more dependent on the lipid–water partition coefficient and the amount of lipid in the organisms (Bryan, 1979). This relationship appears to hold for organic pesticide accumulation in members of the aquatic food chain (Ellgehausen *et al.*, 1980).

This rather limited capacity to regulate accumulation of most toxic pollutants has suggested that significant changes in the level of bioaccumulation be used as an 'early warning' signal of increased pollution stress. Aquatic insects have been found to accumulate metals rapidly and to levels well correlated with those of the water (Nehring, 1976; Nehring *et al.*, 1979), and similar results have been reported for organic contaminants (for example, Kaiser, 1977).

Bioaccumulation of toxic substances can occur to a certain extent before chronic-effects thresholds are reached. The binding of metals to metallothionein-like protein is an example of a protective 'storage function' which keeps the level of metals at the site of action (the enzyme system) below threshold. Pathological changes become apparent if heavy-metal loading exceeds the rate of metallothionein-like protein production (Brown *et al.*, 1977).

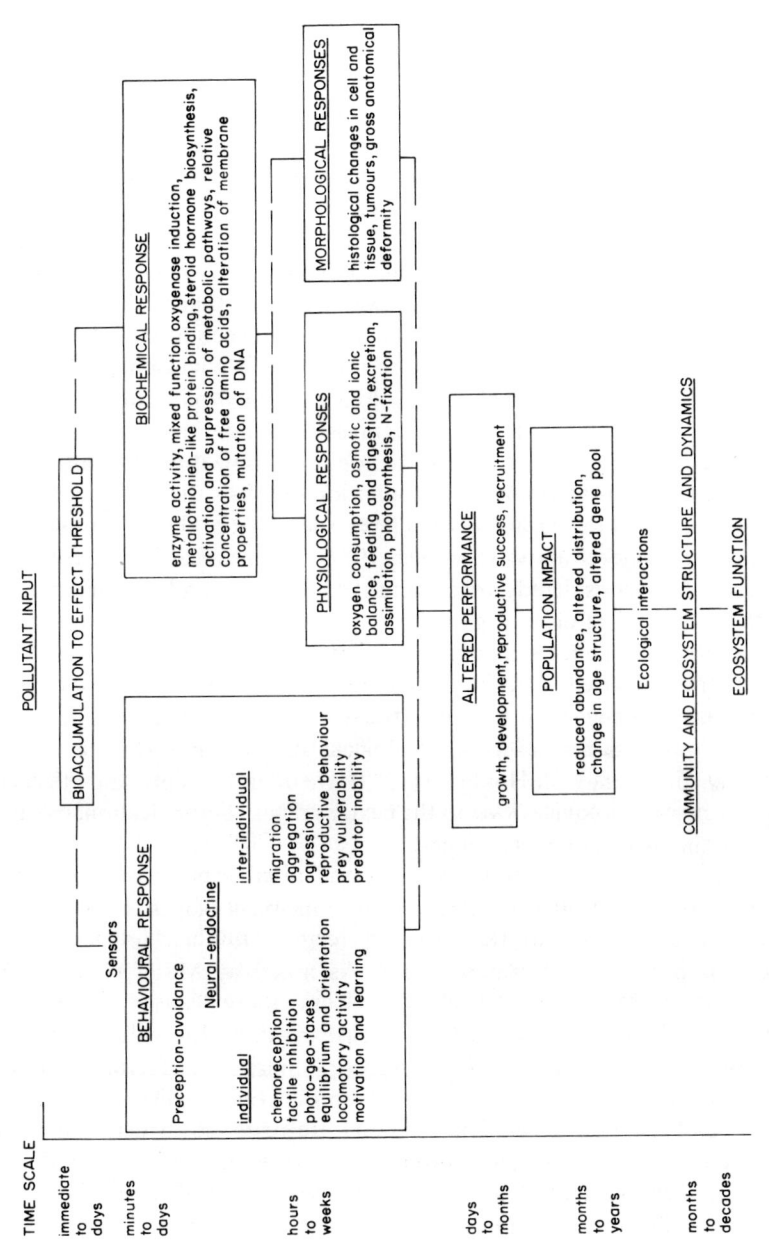

Figure 4.2 A conceptual chronology of induced effects following exposure to toxic pollutants, emphasizing responses in individuals and populations

Increasing tissue concentrations of pollutants through bioaccumulation, particularly organic pollutants such as DDT and PCBs, have led to increased toxic stress (for example, Lincer, 1975; Stickel, 1975; Martin et al., 1976; Roberts et al., 1978). Although there is sometimes little direct association between whole organism or muscle tissue levels of pollutants and the severity of stress response, specific organ concentrations in the brain, liver and kidney have been correlated with stress effects (for example, Hutton, 1980, 1981; Busby et al., 1981). Therefore, abnormally high tissue levels of toxic substances, particularly in critical organs, in wild species should be considered to be indicative of individual contamination and a situation warranting further evaluation of the severity of pollutant stress on exposed populations and ecological interactions.

Chronic effects can be defined as those responses to environmental change whether behavioural, biochemical, morphological, or physiological, that may be induced in one stage of development but expressed at a later time, in a later stage of development, or at a different level of biological organization (Rosenthal and Alderdice, 1976). In assessment of pollution effects, the explanation of mechanism is often at the organizational level below that of the response and its significance, at the level above (Sprague, 1971). Biochemical interactions, for example, should be considered as occurring at a basic level, and are related to the functionality of a tissue or an organ. At higher biological levels, the questions then are whether such effects change the performance of the organisms and, in turn, whether this altered performance can affect the success of assemblages in the ecosystem. The suggested flow of adverse responses induced by pollutant exposure through higher levels of biological organization (Figure 4.2) is conceptually consistent with Holdgate (1979). He described pollutant effects as cascading from one biological level to the next, as repair, detoxification, or other recovery mechanisms are overwhelmed.

The relatively short time between exposure and initial response in an organism can aid in detection of adverse effects in their incipient stages. However, the interpretation of changes at this level, in relation to their impact on the ecosystem, is not always straightforward. As noted by Miller (Chapter 3), distinguishing pollutant-induced changes from those caused by natural environmental changes requires extensive baseline data. Individual organism response may also be influenced by a number of endogenous factors including sex, age, developmental stage, surface area, reproductive condition, nutritional status and biological rhythms. These factors are reviewed in detail in many animal physiology texts and their importance to the monitoring of pollution stress has been emphasized (Uthe et al., 1980; Sastry and Miller, 1981).

Another significant problem in interpreting chronic effects is the organism's ability to compensate for or physiologically acclimate to worsening environmental quality over a short time-course of exposure. Several compensating mechanisms, particularly those involving enzyme rate-function, have been suggested as important to an organism's ability to acclimate to pollution stress (Sastry and Miller, 1981). Therefore, extrapolation of organism response, in an attempt to estimate population success and subsequent community interactions, must depend on an understanding of the organism's physiology and behaviour as well as the ecology of the system. Certainly, indices related to survival, growth, development, reproduction, and recruitment are indicative of potential adverse impacts impinging through population interactions on higher levels of biological organization. However, the interpretation of effects on populations is difficult, as discussed by Moriarty (1978). The complexity of the community within which populations exist does not easily permit a strictly experimental approach, therefore researchers, in forming many of their conclusions, have usually relied heavily on experience and judgement.

4.1.2 Behavioural

Exposure to pollutants, even in exceedingly low concentrations, can elicit behavioural responses (see reviews, Eisler, 1979; Olla et al. 1980). However, behaviour is difficult to assess quantitatively due to variability over time and subject. The proper use of behavioural responses in assessing pollution-induced alterations in field populations is dependent upon quantitatively defining normal behaviour patterns so that either a quantal or gradual change can be demonstrated.

Perception and avoidance of pollutants is the most immediate and perhaps the most important behavioural response for a species exposed to ecosystem contamination (Geckler et al. 1976). Avoidance behaviour has been observed in a number of taxa. Certain fish species have been shown to avoid water polluted with copper at $2.4\,\mu g\,l^{-1}$ or zinc at $54\,\mu g\,l^{-1}$ (Sprague et al., 1965). Similar behaviour has been noted in fish exposed to 1–10 per cent concentrations of pulp mill effluent (Kelso, 1977; Wildish et al., 1977; Lewis and Livingston, 1977) and to several organochlorines (Kynard, 1974). Midge larvae were found to avoid sediments containing more than $400\,mg\,l^{-1}$ Cd or $9000\,mg\,l^{-1}$ Zn (Wentsel et al., 1977). Marine bivalves avoided sediments contaminated by several heavy metals including Pb ($74\,mg\,l^{-1}$) and Cu ($150\,mg\,l^{-1}$) but not sediments contaminated at roughly half these levels (McGreer, 1979). The common mussel *(Mytilus edulis)* closed its shell valves temporarily (5 days) to avoid the detrimental consequences of $0.5\,mg\,l^{-1}$ copper in seawater (Davenport, 1977). Similiarly, the marine snail, *Monodonta articulata*, retracted into its shell at $0.8\,mg\,l^{-1}\,Hg^{2+}$ (Saliba and Vella, 1977). The avoidance of a contaminated environment during reproduction or recruitment would selectively protect a species. This selective behaviour has been documented for postlarval crabs preferentially settling in less oil contaminated sediments (Krebs and Burns, 1977). Although avoidance behaviour would seem beneficial, its environmental significance under specific circumstances has been questioned. It has been suggested that avoidance could be detrimental to a population, should it be unable to reach a spawning or breeding ground, as demonstrated for spawning Atlantic salmon in the copper polluted Northwest Miramichi River (Sprague et al. 1965).

Pollutant interference with sensory perception and capacity is of particular ecological significance when correlated with such functions as feeding, mating and escaping from predators. Inhibition of chemoreceptors by oil, interfering with food location and feeding response, has been documented (FAO, 1977). Number 2 fuel oils at levels of $0.08-0.15\,\text{mg}\,\text{l}^{-1}$ interfered with the ability of the lobster (*Homarus americanus*) to perceive food (Atema, 1977; Atema et al., 1979). Oil also provoked abnormal mouth opening (feeding) responses in exposed corals (Reimer, 1975; Loya and Rinkevich, 1980). Even certain bacteria have been shown to cease feeding when exposed to low concentrations of oils, although their ability to feed was apparently unhampered (Mitchell et al., 1972). The obvious diversity of receptor mechanisms used by this quite wide variety of species indicates that oil interference with chemical feeding cues may be of general significance in spill areas.

Locomotor impairments, as well as sensory perception, are of importance in shelter seeking and construction and prey escape behaviour. The burrowing activity of soft-bottom estuarine invertebrates is essential to their abilities to escape from predators and to create shelter. Chronic levels of toxic metals and phenol increased the time for bivalves to successfully complete burrowing (Stirling, 1975; McGreer, 1979). Fuel oil was reported to produce locomotor impairment in the marsh crab (*Uca pugnax*) as evidenced by its abnormal burrow construction (Krebs and Burns, 1977). The shallow burrows are thought to have contributed to increased over-winter mortality in crab populations.

An effect on locomotor behaviour often contributes to more severe and immediate impacts as in the case where it reduces the species' ability to escape from predators. Controlled field experiments with the marsh crab have demonstrated that the organophosphorous insecticide, temefos, reduced population survival by inhibiting an effective escape response to natural bird predation (Ward and Busch, 1976; Ward et al., 1976).

Abnormal behaviour interfering with reproductive success can have a severe impact on affected species. Avian toxicologists have reported DDE-induced *reduction* in courtship activity of ringed turtle doves (Haegele and Hudson, 1977) and abnormal nesting behaviour in Ontario herring gull populations, probably attributable to organochlorine-induced dysfunctions (Fox et al., 1978). The gull populations feeding on contaminated fish were less attentive and did not defend their nests in a normal manner, accounting for a high incidence of egg loss.

Among the many identified behavioural responses to pollutant stress, those relating to reproduction, migration, shelter construction, and prey vulnerability have been most easily quantified in contaminated ecosystems and can be related to a populations's functional success.

4.1.3 Biochemical

Detection of biochemical response to pollutants has provided much insight into the mechanisms of toxic action and has received considerable attention in the assessment of stress effects in contaminated ecosystems (see reviews, Lee *et al.*, 1980; Sastry and Miller, 1981).

The correlation of chlorinesterase (ChE) inhibition with brain levels of various organophosphorous pesticides has been demonstrated in fish and birds (Verma *et al.*, 1981; Busby *et al.*, 1981). For birds, it has been suggested that brain ChE inhibition exceeding 20 per cent indicated stress, and inhibition greater than 51 per cent caused death attributable to the insecticide (Ludke *et al.*, 1975).

The depression of δ-aminolevulinic acid dehydratase (ALA-D) activity in the blood, kidney and liver of exposed organisms appears to be of value in the assessment of lead intoxication. ALA-D inhibition has been associated with elevated lead accumulation in fish (Jackim, 1973), rats (Mouw *et al.*, 1975), and birds (Hutton, 1980). The marked inhibition of ALA-D in species from lead-contaminated environments may result in significant reductions in haem synthesis, and in neurological consequences. However, some uncertainty exists as to the significance of the ALA-D index to the overall 'fitness' of the exposed populations (Hutton, 1980).

There may also be biochemical warning systems that can, for specific heavy metals, roughly predict the level at which adverse effects on the species may occur. It has been suggested that the pathological effects of heavy metals coincide with the saturation of binding sites on metallothionein-like proteins and the consequent 'spillover' to binding the high molecular weight enzyme fraction (Brown *et al.*, 1977; Brown and Parsons, 1978). The spillover effect in exposed fish and zooplankton was correlated with an environmental mercury concentration around $1 \text{ mg} \text{l}^{-1}$.

Mixed function oxygenase (MFO) reactions may indicate stress from accumulation of certain organic compounds such as oils and some halogenated biphenyls (see Stegeman, 1980). This enzyme 'system' is responsible for the biotransformation of xenobiotics in vertebrates and the active site of catalytic function is associated with the cytochrome P-450 protein (Stegeman, 1977). Results from laboratory studies imply that MFO response to PCB deposition could be used to detect the presence of elevated PCB contamination (Addison *et al.*, 1981). Stegeman (1978) found elevated levels of hepatic cytochrome P-450 and several MFO indicators in fish populations from Wild Harbor. These observations suggested that contaminants from an oil spill 8 years previous were still stressful. This conclusion should be compared to the findings of Teal *et al.* (1978), namely that sediment concentrations of selected aromatics had decreased to about 1 ppm dry weight by 1976, a level thought to be non-toxic. However, some caution must be taken in interpreting MFO results as activity can be related to other forms of chronic stress. Sockeye salmon (*Oncorhynchus nerka*) exposed to low levels of copper showed a rapid concentration-related corticosteroid response (Donaldson and Dye, 1975). Similar corticosteroid responses by fish have been reported for zinc (Watson and McKeown, 1976) and kraft pulp mill effluent (Dye and Donaldson, 1974), and by birds which have ingested oil (Peakall et al., 1981). It was suggested that the disruption of endocrine balance (elevated corticosteroids) is one underlying cause of depressed growth in oil-dosed birds. It has been noted, however, that the corticosteroid response is complex and is affected by acclimation time.

The stability of the lysosomal membrane is important in preventing free hydrolases from causing autolytic cell damage. Lysosomal stability, measured in terms of the latency of activation and release of lysosomal enzymes in the digestive cells of *Mytilus*, was reduced by chronic exposure to crude oil and oil derivatives (reviewed by M. N. Moore, 1980). Lysosomes in the endodermal cells of the hydroid *Companularia flexuosa* were destabilized by exposure thresholds of 1.2–1.9 μg l^{-1} Cu, 40–70 μg l^{-1} Cd and 0.2 μg l^{-1} Hg (Moore and Stebbing, 1976). The assessment of lysosomal stability as a measure of biochemical stress would appear to provide a viable index. It has been shown in the laboratory and in the field to have a statistically significant positive linear relationship with the scope-for-growth index (see altered performance section) for the stressed organism. The basic biological processes underlying this connection are understood (Bayne et al., 1976, 1979).

Certain pollutants can lead to heritable change as a result of chromosomal damage or direct changes in DNA. It is generally accepted that most mutations lessen the capability of a population to cope with its environment. Unfortunately, there is very little information on pollutant-induced genetic changes in wild populations. Davavin et al. (1975) found that several species of algae suffered adverse effects on nucleic acids from oil at levels higher than 100 mg l^{-1}. Longwell (1977) examined the effects of oil-induced chromosome abnormalities on egg development and progeny survival in fish (cod and pollock), finding noticeable effects at No. 2 fuel oil concentrations of the order of 250 μg l^{-1} in the water column. Longwell has also examined chromosomal mutagenesis in developing mackerel eggs, sampled from the pollutant-contaminated New York Bight (Longwell, 1976; Longwell and Hughes, 1980).

In summary, biochemical responses appear to be sensitive to short-term pollutant stress and are often easily associated with the toxic mechanism. Extrapolation to the longer-term 'well-being' of organisms in polluted ecosystems is, in general, poorly defined.

4.1.4 Morphological

Of the various categories of response to pollutants, morphological abnormalities, because of their visibility, can readily serve as definitive evidence of adverse impact. Cell and tissue changes and incidence of gross deformities are particularly suitable for the monitoring of stress effects. Certain tumours and pollutant-potentiated diseases may also be useful indicators (Sindermann, 1979, 1980).

Skeletal anomalies, particularly those of the spinal column of fish and amphibia, have been correlated with pollutant stress and related to reduced individual performance. Long-term lead exposure ($120\,\mu g\,l^{-1}$) of three generations of brook trout (*Slavelinus fontinalis*) results in a greater than 20 per cent increase in skeletal deformities in the second and third generations (Holcombe et al., 1976). Various deformities in tadpoles' anatomy have been described and correlated with chronic pesticide exposure (Cooke, 1972, 1981; Brooks, 1981). Cooke (1981) noted that deformed tadpoles suffered higher juvenile mortality, but survivors showed some recovery from their deformed state at later developmental stages.

Recently reviewed reports of the increased presence of skeletal deformities in fish populations from California, New York, Japan and West Germany indicate that the anomalies are related to heavy metal or chlorinated hydrocarbon pollution (Sindermann et al., 1980). Possible effects of spinal damage such as impaired swimming, feeding and escape ability have been reviewed by Bengtsson (1979). He suggests that with more baseline information on normal incidence of skeletal deformities in fish species, this index may have widespread value in quantifying pollutant stress.

The incidence of shell abnormalities in marine bivalves may have some utility as a similar stress index for intertidal and benthic molluscs. The abnormal thickness and chamber development in the shell of Japanese oysters have been associated with reduced water quality off the French Coast although not directly correlated with levels of any specific toxic pollutant (Heral et al., 1981).

Increased incidence of teratogenic effects found in fish and amphibian larvae from chemical-exposed embryos and spawning adults in laboratory experiments suggests that monitoring the number of abnormal larvae from eggs taken from contaminated aquatic ecosystems may provide an estimate of the severity of environmental stress (Birge et al., 1977, 1978, 1980; Cooke, 1981).

Cell and tissue pathology have often been used to demonstrate incidence and seriousness of pollutant-caused anomalies. In extensive laboratory studies, Gardner (1975) demonstrated that tissues of the sensory organ systems of some marine fishes were damaged by exposure to copper, mercury, silver, the pesticide methoxychlor, crude oil and pulp mill effluent. He concluded that such neurotoxic effects were significant even if they did not cause permanent damage, for even temporary sensory disability can be disastrous in the natural system. Gill histology has also received attention as a means of describing pollutant stress on aquatic species (for example, Bubel, 1976).

The importance of the liver and kidney in metabolizing and eliminating chemical contaminants makes their pathology of particular interest in assessing the stress of exposed vertebrates. Urban pigeons in London having lead levels in the kidney of approximately 720 μg g^{-1} (Hutton, 1981) and wild urban rats with an average of 22.7 μg g^{-1} kidney Pb (Mouw et al., 1975) had pathological symptoms including intranuclear inclusion bodies in cells of the proximal tubules and mitochondrial abnormalities. Liver pathology, increased fat deposition in hepatic cells, and hepatomas have been identified in flounder populations from heavily polluted areas of the Puget Sound (Wellings et al., 1976; McCain et al., 1977) and are not common in populations from uncontaminated regions of the Sound. Liver pathology in fishes has also been related to chronic exposure to crude-oil-contaminated sediments (McCain et al., 1978).

The feasibility of using tumour incidence as an indication of the presence of carcinogens in the sea has been examined (Stich et al., 1976; Bang, 1980; Brown, 1980). The complex aetiology of various tumours and the lack of data on normal incidence confounds the interpretation of such an index. However, certain tumours, for example, liver and epidermal, are prevalent in organisms from polluted waters and are often associated with the bottom-dwelling or feeding habits of the species which place them in contact with the highest local concentrations of these potentially carcinogenic compounds (Sindermann et al., 1980).

The well-understood pathology of the mammalian liver and kidney and the visibility of gross abnormalities provide a strong base for the further study and use of certain morphological indices in assessing pollutant stress in wild populations.

4.1.5 Physiological

Toxic effect on physiological processes provides an important group of endpoints for examination, above the biochemical and sensory levels. The broad categories of physiological response of greatest interest include feeding activity, metabolism, osmotic–ionic balance and photosynthetic activity in plants.

Recognition of an abnormal feeding response provides an initial indication of physiological stress that may lead to eventual growth retardation. Reeve et al. (1977a, b) found that copepods exposed to 10 μg l^{-1} copper demonstrated reduced filtration and ingestion of phytoplankton. Moraitou-Apostolopoulou and Verriopoulos (1979) confirmed that feeding rates of a marine copepod were progressively decreased with increasing copper levels from 1 to 10 μg l^{-1}. Feeding response was reduced in pollution-adapted individuals only at copper concentrations of 5 μg l^{-1} or greater. For yearling brook trout, Drummond et al. (1973) described feeding activity as markedly less aggressive at copper concentrations of 6 μg l^{-1}, and permanently affected at 9 μg l^{-1} and above. Gonzalez et al. (1979) reported that oils reduced the filtration rate of the blue mussel. Exposure to oils has also been demonstrated to inhibit D-glucose uptake and mineralization by bacterial populations (Hodson et al., 1977).

Metabolic processes have been shown to be quite responsive to pollutant stress. Respiratory rate is often used to indicate a stressful environment and can now be accurately measured in the field as well as under laboratory conditions. However, the organism response may be one of either increased or decreased rate, depending upon species, pollutant type and concentration (Bayne, 1980). Stainken (1978) observed that 10 mg l^{-1} oil in water caused the respiratory rate of the soft shell crab to double, while 100 mg l^{-1} caused a significant depression of the rate. Moraitou-Apostolopoulou and Verriopoulos (1979) reported that copper concentrations from 1 to 10 μg l^{-1} produced a continuous increase in oxygen consumption by a marine copepod over a 20-hour period. However, Reeve et al. (1977a, b) did not find a good correlation between zooplankton respiration and copper level in the range of 1 to 10 μg l^{-1}.

PCBs at 1 mg l^{-1} decreased oxygen uptake in fish to 20 per cent of that measured prior to exposure, whereas shrimp exposed to 100 μg l^{-1} increased oxygen uptake by a factor of 3.6 (Anderson et al., 1974). This group also reported species- and age-specific respiratory response of crustaceans to petroleum hydrocarbons. Mussels, taken in samples along a gradient of pollution defined by levels of heavy metals and toxic organics, were found to have increased oxygen consumption rates corresponding to the more polluted sites (Phelps et al., 1981). Gill respiration-rate effects were corroborated by long-term reductions in growth.

The oxygen–nitrogen ratio (O:N) has also been used as an index of substrate utilization for energy production. The ratio of O to N varies naturally with the stage of development and diet and is influenced by the degree of stress. This ratio was reduced in all stages of larval lobsters exposed to 0.25 mg l^{-1} crude oil (Capuzzo and Lancaster, 1981).

Stoner and Livingston (1978) demonstrated that fish exposed to 0.1 and 1.00 per cent bleached kraft mill effluent had elevated ventilation rates and reduced food conversion, suggesting that the pollutant is causing an elevated maintenance metabolism.

With relatively short-term studies, such as those quoted above, there is the danger of misinterpreting a response as permanently damaging when in fact the species may become acclimated. However, increased metabolic demand and reduced food conversion efficiency, even over the short term, may weaken the organism, leaving it susceptible to any additional stresses (Stoner and Livingston, 1978).

The efficient functioning of osmoregulatory processes is essential to aquatic organisms. Preventing the loss of salts in a dilute environment is a life-preserving requirement for a freshwater species. Just as important to the estuarine species is the ability to regulate water–salt balance over a range of salinities, allowing adaptation to a fluctuating salt environment.

Various metals (Cu, Zn, Hg) and chlorinated hydrocarbons (DDT and PCBs) at sublethal concentrations have been found to affect osmoregulatory functions in a variety of estuarine species (Thurberg et al., 1973; Anderson et al., 1974;

Jones, 1975). The effects of mercury on osmoregulatory mechanisms in fish were examined by Renfro et al. (1974).

Presently, a gap remains between laboratory results indicating disrupted osmotic balance and application of this information in assessing the fitness of organisms in contaminated estuarine environments (Bayne et al., 1980).

Water balance and ionic regulation in terrestrial plants is also of great importance to their physiological 'well-being'. Heath (1975) described water loss and ionic alterations caused by ozone exposure of various plant species.

The unique photosynthetic ability of green plants provides an important physiological measure which can be used to assess pollution stress. Adverse pollutant effects on photosynthesis can be estimated for both aquatic and terrestrial species in the laboratory and under natural conditions (Bennett and Hill, 1974; Jensen, 1980). Since the photosynthetic rate is directly related to plant growth, a significant depression or inhibition of photosynthesis will ultimately translate into reduced primary productivity. The important relationship to be established is the degree to which chronic pollutant exposures can repress photosynthetic rates causing a significant retardation in plant growth (Bennett and Hill, 1974).

4.1.6 Altered Performance

Those indices which best reflect individual performance and are most easily related to population fitness are growth and reproductive success.

Growth is the net result of many essential processes such as consumption, excretion and respiration. As a summation of many factors, it is a useful integrated index of physiological status, applicable to multicellular organisms that have not yet reached their maximum biomass. Cell division rate in unicellular species is a useful measure of both growth and reproductive success.

Long-term experimental studies have indicated growth to be quite sensitive to pollutant stress. McKim and Benoit (1971) describe drastic growth-rate reduction in juvenile brook trout exposed to $17-32 \mu g l^{-1}$ copper over a 14-month period, whereas adult fish were unaffected. The rate of growth as compared to control fish was inversely correlated with copper concentration in the range $9-32 \mu g l^{-1}$. They reported that juveniles from unexposed and copper-exposed parents responded identically. The apparent absence of an adaptive process suggests that long-term growth rate may be an effective measure of chronic stress.

Bayne (1975) used the scope-for-growth index to measure the effect of environmental stress on *Mytilus edulis*. This index is an experimentally derived estimate of instantaneous growth rate representing the energy difference between food consumed and loss by excretion and respiration. If scope-for-growth values are positive, the animal has energy for growth and reproduction; when estimates are negative the animal is losing energy. A chronological series of estimates of scope-for-growth allows the estimation of relative growth rate correlated with pollutant concentrations. However, sources of stress other than the pollution must be recognized as affecting this index (Gilfillan, 1980). Investigation of soft shelled clams (*Mya arenaria*), heavily oil contaminated, showed that they gained carbon at 50 per cent the rate of unexposed reference individuals (Gilfillan and Hanson, 1975). The scope-for-growth index was not correlated with body burden of petroleum (Gilfillan *et al.*, 1976) but was related to the fraction of aromatic hydrocarbons (Gilfillan *et al.*, 1977).

A reduction of growth rate appears to be a universal response to chronic exposure to toxic chemicals. Several recent studies have reported reduced growth rates in macroinvertebrates at high but sublethal pollution levels (Percy, 1978, and Gilfillan and Vandermeulen, 1978, for petroleum derivatives; Wu and Levings, 1980, for pulp mill effluent; and Borgmann *et al.*, 1980, for metals). Terrestrial plants are similarly affected by metals and gaseous air pollutants such as sulphur dioxide and ozone (Miller *et al.*, 1977; Constantinidou and Kozlowski, 1979a, b). Cell division and growth rates in mixed microbial populations and algae have been greatly reduced by metals, oils and toxic organic compounds (Hutter and Oldiges, 1980; Gaur and Kumar, 1981; Jensen, 1983; Slater, 1983).

For invertebrates, stress-related decreases in adult body weight have been positively correlated with a reduction in the number of eggs per female, an indicator of reduced reproductive success (Buikema *et al.*, 1980). A mathematical relationship between growth and mortality developed by Borgmann *et al.* (1980) showed that, in general, toxicity was accurately indicated by a drop in growth rate.

Reproduction is the single most important function in the life cycle of an organism. Successful reproduction is essential to the continuation of the species. Therefore, the real test of long-term impact of sublethal pollutant concentrations on an exposed population is whether the population is capable of reproducing successfully. As a means of insuring perpetuation, certain species have even adopted the strategy of shunting a larger than normal proportion of available energy into reproduction under stressful environmental conditions (Bayne, 1975).

The importance of reproductive damage to species survival has stimulated much current research. An evaluation of methods to assess the effects of chemicals on the reproductive function of a number of wild mammalian and nonmammalian taxa was recently prepared (Vouk and Sheehan, 1983) and should be of value in improving the assessment of reproductive damage.

Reproductive failure can occur during a number of processes: courtship, development of gametes, fertilization, embryo development, hatching and early growth. Pollutant effects on these processes are outlined in Table 4.1. Since the reproductive process encompasses all life stages, its successful completion is the basic individual goal. The inability of an organism to successfully complete any

Table 4.1 Some effects of pollutants on reproduction

Vital process	Critical effects of pollutants
Development of gametes	Incomplete or abnormal development of ova or spermatozoa; gene damage
Fertilization	Interference with homing of spermatozoa to the ova; impairment of the ability of spermatozoa to enter the micropyle and successfully fertilize the ova
Embryo development	Cytological and cytogenetic abnormalities including chromosome bridging, breakage and translocation; interference with hardening of the egg; interference with gas and water exchanges; cessation of development
Hatching	Failure to hatch; high mortality of newly hatched larvae; teratogenic abnormalities
Sexual maturation	Histopathological effects on gonads; changes in production and metabolism of gonadotropins
Courting and mating	Destruction of spawning and mating grounds; inappropriate courting or mating behaviour leading to reduced mating success

one stage would indicate a reduced reproductive fitness of the population.

The toxicology of reproductive behaviour is more highly developed for bird species. Behavioural abnormalities, including improper mating responses, reduced nest attentiveness and protection and the chicks unresponsiveness to their mother's call, have been recorded for birds chronically exposed to toxic chemicals (Heinz, 1976, 1979; Fox et al., 1978; Custer and Heinz, 1980). Abnormal reproductive behaviour is often sufficient to cause increased embryonic mortality and has in some cases been correlated with levels of specific pollutants in tissues.

The effects of chemicals on maturation and gamete development have been recently discussed for fish (Donaldson and Scherer, 1983), amphibians (Martin, 1983) and various invertebrate taxa (Dixon, 1983; Davey et al., 1983; Landa et al., 1983). Field evidence indicates that populations exposed to grossly contaminated environments, such as occur after oil spills, are often sexually sterile the next reproductive season (Blumer et al., 1971; Loya, 1975).

For those aquatic species that release gametes directly into the water there is much evidence substantiating the sensitivity of the fertilization phase to pollutant toxicity. Nicol et al. (1977) have shown that extracts of No. 2 fuel oil depress sperm mobility and interfere with fertilization in the sanddollar. The pesticide Lindane, at 25 μg l^{-1}, and sublethal levels of several metals (Hg, Fe, Cy and Cr) have also been shown to be toxic to gametes and to lower the rate of successful fertilization of fish eggs (Billard, 1978).

Recently, Koster and Van den Biggelaar (1980) reported that the development of *Dentalium* (tusk shell) eggs from females collected before an oil spill was significantly better than the development of eggs from contaminated animals, although in early embryonic stages no difference was noted. At time of collection of females, sediment hydrocarbon concentration was about 30 mg kg^{-1} dry weight.

Staveland (1979) noted that one year after an accidental crude oil spill there was no detectable effect in fertilization of *Littorina* eggs; however, hatching success was significantly reduced in 'polluted' populations. Reduced hatch success is perhaps the most common index reported for stressed organisms (e.g., Birge *et al.*, 1979; Peakall, 1983), although egg production (e.g., Reeve *et al.*, 1977b; Wu and Levings, 1980) and the size of brood per female are frequently examined (e.g. Hatakeyama and Yasuno, 1981a). In the special case of bird eggs, shell thickness has been well studied in relation to pollutant stress (review by Cooke, 1973; Stickel, 1973).

The larval stage has also been shown to be quite susceptible to pollutant stress (Birge *et al.*, 1979, 1980). Livingston (1977) reported effects of organochlorine accumulation in fish eggs, noting that the pesticides caused increased embryo and larval mortality. Birge *et al.* (1980) suggest that 10 per cent or greater increase in mortality at the embryonic and larval stage would significantly affect population dynamics in natural communities. Long-term oil stress reduced juvenile settlement in the salt marsh crab at concentrations 5 to 10 times lower than those affecting adults, indicating that the effect need not be increased mortality at the larval phase to adversely affect local recruitment (Krebs and Burns, 1977).

There are relatively few field studies quantifying the impact of pollutants on the reproductive success of wild mammalian populations.

The female ringed seal (*Pusa hispida*) has exhibited pathological changes in the uterus, and an apparently lower reproductive rate associated with high tissue levels of DDT and PCBs (Helle *et al.*, 1976a,b). Pregnant females averaged 75 mg kg^{-1} of DDT in blubber, while non-pregnant females averaged 130 mg kg^{-1}. High premature-birth rates observed in the California sea lion (*Zelophus californianus*) also have been linked to elevated PCBs and DDT levels although an imbalance between mercury–selenium and bromine was also implicated in the etiology (Martin *et al.*, 1976). In an extensive review of residue effects on harbour seals, Reijnders (1980) concluded that the decreases in reproductive success of Dutch seal populations correlates most strongly with high concentrations of PCBs in the tissues.

In his survey of field studies, Schofield (1976) observed a reduction in reproductive success of freshwater fish sampled in low pH lakes. Beamish and Harvey (1972) observed a decrease in pH from 6.8 to 4.4 in a Canadian lake over a period of 10 years, corresponding to a total disappearance of trout, herring and other fish. Pough (1976) has reported similar acidification effects on reproduction in amphibians.

These examples very clearly indicate that chronic pollutant stress, leading to

Table 4.2 Some of the sublethal effects of pollutants on life stages of various animals (modified from Waldichuk, 1979)

Life stage	Vital life process	Critical effects of pollutants
Egg	Meiotic division of cells; fertilization; cleavage mitoses of fertilized egg; hatching; respiration	Gene damage; chromosome abnormalities; damage to egg's membrane; direct toxicity to embryo from pollutant; impaired respiration; reduced hatch
Larva	Metamorphosis; morphological development; feeding; growth; avoidance of predators, parasites and disease	Toxicity from bioaccumulated poisons in yolk sac during early feeding; biochemical changes; physiological damage; deformities; behavioural alterations
Juvenile	Feeding; growth; development of immune systems, endocrine glands; avoidance of predators, parasites and disease	Direct toxicity; reduced feeding and growth; altered predator-prey relations; impaired chemo-reception; reduced resistance to parasites and disease
Adult	Feeding; growth; sexual maturation	Direct toxicity; adverse alteration of environmental conditions, e.g. dissolved oxygen; physiological and biochemical changes; behavioural alterations

reduced reproductive success, can eventually result in the extinction of one or more populations in the exposed ecosystem. Because of the importance of the reproductive process, its investigation in single-species studies of responses to stress is an essential step in determining any long-term pollutant impact.

The effects of exposure to pollutants on the various life stages of organisms have been illustrated. An effect at any stage can reduce the probability of an individual successfully completing its life cycle. A summary of critical behavioural, biochemical, morphological, physiological and integrated effects of pollutants is presented in Table 4.2. The assessment of effects at the various life stages is essential to a complete understanding of a population's susceptibility to toxic stress.

4.2 INTERACTION WITH OTHER ENVIRONMENTAL STRESSES

Each species lives in an environment which exhibits a multitude of physical, chemical and biological constraints. Thus the organism is continually exposed to many factors acting independently or in concert with others. Species are adapted to survive within certain ranges of these factors. All factors together make up Hutchinson's niche concept (Hutchinson, 1944). In general, this means each species exists in an environmental compromise, not always living in the optimal range for all essential functions. Any pollutant stress limits the range of functional response available to the organism prior to its reaching the threshold of damage. Other adverse environmental conditions aggravate the already stressed species. Therefore, combinations of stress can be expected to cause adverse response at lower pollutant concentrations. Multifactor interactions, including pollutant toxicity and the stress response of exposed species, were recently reviewed by Livingston (1979) for a contaminated coastal system and by Babich and Stotzky (1980) for microbial communities.

Examples of pollutant–environment interactions on species response can be found where temperature was examined as the fluctuating environmental factor. The interpretations of the results from interactive stress experiments are complex. Low winter temperatures apparently reduced the toxicity of mining effluent to stream macroinvertebrates (discussed by J. W. Moore, 1980); however, the susceptibility of seabirds to oils is higher at cold temperatures (Levy, 1980). The victims of oil death were found to have only extremely minute oiling of their feathers, below that concentration necessary to inflict mortality under less severe environmental conditions. Atlantic salmon (*Salmo salar*) chronically exposed to 40–60 μg l^{-1} DDT also showed a reduced tolerance to low temperature (Anderson, 1971). Blue crab (*Callinectes sapidus*) mortality in a north Florida estuary contaminated with DDT was seasonally influenced by temperature variation, with the major mortality associated with rapid temperature decreases accompanying cold fronts (Koenig *et al.*, 1976). Survival from hatch to first adult stage of mud crab (*Rhithropanopeus larrisii*) larvae exposed to combinations of 50 or 150 μg l^{-1} Cd and 10, 20 or 30 °/00 (parts per thousand) salinity was significantly reduced at the 20 and 35 °C temperature extremes (Rosenberg and Costlow, 1976). Although the highest temperature (35 °C) caused the greatest decrease in survival, the lower temperature regimes (20, 25 °C) increased development time. Interestingly, results indicated that cycling temperature (20 to 25 °C, 25 to 30 °C) may have a stimulating effect on survival of the larvae compared to constant temperature, both in the presence and the absence of chronic cadmium exposure.

Cadmium at 150 μg l^{-1} in combination with 10 per cent salinity was fatal to the development of the mud crab even within its optimum temperature range (Rosenberg and Costlow, 1976). This response to the combined stresses may be explained by the fact that the exposed mud crab is already under physiological stress, attempting to regulate its body fluid to a hyper-osmotic state. A reverse problem of salinity adjustment must be faced by anadromous juvenile salmonids. Long-term copper exposure at 20–30 μg l^{-1} reduced the tolerance of yearling coho salmon (*Oncorhynchus kisutch*) to increased salinity. Some recovery in ability to survive in seawater was apparent approaching the paar-smolt transformation during the spring migratory period (Lorz and McPherson, 1976). It is suggested that the reduced tolerance to increasing salinity in exposed coho, together with copper-induced behavioural abnormalities, would seriously reduce the chances of successful migration to the ocean and adaptation to seawater.

The pH of the environment into which a pollutant is deposited may influence the chemical form, the solubility, and its toxicity to exposed biota. This is particularly obvious with toxic metals. A decrease in pH of 1 unit from any reference (pH 6–10) resulted in an increase of lead by a factor of 2.1 in the blood of exposed rainbow trout (Hodson et al., 1978). Inhibition of ALA-D followed the same time scale as the change in lead levels in blood, indicating that toxicity increases as pH decreases. Hultberg (1977) reported that there were pronounced declines in sea trout and minnow survival, and reproductive success, associated with the rapid pH drops accompanying the solution of air pollutants in spring snow melt. Most of the mortality was attributable to the short transition period.

An environmental factor of great importance in mediating the stress response of terrestrial species exposed to pollutants is drought. The dry weight yield of two native plant species (*Andropogon scoparius* and *Monarda fistulosa*) was significantly decreased by application of drought stress and 20 μgCd/g soil (Miles and Parker, 1980). The effects of the stresses on weight were determined to be independent and additive.

Most effluents are not composed of a single toxic substance. Therefore, the potential interactions of pollutants in either a synergistic or an antagnostic manner must be considered in evaluating the response of the population to its environment (discussed by Anderson and D'Apollonia, 1978). Recently this aspect of pollution studies has received increased attention among researchers, as ambient levels of pollutants in both urban and rural areas have risen. Separately, injury threshold levels from air pollutants NO_2 and SO_2 were 2 and 0.5 ppm, respectively, whereas together, NO_2 and SO_2 ratios of between 0.1/0.1 and 0.15/0.10 ppm produced maximum injury (Tingey et al., 1971). With some O_3/SO_2 combinations (0.05/0.50, 0.1/0.1, 0.1/0.25, 1.0/0.5 ppm), degrees of injuries were different among various species including radish (more than additive), cabbage (additive) and tomato (less than additive) (Tingey et al. 1973). Often, beneficial interactions may occur; for example, the increased levels of systemic fungicides sprayed on tobacco foliage (benomyl at 1.8 g l^{-1}, diodine at 2.4 g l^{-1}, or maneb at 3.6 g l^{-1}) decreased injury levels by 24 per cent to 60 per cent when combined with continuous ambient O_3 levels above 0.05 ppm (Reinert and Spurr, 1972).

Heavy metals such as Cd, Zn or Ni in the soil from industrial fallout, fertilizers or sewage sludge sources are accumulated differentially by plants and predispose the plants to greater sensitivity and injury at lower O_3 levels (Czuba and Ormrod, 1974; Ormrod, 1977). The Cd–O_3 interaction involved the uptake of cadmium (10–100 μg ml^{-1} in the soil), its redistribution into younger leaves after O_3 exposure (0.05–0.35 ppm for 6 h), and irreversible changes in water content, which resulted in a more than additive injury in plants (Czuba and Ormrod, 1981).

The environment indirectly influences organisms by affecting the availability and nutritional value of foods. Starvation and nutritional deficiencies often cause marked changes in the sensitivity of an animal to toxic chemicals. While it is generally true that starvation enhances toxicity, there are some exceptions. For example, a reduction in caloric intake inhibits the chemically induced tendency for animals to develop tumours (discussed by Uthe *et al.*, 1980). In addition, the nutritional value of the food has been shown to influence the stress response to toxic pollutants. The cladoceran, *Daphnia magna*, fed vitamin-enriched algae, were less sensitive to chronic copper stress than animals fed a trout-granule diet. Sensitivity was based on survival, mean brood size, and r, the instantaneous rate of population increase (Winner *et al.*, 1977). It is of interest to note that there was no significant difference in acute response (72 h LC50) attributable to diet. This situation suggests that the ratio of chronic to acute toxicity would not be constant under varying environmental conditions influencing nutritional state.

4.3 LIFE HISTORY

A thorough knowledge of the life history of a species is critical to the assessment of population responses to chronic pollution stress and to the understanding of altered interaction within the community. It has often been noted that there are life-stage differences in sensitivity to pollutants. The initial fertilization stage in fish reproduction is particularly sensitive to pollutant stress as the sperm and eggs are released directly into the contaminated water (Rosenthal and Alderdice, 1976). McKim (1977), in a review of partial and complete life cycle toxicity tests, found the embryo-larval and early juvenile stages of fish to be the most sensitive to a wide variety of toxic chemicals. Juvenile fiddler crabs are sensitive to oil levels 5 times lower than are adults (Krebs and Burns, 1977), and similar results have been found for gastropods (Staveland, 1979). In addition, Rosenthal and Alderdice (1976) have aptly noted that the response to pollutant stress from exposure at one developmental stage may not be expressed until a later, susceptible stage.

The moulting cycle in arthropods has been shown to be particularly sensitive to pollutant disruption. Metals such as copper and zinc interfered with ecdysis in the shrimp, *Crangon crangon* (Price and Uglow, 1979). Premature mortality in the post-moulting stage was attributed to the large uptake of water containing metals during moulting. A significant increase in the duration of the intermoulting period has been shown to occur with exposure of the marine isopod (*Mesidotea entomon*) to 1 mgl^{-1} crude oil. The increased time between moults was a reflection of the reduced growth rate of exposed animals. As many arthropod species must attain a certain size to reach maturity, pollutant interference with moulting may have both short-term and long-term effects on the exposed populations. In general, pollutant interference at an early life stage can halt development at that point, or can delay maturation. Either effect could mean long-term inhibition of recruitment and loss of sensitive species when the environmental conditions become harsh.

A population is generally well-adapted to its environment. Its fine-tuned response to normal diurnal or seasonal cycles is essential to its well-being and is predictable. Thus, we see that young are born when environmental conditions are suitable, food is available and predators and competitors are of minimal influence. Any interference with the fine-tuning of the organism with respect to its environment can cause severe impacts. An induced prolongation of normal life history through delays in development or metamorphosis would alter population fitness. For example, it has been shown that chronic exposure of frog tadpoles to DDT and organophosphorous pesticides extends the time of larval development to metamorphosis (Cooke, 1970; Mohanty-Hejmadi and Dutta, 1981). A prolongation of the larval phase increases the tadpoles' exposure to both predation and desiccation pressures, leading to a reduced number of survivors reaching a mature adult stage (Dutta and Mohanty-Hejmadi, 1978). Such a series of responses has obvious implications on the population's ability to survive in pesticide-contaminated breeding ponds, common to agricultural regions.

Pollutants can also stimulate behavioural responses uncoordinated with seasonal cycles. Krebs and Burns (1977) described oil-contaminated marsh crabs (*Uca pugnax*) displaying mating colours at a time of year inappropriate for this behaviour. Because of the many other chronic effects these oiled crabs demonstrated, such as larval avoidance of oiled sediments, the impact of the abnormal mating behaviour could not be evaluated in terms of the drastically reduced recruitment noted for the two years following the spill. It is suggested, however, that any induced change in reproductive behaviour which would put mating, egg laying or hatching out of synchrony with normal environmental cues would have a damaging impact on reproductive success. It can be imagined that this would be particularly detrimental to species with a short-lived adult reproductive stage, such as mayflies (Ephemeroptera). These aquatic insects often have less than 24 hours from emergence to mate and oviposit before dying. An uncoordinated male and female emergence or emergence into adverse environmental conditions could drastically reduce local recruitment for the next generation. The issuing forth of larval trout from relatively high pH ground water to the sharply declining pH of streamwater, associated with 'acid' snow melt, is another example of members of a species emerging into environmental conditions detrimental to survival (Schofield, 1976).

4.4 POPULATION INTERACTIONS

It is important to understand how population interactions, as influenced by the effects of sublethal chemical stress, will affect the success of populations and the dynamics of the community as a whole. Toxic pollutants at a sublethal level can impair a species, ability to function normally within the community. Interactions within and among trophic levels can be altered, thereby affecting the structure and function of the ecosystem.

The effects of toxic substances on predator–prey interactions have received considerable attention in the last decade (Cooke, 1971; Kania and O'Hara, 1974; Ward and Busch, 1976; Tagatz, 1976). Sublethal levels of pesticides have been reported to interfere with prey escape and other antipredator behaviour. Ward et al. (1976) found that in in situ tests, the pesticide Temefos reduced the density of marsh crabs (Uca pugnax) in the open test plot by 25 per cent but did not significantly affect the density in the caged plot. Mortality was due to bird predation on the exposed crab population. Farr (1978) demonstrated that killifish (Fundulus grandis), when given a choice of prey, selectively captured the grass shrimp (Palaemonetes pugio) which had been weakened by methylparathion contamination of less than $0.5\,\mu g l^{-1}$. This change in food preference appeared to be effected by a decrease in the grass shrimp's ability to escape predation while impaired by the pesticide. Increased prey vulnerability was also demonstrated for fathead minnows (Pimephales promelas) undergoing acute (0.05–$0.5\,mg l^{-1}$) and chronic (0.013–$0.025\,mg l^{-1}$) cadmium exposure (Sullivan et al., 1978). Schneider et al. (1980) reported increased prey vulnerability of fingerling rainbow trout (Salmo gairdneri) exposed to $7.0\,mg l^{-1}$ phenol but indicated that since this level of exposure was near the lethal concentration, predator–prey interactions did not constitute a sensitive indicator of chronic stress. However, it has been suggested that increased susceptibility to predation be used as a laboratory test of stress effects on trophic interaction (Goodyear, 1972).

Pollutant-caused elimination or drastic reduction of predator populations generally leads to an increased in prey abundance. Hurlbert (1975), in his review of the secondary effects of pesticides, noted that numerous pesticide-induced population increases appeared to have resulted from predator removal. After the application of the fish toxicant, rotenone, to two Canadian lakes to a level of $0.75\,mg l^{-1}$, Anderson (1970) found that the normally benthic-littoral amphipods, Gammarus lacustris and Hyalella azteca, became abundant in the open water. These results suggest that reduced predator pressures allow expansion of prey distribution and abundance.

A reduction in the grazing population can produce a similar response in algae populations. Phytoplankton blooms followed the reduction of zooplankton grazing pressure caused by toxic components of oil following the grounding of the Tsesis in 1977 (Johansson et al., 1980) and the Amoco Cadiz spill in 1978 (Nounou, 1980). A similar bloom response has been noted with application of $10\,\mu g l^{-1}$ of organophosphorous pesticides (Butcher et al., 1977; Hughes et al., 1980a). Again, the zooplankton populations were severally reduced by direct toxic stress.

If a population is not held in check by other factors, an increase in food supply resulting from pollution effects may stimulate population growth of the feeding species. Bacteria and coral normally have a rather symbiotic relationship. The coral, when irritated, produces a mucus which is eaten by the bacteria, thereby cleaning the coral. Apparently, pollution and other stresses increase the mucus-producing response which, in turn, causes increased bacterial feeding, eventually killing the coral. This response has been shown to occur when coral heads were exposed to low concentrations of petroleum hydrocarbons or heavy metals (Mitchell and Chet, 1975; Ducklow and Mitchell, 1979; Loya and Rinkevich, 1980). Like the bacteria, scavenger populations accrue temporary benefit from increased food availability following pesticide kills (see Hurlbert, 1975).

It is also evident that the grazing populations which have the ability to recover rapidly from toxic stress will gain the advantage of utilizing the high algae densities resulting from the initial decimation of grazers. An increased abundance of phytoplankton permitted populations of the cladocern *Moina* when they recovered several weeks after Dursban treatment, to achieve higher densities than they ever had in uncontaminated ponds (Hurlbert *et al.*, 1972). However, the pesticide treatment also killed predators, releasing *Moina* from natural predation stress, which could be a factor contributing to the rapid population increases.

Pollutant-induced reductions in prey abundance and variety may have adverse affects on predator populations. Observations from Södergren (1976) attributed the decline in salmon population to a decrease in prey abundance brought about by the prey's intolerance to metal pollution, in a river containing cobalt at levels of 30–70 μg l^{-1}. This was particularly evident from the decline in large mayfly nymphs in the drift during spring, providing less food for young salmon during this critical growth period. Jeffree and Williams (1980) reported that mining pollution significantly changed the diet of the purple-striped gudgeon, emphasizing that the effects were more complex than simply the depressing or raising of the abundance of dietary species. Similar alterations in diet were also reported for fish exposed to a single treatment (13μg l^{-1}) of diflubenzuron in experimental ponds (Colwell and Schaefer, 1980). Growth rates and condition factors of the fish, however, were similar to control fish 3 months after exposure. Such results imply that switching in prey selection may provide short-term compensation for reduced availability of preferred species.

There is some evidence of food transfer of toxic substances to the predator causing toxic symptoms to appear later. The shrimp *Gammarus pulex*, fed Cd contaminated fungi, showed reduced viability (Duddridge and Wainwright, 1980) and the cladoceran *Moina macrocopa*, fed *Chlorella* containing 340 μg g^{-1} Cd, had reduced reproductive success (Hatakeyama and Yasuno, 1981b). The widely studied case of methylmercury is an example of food-chain bioaccumulation leading toxic effects on top predators (in this case, man). Levels of methylmercury around 0.5 μg l^{-1} in river sediment lead to levels a thousandfold higher in fish (Miller *et al.*, 1977). The much used model of DDT accumulation in both aquatic and terrestrial systems, which can eventually lead to a reduction in reproductive success and, therefore, decreases in populations, is another classic example of trophic interactions with pollutant (Harrison *et al.*,

1970; Edwards, 1973). The effects of DDT, accumulated by fish-eating birds, on the thickness of egg shells and related egg breakage is one of the best-documented examples (see Stickel, 1973). Recently, Beyer and Gish (1980) found that earthworms accumulated organochlorines to tissue levels (32 μg g^{-1} DDT, 8 μg g^{-1} heptachlor and dieldrin) which are potentially hazardous to birds.

Studies have indicated that long-term sublethal toxic stress affects normally healthy organisms, significantly altering their resistance to disease. Plants under the stress of chronic air pollution have been shown to differ in incidence of insect and microbial damage, the incidence either increasing or decreasing in specific cases, depending on the balance of plant and parasite responses (Heagle, 1973; Smith, 1974; Treshow, 1975, 1978).

Within marine fish populations the incidence of various nonspecific disease symptoms such as ulcers and fin rot seem to be clearly associated with degraded estuarine and coastal environments (Sindermann *et al.*, 1980). Hetrick *et al.* (1979) reported increased susceptibility of rainbow trout to infectious hematopoietic necrosis virus after exposure to sublethal (less than 0.01 μg l^{-1}) levels of copper. Mortality was twice as great in stressed as in control groups. Although many of the complexities of the pollutant-disease interaction are not well understood for wild species, their increased susceptibility to infection or infestation while under toxic stress is well documented.

Documenting pollutant effects on competitive interactions has been more difficult than demonstrating feeding interactions. A pollutant-induced reduction in competitive ability could lead to exclusion of the 'weaker' and rapid increase in the more tolerant populations. This type of interaction is a well-known secondary effect of pesticide application (Van den Bosch, 1969). Mosser *et al.* (1972) and Fisher *et al.* (1974) reported that species differences in the toxicity of PCBs and DDT, in pure and mixed diatom cultures, were magnified by interspecies competition. Although the roach (*Leuciscus rutilus*) has disappeared from many acidified lakes (pH below 5.5), in lakes where populations survive, this fish has a considerably higher growth rate and attains a larger size than normal (Almer *et al.*, 1974). Acid waters have apparently decreased intraspecific competition for the available food, thereby allowing survivors to grow faster and larger.

It is becoming increasingly more obvious that population interactions are an essential factor in defining population success as well as changes in ecosystem function and structure, and that they must be considered when evaluating the influence of chemical stress on an ecosystem.

4.5 IMPACT ON POPULATIONS AND ITS RELEVANCE TO ECOSYSTEM RESPONSE

Chemical injury to individuals resulting in premature death and reduced reproductive success and recruitment are ultimately reflected in lower abundance and altered distribution of exposed populations. The description and quantification of abundance and distribution changes in a population are particularly appropriate where pollutants are introduced at a point source so that biological effects may be assessed along a gradient from the source. For example, the abundance and distribution of immature may-flies (*Baetis bicaudatus*) and caddis-flies (*Hydropsyche* sp.) were found to be inversely correlated with soluble copper concentrations along a pollutant gradient in a small mountain stream, during the summer months (Sheehan, 1980). Adult emergence and seasonal increases in heavy metal levels, however, frustrated the use of abundance measures of any single insect population in assessing the effects of copper stress continuously throughout the year. In this case, it was suggested that a series of different species would need to be monitored sequentially throughout the year to account for seasonal effects.

For those populations that have identifiable year classes or generations, the impact of pollutant stress can be assessed through studying changes in population structure. Some animals can be aged rather precisely using such features as annual growth marks (Jones, 1979), while age class in others must be estimated from relative size (Waters, 1977). The influences of natural variability in growth rates, fecundity and recruitment must be considered in any analysis of pollutant-induced changes in age structure. A clear example of pollutant impact on population structure was presented by Notini (1978). He reported that the changes in size distribution of common mussels (*Mytilus edulis*) for the years 1971–1976 reflected the recovery of the population from the severe stress of an oil spill (see discussion, section 5.7). During the first two years following the spill, only a few larger organisms and some newly settled recruits were found. By 1976, the population structure was composed of a complete size range of mussels with the greater proportion in the mid-range, reflecting normal recruitment and age distribution.

A large body of literature exists which indicates that long-term sublethal pollution stress selects for some measure of tolerance (Brown, 1971, 1978; Antonovics, 1975; Ernst, 1975; Bradshaw, 1976; Bryan, 1976). Brown (1976) demonstrated that metal tolerance existed in populations from the polluted River Gannel and persisted in the second generation of amphipods cultured in metal-free waters in the laboratory. Wu and Antonovics (1978) demonstrated that for the grass *Agrostis stolonifera* tolerance to copper and zinc was genetically determined and operated at the cellular level. Tolerance is normally pollutant specific in higher animals, although bacteria (*Bacillus* sp.) from the contaminated New York Bight were found to be resistant to both heavy metals and antibiotics (Timoney *et al.*, 1978). Although tolerance is an expression of change in the population gene pool, there is relatively little data available on actual changes in the gene pools of populations exposed to chronic stress (see review, Beardmore *et at.*, 1980). Genetic structure in a marine snail population was found to exhibit unusually high year-to-year variation in oil-impacted populations (Cole, 1978).

Genetic effects of oil pollution have also been demonstrated in natural populations of *Mytilus* by means of electrophoretic analysis of genotypic frequencies (Battaglia *et al.*, 1980). A higher frequency of the more common allele and a significant drop in average heterozygosities were found in populations from more polluted areas. Luoma (1977) suggested that under the stress of pollution, the probability of an opportunist species developing resistant populations is greater than that of more specialized species. Therefore more simplified ecosystems may result from chronic pollution stress through the elimination of species without resistant genotypes.

4.5.1 Relevance of Population Response Extrapolations

As our ability to detect sublethal toxic injury at the organism and population levels has improved with increased sophistication of scientific methodology and instrumentation, the ultimate question of meaningfulness of extrapolation of this data to natural communities must be considered.

Direct observation of environmental impacts on particular individuals and populations has provided information as to the effects of stress on the endpoints of growth, reproductive success, survival probability and several other characteristics. Such observations can confirm that a single species under specific environmental conditions will be adversely affected by a pollutant, and to take into account natural ecological interaction which cannot generally be done in single species toxicity tests under laboratory conditions (Cairns, 1981; NAS, 1981). Studies of individuals and populations have also provided mechanistic explanations of possible consequences at the community level; for example, the relative sensitivity of key populations to toxic stress should provide insight into changing dominance and trophic patterns. Consequently, it is likely, and desirable, that such studies will remain an integral part of ecotoxicological assessment. Population studies are particularly relevant for those species which play critical roles in the community such as 'keystone' predators (Paine, 1969, 1974) or pollinators (NRCC, 1981b). Also of interest are those populations of economic and aesthetic interest to man.

On the other hand, a population-oriented approach leaves major ecological issues unaddressed. The ecosystem is a complex biogeochemical entity with internal regulating mechanisms. Impacts on a single population may have little effect on the functioning of the ecosystem. If the abundance of preferred prey is reduced by toxic stress, often the populations of predators can switch their choice of food without showing adverse effects or without significantly damaging the food web (Colwell and Schaefer, 1980). Pollutant-sensitive species may be replaced by competitors without affecting biomass or productivity of the system. In a controlled ecosystem experiment, concentrations of 5–10 $\mu g\, l^{-1}$ Cu eliminated sensitive algae populations but biomass and primary productivity were not significantly affected (Thomas and Siebert, 1977; Thomas *et al.*, 1977).

Changes in ecosystem processes, such as decomposition of organic materials and mineral cycling can occur without major effects on the types of organisms normally observed in pollution studies (Van Voris *et al.*, 1980). In the final analysis, Heath *et al.* (1969) pointed out that it was not possible to characterize the response of a system to perturbations solely from a knowledge of the response of component parts. In fact, the state of the 'whole' system must be known in order to understand the collective behaviour of its components (Weiss, 1977).

The limitations of population studies in estimating ecosystem damage from pollutant perturbations suggest that an assessment approach must integrate single species studies with analysis of community structure and dynamics and ecosystem function.

Effects of Pollutants at the Ecosystem Level
Edited by P. J. Sheehan, D. R. Miller, G. C. Butler and Ph. Bourdeau
© 1984 SCOPE. Published by John Wiley & Sons Ltd

CHAPTER 5
Effects on Community and Ecosystem Structure and Dynamics

PATRICK J. SHEEHAN
Division of Biological Sciences
National Research Council of Canada
Ottawa, Ontario, Canada K1A 0R6

5.1	Abundance and Biomass	55
5.2	Reduction in Population Size and Extinction	58
	5.2.1 Loss of Species with Unique Functions	60
	5.2.2 Species Richness	61
5.3	Community Composition and Species Dominance	64
	5.3.1 Species Lists	64
	5.3.2 Indicator Species	65
	5.3.3 Biological Indices	67
	5.3.4 Dominance Patterns	67
5.4	Species Diversity and Similarity Indices	72
	5.4.1 Diversity	72
	5.4.2 Similarity	79
5.5	Spatial Structure	82
5.6	Stability	83
	5.6.1 Inertia	87
	5.6.2 Elasticity	88
	5.6.3 Amplitude	89
	5.6.4 Hysteresis and Malleability	91
	5.6.5 Persistence	91
5.7	Succession and Recovery	92
	5.7.1 Terrestrial Succession	92
	5.7.2 Recovery in Aquatic Ecosystems	93
5.8	An Illustrative Case Study	95

Communities are assemblages of populations structured by biotic interactions and the constraints of their physical and chemical environment. The structure of an ecosystem is defined by the abundance and biomass of all populations and their spatial, taxonomic and trophic organization. The integrated response of the component populations to the presence of toxic pollutants will be reflected in alterations of the structural and dynamic characteristics of the stressed community. It appears that the changes in structural characteristics induced by toxic pollutants may be similar to those caused by natural forms of stress (Gamble *et al.*, 1977).

Ideally, predictions of structural changes should be based on an understanding of the important biological interactions which cause the restructuring; however, ecological theory has not yet been developed to this point. Moreover, Gray *et al.* (1980) emphasized that populations and their environments co-evolve and, within certain limits, resident communities are the products of the history of their environment. As a result, it is difficult to describe a 'typical' ecosystem on which to base a general theory to interpret interactions. Therefore, the approach to assessment of integrated response to pollutants has been the 'after-the-fact' observation of qualitative and quantitative changes in community structure. Cairns *et al.* (1972) suggested that structural changes may be visualized as an information network reflecting environmental conditions but not demonstrating the external mechanisms or internal interactions which brought about the reorganization.

Marine ecologists have argued that structural indices best meet statistical criteria for the monitoring of community response to toxic substances (Heip, 1980: Gray *et al.*, 1980). These indices do not have the daily periodicity of primary productivity or the short-term variability associated with respiration. Community structure, however, is loosely related to ecosystem function, although aspects of structure (e.g. diversity and dominance patterns) may change significantly under stress with no accompanying disruptions of function (e.g. productivity), or inversely, function may be altered without significant changes in composition and diversity (Matthews *et al.* 1982). The lack of a predictable relationship between structural and functional responses to stress suggests the need for a balanced approach in assessing pollution effects at the ecosystem level.

This chapter focuses on structural responses to toxic pollutants. Figure 5.1 depicts a conceptual scheme, fitting the analysis of changes in community structure into the larger framework of integrated ecotoxicological assessment. The importance of developing a greater understanding of the ecological interactions which determine community organization is highlighted. Listed under the heading of 'community structure and dynamics' are several indices which are useful in assessing pollution effects. Although these indices define a fixed structure, no community is static. Stability and succession as influenced by varying community organization, are susceptible to disruption through pollution-induced changes in that organization, and must be considered in assessing long-term effects.

It is the induced changes in structural properties which are of primary interest in pollution assessment, rather than the specific structure of the community itself although, obviously, this is ecologically important as well. To assess change, an appropriate set of structural references is essential. Baseline data on a community prior to pollution exposure can provide an ideal control for comparisons, however, this situation is seldom realized. A time series of changes

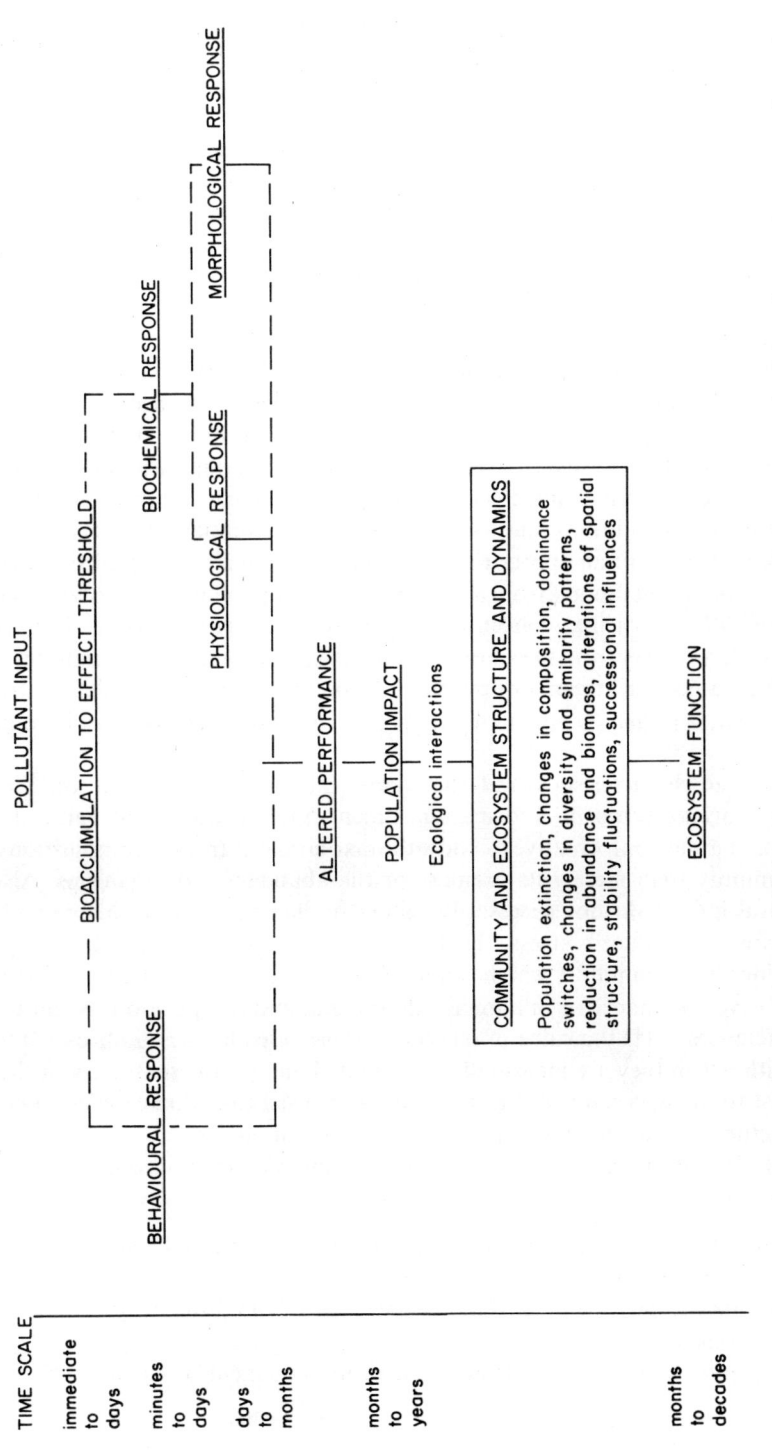

Figure 5.1 A conceptual chronology of induced effects following exposure to toxic pollutants, emphasizing changes in community and ecosystem structure and dynamics

in structural indices after introduction of a pollutant would reflect trends in reorganization. Reductions in numbers, biomass and taxonomic and trophic diversity indicate a disruption in homoeostasis. Conversely, increases in these indices with time would infer at least partial recovery of the system. If baseline data are unavailabe (as is frequently the case), changes can still be monitored in the community over time but obviously no set of values would be available on which to base a definition of an undisturbed community structure. This approach is used frequently to monitor changes occurring in conjunction with pollution abatement.

In cases where there is insufficient baseline data, structural characteristics can be compared among similar communities to provide a measure of relative response to perturbations. However, it must be realized that the immense variability in natural systems makes it impossible to find duplicate communities. Thus any comparison must be based on a limited number of critical environmental conditions and similarities in the composition of available colonizers. Choice of an appropriate reference community is simplified in cases where the pollutant input is from a point source and sampling sites can be defined in terms of decreasing concentration gradient of pollutant at distances progressively further from the source. The location of the reference community is thus designated as upstream or upwind of the pollution source, or is defined by a site along the gradient at which pollutant concentrations approach background levels.

Structural characteristics and the numerical indices dependent on them provide various types of information which differ in ecological value. For instance, it is more informative to know the taxonomic or trophic composition of a community than merely its biomass or the abundance of organisms. Also, structural indices do not necessarily follow similar patterns of change under conditions of induced stress. Hellawell (1977) described several possible alterations in a community which would be reflected only in biomass, in biomass and relative dominance or in biomass, dominance and composition. Because of the differences in (1) the value of information provided by these indices, (2) the ease with which they are measured or calculated and (3) the sensitivity of their response to stress, certain criteria are essential in order to evaluate the usefulness of structural characteristics in monitoring pollutant effects.

Cook (1976) suggested that the ideal community index would have the following properties:

1. Sensitivity to the stressful effects of pollution on the ecosystem.
2. General applicability to various types of ecosystems.
3. Capability to provide a continuous assessment from unpolluted to polluted conditions.
4. Independence of sample size and ease of measurement or calculation.

Two additional properties should be included in the list:

5. The ability to distinguish the cyclical and natural variability of the system.
6. The index should be ecologically meaningful.

The first four criteria, and the sixth, are rather straight-forward, but the fifth requires additional comment. The overall value of using observed changes in structural characteristics to assess pollutant damage is dependent upon distinguishing natural changes (cyclical, successional, stochastic) from those induced by the pollutant (Hellawell, 1977; Cushing, 1979; Heip, 1980). Miller examines this problem (Chapter 3) in terms of statistical criteria. This task requires careful evaluation, particularly in those very important cases where subtle changes may be induced by chronic low level exposure to toxic pollutants. In order to be of utility, an index must meet criteria of minimal variability in time and space. Since most ecosystems are heterogeneous in both time and space, structural characteristics which can be statistically defined by means of a practical number, frequency, and distribution of samplings, must be chosen. When a large part of its variance is found to be associated frequently with time and/or space, a characteristic must be judged as unfit for the monitoring of pollution effects (Gray *et al.*, 1980).

A recent review by Herricks and Cairns (1982) examined aspects of methodology to assess structural changes in aquatic systems but there is no comparable review for hazard assessment in terrestrial systems. In this chapter the effects of toxic pollution on various indices of community structure will be surveyed, with particular emphasis on the utility and sensitivity of each index in stress assessment.

5.1 ABUNDANCE AND BIOMASS

Abundance and biomass are the most simple indices of community structure. Because of their simplicity, these measures do not provide much information on the general ecological character of the system and, therefore, they are often reported only in conjunction with more informative biological data. Both of these characteristics have been shown to vary seasonally in marine benthic communities (Buchanan *et al.*, 1978). In the same study, biomass was not correlated with habitat type (as defined by sediment composition) whereas taxonomic indices did exhibit such correlation. Seasonal fluctuations in abundance would be expected in those communities dominated by species with large numbers of recruits. There is also a great deal of variability inherent in the estimation of abundance. Edwards *et al.* (1975) reported that a minimum of 10 samples were needed to estimate the total number of invertebrates to within ± 40 per cent, with 95 per cent confidence limits (see Chapter 3)

The choice of numerical density or biomass as an appropriate structural characteristic imparts a certain bias to the researcher's approach. Reporting numbers exaggerates the importance of small abundant species while reporting biomass emphasizes those larger organisms usually present in smaller numbers.

This bias becomes less apparent as the size distribution of the community is narrowed. Such a situation is approximated in the freshwater planktonic community, although there are still size differences between bacteria and the larger zooplankton species.

A reduction in abundance and biomass is generally associated with toxic pollutants although in those cases where organic enrichment accompanies toxic substances both abundance and biomass can be increased. Pearson (1975) observed the latter effect in benthic fauna, associated with their exposure to pulp mill effluents. The number of species was reduced in comparison to pre-pollution levels, however biomass rapidly increased to three times 'normal' and then precipitously declined with depression of dissolved oxygen levels.

The response of aquatic fauna to acidification has been shown to vary with community type. This situation is demonstrated in Figure 5.2 (Crisman *et al.*, 1980) where only the abundance of zooplankton is clearly correlated with lake pH, and the number of benthic invertebrates appears to be totally unrelated to acidification. In contrast, Leivestad *et al.* (1976) found that both biomass and density of benthic invertebrates were reduced in acidified Norwegian lakes. Biomass values showed a greater decline with decreasing pH than did abundance, due to the loss of large predators. Others have reported the

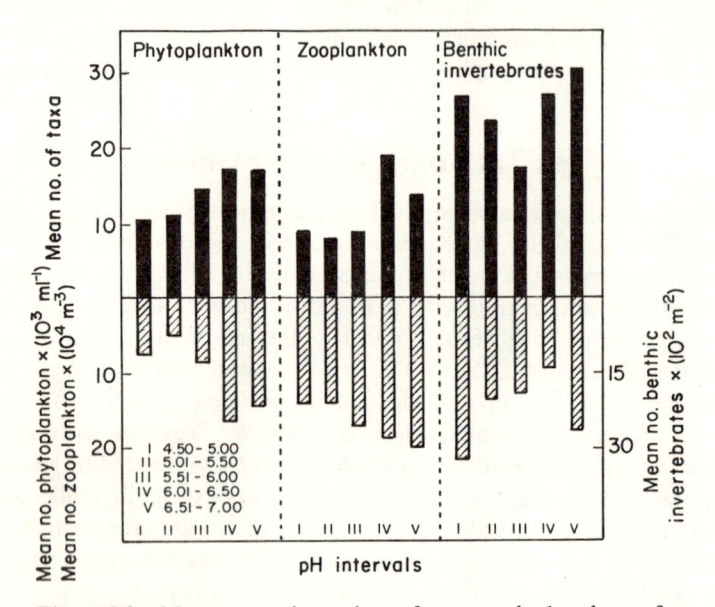

Figure 5.2 Mean annual number of taxa and abundance for phytoplankton, zooplankton and benthic invertebrates from twenty Florida lakes grouped by pH range. (From Crisman *et al.*, 1980. Reproduced by permission of SNSF Project, Box 61, 1432 As-NLH, Norway)

abundance of zooplankton and fish reduced in acidified lakes, but the biomass of periphyton increased (Hendrey et al., 1976). Yan (1979) found that phytoplankton biomass was better correlated with phosphorus concentration than with pH levels in an acid and heavy-metal contaminated lake.

Winner et al. (1975) determined that the abundance of benthic macroinvertebrates in a contaminated stream was inversely related to increasing copper concentrations, although the correlation was not as strong for the numbers of individuals as it was for the numbers of species (Figure 5.3). The variation in abundance was high, making it impossible to determine a graded difference in response between the levels of 38 and 120 $\mu g l^{-1}$ Cu. Reductions in phytoplankton biomass were noted in microcosms after the addition of copper at 50 $\mu g l^{-1}$; but when exposed to less than 10 $\mu g l^{-1}$ Cu, algal standing crop increased slightly (Thomas et al., 1977). In a terrestrial ecosystem, mean plant biomass and total number of species present were found to increase with distance

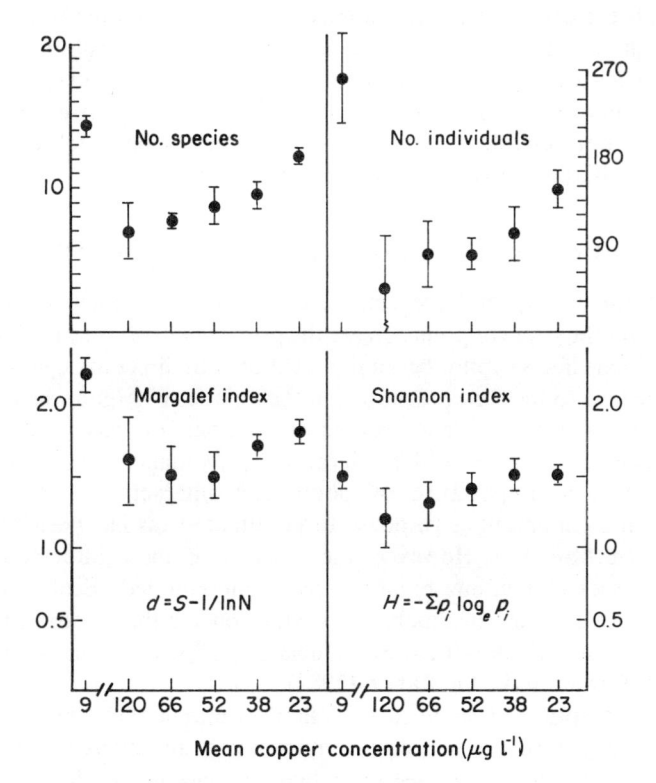

Figure 5.3 Relative changes in four indices of macroinvertebrate community structure at six locations along a copper concentration gradient (from Winner et al., 1975. Reproduced by permission of E. Schweizerbart's che, Verlagsbuchhandlung, Stuttgart)

from a copper smelter but the per cent cover was not correlated with the metal stress (Dawson and Nash, 1980). This situation appears to be explained by the fact that shrubs dominated the vegetation on the sites nearest the smelter. Strojan (1978b) reported that the average numbers of arthropods in soil cores were inversely related to levels of Zn, Pb, Cd and Cu.

Agricultural insecticides, as expected, reduce the abundance and biomass of arthropods and, consequently, can indirectly contribute to increased plant biomass (Shure, 1971).

Oil spills have been reported to depress the biomass of benthic fauna (Jacobs, 1980), intertidal invertebrates (Thomas, 1978), and zooplankton (Johansson *et al.*, 1980; Lindén *et al.*, 1979). However the biomass of phytoplankton may be increased in response to reduced grazing pressure (Nounou, 1980). Oils decrease the per cent cover and number of colonies in coral reef communities (Loya, 1975).

Although the abundance and biomass of specific communities respond to pollutant stress, these indices are lacking in sensitivity and ecological information; in addition, they are highly variable over time and space. The ease with which they are measured is their strongest attribute. These simple indices certainly do not provide, by themselves, a sufficient basis for assessing the severity of pollutant impact on community structure.

5.2 REDUCTION IN POPULATION SIZE AND EXTINCTION

Reductions in density and the complete extinction of sensitive populations resulting from the stress of pollution are the primary factors altering community structure. Such effects cannot be totally attributed to direct toxic mortality, but may be due also to induced reductions in the abilities of organisms to function successfully in competitive and trophic interactions, or may be the result of increased emigration or reduced immigration, in avoidance of the contaminated environment. The importance of population interactions and avoidance behaviour in a community's response to pollutant stress has been emphasized previously (section 4.4). However, the impacts of these indirect effects on localized species extinctions remain largely unquantified. Ecological factors which influence a species' susceptibility to extinction and the consequent effects of selective extinction on the structure and function of the ecosystem were discussed recently by Fowler and MacMahon (1982).

With the exception of some toxic organic contaminants which provide an important energy resource for certain invertebrates and microbes and actually stimulate population growth (see microbial response to oil: Johansson *et al.*, 1980), most toxic pollutants act to reduce the numbers of individuals in tolerant populations as well as the number of species in the community. Concurrent losses in the numbers of individuals and species in an exposed ecosystem are shown in Figure 5.3 (Winner *et al.*, 1975), where both indicators are inversely correlated

EFFECTS ON COMMUNITY AND ECOSYSTEM STRUCTURE AND DYNAMICS 59

BROWN TROUT
Population changes

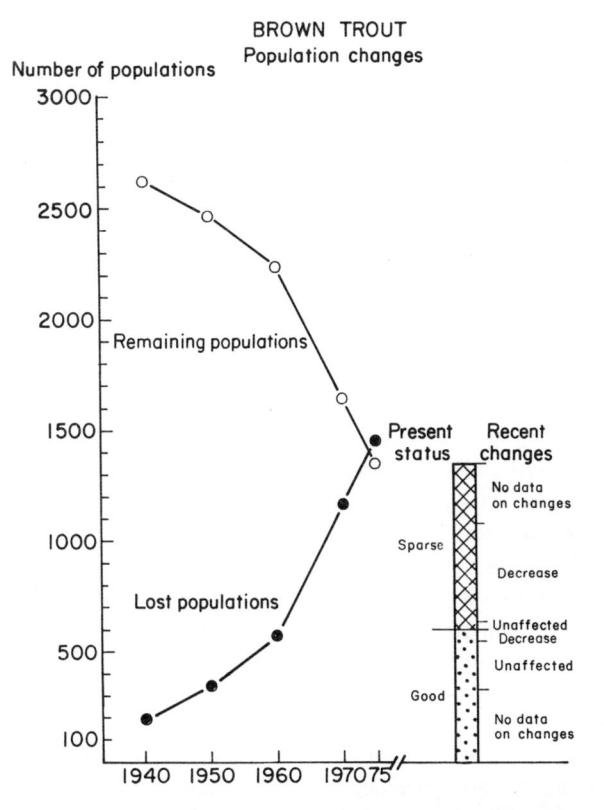

Figure 5.4 Time trend for population losses of brown trout in lakes exposed to acid precipitation in the four southernmost counties in Norway. (From Muniz and Leivestad, 1980. Reproduced by permission of SNSF Project, Box 61, 1432 As-NLH, Norway)

with copper concentrations. It is suggested that the reduced density of stressed populations can accelerate species extinction. Small, severely stressed populations are highly vulnerable to natural forms of biotic stress and environmental fluctuations and, as a result, have a high probability of premature extinction.

One of the better documented time trends in species elimination has been recorded for fish during a period of industrially stimulated acidification of poorly buffered lakes. Data developed from surveys of fisherman and lake owners (Figure 5.4) depicts a decline of nearly 50 per cent in the number of brown trout (*Salmo trutta*) populations in acidified Norwegian lakes during the past forty years. The loss was most exaggerated in lakes with pH < 5.1 (Wright and Snekvik, 1978; Muniz and Leivestad, 1980). Although the use of this type of survey data to establish a relationship between fisheries status and acidification has been questioned (EPRI, 1979; Brown and Sadler, 1981), there appears to be little doubt that a large number of poorly buffered lakes in southern Scandinavia have lost most, if not all, of their fish populations. Declining fish populations in lakes exposed to acid precipitation inputs have been reported in Canada and the United States as well, and decreased recruitment of young fish has been cited as the primary factor leading to the gradual extinctions (Beamish, 1974; Beamish et al., 1975; Schofield, 1976; NRCC, 1981a; Haines, 1981). A survey of 150 lakes in Ontario, Canada revealed that 70 had a pH of 5.5 or lower, with acid levels having increased more than a hundredfold in the 1960s. Beamish and Harvey (1972) reported that six of seven fish species had disappeared from Lumsden Lake by 1971. The loss of sports fish populations from other acidified lakes in the Sudbury district was confirmed by Beamish (1974). Further reductions in fish species have occurred in this region in the past decade (Harvey and Lee, 1980). More than 100 lakes in the Sudbury area, with pH less than 5, have lost all fish populations (OME, 1979; Harvey, 1980). Documentation of acidification in waters of the northeastern United States has shown that almost 50 per cent of the lakes surveyed, in which pH was below 5, contained no fish (Schofield, 1976).

The loss of fish populations has had profound effects on the lake ecosystems since these top consumers contribute to the regulation of energy and nutrient turnover, and to the maintenance of the community composition of zooplankton and benthic invertebrates (Muniz and Leivestad, 1980; Økland and Økland, 1980; Raddum, 1980).

Although the uniqueness in role and in sensitivity of individual species has been stressed (Resh and Unzicker, 1975), difficulties in species identification, particularly of the larval stages, has led to the use of higher taxonomic categories in pollution assessment. Unfortunately, a comparison of data from different levels suggests that only species and generic numbers provide sufficient sensitivity to monitor changes in pollutant stress. Furthermore, even though these two taxonomic levels follow similiar trends (see Hellawell, 1977), there is a significant loss of information when identification is made only to the generic level.

5.2.1. Loss of Species with Unique Functions

Although it is very difficult to evaluate the loss of any one species, it is easier to appreciate the disappearance of those that fulfill unique roles within the community. These can be species that perform unique individual actions (e.g. pollination) or 'key' species which have an important role in controlling ecosystem function and stability (Lewis, 1978).

A reduction in natural pollinator populations through pesticide application can affect fertilization and seed set in natural plant communities. The accidental killing of bees resulting from the use of insecticides has been a problem in the ited States for nearly a century. Johansen (1977) concluded that the long-term

sample variability in the proportions of epiphytic worms associated with the alga *Ulva lactuga*. In other experiments he demonstrated that the diversity of organisms which colonized coral skeletons implanted in oil-contaminated habitats was lower and more variable than that of the colonizers in uncontaminated areas (Figure 5.10). The disturbed micro-ecosystems did not achieve a stable pattern of fluctuations and, in addition, they displayed wide variation in species composition, reflected in the range of diversity values measured. Jacobs (1980) reported that both diversity and evenness of littoral assemblages fluctuated abnormally after the *Amoco Cadiz* spill. The increased fluctuations were described as rapid changes in dominance associated with the expansion of opportunistic species.

There is some evidence, however, which conflicts with the postulate that stress increases the variability in structural indices. Winner *et al.* (1980) found that copper pollution damped the variability in community composition, over time, among sequential macroinvertebrate samples. In fact, they reported that the affinity among samples increased with the growing dominance of chironomids at the higher concentrations of copper. Read *et al.* (1978) observed that the smallest temporal changes in PIE (see Table 5.2) were found in assemblages inhabiting a grossly polluted beach, whereas the largest changes occurred in a relatively unstressed beach community. It is not clear to what extent these conflicting reports can be explained by a difference in the severity of stress. The great simplification in structure of highly stressed communities suggests that temporal variability would be damped as resident species come to be controlled more by the pollutant than by natural biological factors.

5.6.1 Inertia

The inertia of an ecosystem, i.e. its capacity to resist change, is a measure of that system's resistance to pollution. In the context of ecological modelling, inertia has been called buffering capacity (Jørgensen and Mejer, 1977, 1979) and the inverse of inducible adaptability (Conrad, 1976). An ecosystem's inertia is a function of several basic properties:

1. The resistance of the organisms to environmental fluctuations.
2. The degree of functional redundancy within the system.
3. The self-cleansing capacity.
4. The chemical buffering capacity.
5. The proximity of its parameters to ecological threshold values for factors such as temperature and salinity (Cairns and Dickson, 1977).

Accordingly, it has been suggested that ecosystems with a high degree of specialization and little history of fluctuating environmental parameters have low inertia and are highly vulnerable to pollution damage (Boesch, 1974; Jernelov and Rosenberg, 1976).

To assess the inertia of an ecosystem, the degree of chemical contamination must be related to some measure of change. Westman (1978) proposed a measure of damage of 50 per cent change in composition and relative abundance of species, using the percentage similarity index (PS). This measure would be patterned after well-known bioassay techniques. The quantity of a chemical which causes a 50 per cent change in PS (between the before-exposure and after-exposure values) would be the measure of inertia. This technique has been used to measure the inertia of an oak–pine forest exposed to chronic gamma radiation (Woodwell, 1967), although the characteristic chosen was species richness. The criterion of a 50 per cent reduction in macroinvertebrate richness (excluding the chironomid taxa) was used by Sheehan (1980) to estimate the impact of metal pollution on the structure of a stream system. This analysis demonstrated that high seasonal influxes of copper were annually effective in overwhelming the inertia of the system, depressing species richness to a level below 50 per cent of that in the reference community.

5.6.2 Elasticity

The elasticity of an ecosystem is a measure of its ability to recover, within acceptable limits, after being perturbed. It is best related to restoration time, and is affected by a variety of factors.

The relationships of species abundance interactions, and vegetative structure to community elasticity have been examined experimentally, through selective plant removals from an old field community (Allen and Forman, 1976). Recovery was inversely correlated with the per cent cover of species removed. The recovery response was also highly dependent on the particular species removed, emphasizing the importance of species interactions and of the species' ability to reproduce rapidly and fill gaps in community function. High elasticity was positively correlated with vertical stratification (belayered with dense ground cover), horizontal patchiness (many low diversity patches), and the abundance of species which were capable of rapid vegetative reproduction. In particular, this study emphasized that spatial heterogeneity of biota was essential to ensure community recovery from perturbation. This characteristic has been linked also to the maintenance of stable population interactions through its provision of cover, refuge and patches of potential immigrants for dispersal (Stenseth, 1980).

The mobility of juvenile recruits and potential immigrants is a major factor in the rapid recovery of animal communities. Leppäkoski (1975) suggested that migrating populations can contribute significantly to ecosystem elasticity. When exposed to increasing pollution, such a group can move away from the source while maintaining its structural integrity. Thus, these populations are always available to contribute immigrants to the rapid recovery of a system when the toxic stress has abated. This idea developed from observations of the effects of pollution in estuaries; it is not known yet to what extent it can be applied to other situations.

Elasticity, as measured by restoration time is not a simple concept. It can be defined in terms of various characteristics each of which will provide a different estimate of community recovery. Rosenberg (1976) showed that baseline levels of faunal density and biomass were regained more rapidly than were the numbers of species, but both the biomass and density of the fauna showed erratic fluctuations during the early recovery period. Therefore, the predictability of the recovered state must be considered. Westman (1978) proposed that recovery time to reach 85 per cent similarity to the original composition be used, emphasizing that achievement of 100 per cent agreement is an unrealistic expectation. Restoration rates have also been assessed in terms of species-shifts from reproduction-oriented to competition-oriented strategists (Opler et al., 1977). The use of colonization data and appropriate mathematical model curves (Sheldon, 1977) might offer a rapid appraisal of stream recovery. However, care must be taken to confirm that the colonizers do indeed survive throughout their vulnerable life history stages and then reproduce successfully within the recovering system.

There are several obvious difficulties in estimating elasticity. Stochastic as well as deterministic processes contribute to successional recovery and, thus, to the nature of the climactic structure. Recovery time may be long, and the rate of change might not be linear, so there is no guarantee that early estimates can be extrapolated to predict long-term progress towards restoration of either structure or function. Also, as previously mentioned, ecosystem characteristics do not necessarily recover at concurrent rates.

5.6.3 Amplitude

Amplitude, as used by Westman (1978), is that maximum amount of perturbation from which an ecosystem can still recover. The critical question with regard to the 'brittleness' of a polluted system is whether sufficient structural organization and biotic interaction has been maintained to allow it to persist. This question is far from simple and forms the crux of the problem of establishing protective standards for an ecosystem.

A number of environmental factors and species characteristics have been shown to increase the amount of stress that an ecosystem can bear before becoming permanently impaired (Orians, 1975). These include intraspecific variability of component species, capacity for long-distance dispersal, broad physical tolerance limits, and low density-dependence in birth rates.

The measurement of amplitude involves the determination of a threshold, if there is one, beyond which recovery to the original state can no longer occur (Westman, 1978). The necessity of estimating amplitude has long been recognized with respect to recovery in agricultural systems from over-grazing or

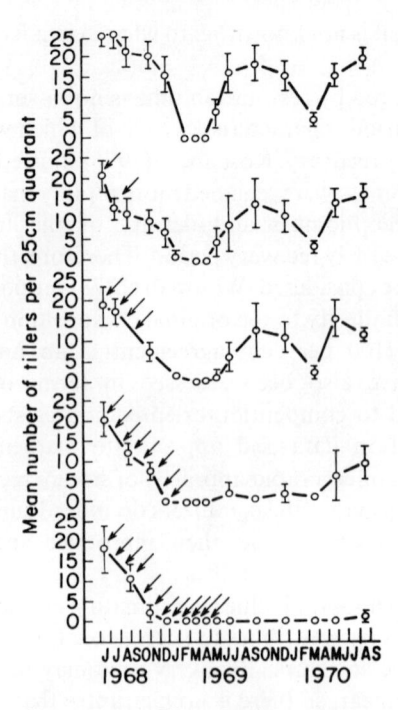

Figure 5.11 The results of successive oilings of plots of *Spartina anglica*. Oil was applied on the dates arrowed; 95 per cent confidence limits (*t*-test) are indicated. (from Baker, 1971. Reproduced by Permission of the Institute of Petroleum, London)

nutrient stresses. The relationship between system amplitude and chronic or successive inputs of toxic pollutants is less well documented. Baker (1971) examined the successive inhibition of recovery with increasing duration of oil stress in the salt marsh grass *Spartina anglica* (Figure 5.11). From her data, she concluded that approximately 12 successive oilings, with 90 per cent crude oil over a 14-month period, would exceed the system's threshold of self-repair. The application of repeated doses of No. 2 fuel oil as slicks on a salt marsh grass community resulted in similar long-term reductions in plant vigour and productivity (Hershner and Lake, 1980). Although neither of these studies was of sufficient duration to demonstrate long-term inhibition of recovery, field observations by Hampson and Moul (1978), following a No. 2 fuel oil spill, indicated that the inability of grasses to reestablish themselves either by reseeding or by rhizome growth would be aggravated by increased erosion, preventing any likely short-term recovery. There is little data on the amplitude of pollutant damage which a system can sustain for periods approximating those of normal succession. Examples of continuous disruption by fire preventing reestablishment of normal composition patterns in forests have been documented for extended periods (Vale, 1977).

5.6.4 Hysteresis and Malleability

Hysteresis is a measure of the degree to which an ecosystem's pattern of recovery is not the reversal of the pattern of species loss. There is very little data comparing the disappearance and reappearance of species in relation to effluent release and abatement. Westman (1978) suggested that hysteresis could be measured by such a comparison, using techniques such as Spearman's rank correlation coefficient, and that this information should be of considerable interest to environmental managers.

Malleability is a measure of the ease with which a system is permanently altered to a new stable state. An ecosystem can possess a multiplicity of stable states (May, 1977). The problem facing the ecotoxicologist is demonstrating that the system is indeed new and stable. In the initial stages of recovery from a disruption, a system might be expected to display large differences in community composition (PS) from year to year. Small changes in PS over several years would indicate that a stable configuration had been achieved. Malleability could then be assessed by analysis of the differences between characteristics of the original and the new states, in relation to the extent of pollution.

5.6.5 Persistence

The ultimate criterion in pollution assessment is the persistence of a functioning ecosystem. It would appear to be more practical to define persistence in terms of primary productivity, organic decomposition rate, or nutrient recycling capability, since these processes are central to ecosystem function. Studies of ecosystem restoration have indicated that essential processes involving nutrient conservation and cycling are exceptionally difficult to reestablish. Bradshaw *et al.* (1978) reported that the overriding problem in restoration of a terrestrial ecosystem is the provision of adequate inorganic nitrogen. Therefore, the breakdown of essential nutrient cycles may provide a key to evaluating a system's persistence under stress. Likens *et al.* (1978) noted that the recovery of deforested ecosystems was dependent upon nutrient availability and the reestablishment of biological regulation of nutrient cycles. In the experimentally logged forest, 28 per cent of the total pool of nitrogen was lost, and it was estimated that this would require 20 years to be replaced by natural processes (mineralization and N-fixation).

The above field observations on the sensitivity of nutrient cycles to system stability are generally in agreement with predictions using the Liapunov Direct

Method (mathematical stability model). This method was used by Lasalle and Lefschetz (1961) to predict that damage to decomposers or to nutrient pools is a potentially stronger cause of instability than are disturbances to predator-prey components of the system. Similarly, the relative stability of a generalized linear ecosystem model was found to be influenced by the efficiency of nutrient recycling processes (Webster *et al.*, 1975; Halfon, 1976). Closed-looped nutrient cycling was also shown to have great influence on the homeostasis of a modelled ecosystem (Waide *et al.*, 1974).

Although there appears to be a growing interest in the practical assessment and modelling of ecosystem resilience and persistence under various conditions of stress, the use of stability criteria in pollution-effects research is still in its infancy.

5.7 SUCCESSION AND RECOVERY

5.7.1 Terrestrial Succession

Ecological succession has recently been defined as a less directed process than originally conceived. MacMahon (1980) defined succession as merely the change in an area of the earth's surface and its inhabitants over a moderate period of time (i.e. tens to hundreds of years), during which environmental conditions remain relatively constant. He assumed no inherent order to the process and no precisely defined time schedule. Succession is usually characterized as changes through time in species composition and community structure, but other ecosystem attributes such as community energetics (Odum, 1969) and chemical budgets (Woodwell *et al.*, 1975; Vitousek and Reiners, 1975; Gorham *et al.*, 1979) may respond to seral change.

Although some changes in species composition take place in all systems after a disturbance, succession is most obvious in terrestrial systems which are not under severe environmental stress. MacMahon (1980) concluded that in harsh environments such as desert or tundra, following a disturbance, the only viable colonists were the limited numbers of tolerant species previously resident. Even for forest ecosystems, for which successional changes in vegetative structure have been well documented, there have been few studies encompassing changes in both plants and animals.

The successional state of a system has been described as an inverse function of stress (Regier and Cowell, 1972). The sequence through which community structure is simplified by chronic stress has been suggested to be the reverse of succession, that is, retrogression (Whittaker and Woodwell, 1973). Odum (1969) listed several trends descriptive of developmental and mature stages. The usefulness of many of these general descriptors has been questioned. However, those specifically dealing with the decrease in the productivity to biomass ratio (P/B), the accumulation of organic matter, heightened nutrient retention, and increases in species richness, with time, seem to have general applicability at the ecosystem level (MacMahon, 1980). The reverse of these trends would then be evidence of retrogression.

A pattern of simplification has been documented for forested ecosystems in proximity to SO_2 emission sources (Guderian and Kueppers, 1980). Under high dosage, near the source of contamination, the tree stratum is destroyed, followed by subsequent die-back of shrubs, herbs, mosses or lichens, one after another until the zone is barren. The denuded zone is surrounded by a transition zone containing isolated clusters of grass and resistant ground cover. This area adjoins the outer stunted forest zone. Less severe changes in composition are found under low to intermediate pollutant dosage. Kercher *et al.* (1980) have forecast from models that SO_2 pollution would increase white fir (*Abies conclor*) dominance at the expense of ponderosa pine (*Pinus ponderosa*) stands in a western coniferous forest. The reduced vigour of the stressed pines would provide the fir with a substantial advantage even at non-necrotic SO_2 levels. It is also evident that air pollutants influence the community makeup of later series by influencing reproductive success (Keller, 1976).

Gradient studies in forested systems have demonstrated retrogressive trends in biomass and species richness associated with increasing levels of pollutants (e.g. McClenahen, 1978; Freedman and Hutchinson, 1980a; Scale, 1982). There is less definitive evidence on P/B trends. The elimination of large tree species, as frequently reported in cases of severe stress, could raise the ratio of productivity to biomass in the remaining undergrowth community. However, the effects of the pollutant on photosynthesis might significantly retard primary productivity in these normally more rapid growing species.

5.7.2 Recovery in Aquatic Ecosystems

There is little defined theory describing successional patterns in freshwater and marine systems. However, there exists very good information on the successive recovery of ecosystems following pollution abatement and oil spill cleanups (e.g. Rosenberg, 1976; JFRBC, Special Issue, 1978; Turnpenny and Williams, 1981).

Much of the emphasis in recovery studies has been on the sequence of species recolonization. The general pattern of rocky shore community reestablishment after the *Torrey Canyon* oil spill was observed to be similar to that found after small scale removal experiments (Southward and Southward, 1978). However, the time scale of recovery of the oil-polluted system was significantly longer due to: (1) residual toxicity, (2) the extent of widespread mortality which restricted potential adult immigration and (3) the nearly complete removal of the herbivore community. Loya (1976) found that low level residual oil toxicity slowed the rate of recovery for a Red Sea coral community and may, in fact, have prevented the community from returning to its former configuration. Climate also affects recovery time. For example, oil spilled in colder water degrades more slowly, thereby retarding recolonization (Vandermeulen *et al.*, 1977). Conversely, spills in the tropics, even though the oil may degrade more rapidly, can lead to higher water temperatures through increased absorption of solar radiation by the contaminated ocean surface (Chan, 1977). The compounded effects of temperature and oil toxicity can further stress intertidal communities. Southward and Southward (1978) noted that the temperate climate of Cornwall may have mitigated the effects of the *Torrey Canyon* spill on marine life.

Recovery from pollution can be separated into two phases: (1) colonization, (2) reestablishment of biotic interaction. During the colonization phase a species' capacity to immigrate, rapidly reproduce, and disperse young are essential attributes (Rosenberg, 1976), along with a resistance to persistent toxicity. The early colonizing community therefore consists of opportunists and forms highly resistant to pollution (Grassle and Grassle, 1974; Leppäkoski, 1975; Leppäkoski and Lindström, 1978). These species display many of the characteristics attributed to the reproduction-oriented strategist and may have been members of the original assemblage, although at relatively low population densities. The polychaete *Capitella capitata* exemplifies such a species in temperate estuarine environments, having widely dispersing pelagic larvae, an opportunistic benthic stage, and a short life cycle (Grassle and Grassle, 1974). Rosenberg (1976) reported *C. capitata* to be the dominant early successional colonizer, following abatement of pulp mill effluent. This 'opportunist' colonized unoccupied space rapidly under conditions of reduced competition, and comprised more than 90 per cent of the individuals in the benthic community during the first year of recovery.

In intertidal systems, shifts in the spatial patterns of species are common for the period prior to the development of strong biotic interactions. This is due primarily to the differing time scales of resettlement for different species. Southward and Southward (1978) reported that the rapid regrowth of macroalgae high in the intertidal splash zone was due to the absence of grazing pressure by limpets, which were drastically reduced in numbers by the *Tsesis* spill and cleanup. The lack of biological interaction appeared to be more important than was physical dessication in establishing the upper limit of algal growth on rocks.

The latter recovery phase is dominated by density-dependent population regulation. Specialists become more common following the displacement of early dominant opportunists. An obvious indication of this phase is the reestablishement of population equilibriums, through trophic and competitive interactions. Fluctuations in population densities of predators and prey are slowly damped as balance is restored in the pollution-disrupted cycle. Figure 5.12 demonstrates the restoration of these relationships for the limpet *Patella*, one of its major food sources, the alga *Fucus*, and its major competitors for space, barnacles. A decrease in both herbivore and algal density occurred between 1975 and 1976, at which time *Fucus* cover was reduced and *Patella* populations stabilized at a relatively lower density. With *Fucus* removed, competition for

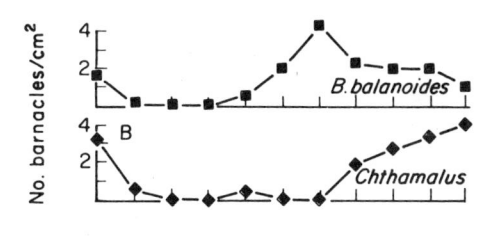

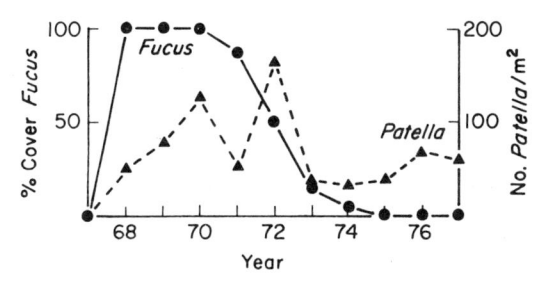

EFFECTS ON COMMUNITY AND ECOSYSTEM STRUCTURE AND DYNAMICS 95

Figure 5.12 Population fluctuations during recolonization of flat rocks at mean tide level, at Trevone, from 1968 to 1977. Lower section: circles indicate percentage cover by *Fucus*; triangles, no. of *Patella vulgata* m^{-2}. Upper section: squares indicate no. of *Balanus balanoides* cm^{-2}; diamonds, no. of *Chthamalus stellaius* cm^{-2}. (From Southward and Southward, 1978. Reproduced by permission of the *Journal of the Fisheries Research Board of Canada*)

attachment space increased between the barnacles, *Balanus* and *Chthamalus*. The recovery of the barnacle population when, by 1973, *Fucus* cover had been reduced by 75 per cent from the maximum levels attained during the early recovery phase, can be seen in Figure 5.12. This figure exemplifies not only the restoration of balance among populations, but also the effects of earlier stages on establishment of the eventual equilibrium. The importance of assessing recovery through both the recolonization and interaction phases has been repeatedly emphasized (Grassle and Grassle, 1974; Rosenberg, 1976; Leppäkoski and Lindström, 1978).

5.8 AN ILLUSTRATIVE CASE STUDY

Ecosystem recovery following the introduction of abatement measure in chronically polluted systems can illustrate many changes in structural charateristics common to the original sequence of degeneration. Many of these principles of structural destabilization were exemplified in the comprehensive study

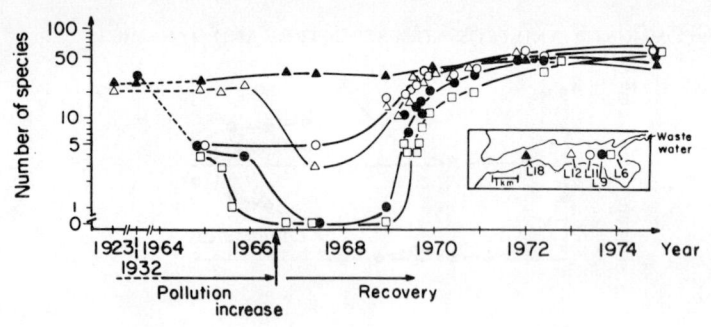

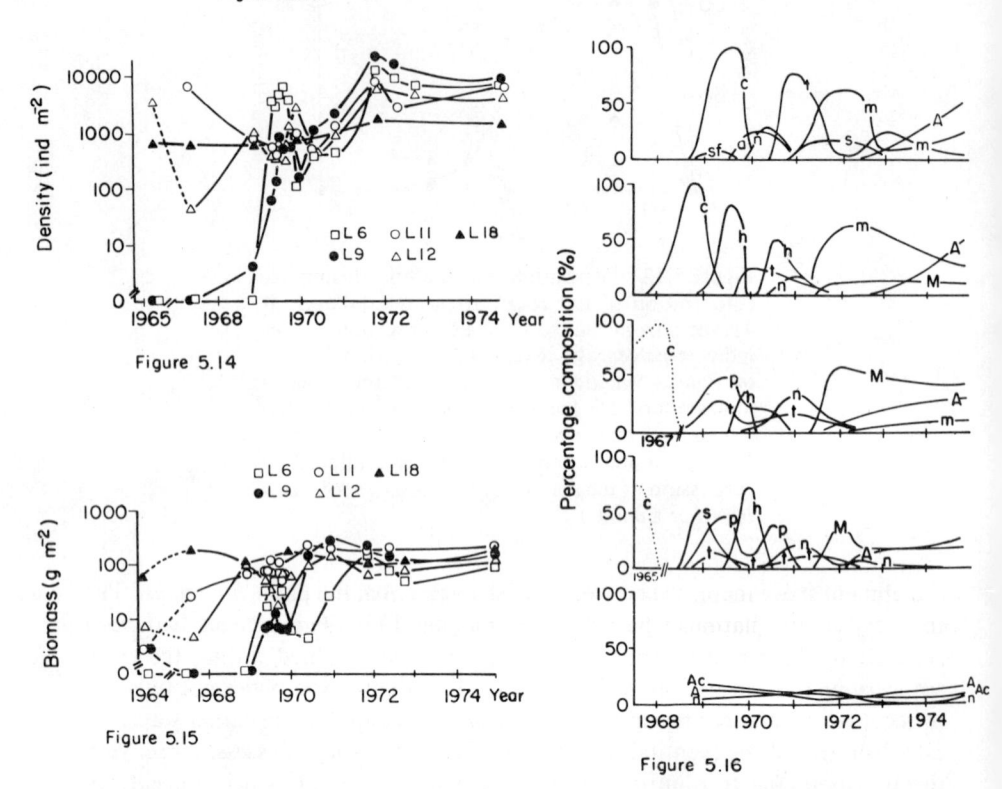

Structural characteristics describing the recovery of a pollution-perturbed estuarine system. (From Rosenberg, 1976, by permission of *Oikos*, Munksgaard, Copenhagen)

Figure 5.13 The succession of benthic species at five stations in the Saltkallegjord. The arrow indicates the time of pollution abatement

Figure 5.14 Dynamics of faunal density during 1965 to 1974

Figure 5.15 Benthic faunal dynamics of biomass (wet weight).

Figure 5.16 Percentage composition of some abundant populations during succession in the Saltkallefjord. c: *Capitella capitata*, a: *Abra alba*, S: *Scalibregma inflatum*, A: *Amphiura filiformis*, M: *Myriochele oculata*, Ac: *Amphiura chiajei*, n: *Abra nitida*, sf: *Scolelepis fuliginosa*, t: *Thyasira spp.*, m: *Mysela bidentata*, h: *Heteromastus filiformis*, p: *Polyphysia crassa*

(Rosenberg, 1976) of benthic faunal dynamics following abatement of sulphite and pulp mill effluent into the Staltkallefjord in Sweden. Unfortunately, as with many pollution studies, good records of community characteristics were unavailable for the periods prior to and during much of the degeneration. During the period of severe pollution stress, the number of macroinvertebrate species was inversely related to the pollutant concentration, which decreased with distance from the effluent source. Figure 5.13 illustrates the increased time delay prior to recovery in species number at the more severely polluted sites. The indices of density and biomass (in Figures 5.14 and 5.15, respectively) show heterogencity and discontinuity in the development of early recovery assemblages. Faunal density rose rapidly at all sample stations but there was a delay of approximately 2 years before species recruitment began at the most polluted sites. The graph of the number of individuals shows two peaks, 1969 and 1972, which appear to be representative of an estuarine recovery pattern (Leppäkoski, 1975). The data showed a primary maximum indicative of marine assemblages and a secondary maximum for freshwater-tolerant groups. Biomass (wet weight) also increased rapidly after an initial delay but continued to fluctuate throughout the colonizing phase of recovery. Biomass levels appeared to stabilize after 1971. The fluctuations in density and biomass charateristics were in contrast to the continuous trend of species recruitment in the early recolonizing phase. This emphasizes the predictability of the latter of these simple indices and is in accord with earlier discussions (sections 5.1 and 5.2).

Successive changes in dominance are displayed in Figure 5.16. With decreasing distance from the pollution source, the early dominance of the opportunistic polychaete, *Capitella capitata*, was evident from its higher percentage of relative abundance. In general, amplitude changes in relative abundance are larger and take longer to approach equilibrium in more severly polluted systems. The short duration of dominance displayed at site L12 is typical of the pattern expected for opportunistic populations which are able to expand rapidly into 'empty' niches created by pollution stress (Grassle and Grassle, 1974).

The information provided by the more complex indices of diversity and evenness, H' and J in this case, was somewhat confusing (Figure 5.17). An understanding of the biology of the estuarine system, and a great deal of care, must be exercised in interpreting these diversity trends. Although the species diversity at the reference site (L18) was high and relatively stable, as expected, the values calculated for polluted sites did not show a regular trend. Diversity at these sites peaked early in the recovery period and then declined by the third year. There was also some unexpected ordering of sites according to diversity during the early recolonization phase. For example, site L6, closest to the pollutant source, was more diverse than the less polluted site L12, in 1970. Some of the contradictions contained in the Shannon diversity data were associated with the equitability component. The evenness data showed that the relative abundance at several polluted sites was higher than that of the reference community during

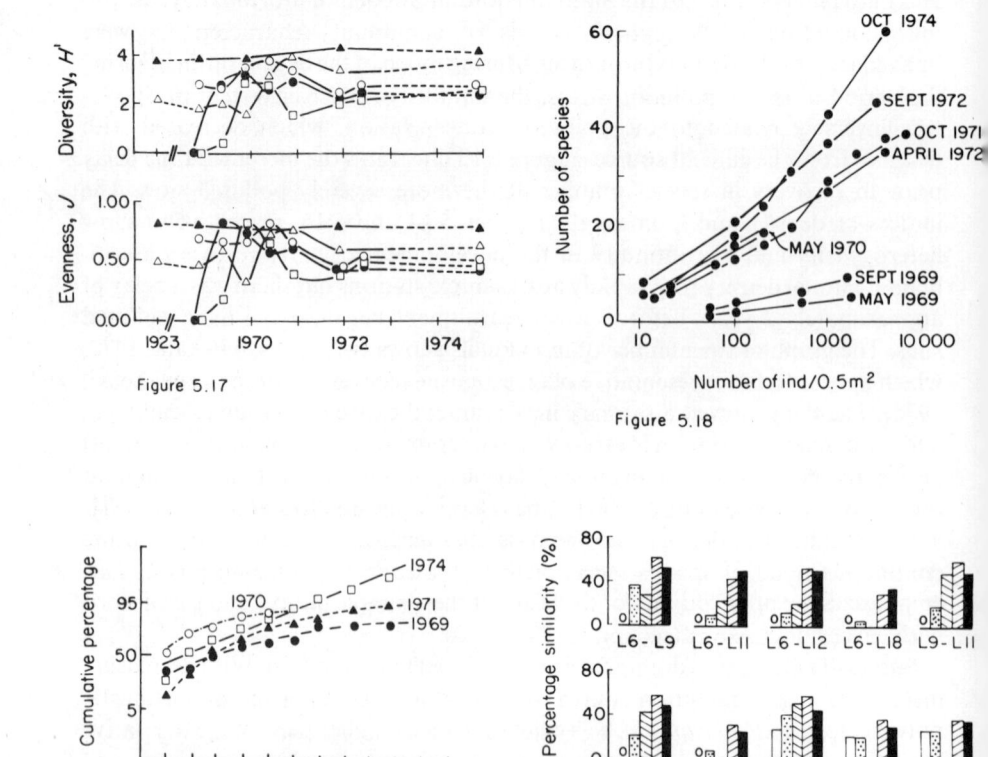

Structural characteristics describing the recovery of a pollution-perturbed estuarine system. (Figures 5.17, 5.18 and 5.20 from Rosenberg, 1976, by permission of *Oikos*, Munksgaard, Copenhagen; Figure 5.19 from Gray, 1979, by permission of The Royal Society)

Figure 5.17 Diversity according to Shannon (H') and evenness (J), at five stations in the Saltkallefjord

Figure 5.18 Diversity at station L6 according to Sanders' rarefaction technique

Figure 5.19 Log-normal plots of individuals per species during recovery from pollution in the Saltkallefjord. Reproduced with permission from Gray (1979)

Figure 5.20 Percentage species similarity of benthic fauna between pairs of stations in the Saltkallefjord. Data from L18 in 1970 is missing. 0 indicates no similarity

the early abatement period. High evenness values are frequently found for moderately stressed communities containing reduced numbers of species and individuals (Gray, 1979).

Rosenberg (1976) suggested that the Sanders' rarefaction technique provided a clearer picture of diversity trends (Figure 5.18) than the more complicated Shannon index. Gray (1979) reported that log-normal plots of Rosenberg's data (Figure 5.19) demonstrated a continuous pattern of recovery. The shape of the curves changed from a polluted pattern, through the transition form, to a typical clean-water distribution, through five years following abatement (see Gray and Mirza, 1979, for an explanation of log-normal plots).

Similarity measures such as percentage similarity (*PS*) were used both to order communities according to the magnitude of change due to the severity of sulphite pulp effluent stress and to establish community elasticity (Figure 5.20). In 1968 no similarity was found between communities at the two most polluted sites (L6, L9) and any other community. There was successive increase in PS from 1968 to 1970, and a specific trend towards increasing similarity was apparent during the later seral stages (1970–1971) as compared to the previous period. However, the pattern was not one of successive step increases as with the earlier recovery phase. The inability to define succession stages from 1971 to 1974, using the PS index, probably indicates that either (1) natural fluctuations were too large to permit separation of successive change in species number and abundance or (2) the pollutant concentration was no longer the factor controlling relative population abundance. This would infer that biological regulation of community processes had been reestablished.

The usefulness of similarity measures for distinguishing seral recovery through the intial recolonization phase appears to be greater than that of the complex diversity indices and, therefore, it complements the simple characteristic of species richness, which also shows an unambiguous trend with ecosystem recovery. The Rosenberg study also provided insight into several aspects of the stability–pollution relationship in an estuarine ecosystem. The reduced fluctuations in species number and fauna density from 1972 to 1974 indicated that the system had returned to a state of 'constancy'. The rather rapid recovery (5–7 years) and the fact that eleven of the fifteen species recorded at site L9 before pollution had reappeared in significant numbers by 1974, suggests that the system was highly elastic. High inertia and elasticity are properties expected of a physically and chemically varying ecosystem with an extensive history of pollutant stress. This example is an effective summary of the usefulness of structural characteristics in describing the ecosystem-level effects of pollution.

Effects of Pollutants at the Ecosystem Level
Edited by P. J. Sheehan, D. R. Miller, G. C. Butler and Ph. Bourdeau
© 1984 SCOPE. Published by John Wiley & Sons Ltd

CHAPTER 6
Functional Changes in the Ecosystem

PATRICK J. SHEEHAN
Division of Biological Sciences
National Research Council of Canada
Ottawa, Ontario, Canada K1A OR6

6.1 Material and Energy Movement.	103
6.2 Decomposition and Element Cycles.	107
6.2.1 Acid Precipitation.	109
6.2.2 Heavy Metals.	118
6.2.3 Pesticides and Oils.	124
6.3 Productivity and Respiration.	130
6.3.1 Acid Rain.	131
6.3.2 Heavy Metals.	136
6.3.3 Pesticides and Oils.	139
6.4 Food Web and Functional Regulation.	141
6.5 Assessing Changes in Energy Flow and Nutrient Cycling.	144

The ecosystem is the fundamental unit of ecology. Yet the identification and use of appropriate indices to assess the effects of perturbations on 'whole' system functioning have been slow to develop. The focus of ecological research has been on the component populations rather than on the system. This fact is particularly evident in the assessment of toxic pollutants where a preponderance of effort has been spent to identify individual and population effects which can seldom be extrapolated to describe changes at the ecosystem level (Hammons, 1980; NAS, 1981).

The ecosystem is an integrated system with fundamental characteristics which transcend the simple summation of component processes (e.g. Schindler *et al.*, 1980b; O'Neill and Reichle, 1980). Therefore, analysis of the ecosystem as a unit in ecotoxicology is based upon the premise that the system as a whole possesses characteristics which not only reflect the integrated response of component populations to perturbation but, in addition, provide a more comprehensive picture of ecosystem 'status'. The ecosystem is a viable unit and tends to persist through adverse environmental fluctuations, often reflecting changes in structure and efficiency of function, whereas certain individual populations within the

EFFECTS OF POLLUTANTS AT THE ECOSYSTEM LEVEL

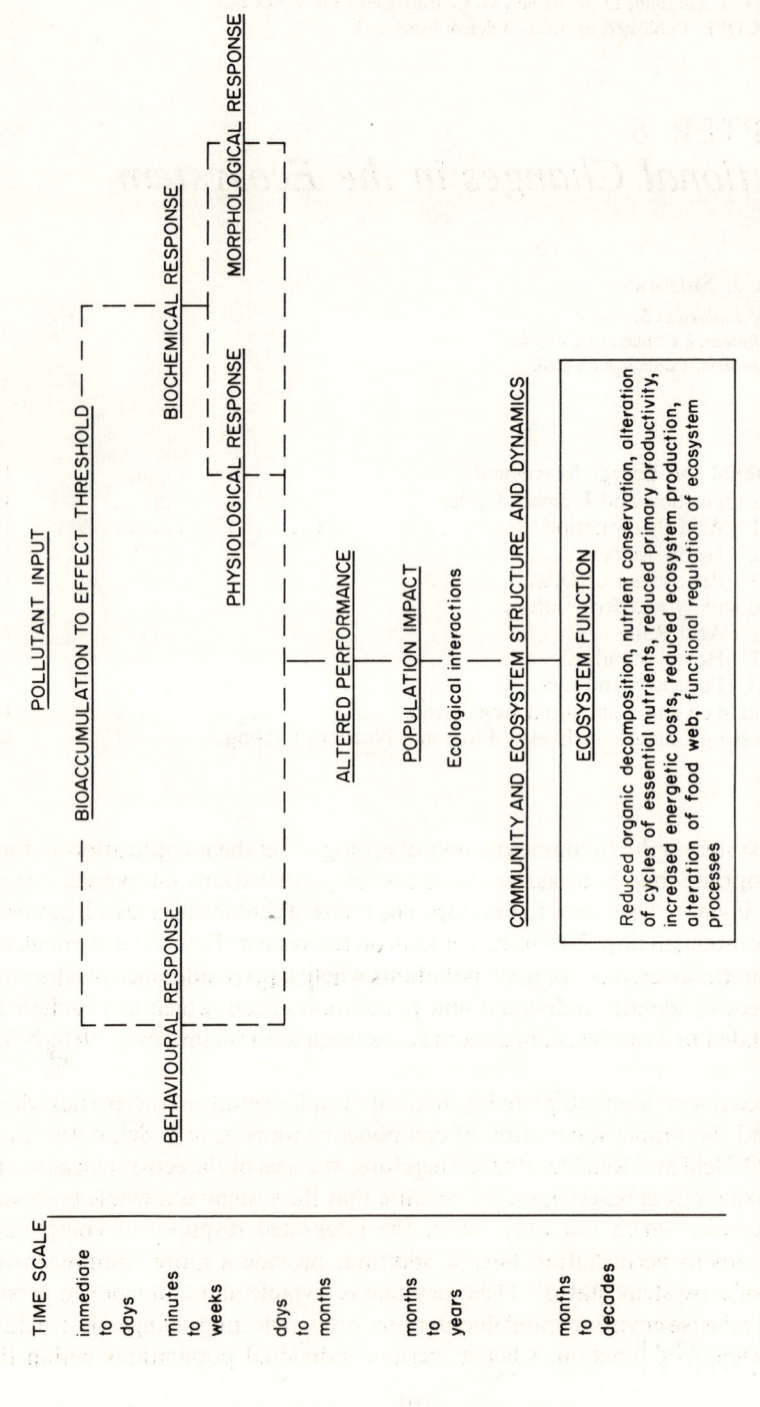

Figure 6.1 A conceptual chronology of induced effects following exposure to toxic pollutants, emphasizing changes in ecosystem function

disturbed community do not survive. Although pollutant-caused perturbations have the potential to influence all components of the ecosystem, the impacts on a single species may have negligible effects on system function. Complex internal feedbacks and controls can minimize overall impact if some form of redundancy exists (see O'Neill and Waide, 1982). For example, sensitive species may be replaced by competitors without affecting the system, or regulation of trophic functions may be maintained by predator flexibility in prey choice. O'Neill and Giddings (1979) have arugued that under conditions of nutrient limitation, major shifts in phytoplankton populations can occur, including the loss of sensitive species but without detectable changes in primary production. On the other hand, critical ecosystem processes such as litter decomposition are sensitive to pollutant stress and display measurable changes without obvious impacts on the highly visible populations normally monitored (Van Voris *et al.*, 1980).

Although there has been a tendencey to emphasize structural characteristics (see Chapter 5) in ecosystem pollution studies, the assessment of effects on the fundamental functional characteristics of biomass production, trophic regulation and nutrient cycling must be encouraged (Barrett *et al.*, 1976; Matthews *et al.*, 1982). Figure 6.1 lists several critical ecosystem processes which can be assessed in response to long-term pollution stress. Changes in primary productivity, energy flow, decomposition, material cycling, and internal regulatory processes provide insight into the dynamics of stress response at the ecosystem level rather than providing a measure of eventual outcome. It is the investigation of these processes which promises the greatest hope for the development of dynamic simulation models which can predict ecosystem level response to loading of toxic pollutants.

6.1 MATERIAL AND ENERGY MOVEMENT

In the early stages, ecosystem analysis did not encourage many investigators to employ a process-oriented approach to pollution studies (see Levins, 1975). A dichotomy of approaches to energy and material analysis were followed (Rigler, 1975). Energy studies were primarily based on trophic models. Mineral cycling investigations explored element availability in environmental pools not generally representative of either trophic levels or populations. However, advances in theory and analysis have encouraged a more holistic framework for assessing ecosystem nutrient and energy processing. Odum (1969) discussed the strategies of ecosystem development in relation to bioenergetic processes, nutrient cycling and regulation of internal homeostasis. Reichle (1975) indicated that one analytical approach is to characterize the nutrient, energy and water budgets of the ecosystem and explain their dynamic behaviour through an understanding of the basic mechanisms governing the internal processes of the system. Specific hypotheses have been proposed that integrate ecosystem energy and material metabolism (Cummins, 1974; O'Neill *et al.*, 1975; Reichle *et al.*, 1975; O'Neill and Reichle, 1980) and a number of authors have proposed the use of functional charateristics in stress assessment (Barrett *et al.*, 1976; Odum, 1977; O'Neill *et al.*, 1977; O'Neill and Reichle, 1980; O'Neill and Waide, 1982).

Ecosystems are essentially energy processing units. The assemblage of organisms and their associated environment interact collectively through the exchange and incorporation of energy and materials. Although there does not appear to be one generally accepted theory of energy maximization in terms of ecosystem persistence, it has been hypothesized that the ecosystem processes energy in a fashion that promotes maximum persistent biomass within the physical and chemical constraints of the environment (Reichle *et al.*, 1975; O'Neill *et al.*, 1975). This hypothesis may be particularly applicable to terrestrial forest ecosystems which generally support a large biomass. Whittaker and Woodwell (1971) indicated that it is biomass accumulation in relation to productivity that reaches a maximum in the climax community of most terrestrial forests.

Application of the above hypothesis to phytoplankton communities in aquatic ecosystems is less certain. In general the biomass accumulation ratio (total organic biomass over net annual productivity) for aquatic systems is less than one, while it is greater than one in terrestrial systems, reaching 2–5 to 30–50 for perennial herbs and mature forests, respectively (Whittaker and Woodwell, 1971). Most aquatic ecologists would certainly support an alternative hypothesis that points to maximization of productivity rather than biomass in plankton-dominated systems (see Hutchinson, 1973). However, under the limitations imparted to the biomass hypothesis by the phrase 'within the physical and chemical constraints of the environment', it would appear to be generally applicable as a basis for analysis of any ecosystem.

The ecosystem can minimize and counteract the influence of environmental stress through population shifts and interactions. According to the hypothesis, unless a population contributes to the maintenance of persistent maximum biomass either through efficient photosynthetic energy conversion at the autotrophic level or through the efficient upgrading of energy quality, regulation of internal homeostasis, and/or cycling of essential nutrient at the heterotrophic level, there is strong negative selection against that population. Therefore, the ecosystem as a unit of investigation, is characterized as persisting in spite of perturbation, through dynamic shifts in its nutrient and energy metabolism. Populations are 'sacrificed' to preserve the integrity of the ecosystem, much as cells grow and die even as the organism persists. Of course, we cannot assume the same level of integration in an ecosystem as exists in the centralized organization of an organism.

As functional processes within the ecosystem, energy and material movements are intimately related. Within the constraints of the physical environment, the rate of the primary energy transformation process, photosynthesis, is regulated by essential element availability. There is an array of elements required for growth, reproduction, replacement of structural parts and general maintenance of all plants and animals. Mineral nutrients and food are often exchanged in the same community between producer and consumer levels, although the relative rates of usage are not generally equivalent. Therefore, a fraction of the energy processed by the ecosystem is expended in nutrient recycling and biomass retention (O'Neill and Reichle, 1980).

The cycling of carbon is most closely linked to energy flow since primary production is usually measured as organic carbon accumulation and respiration cost as carbon dioxide release. Other macronutrients and certain essential trace elements are selectively retained by the biota, thus these elements may pass through the food chain more slowly than bound chemical energy. Species retain elements and energy in their biomass. Maximum development of biomass storage is best characterized by the vast structural pool in forested systems. These huge biotic structures represent a living reservoir of energy and materials to the ecosystem; however, this reservoir is maintained only at a significant energy cost. Systems also have a large alternative material and energy base consisting of detrital organic material.

However, it is not merely the retention of elements which insures ecosystem persistence. Both a means to regulate remobilization of these nutrient resources (a major function of the consumer communities, in particular, decomposers) and a mechanism to regulate functions within the consumer community are essential. Consumer interactions thus contribute to the overall homeostasis of the ecosystem. Efficient remobilization of nutrients requires energy expenditure, however, and there are costs in regulation and homeostasis.

In all but plankton dominated ecosystems, the bulk of the net primary productivity enters the decomposition subsystem directly as dead organic matter (detritus). The faeces and carcasses of the consumer community also contribute to the detrital input to decomposition. The decomposition system is immensely important to both energy metabolism and efficient nutrient mobilization and conservation. Decomposers which utilize the organic energy of detritus also function to maintain essential nutrient elements in forms available for plant uptake.

Consumer species function as regulators of the major processes of primary productivity and decomposition, either through feeding interaction, competition, alteration of the environment or through performance of some unique functional role such as pollination. Within the ecosystem, such regulation is achieved through the diversity of the controlling species, the complexity of interactions or the unique and specific character of the feedback. Host–parasite and symbiotic interactions exemplify the latter control mechanism. The importance of any one mechanism of consumer control would depend on the system itself and on the species' evolution of adaptations to cope with fluctuations in the environment (Ricklefs, 1973). Species interactions which contribute to ecosystem regulation often have characteristic response time lags which increase the instability of the interaction. Therefore, natural and man-related perturbations to the system may not be recognized until some time after their introduction, as noted by the time scale of responses in Figure 6.1.

The interrelationship of energy and material movements can be depicted simply, as in Figure 6.2. Although this representation was developed for a terrestrial ecosystem, it is generally applicable to aquatic systems, given that there has been no demonstration of an equivalent process for withdrawal of nutrients from senescent leaves, and there is little or no dissolved pool of refractory organic compounds which can be classified as humic. There are also some

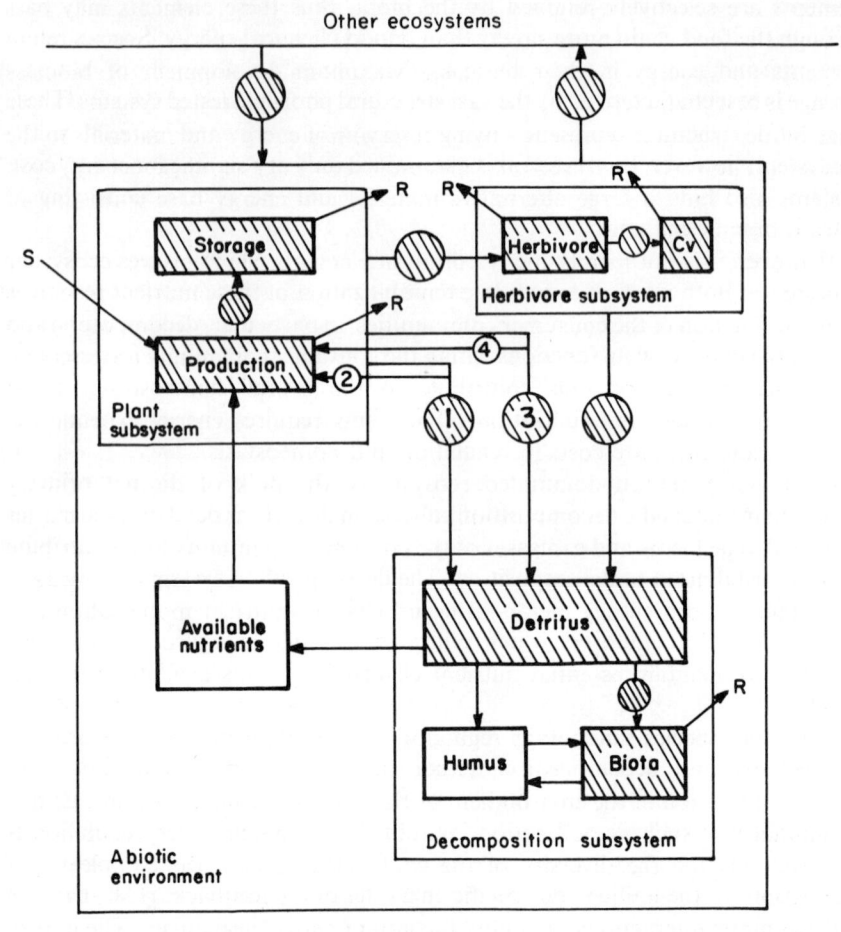

Figure 6.2 Nutrient and energy pools and fluxes within ecosystems. Pools and fluxes of both energy and nutrients are shaded. 1: litter production; 2: withdrawal from senescent leaves; 3: leaching losses; 4: reuse of dissolved organic carbon (DOC) by algae; S: solar radiation; R: respiration losses. The uptake and release of gases by all three subsystems are implicit

obvious differences in energy and material processing in terrestrial and aquatic ecosystems. Terrestrial systems are based on structurally complex vascular plants where the 'detritus' food chains are quantitatively more prominent. Biomass storage and efficient nutrient recycling are highly developed. Lakes and marine ecosystems are based on rapidly reproducing autotrophic populations supporting well-developed 'grazing' food chains (Wiegart and Owens, 1971). Most aquatic systems, particularly streams, can be conceptualized as being greatly dependent on the surrounding watershed for nutrient input (Likens and Bormann, 1975) although there is some evidence of the importance of internal nutrient recycling in freshwater lakes. Axler *et al.* (1981) reported that during most of the growing season, nitrogen regeneration via zooplankton excretion and microbial mineralization was critical to phytoplankton growth in Castle Lake, since levels of dissolved inorganic nitrogen were insufficient to support the rates of primary production measured. These differences emphasize the importance of examining nutrient conservation processes in terrestrial systems under pollution stress, while considering changes in primary production and consumer regulating processes in perturbed phytoplankton-based aquatic systems. This would not ignore the importance of recycling of essential elements in lake and marine systems, but would place emphasis on those specific processes governing essential element availablity such as nitrogen fixation, nitrification, ammonification, consumer regeneration of ammonia and sediment–water interchanges. Nor would such an approach ignore direct pollutant interference with terrestrial productivity or yield when such effects are measureable.

This survey builds a case for examining toxic pollutant interference with ecosystem function. (It would appear that the theory and methodology would be quite applicable to physical perturbations as well.) The focus will be on productivity and decomposition subsystem processes, and the ways in which these processes are regulated.

6.2 DECOMPOSITION AND ELEMENT CYCLES

Toxic pollutants can interfere with the decomposition process itself and interrupt internal recycling of essential elements, particularly nitrogen. The decomposition process forms the major link between nutrient availability and primary production through its control of the breakdown of organic matter, essential element immobilization and release, and soil humus formation. Decomposition is also the process through which a major portion of all heterotrophic energy is expended (Heal and MacLean, 1975). In unperturbed systems where quantities of 'available' inorganic forms of nutrients are seldom in excess of plant demand, essential element cycles, such as that for nitrogen, are particularly vulnerable to pollution disruption. Aside from their role as mineralizers, microorganisms contribute substantially to nutrient conservation through nitrogen fixation and essential element immobilization in their cells.

Decomposition processes in terrestrial ecosystems have been extensively reviewed (Witkamp and Ausmus, 1976; Lohm and Persson, 1977; Swift et al., 1979). A number of reports have detailed the decomposition process in aquatic systems, emphasizing its importance in energy movement rather than nutrient cycling (Fisher and Likens, 1973; Cummins, 1974; Boling et al., 1975; Saunders, 1976). Woodall and Wallace (1975) indicated that streams function primarily through the use and transport of nutrients from the watershed. However, Likens (1975) suggested that within stream boundaries, nutrients may be cycled very rapidly among living components, nonliving components or both. Filter feeders may play a significant role in such recycling processes (Wallace et al., 1977). The study of critical regeneration of nitrogen in Castle Lake, previously cited (Axler et al., 1981), is an example of current research to determine the importance of element recycling by biota in lake ecosystems, in relation to primary productivity.

Decomposition can be divided into three processes: (1) leaching, (2) faunal breakdown, (3) microbial mineralization. Leaching processes affect element movement from live plant tissue and detrital materials. The process is primarily physical in nature (Witkamp and Frank, 1970) although the rate of leaching can be influenced by the extent of fragmentation and breakdown of leaf litter by invertebrate and microbial feeders. Extracellular release by phytoplankton in aquatic systems provides a biological mechanism for the transfer of organics to the dissolved pool. The largest portion of dissolved organics are leached early in the decomposition process. A diverse community of decomposer organisms effects element cycling by means of mechanical comminution of detritus, enzymatic breakdown of ingested matter, channelization of woody substrates and colonization of detritus by microbes (Witkamp and Ausmus, 1976). Microbial decomposition, which generally begins before leaf senescence, leads to the release of inorganic elements. Microbial cells immobilize considerable amounts of N, P and K, retaining them against leaching loss (Ausmus et al., 1976).

The rate of decomposition is regulated by three controlling factors: (1) the physico-chemical environment, (2) the available resource (detritus) quantity, (3) the resource quality. Pollution input generally affects decomposer activity through direct toxic stress (i.e. changes in the chemical environment), although toxic organic pollutants (oil, industrial organic effluents, pesticides) can play an additional role by affecting the other controlling variables.

There have been few ecosystem studies directed specifically at determining the effects of toxic pollutants on the decomposition process and the efficiency with which essential elements are conserved and cycled. However, there presently appears to be a theoretical base and a developed experimental technique to support such studies (Witkamp and Ausmus, 1976). It is important to consider the fact that in terrestrial systems, during the early successional stages, nutrient recyling is only minimally developed and the systems are 'leaky'. Reversion of normally element-conservative, mature systems to a state in which they export excessive levels of calcium, magnesium and inorganic nitrogen, is convincing evidence of the breakdown of their characteristic internal regulatory mechanisms. This type of retrogression infers a reduction in the biotic control of nutrients. The investigations at the Hubbard Brook Forest watershed have provided much experimental impetus for the study of nutrient dynamics and their control processes as a means of determining the effects of perturbations on ecosystem function (Likens et al., 1970; Gosz et al., 1973; Likens and Bormann, 1975). O'Neill et al. (1977), using a series of terrestrial microcosm experiments involving additions of either 100 mg cm^{-2} Na$_2$HAsO$_4$ or 11 mg cm^{-2} Pb, demonstrated that pollutant disruption of the nutrient cycle, as measured by the net rate of leached soil nutrients (Ca and NO$_3$—N), provided a more sensitive and reliable measure of system disturbance than did population numbers, biomass or microbial diversity.

Several recent studies of specific ecosystem pollutant problems such as acid precipitation (Abrahamsen, 1980; Bjor and Teigen, 1980; Norton et al., 1980; Wright and Johannessen, 1980; NRCC, 1981a) and heavy metals (Jackson and Watson, 1977; Jackson et al., 1978; Hughes et al., 1980b; Lepp, 1981) have provided additional evidence of decomposition and nutrient cycling disruptions, further illustrating the implications of long-term disturbances on the metabolic functioning of a system. Although a preponderance of the information on the disruption of decomposition processes and their function in material conservation and recycling is from laboratory or microcosm and short-term small-scale field investigations, it seems clear that such processes should be examined when evaluating ecosystem perturbations.

6.2.1 Acid Precipitation

Acid precipitation is a very topical issue and a damaging pollution problem. There is considerable evidence that directly through toxic effects and indirectly through lowering of environmental pH, atmospherically borne H$_2$SO$_4$ and HNO$_3$ have caused extensive disruption of material conservation and cycling in both terrestrial and aquatic systems.

Through-fall studies have indicated increased foliar leaching with increased incidence of acidic rain (Abrahamsen et al., 1976; Cronan, 1980; Cronan et al., 1980). Abrahamsen et al. (1976) showed that concentrations of metal cations increased with the fall of acid rain. Previously, cation exchange for hydrogen ion in the tree crown had been postulated to explain foliar leaching (Eaton et al., 1973), although processes such as foliar exudation are likely of importance (Cronan, 1980). Potassium (K$^+$) efflux from lichens has been correlated with SO$_2$ levels and the acidity of precipitation (Nieboer et al., 1976; Tomassini et al., 1977). Acidification experiments in Norway demonstrated that rain pH significantly influenced the amount of calcium, magnesium and potassium in through-fall (Figure 6.3). It is probable that a large portion of the through-fall enrichment

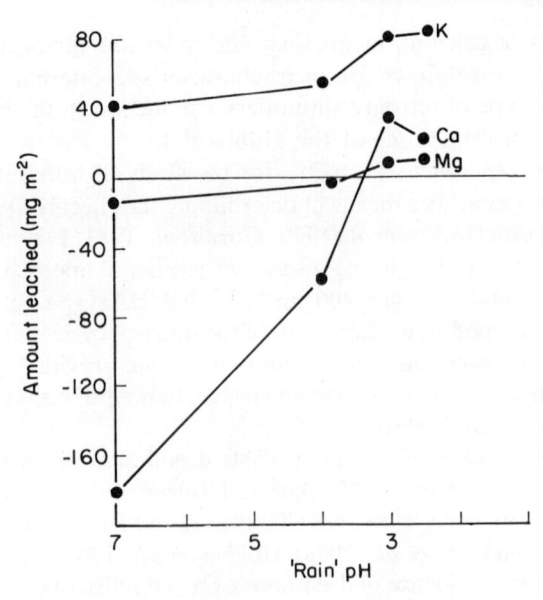

Figure 6.3 Effect of pH of simulated rain on leaching of Ca, Mg and K from spruce crowns. Average of two treatments of 50 mm 'rain'. Before passing the tree crowns the 'rain' contained 315, 35 and 35 mg m^{-2} of Ca, Mg and K, respectively. (Reproduced by permission from Abrahamsen et al., 1976)

in Cl$^-$, SO$_4^{2-}$, Na$^+$ and H$^+$ is derived from dry deposition rather than from leaching (Abrahamsen et al., 1977). Wood and Bormann (1975) demonstrated increased foliar leaching of nutrient cations (Ca^{2+}, Mg^{2+}, K$^+$) to be a function of increases in the acidity of an artificial mist applied to greenhouse plants. Increased leaching was demonstrated with no visible foliar damage at pH 3.3 but cation levels in the leachate were considerably elevated with tissue damage below pH 3.0. Leaf cuticle erosion, accelerated by acids (Shriner, 1976), and cuticular abrasion by other environmental stresses (Tukey and Morgan, 1963), may account in part for accelerated foliar leakage in acid rain disturbed systems. Mathematical techniques to separate crown leaching components were recently described by Lakhani and Miller (1980).

There is also evidence suggestive of accelerated nutrient loss due to structural damage and root leakage accompanying decreases in plant resistance to pest infestation under SO$_2$ stress (Knabe, 1976). Pest invasion can eventually kill trees, reducing their long-term effectiveness as nutrient-conserving units. Polish forest soils exposed to average annual SO$_2$ concentrations greater than

$0.8\,\mathrm{mg\,m^{-3}}$ are becoming acidic, favouring establishment of root-destroying fungi such as *Armillaria mellea*, bringing about short-term increases in root leakage which eventually lead to disruption of the ability of plants to take up nutrients (Grzywacz and Wazny, 1973).

Perhaps more detrimental to plant productivity is the increased acidic leaching of mineral and detrital elements from soils (Rorison, 1980), which results also in the release of potentially toxic heavy metals into ground and surface waters (Beamish and Van Loon, 1977; Cronan *et al.*, 1978; Hutchinson, 1980). This loss is particularly damaging in the nutrient deficient soils of many forested ecosystems (Engstrom, 1971). Overrein (1972) observed dramatic releases of calcium from the root zone in forest soils exposed to precipitation with a pH of 3.0. Lysimeter studies have demonstrated that simulated acid rain (pH 2.5) induced net losses of K, Ca, Na, Mg, Mn, Al and NO_3 and NH_4 from podzol-type soils (Bjor and Teigen, 1980). It was the leaching of calcium and magnesium which was most affected by acid rain with a pH in the range of 3–4, and this effect was most noticeable in the surface layer of the soil profile. Figure 6.4 illustrates the effects of acidification on levels of calcium, magnesium and potassium in soil, leachate and vegetation. The leaching of nitrogen is generally

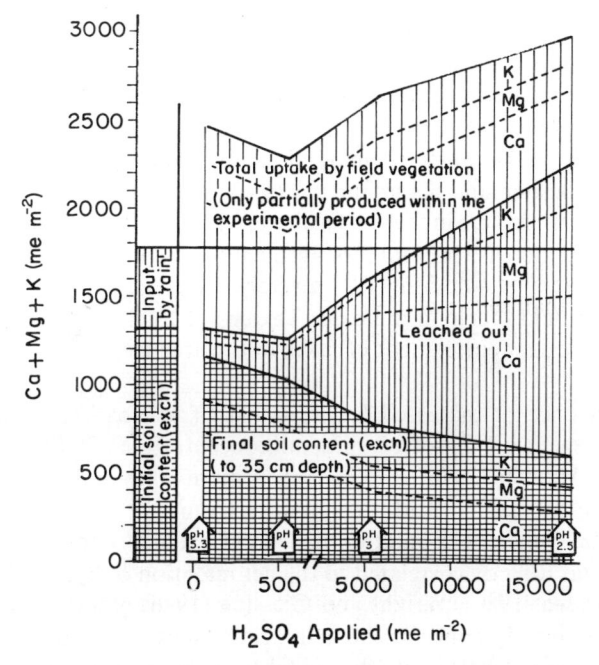

Figure 6.4 Acid rain treatment effects on calcium, magnesium and potassium in soil, leachate and vegetation. (Reproduced by permission from Bjor and Teigen, 1980)

much less than the input by precipitation (Abrahamsen, 1980; Cronan, 1980) and, therefore, the net losses of Ca, Mg and Mn from the soil nutrient pool are of greater concern. The crucial question is whether or not the increases in weathering of minerals due to acidification can compensate for the increased leaching losses in surface and ground-water runoff.

The degree of leaching increases considerably as a function of decreases in pH of the precipitation (Abrahamsen et al., 1977; Abrahamsen, 1980). However, the quantity of leached cations that were measured were not found to be equivalent to the amount of acid added (H^+). These results lend support to the theories advanced by Wiklander (1975, 1980) who called attention to the low efficiency of hydrogen ions in exchanging metal cations in acid soils. Application of 50 mm of 'rain' (pH < 3) to soils significantly reduced their base saturation and lowered soil pH (Abrahamsen et al., 1976). Aluminium, because it is common in soil minerals and is chemically active, plays a dominant role in regulating soil solution acidity. High levels of Al may be brought into soil solution to buffer soils receiving acid rain (Voigt, 1980; Johnson et al., 1981). Abrahamsen and Stuanes (1980) reported that simulated acid rain produced effluent solutions differing in pH, SO_4^{2-} and Al concentrations from soils of the same type as one another but dominated by different plants. Leaching studies have consistently demonstrated potential nutrient loss due to increased acid leaching, but their small scale has raised questions about the application of these experimental results to the more extensive, internally regulated plant and soil nutrient pools of forests. However, data from Cronan (1980) indicated that in the subalpine zone of the White Mountains, New Hampshire, mineral acids dominate solution chemistry. Calcium, magnesium and potassium increase in the through-fall. Levels of these cations gradually decline in the soil leachate, with soluble aluminium accounting for most of the increase in total dissolved cations between the through-fall and forest floor percolate solutions. In a severely disrupted forest ecosystem near a metal smelter, acid precipitation caused extensive leaching of Ca from the soil, with the consequent enrichment of lakes in the area (reviewed by Gorham and McFee, 1980).

The occurrence of increased leaching loss of soil cations is supported by evidence of increased dissolved mineral concentrations in Swedish rivers, as discussed by Malmer (1976), and in some remote lakes in Sweden and Canada which receive substantial acid precipitation (Almer et al., 1978; Dillon et al., 1980). In small watersheds on granitic bedrock the net output of combined Ca, Mg and Al was directly correlated to the net retention of hydrogen ion (Wright and Johannessen, 1980). Wright and Gjessing (1976) reported that there were increased Ca and Mg concentrations in acidic lakes in Scandinavia and North America relative to levels in unaffected lakes in the same geographic areas. The net export of calcium, magnesium and potassium as previously described is due, in part, to natural weathering, but a major fraction results from acid-induced leaching from normally element-conservative plant and soil systems. In addition, the base-impoverished surface sediments in acid lakes in the Adirondack Mountains indicates increased leaching within an aquatic system (Galloway et al., 1976).

The sensitivity of certain decomposer organisms to an acid pH environment has been established, but the effect of precipitation-induced acidification on the functional effectiveness of the decomposer community is less well defined. Groups like protozoa and earthworms are very rare in most soils with pH below 4 (Stout and Heal, 1967; Satchell, 1967), however, Wood (1974) found extensive earthworm colonization in the organic and mineral-rich but acidic (pH ≤ 4.5) alpine soils of Mount Kosciusko, Australia. Although Abrahamsen et al. reported in 1976 that they had found no significant effects of acid rain on soil fauna, later studies indicated that populations of *Cognettia sphagnetorum*, which feed primarily on litter and associated microbes, were significantly reduced in experimentally acidified plots with litter-humus pH generally less than 5 (Lundkvist, 1977; Abrahamsen et al., 1980).

In preference experiments, Lumbricid and Enchytraeid worms either selectively colonized basic soils or avoided humus pH < 3.9 (Hagvar, 1980). The abundance of the total Enchytraeid group was significantly reduced in acified forest plots where pH had been lowered to 4.1, 4.2 and 4.7, respectively, in the top three soil profiles. The vertical distribution of *Cognettia sphagnetorum* also changed in acidified plots, as the main part of the population was found not in the humus-litter layer, but rather, in a lower profile (Bååth et al., 1980a, b). Other important soil invertebrates such as mites (Acari) and spring tails (Collembola) are affected differently by acid precipitation. The total abundance of all mites appeared to be unaffected by rain acidity, pH 4 to 2.5, in experimentally acidified soil plots (Bååth et al., 1980a, b; Abrahamsen et al., 1980). However, in similar experiments, the total numbers of Acari were decreased in soil with pH 2.5 to 4, as was the vertical distribution of the Oribatid group (Hågvar and Amundsen, 1981). It was suggested that reduced abundance of soil invertebrates was best explained by the correlation between soil pH and reproductive success. On the other hand, the total abundance of Collembola increased in acidified plots, pH 2.5 to 4 (Abrahamsen et al., 1980), suggesting release from competitive or trophic constraints. The lengths of FDA-active fungi and the numbers, biomass and cell size of soil bacteria have been shown to be reduced by experimental acidification of soils to a pH of 4.2, while total fungal mycelium appear to increase in abundance (Bååth et al., 1980a, b; Lohm, 1980). Isolated soil bacteria were unable to grow at pH values less than 4. Acid treatment also caused marked changes in the physiological abilities of soil bacteria. In acetate and in 'nitrogen free' media, growth was impaired (Bååth et al., 1980a).

Aquatic crustaceans such as the detritus-feeding amphipod, *Gammarus lacustris*, are not found in Norwegian lakes below pH 6, nor are oligochaete worms abundant in these lakes' deep sediments (Leivested et al., 1976). Minshall and Minshall (1978) found that the mortality rate for *Gammarus* in water with a pH of 5–6 was twice as high as in pH 6–7 water. In general, the number of species in acid lakes decreases with pH (Almer et al., 1978; Økland and Økland, 1980) although it is assumed that pollutants associated with acidification (e.g. metals, particularly aluminium) may also restrict faunal abundance. Mossberg and Nyberg (1979) reported that the number of species declined at a pH of 4.2 to 5 in Norwegian lakes. Reducing the pH in a stream from 5.4 to 4.0 reduced macroinvertebrate standing crop (Hall et al., 1980). Macroinvertebrates are less active in leaf decomposition in naturally acidic streams (in Hendrey et al., 1976). Although no marked differences in the content of fine particulate organic matter (FPOM) or dissolved organic material (DOM) were detected between an acidic and a non-acidic stream, the breakdown of leaf material was much slower in the former system (Friberg et al., 1980). The biomass of invertebrates per leaf pack was less, the number of species was less and the community of functional feeders contained far more 'shredders' and fewer 'scrapers' in the acidified stream (pH 4.3–5.9) than in the nonacidic one (pH 6.5–7.3). The reduced breakdown of leaf material could be the direct result of organism sensitivity to low pH or an indirect response to a reduced microbial community associated with detritus. Boling et al. (1975) found that certain macroinvertebrate detrital feeders in streams prefer 'conditioned' (microbially colonized) leaves, which are of increased nutritional value. Physiological studies of decomposer bacteria indicate that very few kinds function actively below pH 4 (Doetsch and Cook, 1973). Long- and short-term studies of the effects of soil acidification showed that heterotrophic bacterial counts were severely reduced, resulting in decreased microbial respiration (Bryant et al., 1979). The effects of aquatic acidification on microbial activity were determined in laboratory experiments by examining the decomposition of peptones (Bick and Drews, 1973). Total bacterial cell counts and the numbers of ciliated protozoa decreased with decreasing pH. The investigators noted that for the peptone substrate, decomposition and nitrification were also reduced as the pH decreased, and oxidation of ammonia ceased below pH 5. At pH 4 and lower, the numbers of fungi increased, however. In acidified lakes the effect of reduced pH on decomposition processes appears to be greater in the littoral rather than in the profundal sediments (Gahnström et al., 1980). Sphagnum mats accompanied by dense filamentous algae have been observed covering acid lake sediments, restricting element movement from the sediments (Hendrey and Vertucci, 1980).

Laboratory studies have provided further evidence for reduction of decomposer metabolic activity under acidified conditions. In soil tests, a highly significant correlation was found between the relative amount of CO_2 produced and the exchangeable hydrogen ion content of the soil (Francis et al., 1980). Experiments using glucose and glutamic acid substrates demonstrated that reduced pH led to a shift in the dominant organism from bacteria to fungi, with an accompanying decrease in zooflagellate fauna and consequent decrease in the consumption of oxygen (Leivestad et al., 1976). The extent of the inhibition of glucose oxidation and the fall in pH were found to be proportional to the period of exposure to SO_2 (Grant et al., 1979). Oxygen consumption, used as a measure of the rate of decomposition of homogenized leaf litter, was reduced by 50 per cent when the pH decreased from 7.0 to 5.2, and no adaptation to lower pH was achieved over a 3-week period (Traaen, 1974). In both natural and artificial sediment cores, the rate of oxygen uptake indicated that heterotrophic microbial activity was reduced when the pH of overlying water was decreased to below the neutral range (Leivestad et al., 1976).

Although the effects of acid rain on decomposition may not be obvious for some time after exposure (Johnson and Shriner, 1980; Roberts et al., 1980), the end results of decreased number and activity of decomposer organisms are a reduction in the decomposition rate and the accumulation of coarse detrital material in the ecosystem. The decomposition rate of lodgepole pine needles increased significantly above pH 3.5, whereas at a pH of 1, no decomposition took place (Abrahamsen et al., 1977). The rate of decomposition of Scots pine needle and root litter was significantly reduced by application of acid rain for a period greater than one year, which lowered soil pH by 0.5 units to 4.2 and 4.1 in the top horizon (Bååth et al., 1980a; Lohm, 1980). There was also a reduction in nitrogen retention in the needle litter of the acidified plots. Decomposition of cellulose was significantly retarded with increased rain acidity in some experiments (Ruschmeyer and Schmidt, 1958; Baath et al., 1980a); however, this finding has not been consistently duplicated (Hovland and Abrahamsen, 1976). Recently, Killham and Wainright (1981) reported cellulose degradation to be only marginally inhibited by atmospheric pollution from a coke plant which reduced decomposition of deciduous leaf litter by 35 per cent. There are few data which document increased litter accumulation associated with pH decreases. Freedman and Hutchinson (1980b) observed a trend toward higher standing crops of litter at sites closer to a nickel smelter, but heavy metal levels as well as acidity in the litter were significantly higher than normal. Litter-bag experiments in acidic stream waters have indicated reduced decomposition (Hendrey et al., 1976). Similar results have been reported for acidic lakes (Traaen, 1980; Francis et al., in press). Figure 6.5 illustrates decreases in the per cent weight loss of leaf litter with increasing lake acidity and an associated decrease in bacterial colonization. There were species-specific differences in decomposition rate which appear to be associated with microbial colonization and activity. Abnormal accumulations of coarse organic detritus have been observed on the bottoms of six Swedish lakes where pH has decreased by more than 1.5 units in the past 30 to 40 years (Grahn et al., 1974).

In both Sweden and Canada, acidified lakes have been treated with alkaline substances to determine the effect of neutralizing pH on the decomposition processes. Such treatments have complex effects on water chemistry, but have resulted in increases in heterotrophic bacteria, increased microbial activity and an acceleration of organic decomposition processes (Andersson et al., 1975; Scheider et al., 1975).

Acidification of the environment directly and indirectly affects the nitrogen

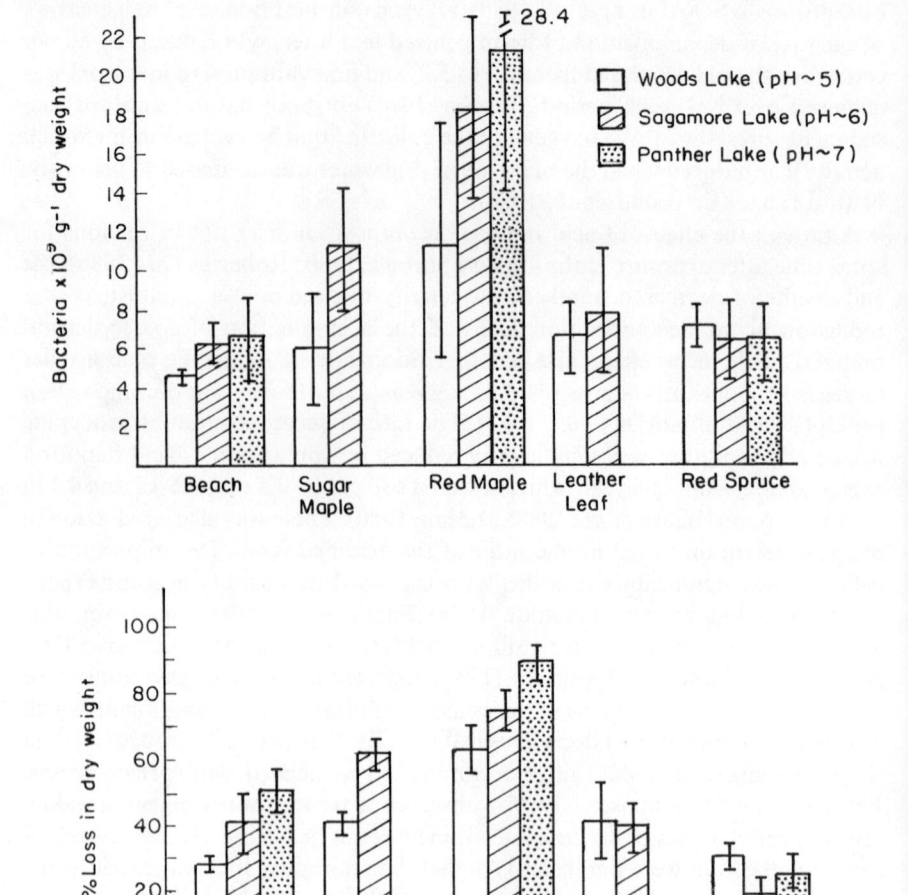

Figure 6.5 pH influence on microbial colonization of leaf litter and on leaf decomposition in three Adirondack lakes. (From Francis *et al.*, in press. Reproduced by permission of Ann Arbor Science Publishers)

cycle. Increases in soil acidity adversely affect the root–mycorrhiza relationship, presumably disrupting the nitrogen fixing activities of associated bacteria. The root nodulation process which is the prelude to nitrogen fixation is particularly sensitive to soil acidification (Alexander, 1980a). In addition, the numbers of infective root-nodule bacteria are often reduced to exceedingly small populations as the pH value falls below 5.0 (Rice *et al.*, 1977). Alexander (1980a, b) concluded that the nitrogen inputs resulting from the legume-*Rhizobium* symbiosis would likely be greatly lowered as soil acidity increases and the same effects would reduce nonsymbiotic fixation. Nitrogen fixation by free-living bacteria in naturally acid and pH-adjusted soils did not take place at or below pH 4.7 and at pH 5.8 the amount of nitrogen fixed was 1000 times less than in soils at pH 6.4 (Francis et al., 1980). Acid water (pH 5) also markedly reduced the rate of fixation in litter samples (Denison et al., 1977). Some nitrogen-fixing lichens such as *Lobaria oregana*, very important to the nitrogen balance of mature Douglas fir forests, are particularly sensitive to acid rain (Tamm, 1976). Wodzinski et al. (1977) demonstrated that nitrogen-fixing blue-green algae were one of the microbial groups particularly sensitive to acidification.

Several investigators have reported reductions in nitrification rates in acidified artificial and natural soils (Hovland and Ishac, 1975; Tamm, 1976). Autotrophic nitrifiers were demonstrated to be very sensitive to acidic conditions (Francis et al., 1980). Remacle (1977) reported that nitrification rates decreased with decreasing pH while ammonia production increased until pH fell below 5.1. Tamm et al. (1977) observed that acidification of soils decreased carbon dioxide evolution, with accompanying increases in ammonification and decreases in nitrification. They concluded that the decreased microbial activity was more pronounced for nitrogen immobilization processes than for decomposition processes in general. However, it should be noted that the importance of reduced nitrification in acidified systems is not totally understood. Indeed, less nitrogen in the leachable nitrate form may in fact reduce net nitrogen loss. Tamm (1976) found that repeated acid treatments of soils reduces both nitrification and ammonification, indicating a severe disruption in organic nitrogen mineralization processes. Inhibition of denitrification has been reported for bacteria in soil acidified to a pH of 4.1–4.2 (Bååth et al., 1980a). One complication that arises when interpreting changes in the nitrogen cycle is the input of nitrogen compounds from the HNO_3 portion of the acid precipitation. The impact of these compounds as fertilizer sources is not yet fully understood, although Bengtson et al. (1980) reported that nitrogen oxides were taken up by Scots pines in amounts linearly related to NO_x concentrations in the field.

Little is known about the effects of acidification on sulphur-cycle bacteria although Schindler et al. (1980a) found that sulphate-reducing bacteria neutralize a considerable level of acid input to lakes, and sulphate reduction to sulphide in soils is markedly inhibited below pH values of 6 (Connel and Patrick, 1968). Both the chemistry of phosphorus and its uptake by algae are pH sensitive, and the solubility of heavy metal ions increases at low pH, suggesting that synergistic adverse effects may accompany ecosystem acidification. Very little is known about the consequences of the loss of long-term essential element storage in the biomass of large consumers (fish), eliminated from acidified aquatic systems.

In summary, field and laboratory experiments have indicated that acid precipitation interferes with ecosystem nutrient conservation and cycling processes through several mechanisms:

1. Increasing leaching export of essential cations.
2. Bringing heavy metal ions into solution.
3. Inhibiting organic decomposition and shrinking the available nutrient pool.
4. Inhibiting N_2 fixation and reducing nitrification and ammonification (under severely reduced pH conditions).

6.2.2 Heavy Metals

The reported effects of metal pollution on decomposition and element cycling are quite similar to those previously discussed for acid precipitation. There is no reason to suggest that low pH need accompany metal pollution in explaining the similarity in patterns of disruption; however, metals and SO_2 are at times closely interrelated (see review by Haines, 1981). They can be derived from the same industrial source. Acidity can enhance metal toxicity (Skidmore, 1964) and increase metal solubility (Stumm and Morgan, 1981), bringing toxic concentrations of certain metals (e.g. aluminium) into solution even at pH values not considered harmful (Cronan and Schofield, 1979).

O'Neill et al. (1977) reported that lead-contaminated (11 mg cm^{-2}) terrestrial microcosms showed no signs of growth sensitivity even after 9 months of exposure. However, the mean calcium level in the soil leachate, 29.4 mg l^{-1}, was significantly higher than the mean of 20.3 μg l^{-1} measured for unexposed microcosms. Leached nitrate levels were also elevated but were not statistically higher than for control plots. The treated microcosms continued to lose nutrients long after exchange processes would have equilibrated, indicating a disruption of calcium immobilization processes, accelerating Ca export from the system. Jackson et al. (1978) found a similar sustained leaching effect for Ca and NO_3-N due to synergistic heavy metal contamination in an intact forest microcosm exposed to Pb, Cd, Zn and Cu at levels of 11.0, 0.128, 0.748 and 0.161 mg cm^{-2}, respectively. The soil pool of extractable nutrients showed signs of depletion after 20 months of contamination. Arsenic-induced stress stimulated loss of PO_4-P, NO_3-N, NH_4-N and DOC from a grassland microcosm (Jackson et al., 1979). The advanced stage of pollution response near a lead smelter is characterized by nutrient depletion in the soil and litter pools (Figure 6.6) (Jackson and Watson, 1977). Depletion of Ca, Mg and K is correlated with distance from the metal source. Severely metal-stressed terrestrial ecosystems can support only sparse vegetation and later show signs of general material erosion, emphasizing the deterioration of nutrient conservation mechanisms (Jordan, 1975).

There is little evidence to suggest that aerial metal pollution stimulates foliar leaching of nutrients as was documented previously for acid precipitation. Severe heavy metal dusting can cause foliar injury to plants (Kraus and Kaiser, 1977), which could eventually contribute to element export. There is some information suggesting that toxic metals stimulate extracellular release of dissolved organic

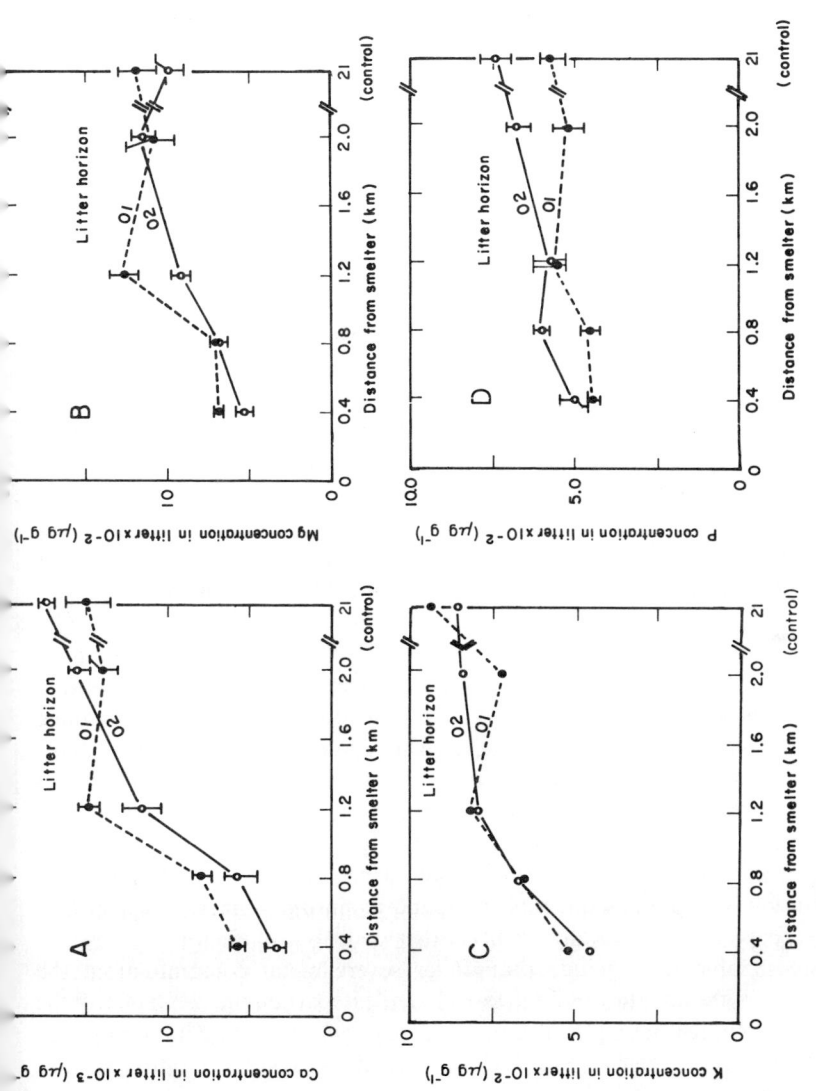

Figure 6.6 Macronutrient concentrations of (A) Ca, (B) Mg, (C) K, and (D) P(± 1 SE, $n=12$) in forest floor litter as a function of distance from a lead smelter. (Reproduced from Jackson and Watson, 1977, *Journal of Environmental Quality*, volume 6, page 336, by permission of the American Society of Agronomy, Crop Science Society of America and the Soil Science Society of America)

carbon (DOC) from phytoplankton species. Wetzel (1975) suggested that membrane damage would increase the loss of DOC. Thomas *et al.* (1977) noted increased extracellular release with a copper concentration of 10–50 μg l^{-1}. A precipitous release of DOC with the addition of 50 μg l^{-1} copper has been demonstrated, indicating that cellular disruption is probably occurring in some phytoplankton species (Vaccaro *et al.*, 1977).

Decomposer populations and their activities are adversely affected by toxic metal concentrations. Although there is a rapidly growing body of literature on metal accumulation by litter-feeding invertebrates in contaminated soils, there is little definitive information on the detrimental effects of metal accumulation on soil community processes. Williamson and Evans (1972, 1973) found that a lead content of between 165 and 19,000 μg g^{-1} had little influence on soil invertebrates collected near lead-contaminated roadsides and around mining operations. However, the reported Pb concentrations in their study were far less than those measured for industrially impacted areas. Watson *et al.* (1976) reported reductions in total arthropod numbers and biomass in contaminated litter near a lead smelting–mining complex in Missouri where soil-lead levels were greater than 88 000 μg g^{-1}. Table 6.1 indicates that the biomass of invertebrates in all trophic categories was decreased significantly, relative to the proximity to the pollution source. Strojan (1978a, b) found the density of all major taxonomic groups to be lower near a zinc smelter (Zn $\geq$ 26 000 μg g^{-1}), with the orbatid mite populations, important leaf litter feeders, most severely reduced.

The adverse effects of metals on aquatic macroinvertebrates (primarily a detritus-based community) encompasses all taxonomic groups; the trend is towards reduced numbers of species and individuals. In a copper-polluted mountain stream (20–630 μg l^{-1} Cu), Sheehan (1980) observed abnormally small macroinvertebrate populations and alterations in the distribution of types of feeders. The diverse detrital feeding family Chironomidae appears to have several metal-tolerant species (Winner *et al.*, 1975, 1980). Occhiogrosso *et al.* (1979) proposed a direct cause–effect relationship between high heavy metal concentrations in aquatic sediments and reduced density of macroinvertebrates, although they could not totally dismiss other contributing factors.

It is reasonable to conclude that under severe metal contamination, the numbers of both aquatic and terrestrial detrital particulate feeders will be severely reduced, restricting their important contribution to the comminution of litter. The nature of their role in moderately polluted systems requires further investigation.

Excessive metal accumulation leads to a restructuring of microbial community composition. Jensen (1977) described the response of soil microbes to lead additions of 5000 μg g^{-1} as reflecting a reduction in bacterial density with corresponding proliferation of a few tolerant fungal species. Metal-resistant fungal species found in zinc-contaminated soils appear to grow normally (Jordan

Table 6.1 Mean biomass of arthroped predators, detritivores and fungivores in litter from the 02 horizon on Crooked Creek Watershed ($n = 16$). (Reproduced from Jackson and Watson, 1977, *Journal of Environmental Quality*, volume 6, page 334, by permission of the American Society of Agronomy, Crop Science Society of America and Soil Science Society of America

Distance from smelter (km)	Biomass (mg m^{-2})		
	Predator	Detritivore	Fungivore
0.4	2.1[1]	2.3[1]	0.8[1]
0.8	5.8[1]	16.6[1]	2.4[1]
1.2	14.0	12.6[1]	5.8
2.0	87.6	92.6	16.1
21.0 (control)	17.3	61.1	22.1

[1] Significantly different from control ($P \leq 0.1$).

and Lechevalier, 1975). In addition, some zinc-tolerant actinomycetes and bacteria were present in the same soils, containing up to 13.5 per cent Zn in the upper horizon. Tolerant bacterial strains have been shown to display physiological characteristics different from those of sensitive species. The density of resistant strains has been shown to be inversely related to the level of metal concentration in the environment (Houba and Remacle, 1980). The work of Jordan and Lechevalier (1975) and Strojan (1978a) clearly indicated that both microflora and invertebrate activities in soil had been adversely affected near metal smelting operations, where concentrations of 900 μg g^{-1} Cd, 2300 μg g^{-1} Pb and 26,000 μg g^{-1} Zn in litter were 100 times greater than in samples from a control site. Jackson and Watson (1977) found significantly high C/N ratios in litter near smelters (where levels of 88 348 μg g^{-1} Pb, 128 μg g^{-1} Cd, 2189 μg g^{-1} Zn and 1315 μg g^{-1} Cu were 50 to 500 times those of control samples), indirectly indicating a reduction in microbial activity. The C/N ratio is inversely related to microbial population density (Witkamp, 1966). The microbial colonization of leaf litter in an artificial stream system was shown to be inhibited by relatively low levels of cadmium, 5 and 10 μg l^{-1} (Giesy, 1978). Reduced microbial colonization of litter has been shown to affect invertebrate feeding (Boling *et al.*, 1975). If these results can be applied generally, macroinvertebrates requiring leaf conditioning could be severely impacted at metal concentrations which inhibit microbial colonization but do not directly affect the feeding invertebrates.

Bond *et al.* (1976) found that Cd at 10 μg g^{-1} in a forest soil and litter microcosm reduced oxygen consumption and respiration by 40 per cent. A number of investigators have demonstrated reduced decomposition activity as measured by carbon dioxide evolution in artifically metal-amended soils (Ebregt and Boldewijn, 1977; Doelman and Haanstra, 1979a) and in contaminated litter-soil samples (Jackson and Watson, 1977 Chaney *et al.*, 1978; Strojan, 1978b).

Metal effects on soil enzyme activity indicate inhibition of essential reactions. Rühling and Tyler (1973) showed a highly significant negative correlation between total metal concentration (Zn + Cu + Cd + Ni) and dehydrogenase activity. Amylase activity in metal-contaminated litter was found to decrease linearly with the log concentration of total Cu + Zn + Pb + Cd (Ebregt and Boldewijn, 1977). A concentration of 1500 μg g^{-1} Pb considerably increased the lag time to reach the peak rates of oxidation of starch and cellulose, and reduced the peak rates of oxidation of glucose, starch and glutamic acid, in soils (Doelman and Haanstra, 1979b). Tyler (1974, 1976) found a partial but immediate inhibition of phosphatase and urease enzymes due to copper and zinc salts, whereas β-glucosidase activity was not measureably affected. In Douglas fir needle litter, invertase activity was immediately inhibited and cellulase activity was depressed after four weeks, with treatments of 1000 μg g^{-1} Hg or Cd (Spalding, 1979). Houba and Remacle (1980) noted that metal inhibition of certain essential enzymes may explain why metal-tolerant bacteria are unable to perform complete cellulolysis, pectinolysis or hydrolysis of starch under stressed conditions.

Whole-lake experiments with arsenic showed that seasonal influences mediate the impact of this pollutant on the microbial degradation of organic matter (Brunskill et al., 1980). Under winter ice, 40 μmol l^{-1} arsenate or arseniate inhibited degradation by 50 per cent, while during the ice-free season, there was little apparent effect on degradation rates.

The findings reviewed above lead to the conclusion that excess metal loading can significantly decrease the organic decomposition rate and consequently lead to increased litter accumulation. Rühling and Tyler (1973) concluded that even a moderate quantity of metal pollution would depress decomposition in Swedish forests. Strojan (1978a) observed progressively less weight loss from litter bags with increasing proximity to a zinc smelter and a significantly large ground accumulation of litter at 1 km from the smelter where levels of Zn, Pb and Cd were 26 000 μg g^{-1}, 2300 μg g^{-1}, and 900 μg g^{-1}, respectively. For metal-tolerant vegetation, growing on metal mine wastes containing high concentrations of lead ($\sim$ 14,000 μg g^{-1}) and zinc (400 μg g^{-1}), the rate of decomposition was retarded (Williams et al., 1977). Even in less contaminated urban areas, black oak litter having levels of Pb from 33 to 305 μg g^{-1} and Zn from 146 to 382 μg g^{-1}, decomposed at a slower rate than that on rural sites where metal concentrations were at least 75 per cent lower (Inman and Parker, 1978). Dixon et al. (1978) predicted that a 50 per cent increase in litter mass would occur in a simulated system with gross metal inputs, in less than six years. Jackson and Watson (1977) demonstrated an inverse distance–litter accumulation relationship near a lead smelter, as shown in Figure 6.7. Using regression analysis, Coughtrey et al. (1979) demonstrated that litter accumulation was highly dependent on cadmium levels and was associated with a particular size range of material (0.5–2 mm), suggesting that the effects occur during the later stages of breakdown. Reduced

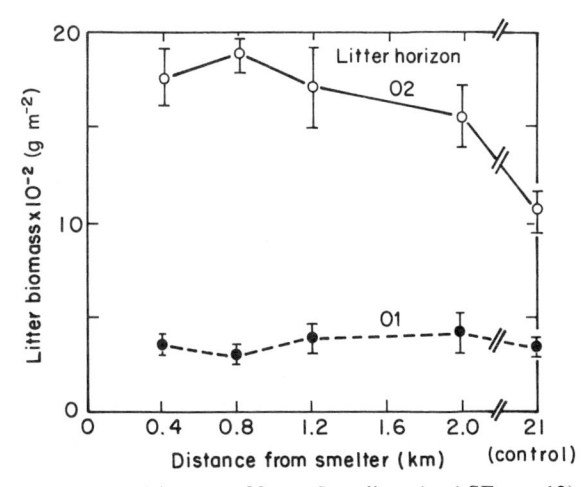

Figure 6.7 Biomass of forest floor litter ($\pm$ 1 SE, $n = 12$) as a function of distance from a lead smelter. (Reproduced from Jackson and Watson, 1977, *Journal of Environmental Quality*, volume 6, page 335, by permission of the American Society of Agronomy, Crop Science Society of America and the Soil Science Society of America)

leaf decomposition has also been demonstrated for a cadmium-stressed (5–10 μg l^{-1}) artificial stream (Giesy, 1978). The probable consequences of reduced litter decomposition on ecosystem function were summarized by Tyler (1972). Initially, some increase in soil litter occurs, with both the new fall and existing litter incompletely decomposed. This litter accretion can occur for only a limited time as more and more mineral nutrients become bound in forms unavailable to plants. The turnover of phosphorus, particularly, is retarded. Consequently, plant productivity decreases as does further litter fall. Even though litter mass is increased, concentrations of the macronutrients Ca, Mg, K and P are significantly depressed near metal smelters, probably as a result of an increase in acidic leaching accompanied by a decrease in microbial immobilization of these elements (Jackson and Watson, 1977).

As demonstrated for acid rain, excessive metal concentration interrupts the delicate ecosystem nitrogen cycle. Horne and Goldman (1974) observed the suppression of nitrogen fixation by blue-green algae with the addition of low levels of copper. They estimated that if ambient copper levels increased by 10 μg l^{-1} this could reduce the lake nitrogen budget by 40–50 per cent. Cadmium at 1.1 mg kg^{-1} has been shown to inhibit nitrogen fixation in Douglas fir needle litter (Lighthart, 1980). With increasing Ni levels, soil nitrification reactions were inhibited to a greater extent than were nitrogen and carbon mineralization (Giashuddin and Cornfield, 1978). Among the other heavy metals which are effective inhibitors are Hg, Cd and Cr (Liang and Tabatabai, 1978). Although zinc is not as potent an inhibitor as other heavy metals, $100\,\mu g\,g^{-1}$ Zn significantly reduced nitrification in amended soils (Wilson, 1977). The ammonification rate in some vineyard soils was retarded by the accumulated levels of copper-base fungicides (Baroux, 1972). Harrison *et al.* (1977) reported that inhibition of nitrate uptake, synthesis of nitrate reductase and loss of accumulated ammonia, were some of the acute effects of copper on phytoplankton. Copper, at a concentration of 10 to $25\,\mu g\,l^{-1}$, also inhibited uptake of silicic acid, essential to diatom growth (Goering *et al.*, 1977). In a series of microcosm studies, Heath (1979) demonstrated that cadmium stress adversely affected nutrient assimilation by microbes. Uptake of nitrogen and phosphorus occurred more slowly at $1\,mg\,l^{-1}$ Cd than at natural levels; these elements were barely assimilated during the course of the experiment at $10\,mg\,l^{-1}$ Cd. Phosphorus assimilation was the more severely affected process.

A summary of the effects of metal loading on decomposition processes includes numerous indications of disruption. Soil leaching is increased, and the decomposition rate decreased, contributing to an accumulation of nutrient-depleted litter.

6.2.3 Pesticides and Oils

Organic pesticides and oils cause a more specialized set of environmental disturbances than those discussed for either SO_2 or metals. The influence of these pollutants on decomposition processes and nutrient conservation has been less thoroughly investigated and, therefore, is more poorly defined.

Pesticides are used to restrict the numbers of certain undesirable species; however, because of their general lack of target specificity they can also reduce the populations of essential decomposer species, thus disrupting the balance within the decomposer community. Oil spills, essentially an aquatic problem, lead to severe impacts on nearshore marine ecosystems. The organic nature of these pollutants results in their providing additional organic energy to the affected system. This energy source can supply positive feedback to tolerant species, whereas sensitive and sometimes ecologically critical groups, such as those necessary for complete mineralization of normal litter, suffer extinction or reduction in numbers. Since there have been few studies on the influence of these pollutants on litter decomposition, the adverse effects of pesticides and oil spills are best surveyed through their direct impact on microbial and macroinvertebrate decomposers.

A comprehensive review of the impact of pesticides on soil fauna was presented by Edwards and Thompson (1973). These authors indicated that pesticides often affect nontarget species, many of which are saprophagous invertebrates (e.g. earthworms, Enchytraeid worms, Collembola, some mites and Diptera larvae) essential to the breakdown of leaf litter. Insecticides and nematicides are obviously toxic to many exposed soil and aquatic invertebrates.

Barrett (1968), examining the effects of the pesticide Sevin on a grassland ecosystem, reported that surface-dwelling invertebrates were reduced by 95 per cent in treated areas (2.5 kg ha^{-1} Sevin) and populations remained lower than control levels for more than 7 weeks. After 3 weeks, a significant reduction in litter decomposition rate was measured, presumably resulting from the earlier reduction in soil microarthropods and/or microbial decomposers. Brown (1978) reported that several carbamate pesticides (such as Sevin) were toxic to Collembola and mites at application levels of 1.4 to 11 kg ha^{-1}, commonly used for gypsy moth control. Most pesticides were also highly toxic to the freshwater macroinvertebrate decomposer community. Sanders and Cope (1968) reported that of thirty-nine pesticides tested all but four had a 96-hour LC50 for stonefly nymphs, of less than 50 μg l^{-1}, with the majority of LC50s less than 10 μg l^{-1}. The spraying of organochlorines, organophosphates and carbamates for the control of forest pests (e.g. gypsy moths) has been found to have a severe impact on stream macroinvertebrates (reviewed by Brown, 1978).

There has been less research on the effects of oil on marine macro- and microinvertebrate decomposers. The activities of benthic deposit-feeders have been shown to be important in accelerating the degradation of sediment-bound oil (Gordon et al., 1978). However Lindén et al. (1979) found that detrital feeding amphipods were reduced in oil spill areas, possibly through emigration as well as mortality. After ten months no significant recovery of amphipods was noted. As most other members of the soft-bottom community were also adversely impacted, it is difficult to imagine the litter-processing role of organisms like amphipods being successfully filled until the effects of the spill had subsided. Hampson and Moul (1978) reported that a fuel oil spill caused mortality of marine polychaetes, amphipods, decapods and isopods, and a significant reduction in faunal populations in marsh sediments was still evident three years later.

Although there is little information available, it has been noted that the effects of oils and pesticides on the microbial community deviate somewhat from those discussed for metals or acid rain. Pesticides and oils are toxic to certain microbial species. However, the most striking feature of microbial community response is the rapid population growth of those species directly involved in mineralization of the pollutant itself. Microbial responses to an oil spill are of two types: an inhibition of some naturally occuring groups of bacteria that may be critical in maintaining ecosystem balance and/or the development of a petroleum-degrading population that contributes to the removal of the oil (FAO, 1977). Growth-limiting effects of fuel oil No. 2 and Louisiana crude have been noted for ecologically important proteinolytic, cellulolytic, lypolytic and chitinolytic populations (Walker et al., 1975). This study could have far-reaching implications relating to the completeness of litter decomposition which might be achieved under oil stress. Colwell et al. (1978) observed proportional decreases in starch hydrolysers and chitin digesters with parallel increases in petroleum

Table 6.2 Effect of *Metula* oil on representation of physiological groups of microorganisms comprising the heterotrophic populations of beach sand, given as per cent of the total viable count (*TVC*). Sd = standard deviation. (From Colwell *et al.*, 1978. Reproduced by permission of the *Journal of the Fisheries Research Board of Canada*.)

Physiological group	Clean sand			Oil-impacted sand		
	Mean % of TVC	No. of samples examined	sd	Mean % of TVC	No. of samples examined	sd
Starch	17.1	9	8.7	3.7	13	6.2
Chitin	6.6	19	8.1	1.8	15	1.6
Petroleum	0.018	17	0.043	0.050	19	0.145

degraders (Table 6.2). The altered balance was still evident 2 years after the *Metula* spill. Hodson *et al.* (1977) noted a reduction in microbial activity, as defined by glucose uptake, due to several crude and refined oils, an effect which could be interpreted as microbial switching from a natural substrate to oil.

Pesticide application appears to cause similar changes in soil microbial communities. Although reviews by Pfister (1972) and Tu and Miles (1976) indicated that halogenated pesticides and organophosphorous and carbamate insecticides did not significantly inhibit most microbial populations at normal application rates, certain pesticide compounds were selective for important decomposer groups such as fungi. Brown (1978) cited studies indicating that fungicides severly reduced fungi and nematodes, but often stimulated bacterial growth through elimination of competition and/or the utilization of dead fungi or the fungicide as an additional energy source. Mirex, at 100 μg g^{-1} of soil, has also been shown to stimulate bacterial activity (Lue and de la Cruz, 1978). Fumigation of Douglas fir needle litter with a mixture of 98 per cent methyl bromide and 2 per cent chloropicrin caused increased microbial respiration (after a lag of nine days) which coincided with an increase in cellulase, xylanase, mannase and amylase activities (Spalding, 1979). Exposure of forest microcosm soils to 0.73 and 7.42 mg cm^{-2} hexachlorobenzene (HCB), on the other hand, caused increased Ca loss and decreased CO_2 efflux (Ausmus *et al.*, 1979). Initial, rapid microbial decomposition of organic pollutants has been shown to deplete essential elements such as oxygen, nitrogen and phosphorus (Aminot, 1981), restricting subsequent mineralization rates due to nutrient limitations. The rates of mineralization of certain natural substrates (cellulose, chitin, starch,) may also suffer due to changes in the numbers and types of specialized decomposers, with pollutant degraders becoming dominant.

Overall, the rates of decomposition of soil litter in pesticide-polluted systems are generally lowered (Barrett, 1968; Perfect, 1979), the reduction of invertebrate activity being the most significant factor. The impact of oil on natural litter degradation remains undemonstrated in microcosm or ecosystem studies; a reduction in decomposition can only be inferred from the indirect evidence as presented above.

There is conflicting evidence also with regard to pesticide interference with nitrogen-cycling processes. A review of pesticide studies revealed that nitrogen fixation, ammonification and nitrification were mildly affected by pesticide residue concentrations, but that the severity of the impact was pesticide, soil-type and process dependent (Tu and Miles, 1976). Of these processes, nitrification appeared to be the most sensitive to initial exposure. Winteringham (1977) indicated that nitrification showed more sensitivity than ammonification to organochlorines such as DDT. Reduction in root nodules has been demonstrated for DDT at 100 kg ha^{-1}. Herbicides such as PCP and DNOC decrease introgen fixation by *Azotobacter*, and phenoxy herbicides can suppress legume nodulation (Brown, 1978). Pentachlorophenate, at concentrations of 100 μg g^{-1} and above, strongly inhibited N-fixation in soils (Tam and Trevors, 1981). In addition, PCP was found to decrease ammonification, while most other herbicides increased the formation of NH_4-N.

The importance of such disruptions to the availability of nitrogen to primary producers has not been clearly demonstrated. It is suggested that inhibition of nitrogen fixation could be particularly critical in nutrient-deficient natural ecosystems. Shindler *et al.* (1975) hypothesized that *Azotobacter* species were stimulated to fix atmospheric nitrogen in oil-polluted freshwater ponds, thereby contributing to the heightened N/P ratio observed, in comparison to unpolluted controls.

There is no evidence that oils or pesticides promote direct leaching of nutrients from soils or sediments as previously noted for metals and acid rain. However, in cases of severe exposure, nutrient export and increased erosion have occurred. Likens and Bormann (1975) reported that herbicides reinforced the already established trend of nitrate loss from felled vegetation. Pesticides may also accelerate nitrogen loss in agricultural systems (Winteringham, 1980). In the absence of active plant uptake, the very mobile nitrate is flushed from the ecosystem. Woodwell (1970) theorized that herbicides cause 'leaks' in the nutrient conservation system, presumably through elimination of biomass storage capacity. Perfect (1979) and Perfect *et al.* (1979) suggested that there was a link between DDT interference with nutrient cycling and subsequent decreased crop yield in agro-ecosystems, although they were unable to establish the exact nature of this disruption. Hampson and Moul (1978) showed that erosion of marsh soils was increased by the long-term damaging effects of fuel oil on the marsh vegetation. After 3 years they found that erosion rates were 24 times greater than those measured at an undisturbed control site. The increase correlated well with the degeneration of the intertidal marsh root system which normally acts as a sediment-binding mechanism. Zeiman (1975) concluded that the denuding effect of oils on sea-grass systems led to increased erosion and that oil,

Table 6.3 Summary of pollutant effects on decomposition processes and element conservation in ecosystems

Pollutant	Decomposer organisms — Invertebrates	Decomposer organisms — Microbes	Litter accumulation	Litter decomposition rate	Leaching	Nutrient export	Nitrogen cycling N_2-fixation	NO_2, NO_3
Acid precipitation	Decreased	Decreased; favoured fungi	(+)	Decreased	Increased soil and foliar loss	Accelerated net loss	Decreased	Decreased
Heavy metals	Decreased	Decreased; may favour fungi	(+)	Decreased	Increased soil loss	Accelerated net loss	Decreased	Decreased
Pesticides	Decreased	Species composition changes favouring pollutant-mineralizing species at the expense of ecologically important groups; fungicides are particularly harmful to fungi	(+)	Decreased	Effects unknown; herbicides can stimulate vegetative release of dissolved organics; HCB induced Ca leaching	Accelerated net loss particularly with herbicides	No effect?	Decreased?

FUNCTIONAL CHANGES IN THE ECOSYSTEM

| Oil | Decreased | Species composition changes favouring oil mineralizers; possible decrease in the proportion of chitin and starch degraders | (?) | Effects unknown; rates assumed to be depressed; decomposition may be at some point nutrient limited | Effects unknown | Erosion has been demonstrated for Marsh Grass systems | Possibly stimulated in some cases | Effects unknown |

coating the sediment, inhibited rapid recolonization, thereby accelerating the erosion process.

Although there are some obvious differences in the effects of each pollutant discussed, on decomposition subsystems, a summary of their effects on component organisms and selected decompostion processes shows that there is some similarity in ecosystem response (Table 6.3). The rate of invertebrate comminution of litter is often drastically reduced as is the general rate of litter decomposition. Material export, resulting from leaching of inorganic nutrients or from general soil erosion, is associated with pollution loading at subacute levels and is accelerated with acute long-term or gross exposure. Inhibition of conservative nitrogen cycling has frequently been found to occur, with the nitrification process being particularly sensitive. The similarities among decomposer subsystem responses to pollution strongly suggest the feasibility of using decomposition and element cycling criteria to measure pollution stress at the ecosystem level. This approach should be particularly effective in cases of terrestrial and flowing-water systems, where a large fraction of the organic matter is processed as detrital material.

6.3 PRODUCTIVITY AND RESPIRATION

Solar radiation is the ultimate source of energy for the living system; however, it must be concentrated and stored by the photosynthetic process. This process is the critical link in the energy base, fundamental to both autotrophic maintenance and heterotrophic development. Inhibition of photosynthesis due to pollution stops ecosystem energy movement at this most critical initial conversion. The impact of a toxic pollutant on photosynthesis may be effected through death of the plant or necrosis of photosynthetic tissue, through direct inhibition of photosynthetic reactions, or through indirect rate limitation brought about by a pollutant's effect on the availability of essential elements for plant growth.

The primary result of partial inhibition of photosynthesis is a progressive reduction in the size of the available energy pool, and an increase in the turnover time in which the pool is replenished. This decrease in gross primary production, even if accompanied by an equivalent reduction in autotrophic respiration, will still significantly diminish net primary production. This would mean that less energy would be available to heterotrophs. Since pollutant stress often, in fact, increases total ecosystem respiration while decreasing productivity, the available energy pool could be even further reduced. A diminished net energy reservoir is often accompanied by a reduction in species variety. A simplified plant community can mean reduced availability of preferred species and changes in the digestibility and nutrition of available food. The availability of essential 'quality' energy to heterotrophs then, is subject to the additional constraint of changed nutritional value.

Another aspect of primary productivity is the relationship of the pollutant-generated perturbation to the seasonality of ecosystem growth. The terrestrial plant community is most sensitive to stress during certain vegetative stages (germination and bud formation). Exposure to toxic substances during these critical phases enhances the potential for plant damage. Other pollutant-sensitive factors governing plant growth such as nutrient availability and grazing pressure are often correlated with seasonality, as well. Hunding and Lange (1978) stated that the developmental stage and physiological states of individual species must be considered in the study of ecotoxicology of aquatic plant communities. The vulnerability of terrestrial productivity to pollution stress is also influenced by the stage of successional development of the ecosystem. Vitousek and Reiners (1975) argued that in the course of succession, element outputs are initially high, that they later drop to a minimum due to accumulation in biomass and detritus when net ecosystem production is highest and that, eventually, outputs rise to balance inputs. Therefore, late successional plants are less well adapted to conditions of nutrient flux which often accompany pollutant disturbances than are species from the early colonizing community.

Smith (1974) divided the impact of pollutants on plants into three major classes based on the severity of response. A class I response represents low pollutant loading with the plant community acting as a sink for the contaminants. This type of response is subtle and can only be estimated through chemical extraction techniques; however, it often provides the initial evidence of pollutant accumulation and movement from the abiotic to biotic components of the ecosystem. Class II responses are found with intermediate pollutant loading. At this stage, plants show reduced growth, reduced reproduction and increased morbidity of individual species, leading to reduced productivity and biomass, altered species composition and increased vulnerability at the ecosystem level. This set of reactions is based primarily upon extrapolation of controlled laboratory and small field experiments with limited microcosm and ecosystem data. Class III responses represent acute damage to the plant community; resultant simplifications in the biotic structure of the ecosystem infer reduced stability. Extensive acute damage is readily observable in the field.

The development of an understanding of ecosystems functioning at the class II response level is of particular interest to ecotoxicologists since the reduced 'health' of the system can often be reversed if correct diagnosis leads to decreased contamination.

6.3.1 Acid Rain

There are a variety of experimental studies demonstrating that air pollutants such as acid precipitation reduce photosynthetic rate, plant growth and reproductive success, and consequently reduced plant yield. Bennett and Hill (1974) demonstrated that chronic exposure of plants to a number of phytotoxic air pollutants (HF, Cl_2, O_3, SO_2, NO_2, NO) can cause a reversible suppression of photosynthesis. Of the phytotoxic air pollutants tested, combinations of SO_2-NO_2, major constituents of acid rain and ozone, are the most likely to occur in the ambient atmosphere in sufficiently high concentrations to depress this process. Bennett and Hill concluded that although plants generally recover from recurring episodes of short-term pollutant exposure, when sufficient recovery time is allowed, it is probable that prolonged exposure chronically depresses photosynthesis, leading to visible symptoms of injury and reduced growth.

Much of what we know of the effects of SO_2 on primary producers has come from lichen investigations. Although the aim of much of this research has been to discover why this unique group is sensitive to SO_2, it has produced some general understanding of the mechanisms of SO_2 damage to plants and of their ability to recover under various stress conditions. With short-term exposure, lichens are no more sensitive to SO_2 than are higher plants; however, they have the ability to efficiently absorb the pollutant, accounting for their disappearance from contaminated areas (Nash, 1973). Lichens are more sensitive to sulphur in simulated acid rain than in its dry forms (Türk et al., 1974). There is a very rapid (within 15 minutes) decrease in photosynthetic ^{14}C fixation with exposure to aqueous sulphur dioxide, and repeat exposures reduce the plants' ability to recover a normal photosynthetic rate (Puckett et al., 1974). After a slight increase in photosynthesis at low SO_2 concentrations, the reduction in photosynthetic ^{14}C fixation becomes pronounced with lower pH and increased sulphur dioxide concentrations, as shown in Figure 6.8 (Nieboer et al., 1976). Tomassini et al. (1977) demonstrated that the per cent total ^{14}C fixation decreased expontially with increasing concentrations of SO_2 at gross exposure levels with short exposure time. The fixation rate–exposure time relationship became linear with low level exposure for 6 to 12 hours. Through extrapolation, they estimated the injury threshold concentration of SO_2 as acid precipitation for the lichen *Cladonia rangiferina* to be $10 \mu g l^{-1}$ for six months.

Researchers have found that limited exposure to SO_2 not only reduces gross carbon assimilation but also influences metabolic expenditures. Respiration rates were depressed during incubation in sulphur dioxide solutions (Baddeley et al., 1973; Tomassini et al., 1977). Differences in lichen respiratory response which are not specifically related to SO_2 concentration may be due to differences in the sensitivity of the fungal partner; however, the algal symbiont has been shown to be the more sensitive to SO_2. In fact, Hällgren and Huss (1975) showed that the nitrogen-fixing capability of lichens and blue-green algae was more sensitive than their photosynthetic capacity to SO_2 exposure. Thus, SO_2 pollution has very complex metabolic effects on the exposed lichen, effects which either directly or indirectly reduce its productive capacity.

The foliar efflux of potassium ions and the bleaching of pigment have been found to result from extended SO_2 exposure, and these injurious effects culminate in permanent reduction of net photosynthesis and, presumably, permanent impairment of the lichen (Nieboer et al. 1976). Excessive foliar

FUNCTIONAL CHANGES IN THE ECOSYSTEM 133

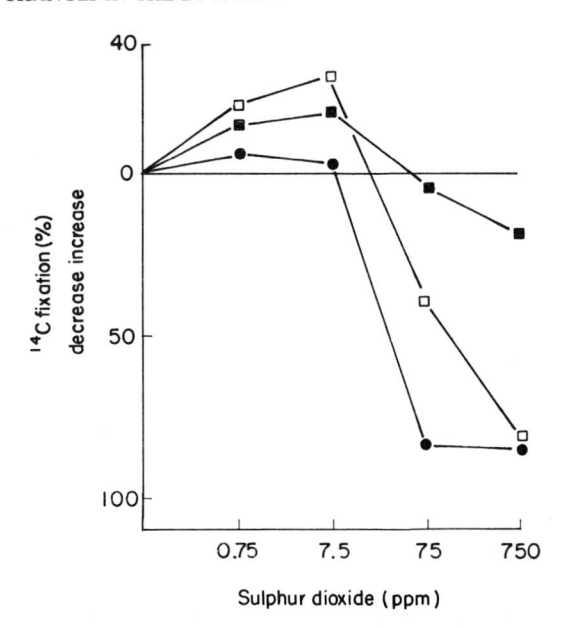

[Figure: graph of ^{14}C fixation (%) decrease/increase vs Sulphur dioxide (ppm), x-axis values 0.75, 7.5, 75, 750]

Figure 6.8 The percentage reduction in net ^{14}C fixation by *Stereocaulon paschale*, incubated for 6 h in solutions of sulphur dioxide buffered at various pH values, as compared with control samples without sulphur dioxide. Each point represents the mean of six replicates. ■ pH 6.6; □ pH 4.4; ● pH 3.2. (Reproduced with permission from Nieboer *et al.*, 1976. In Mansfield, *Effect of Air Pollutants on Plants*, Cambridge University Press)

leaching represents severe disruption of the plant community and infers photosynthetic impairment on a broad scale.

Feder (1973) noted that the line of reasoning developed from SO_2-lichen studies may be quite applicable to much of the terrestrial plant community. He indicated that the hidden injuries to plants from chronic exposure are reflected in reduced photosynthesis, decreased growth, and fast aging of foliage, which cumulatively result in reduced system viability and yield. Increased transpiration and dark respiration resulting from exposure to elevated SO_2 levels of > 3.5 mg m^{-3}, for the initial few days, constitute heightened metabolic expenditures (Ziegler, 1975) and consequently reduce productivity. Recent work by Constantinidou and Kozlowski (1979a, b) revealed additional indirect evidence of SO_2-induced loss in primary productivity as measured by the following parameters: (1) diminished leaf expansion, (2) fewer emerging leaves, (3) a decrease in total non-structural proteins and carbohydrates in new leaves, (4) reduction in root weight. Marshall and Furnier (1981) reported that root growth was inhibited more than shoot growth in tree seedlings, and the reduction in growth was negatively correlated with the level of SO_2 exposure.

Although air pollutants such as SO_2 can directly inhibit photosynthesis, the reduction of vegetative growth may occur primarily through stimulation of essential element leaching as suggested by Tyler (1972). Reduced growth might not be immediately obvious and there could in fact be a substantial delay, encompassing a period of nutrient depletion in the soil through leaching and element immobilization as a result of reduced decomposition, plus draw-down time for the tree's own substantial storage of materials in its structural biomass.

An additional consideration is the adverse effect of SO_2 on plant reproduction with respect to long-term yield. In spruce species, cone yield was shown to be smaller (Pelz, 1963) and in pine species, cone and seed weight were decreased, and seed viability was diminished, as a result of exposure to $0.06\,mg\ SO_4^{2-}\ cm^{-2}\,month^{-1}$, near a coal-burning power generating plant (Houston and Dochinger, 1977). Pollen grains were smaller, with a higher proportion of underdevelopment (Shkarlet, 1972), and a decreased ability to develop tubes (Varshney and Varshney, 1981). Small amounts of simulated acid rain applied to soybeans decreased the number of pods per plant (Evans *et al.*, 1980) and significantly reduced seed yield (Irving and Miller, 1980). Possible harmful influences of acid precipitation on germination and seedling establishment in mineral soils have been pointed out as well (Teigen, 1975; Abrahamsen *et al.*, 1977). About 80 per cent of Norway spruce and Scots pine seeds did not develop normal seedlings at pH 3.8 (Teigen, 1975).

Although there is considerable evidence of SO_2-induced reductions in the short-term growth rates of plants potentially affecting agricultural yields (e.g. Reich *et al.*, 1982), long-term reductions have not been as clearly demonstrated. However, in a forested area surrounding a 'sour gas' processing plant, sulphur emissions reduced the woody production of hybrid lodgepole × jack pines (Figure 6.9). The maximum depression of basal area incrementation occurred at sites approximately 1 and 3 km from the gas plant and decreased to minimal or no effect at a distance of 9.6 km. A difference of approximately 2 per cent in basal area growth was evident between trees at 1 and 9.6 km over a 14 year period following the start-up of the gas plant (Legge, 1980). Results of field experiments using artificial acid rain as the treatment on Norway spruce and Scots pine showed that growth effects were most obvious at pH 2.0 and 2.5 (Tveite, 1980; Tveite and Abrahamsen, 1980). Growth in height of Scots pines showed an increase with acid rain treatment during a 4-year period. Tveite and Abrahamsen (1980) suggested that the stimulated growth was probably the result of increased nitrogen or sulphur availability, or both. On the other hand, in the Norway spruce stand, height and girth growth were reduced by the low pH treatments (Tveite, 1980). A regional tree-ring study in Norway also yielded inconsistent growth results associated with acid precipitation (Abrahamsen *et al.*, 1976). The conference summary from the recent SNSF international meeting on the

FUNCTIONAL CHANGES IN THE ECOSYSTEM 135

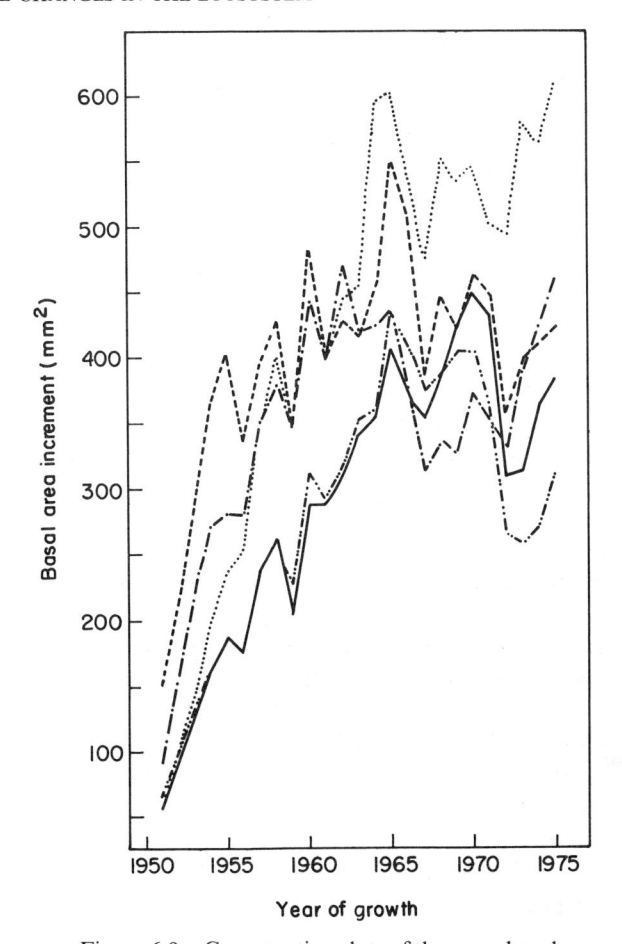

Figure 6.9 Comparative plots of the mean basal area increments from lodgepole x jack pine trees at each of five ecologically analogous sampling sites approximately 1.2 (———), 2.8 (———..———), 6.0 (———.———), 7.6 (----) and 9.6 (.....) km, respectively, from a 'sour gas' processing plant source of sulphur gas emissions. (Modified from Legge, 1980. Reproduced by permission of USDA Forest Service)

ecological impacts of acid precipitation stated that it was not yet possible to draw a definitive conclusion on the effects of acidification on forest growth, although induced deficiencies of K, Ca and Mg may have long-term effects on productivity (Last et al., 1980).

Some of the symptoms of reduced productivity have been noted for aquatic systems although the primary producer response is less clear. Production at the macro- and microconsumer levels was reduced by experimental acidification of a stream ecosystem (Hall and Likens, 1980). Acidified lakes are reported to have fewer species of phytoplankton (Almer et al., 1974; Dillon et al., 1979; Yan, 1979), and lower primary production has been attributed to reduced nutrient supply (Grahn et al., 1974). However, in some acidified lakes and streams the growth of mosses and filamentous algae has been greatly increased, with a corresponding reduction in macrophyte density (Leivestad et al., 1976; Hendrey et al., 1976; Grahn, 1977). In a whole-lake study, Müller (1980) found that artificial acidification caused an increase in periphyton growth in the littoral zone; however, there was no corresponding increase in productivity. It has been suggested that as lakes are acidified they become more oligotrophic (Grahn and Hultberg, 1974; Grahn et al., 1974). The process was termed self-accelerating oligotrophication since acid-induced decreases in productivity could cause starvation or reduced 'fitness' in consumers, leading to further disruptions of the food web and reduced biomass. Schindler (1980) and Schindler et al. (1980a) conducted experimental acidification of a whole lake to test the oligotrophication hypothesis. They reported that phytoplankton biomass and chlorophyll a increased as the lake became more acid. Phytoplankton production per unit area also increased and there was no evidence of declining production in the food chain. However, the pH in the experimental lake was only lowered to 5.6, above the value of 5 at which chronic effects were observed in Swedish lakes (Wright and Henriksen, 1978). Schindler (1980) also questioned whether primary productivity increases would be a transitory phase in the acidification process. An NRCC report (1981a) concluded from the data available that productivity did not appear to be significantly reduced despite the very low concentrations of dissolved inorganic carbon characteristic of acidic lakes.

6.3.2 Heavy Metals

Contrary to the reports of direct SO_2 effects on photosynthesis in terrestrial plants, there is little information indicating that metals directly affect the photosynthetic process (Foy et al., 1978). However, the end result of medium to high levels of heavy metal pollution appears to be reduced primary production. Cupric ions have been shown to inhibit photosynthetic electron transport in isolated chloroplasts, but this effect has not been demonstrated in whole-plant studies (Cedeno-Maldonado et al., 1972). Reductions in ^{14}C fixation can be induced by a number of heavy metals (Ni, Cu, Zn, Pb) in both algal cultures and natural phytoplankton populations (Erickson, 1972; Bartlett et al., 1974; Steemann Nielsen and Bruun Laursen, 1976). It has been suggested that lead may inhibit plant growth near major highways (Coello et al., 1974). It is possible that where heavy metals have reduced terrestrial plant productivity, the effects may have been indirect and may have been manifested through reduction of photosynthetic surfaces, inhibition of metabolic enzyme systems or interference with nutrient uptake. Furthermore, damage to a forest ecosystem, due to other stresses such as fire, is aggravated by high metal levels. Toxic metals apparently inhibit seed production, preventing vegetative reestablishment.

The most general symptoms of excessive metal levels are stunting and chlorosis. The latter represents tissue injury and may be due to an iron deficiency caused by competitive interaction with multivalent heavy metal ions. Chlorosis from excess Zn, Cu, Ni and Cd appears to be due to direct or indirect interaction with foliar Fe. This symptom is aggravated by low pH and iron-depleted soils (Foy et al., 1978). White et al. (1974) observed that excessive zinc may also increase manganese transport levels in plant tops, interfering with chlorophyll synthesis. Inhibition of enzyme activity has long been cited as a major biochemical mechanism associated with metal toxity. Competitive inhibition of metalloenzymes may be of particular importance. Increased metabolic costs and indirect reduction of photosynthetic capability would eventually lead to reductions in whole-plant growth. Stunting can result from the interaction of several factors including specific toxicity, antagonism with nutrients and inhibition of root penetration and growth.

Toxicity is first evident in root tips, followed by subsequent inhibition of lateral root development, resulting in uptake of phosphorous, potassium and iron at rates insufficient for normal growth (Barber, 1974). Whitby and Hutchinson (1974) demonstrated a relative reduction of root elongation by tomato plants in solutions of nickel, copper, cobalt and aluminium, between $2-10\,mg\,l^{-1}$. They noted this as significant because water extracts often yield levels in excess of $10\,mg\,l^{-1}$ in soils from contaminated areas. Plants grown in soil-water extracts taken from sites closer than 3.8 km to a smelter (Sudbury) showed almost total inhibition of radicle elongation. At distances of 3.8 to 10.4 km, root growth, when compared with that of controls, was found to be related to proximity to the smelter and corresponding concentrations of Ni, Cu, Al and Zn. Observers noted a depletion in fine roots in litter polluted by lead smelter emissions ($30\,000-88\,000\,\mu g\,g^{-1}$ Pb), and found extremely high concentrations of heavy metals in root tissues (Jackson and Watson, 1977). Seedlings in zinc-contaminated soils, collected near a smelter operation, produced significantly less root and shoot growth than did controls: Figure 6.10 (Jordan, 1975). Presumably, inhibition of decomposition and elevation of soil nutrient leaching by metals would aggravate any difficulty in the uptake of essential elements by unhealthy roots. Jackson and Watson (1977) found depressed nutrient concentrations in leaf tissue sampled from plants near a lead smelter. It might require several years of inadequate nutrient uptake before such a deficiency becomes noticeable in plant leaves (O'Neill et al., 1977; Jackson et al., 1978). Reduced nutrient levels in leaves should be considered as indirect evidence of a corresponding reduction in structural growth.

Tyler (1972) suggested that metal-retarded mineralization of litter, with subsequent reductions in the availability of nutrients, contributes to reduced

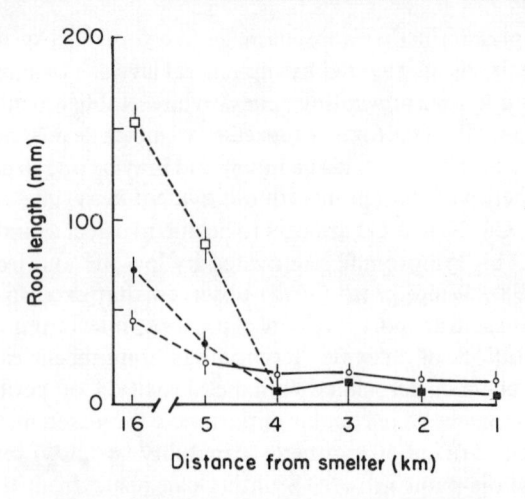

Figure 6.10 Root length of *Quercus rubra* seedlings germinated on soils collected various distances from a metal smelter. Means ± SE are plotted. Open circles indicate 42 days, closed circles 60 days, and squares 69 days growth. Means differing significantly at the $P \leq 0.05$ level are connected by dashed lines. (From Jordan, 1975. Copyright © 1975 Ecological Society of America)

ecosystem productivity. This series of events may act as a self-stimulating cycle. If pollutant loading were continued, the cycle could only be reversed through the genetic adjustment of organisms at all important links in the decomposition chain. Bradshaw et al. (1978) indicated that restoration of vegetation on metal mining waste required application of organic matter, to reduce direct metal toxicity through adsorption and chelation, and fertilizers containing N, Ca, P, K and Mg, to overcome very depressed nutrient levels in the soils.

Although cases of excessive metal concentrations in soils or irrigation water reducing crop productivity have been documented (Huisingh, 1974; Bingham et al., 1980), there are few examples of long-term reductions in productivity in nonagricultural ecosystems. Rather, it is thought that for forested systems, a gradual reduction in plant growth does occur but is not easily quantifiable (Hughes et al., 1980b). Hutchinson and Whitby (1974, 1977) reported a considerable reduction in forest productivity due to a combination of SO_2 acidity and metals.

Reduction of net primary productivity and phytoplankton biomass due to metal contamination has been observed more frequently in aquatic systems, where the producers' small size and rapid turnover rate elicit a more rapid response to perturbation and facilitate productivity measurements. The microcosm studies by Gächter and Máres (1979) demonstrated that an initial combined loading of heavy metals ($\leq 1\,\mu g l^{-1}$ Hg, $5-7\,\mu g l^{-1}$ Cu, $2-3\,\mu g l^{-1}$ Cd, $98-196\,\mu g l^{-1}$ Zn and $4-8\,\mu g l^{-1}$ Pb) lowered phytoplankton biomass as well as species number and photosynthetic activity. Comparable results have been reported for single additions of 10 to 50 $\mu g l^{-1}$ copper (Thomas et al., 1977; Harrison et al., 1977). Levels of copper as low as 5 to 10 $\mu g l^{-1}$ affect algal growth, indirectly, through inhibition of nitrogen fixation (Horne and Goldman, 1974), and at 25 $\mu g l^{-1}$, through interference with silicic acid uptake by diatoms (Goering et al., 1977). Mercury, at levels greater than 30 $\mu g l^{-1}$, lowers the rate of algal cell division and induces cellular losses of nitrogen and lipids (Sick and Windom, 1975). Growth of *Selanastrus capricornutum* was inhibited in assays with contaminated Coeur d'Alene River water containing Zn levels greater than 0.5 $mg l^{-1}$ (Bartelett et al., 1974). In a further study of heavy-metal effects, Wissmar (1972) reported that in the highly polluted Coeur d'Alene Lake, the combined effects of Zn, Cu and Cd reduced carbon fixation by lake phytoplankton.

Aquatic microcosm studies have shown that toxic metal levels affect the ratio of gross primary production (GPP or often termed simply P) to total ecosystem respiration (R). Odum (1956) proposed the use of the P/R ratio to classify systems as autotrophic ($P/R > 1$) or heterotropic ($P/R < 1$), and suggested that mature aquatic ecosystems tend to approach a P/R ratio of 1. This approximate value has been measured in a whole-lake ecosystem (Jordan and Likens, 1975). With the addition of 11.5 $mg l^{-1}$ arsenic to microcosms, P/R dropped to nearly zero and did not recover for 2 weeks (Giddings and Eddlemon, 1978). An input of 10 $mg l^{-1}$ cadmium (Heath, 1979) or 0.5 $mg l^{-1}$ copper (Whitworth and Lane, 1969) significantly depressed P/R in similar microcosm studies. A ratio of less than 1 would indicate a negative energy balance which could be detrimental if allochthonous sources were unavailable or the system was unable to recover rapidly. However, Fisher and Likens (1973) found $P/R < 1$ for streams and argued that most mature aquatic ecosystems would be heterotrophic and open to allochthonous inputs. The applicability of the P/R ratio in evaluating stress would be doubtful for ecosystems that were already primarily heterotrophic ($P/R < 1$).

Evidence of loss of net productivity at intermediate levels of metal stress is substantially lacking for forested ecosystems although a good case for its existence has been argued. Metal-induced depression of productivity most certainly occurs and may persist in polluted aquatic systems; therefore, its use as a dynamic measure of ecosystem response to metal stress should be encouraged.

6.3.3 Pesticides and Oils

Generalizations concerning the response of the primary production processes to either pesticides or oil pollution are nearly impossible. Productivity is either stimulated or depressed depending on the type of pesticide or oil, as well as its concentration, and a number of other factors.

Herbicides in general are quite disruptive to production processes and cause widespread mortality. The product 2,4-D applied to Colorado rangeland reduced forbes on a fairly permanent basis. Use of 2,4-D and 2,4,5-T in Vietnam showed that a double spraying of 28 kg ha^{-1} would reduce triple canopy forest to a grass–sedge system quite rapidly (in Brown, 1978).

Pesticides are often applied to increase crop yield through the reduction of pest species. However, continuous application under nonagricultural conditions (without fertilizer addition) has led to an eventual decrease in productivity (Perfect, 1979; Perfect et al., 1979). It has been shown that DDT-treated plots had higher plant yield than untreated plots for the first few years of application, after which the trend reversed, and the decline in the rate of yield as function of time was accelerated in the treated plots. The reduction in yield was associated with DDT's effects on soil biota and nutrient cycling processes.

Organochlorines generally reduce photosynthesis in aquatic ecosystems (Wurster, 1968; Johannes, 1975), although 0.02 mg l^{-1} DDT did not significantly depress P/R in artificial pool studies (Whitworth and Lane, 1969). A stream ecosystem demonstrated the capacity to adjust P/R rapidly in response to 9 mg l^{-1} of the lampricide TMF (Maki and johnson, 1976). Organophosphorous compounds often stimulate primary productivity either through addition of phosphorous or through reducing grazing pressure (Butcher et al., 1975, 1977). Increases in photosynthetic activity were found to correlate with levels of pesticide application, giving additional evidence of the phenomenon of nutrient stimulation.

Very few studies have been made of oil-pollution effects on terrestrial ecosystems. An oil spill in a grass–herb community was found to reduce total plant production by 74 per cent over a 6-month period (Kinako, 1981). The inhibitory effects of oils on the aquatic plant community are most obvious for rooted species. Marsh grasses have been shown to be severely reduced by No. 2 fuel oil, and recovery of the community either through reseeding or rhizome growth was hampered (Hampson and Moul, 1978). Baker (1971) found that the recovery of salt marsh grass was reduced by successive spills. Hanna et al. (1975) showed that freshwater macrophytes were greatly reduced in arctic lakes under oil stress, although the productivity of the lakes was not significantly reduced. This finding infers that perhaps phytoplankton are stimulated by the addition of oil. Algal blooms have been reported after major oil spills (Lindén et al., 1979; Johansson et al., 1980); the investigators hypothesized that reduced grazing pressure was the primary reason for these blooms. Shindler et al. (1975) suggested that nitrogen fixation might be stimulated by oil in freshwater systems, also contributing to increased productivity. Short-term exposure to No. 2 fuel oil at 1.2 to 15.3 mg hydrocarbon l^{-1}, however, depressed photosynthetic activity and phytoplankton biomass in *Vaucheria* (Bott and Rogenmuser, 1978). The general consensus emerging from oil pollution research on pelagic systems is that although natural phytoplankton communities are sometimes adversely affected by oil, the organisms are sufficiently prolific that individual oil spills will have only transitory effects on overall community productivity (FAO, 1977).

The redundancy existing in the functional and regulatory processes of ecosystems, and the large stored biomass inherent in forest systems, tend to mask short-term pollutant effects on the producer communities. Long-term studies of acid precipitation effects on forests and of the influence of metals on productivity adjacent to smelter operations may clarify our understanding of the effects of pollution on long-term energy processing.

6.4 FOOD WEB AND FUNCTIONAL REGULATION

One of the most poorly understood yet important features of ecosystem function is the role that consumer organisms play in regulating energy and nutrient dynamics. Owen and Weigert (1976) suggested that the mutualistic relationship between some plants and their consumers promotes maximum fitness in the plants. The important role of herbivores in recirculating nutrients contributes to an ecosystem's resilience to perturbations (Webster et al., 1975).

The consumer complex acts to control the producer and decomposer components of the system (Golley, 1973). Predator and parasite populations are limited by the reductions in energy as it passes through the food chain and by the availability of essential materials. Thus, there are physical and chemical limits to the length of the food chain, and any further trophic elaboration is biologically regulated. This control can be achieved through a variety of competitive and trophic interactions and through the feedback from specific interdependent relationships.

Golley et al. (1975) grouped the various impacts of regulators into categories relating to : (1) destruction of a component by organisms, (2) the movement of materials, (3) alteration of the environment, (4) interactions with other consumers, particularly predators. Each of these categories involves direct effects on biological components and on the rates of transfer of energy and materials between components. These impacts may have hierarchical and indirect effects on many components within the system at the same time. Also, they can change the abiotic environment.

The role of consumers in functionally regulating ecosystems has been most frequently discussed in terms of energy flow (see Teal, 1957; Fisher and Likens, 1973; Reichle et al., 1973a, b). The impact of consumer regulation on energy movement has been generally assumed to be equivalent to the magnitude of energy flowing through the population. For example, the effect of phytophagous insects on forest production was related to their consumption of foliage (Rafes, 1970; Mattson and Addy, 1975). Cummins (1974) associated energy processing in a heterotrophic stream with the breakdown of leaf litter, thereby accounting in part for the flow of both energy and nutrient materials.

More recently, analysis of consumer regulation of nutrient cycles through

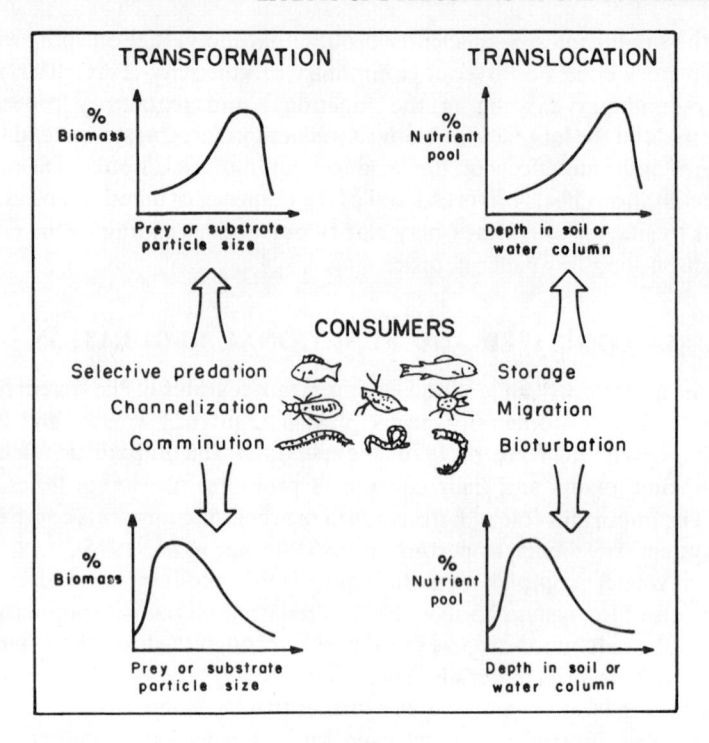

Figure 6.11 Graphic representation of consumer effects on nutrient cycling processes in ecosystems. Translocation or redistribution of nutrients occurs when mobile consumers cross subsystem boundaries. (From Kitchell et al., 1979. Copyright © American Institute of Biological Sciences)

processes not directly reflected in energy flow has been proposed. Kitchell et al. (1979) suggested that regulation at the physico-chemical level occurs through consumer impact on translocation and transformation processes. Consumer effects on the cycling process are summarized in Figure 6.11. Translocation processes would include migration, storage in biomass, bioturbation (mixing of sediments) and movement of litter and faeces within the system. Transformation processes would include selective predation, the comminution or agglutination of litter particles and chemical transformation of substrates. Translocation by mobile consumers redistributes nutrients, retards their loss to soils and sediment and facilitates nutrient availability for primary producers, particularly during critical growth periods, through control of the rate of excretion. Transformation of the size of organic units alters cycling rates through changes in surface–volume relationships. Nutrient pools composed of large biomass units (e.g. tree stumps, leaf litter) may be broken into smaller units, increasing nutrient release. On the other hand, nutrients may be accumulated in consumers of large biomass, resulting in conservation. Although there appears to be a framework for the examination of consumer regulation of both energy and nutrient cycling processes, few details of individual consumer impact have been quantitatively described. Given the complexity of functional regulation, there is little wonder that the effects of pollutants on regulatory interactions have not been properly investigated.

The simplest example of disruption of regulatory processes by pollution is the elimination of grazing species by increased mortality or by stress-induced emmigration. In aquatic systems, pollutant reduction of grazers is frequently followed by phytoplankton increases, often to bloom proportions (previously discussed in section 6.3). The control of primary production by grazers in aquatic systems is also diminished through pollution-induced composition changes in the producer community towards less preferable and less edible species. Large populations of mosses and filamentous algae are sometimes established in acidified lakes and streams, and the existing plant community becomes less diverse, thereby reducing the variety of species available to consumer levels, and hence, reducing the food supply (Hendrey *et al.*, 1976). The long-term input of a variety of pollutants into areas of the New York Bight has altered macrobenthic communities in the Christiaensen Basin (Boesch, 1979). Populations of species which are the predominant food for fish have been reduced. Therefore, although this basin supports a dense and productive benthic community, little of this productivity appears to be transferred up the food web.

Changed predator–prey relations in acidified lakes may be responsible for many of the biological changes reported (Eriksson *et al.*, 1980; Henrikson *et al.*, 1980). In acid lakes where fish populations have disappeared, they have been replaced by invertebrate top predators with different feeding strategies. Invertebrate zooplankton feeders, unlike fish, generally take smaller prey, which results in the proliferation of the larger species. The new grazer community of predominantly large species may, in turn, have an impact on phytoplankton productivity through changes in grazing pressure.

There is some evidence that simplification has potentially greater impact on the upper links in the food chain. Several investigators have suggested that there are significant differences in dietary strategy, in polluted versus unpolluted ecosystems, which reflect either changes in consumer behaviour or prey abundance, or perhaps both (Södergren, 1976; Jeffree and Williams, 1980). Predators may be eliminated for lack of quantity and quality of prey, as well as through direct toxic mortality (Pimentel, 1971). Hurlbert (1971) found that an organophosphorous insecticide had a more severe impact on primarily predaceous insects than on primarily grazing or collecting species. This was also noted by Stickel (1975) for other toxic compounds. Södergren (1976) attributed the decline in a salmon population to reduced prey abundance in metal-and acid-polluted waters.

Consumer–consumer interactions amended by pollutant introduction are exemplified by species population explosions upon elimination of natural predators due to pesticides (reviewed by Pimentel, 1972). In addition, Nounou (1980) showed that oil induced a series of relatively rapid oscillations in some consumer populations through elimination of certain species.

Pollution-induced perturbations have been observed to be passed up the food chain. Saward *et al.* (1975), in a microcosm study, showed that sublethal concentrations of copper (10, 30 and 100 μg l^{-1}) reduced algal productivity, which reduced the growth of grazing snails and, in turn, reduced the available food and the growth of young snail-feeding fish. Certainly, the simplicity of the microcosm community did not allow for prey-switching behaviour to supplement consumer diets, but studies of this type do indicate the functional importance of specialized feeding relationships. Holdgate (1979) discussed the SO_2-induced elimination of lichens with a parallel decline in a lichen-specific feeding insect (Psocidae) as exemplifying this type of trophic repercussion.

As an indirect result of normal consumer activity, toxic materials are accumulated and transported within polluted ecosystems. Most toxic pollutants are accumulated in biota at higher than ambient levels (Hutchinson *et al.*, 1975). There is evidence that some pollutants (organic pesticides and certain metals) are progressively accumulated up through the food chain. Trophic accumulation has been demonstrated for DDT in flesh-eating birds such as ospreys, peregrine falcons, Cooper's hawks, pelicans and a number of others (Ratcliffe, 1973; Snyder *et al.*, 1973; Stickel, 1973). Methylmercury is a pollutant which is bioaccumulated through food (Armstrong and Hamilton, 1973) and has caused human sickness and death; the Minamata incident is the most notorious example, with respect to man, of trophic accumulation and consequent toxicity. (See NRCC, 1979, for a review of mercury bioaccumulation and toxicity.)

6.5 ASSESSING CHANGES IN ENERGY FLOW AND NUTRIENT CYCLING

Changes in the characteristics of ecosystem function listed in Figure 6.1, and discussed in the previous sections, can provide quantitative data on ecosystem-level response to pollutant perturbations. However, the analysis of functional processes in pollution studies has had a relatively short history and is dependent upon the development of ecosystem theory and the collection of sufficient baseline data to establish 'normal' process rates.

With the development of ecosystem theory and methodologies, a number of more sophisticated functional indices have been proposed for pollution evaluation (O'Neill *et al.*, 1981). Power, defined as the total energy movement through the ecosystem per unit biomass, is a metabolic index which should respond to perturbation and which is related to the maximization of persistent biomass. Odum and Pinkerton (1955) and Odum *et al.* (1977) have argued that through the interaction of low quality energy (solar radiation) and high quality energy (biomass), power influences the capability of the ecosystem to respond to perturbations. Power has been related to both the ability of the system to recover from disturbance (O'Neill, 1976) and the time between perturbation and recovery (DeAngelis, 1980).

The characteristic of phase space has been used to describe an ecosystem in terms of a matrix of redox reactions (Schindler et al., 1980b). The phase-space response (pH–dissolved oxygen) of an aquatic microcosm has been demonstrated for physical (stirring), chemical (acidification) and biological (species introduction) perturbations (Waide et al., 1980).

The analysis of frequencies of response has been used to characterize periodicities in metabolism, biomass and nutrient stability (Emanuel et al., 1978a, b; Van Voris et al., 1980). The number of low-frequency peaks in CO_2 efflux was suggested as an index of functional complexity of the ecosystem, with the assumption that the periodicity of peaks must be related to functional interactions among components, as influenced by feedbacks and time delays. Recently, Dwyer and Perez (1983) reported that the number of functional periodicities was reduced when the complexity of the microcosm community was simplified.

A cycling index quantifying energy and material flows has been suggested by Finn (1976, 1980). The path length (PL) is defined as the number of compartments through which the inflow passes, and the cycling index (CI) is the proportion of flow that cycles through components of the system. Flow indices can characterize the nutrient cycling in different ecosystems, however, the relationship of CI to other measures of cycling (leachability and turnover rate) is presently unclear (Finn, 1980). Although these indices offer an elegant mechanism for representing the complex processes of cycling through a single numeric quantity, their value in assessing the impact of disturbances to ecosystems has not yet been demonstrated (May, 1981).

In streams where nutrient retention processes are competing with downstream transport, recycling has been described in terms of the concept of spiralling length (Wallace et al., 1977; Webster and Patten, 1979). Spiralling can be visualized as the number of times particulate organic matter is taken up by feeding organisms as it is transported downstream, i.e. the shorter the distance between reingestions, the tighter the spiral. Using radiotracers, spiralling length can be measured in natural ecosystems (Newbold et al., 1981), although the effects of pollutants on this process have not yet been evaluated.

The ecosystem is the fundamental unit in ecology and must be recognized as such in ecotoxicology. There is a theoretical base and a developing research methodology for the analysis of changes in energy and cycling processes in perturbed ecosystems. The complexities of natural systems should not dissuade scientists from focusing their efforts on understanding the effects of toxic pollutants on the state of the whole ecosystem.

Effects of Pollutants at the Ecosystem Level
Edited by P. J. Sheehan, D. R. Miller, G. C. Butler and Ph. Bourdeau
© 1984 SCOPE. Published by John Wiley & Sons Ltd

Reference List for Chapters 2 to 6

JOANNE M. RIDGEWAY
Division of Biological Sciences
National Research Council of Canada
Ottawa, Ontario, Canada K1A 0R6

Abrahamsen, G. (1980). Acid precipitation, plant nutrients and forest growth. In Drabløs, D. and Tollan, A. (Eds.), *Ecological Impact of Acid Precipitation*—Proceedings of an international conference, Sandefjord, Norway, The SNSF project, pp. 58–63.

Abrahamsen, G. and Stuanes, A. O. (1980). Effects of simulated rain on the effluent from lysimeters with acid, shallow soil, rich in organic matter. In Drabløs, D. and Tollan, A. (Eds.), *Ecological Impact of Acid Precipitation*—Proceedings of an international conference, Sandefjord, Norway, The SNSF project, pp. 152–153.

Abrahamsen, G., Bjor, K., Horntvedt, R. and Tveite, B. (1976). Effects of acid precipitation on coniferous forest. In Braekke, F. H. (Ed.), *Impact of Acid Precipitation on Forest and Freshwater Ecosystems in Norway*, Research Report 6/76, The SNSF project, pp. 37–63.

Abrahamsen, G., Horntvedt, R. and Tveite, B. (1977). Impacts of acid precipitation on coniferous forest ecosystems. *Water Air Soil Pollut.*, **8**, 57–73.

Abrahamsen, G., Hovland, J. and Hågvar, S. (1980). Effects of artificial acid rain and liming on soil organisms and the decomposition of organic matter. In Hutchinson, T. C. and Havas, M. (Eds.), *Effects of Acid Precipitation on Terrestrial Ecosystems*, Plenum Press, New York and London, pp. 341–362.

Addison, R. F., Zinck, M. E. and Willis, D. E. (1981). Time- and dose-dependence of hepatic mixed function oxidase activity in brook trout *Salvelinus fontinalis* on polychlorinated bi-phenyl residues: implications for biological effects monitoring. *Environ. Pollut. Ser. A*, **25**, 211–218.

Åkesson, B. (1983). Methods for assessing the effects of chemicals on reproduction in marine worms. In Vouk. V. B. and Sheehan, P. J. (Eds.), *Methods for Assessing the Effects of Chemicals on Reproductive Functions*, SCOPE 20, John Wiley and Sons, Chichester, New York, Brisbane, Toronto, pp. 459–482.

Alexander, M. (1980a). Effects of acid precipitation on biochemical activities in soil. In Drabløs, D. and Tollan, A. (Eds.), *Ecological Impact of Acid Precipitation*—Proceedings of an international conference, Sandefjord, Norway, The SNSF project, pp. 47–52.

Alexander, M. (1980b). Effects of acidity on microorganisms and microbial processes in soil. In Hutchinson, T. C. and Havas, M. (Eds.), *Effects of Acid Precipitation on Terrestrial Ecosystems*, Plenum Press, New York and London, pp. 363–374.

Allen, E. B. and Forman, R. T. T. (1976). Plant species removals and old-field community structure and stability. *Ecology*, **57**, 1233–1243.

Almer, B., Dickson, W., Ekström, C., Hörnström, E. and Miller, U. (1974). Effects of acidification on Swedish lakes, *Ambio*, **3**, 30–36.
Almer, B., Dickson, W., Ekström, C. and Hörnström, E. (1978). Sulfur pollution and the aquatic ecosystem. In Nriagu, J. (Ed.), *Sulfur in the Environment*, Part II: *Ecological Impacts*, John Wiley and Sons, New York, pp. 273–311.
Aminot, A. (1981). Abnormalities of coastal hydrobiological system following the Amoco Cadiz oil spill. Qualitative and quantitative studies on the *in situ* biodegradation of hydrocarbons. In *Amoco Cadiz: Fates and Effects of the Oil Spill, Proceedings of an International Symposium*. Centre for National Exploitation of Oceans, Paris, France, pp. 223–242 (in French).
Anderson, J. M. (1971). Assessment of the effects of pollutants on physiology and behavior. *Proc. R. Soc. Lond. B*, **177**, 307–320.
Anderson, J. W., Neff, J. M. and Petrocelli, S. R. (1974). Sublethal effects of oil, heavy metals and PCBs on marine organisms. In Khan, M. A. Q. and Bederka, J. P. Jr. (Eds.), *Survival in Toxic Environments*, Academic Press, Inc., New York, San Francisco, London, pp. 83–121.
Anderson, P. D. and d'Apollonia, S. (1978). Aquatic animals. In Butler, G. C. (Ed.), *Principles of Ecotoxicology*, SCOPE 12, John Wiley and Sons, Chichester, New York, Brisbane, Toronto, pp. 187–221.
Anderson, R. S. (1970). Effects of rotenone on zooplankton communities and a study of their recovery patterns in two mountain lakes in Alberta. *J. Fish. Res. Board Can.*, **27**, 1335–1356.
Andersson, I., Grahn, O., Hultberg, H. and Landner, L. (1975). Comparative studies of different techniques for the restoration of acidified lakes. STU-report 73–3651, Institute for Water and Air Research, Stockholm, 48 pages.
Anonymous (1980). Bulletin of the Man and the Biosphere Program. *Nat. Resour.*, **16**.
Antonovics, J. (1975). Metal tolerance in plants: perfecting an evolutionary paradigm. In *International Conference on Heavy Metals in the Environment*, Toronto, Canada. Symposium Proceedings, vol. II, pp. 169–186, Institute for Environmental Studies, Toronto, Canada.
Antonovics, J., Bradshaw, A. D. and Turner, R. G. (1971). Heavy metal tolerance in plants. *Adv. Ecol. Res.*, **7**: 1–85.
Archibald, R. E. M. (1972). Diversity in some South African diatom associations and its relation to water quality. *Water Res.*, **6**, 1229–1238.
Armitage, P. D. (1980). The effects of mine drainage and organic enrichment on benthos in the River Nent system, Northern Pennines. *Hydrobiologia*, **74**, 119–128.
Armstrong, F. A. J. and Hamilton, A. L. (1973). Pathways of mercury in a polluted northwestern Ontario lake. In Singer, P. C. (Ed.), *Trace Metals and Metal-Organic Interactions in Natural Waters*, Ann Arbor Science Publishers Inc., Ann Arbor, Michigan, pp. 131–156.
Atema, J. (1977). The effects of oil on lobsters. *Oceanus*, **20**, 67–73.
Atema, J., Karnofsky, E. B. and Oleszko-Szuts, S. (1979). Lobster behavior and chemoreception: sublethal effects of Number 2 fuel oil. In Jacoff, F. (Ed.), *Advances in Marine Environmental Research*, U.S. Environmental Protection Agency, EPA-600/9-79-035, Cincinnati, Ohio, pp. 122–134.
Ausmus, B. S., Edwards, N. T. and Witkamp, M. (1976). Microbial immobilization of carbon, nitrogen, phosphorus and potassium: implications for forest ecosystem processes. In Anderson, J. M. and Macfadyen, A. (Eds.), *The Role of Terrestrial and Aquatic Organisms in Decomposition Processes*, Blackwell Scientific Publications, Oxford, London, Edinburgh, Melbourne, pp. 397–416.

Ausmus, B. S., Kimbrough, S., Jackson, D. R. and Lindberg, S. (1979). The behavior of hexachlorobenzene in pine forest microcosms: transport and effects on soil processes. *Environ. Pollut.*, **20**, 103–111.
Axler, R. P., Redfield, G. W. and Goldman, C. R. (1981). The importance of regenerated nitrogen to phytoplankton productivity in a subalpine lake. *Ecology*, **62**: 345–354.
Bååth, E., Berg, B., Lohm, U., Lundgren, B., Lundkvist, H., Rosswall, T., Söderström, B. and Wiren, A. (1980a). Effects of experimental acidification and liming on soil organisms and decomposition in Scots pine forest. *Pedobiologia*, **20**, 85–100.
Bååth, E., Berg, B., Lohm, U., Lundgren, B., Lundkvist, H., Rosswall, T., Söderström, B. and Wiren, A. (1980b). Soil organisms and litter decomposition in a Scots pine forest—effects of experimental acidification. In Hutchinson, T. C. and Havas, M. (Eds.), *Effects of Acid Precipitation on Terrestrial Ecosystems*, Plenum Press, New York and London, pp. 375–380.
Babich, H. and Stotzky, G. (1980). Environmental factors that influence the toxicity of heavy metal and gaseous pollutants to microorganisms. *Crit. Rev. Microbiol.* **8**, 99–145.
Baddeley, M. S., Ferry, B. W. and Finegan, E. I. (1973). Sulphur dioxide and respiration in Lichens. In Ferry, B. W., Baddeley, M. S. and Hawksworth, D. L. (Eds.), *Air Pollution and Lichens*, University of Toronto Press, Toronto and Buffalo, pp. 299–313.
Baker, J. M. (1971). Studies on salt marsh communities—successive spillages. In Cowell, E. B. (Ed.), *The Ecological Effects of Oil Pollution on Littoral Communities*, Proceeding of a symposium organized by the Institute of Petroleum, London, 30 November–1 December 1970, Elsevier Publishing Co. Ltd, Amsterdam, London, New York, pp. 21–32.
Bang, F. B. (1980). Monitoring pathological changes as they occur in estuaries and in the ocean in order to measure pollution (with special reference to invertebrates). In McIntyre, A. D. and Pearce, J. B. (Eds.), *Biological Effects of Marine Pollution and the Problems of Monitoring*, Rapp., P.-v. Réun. Cons. Int. Explor. Mer, **179**, 118–124.
Barber, S. A. (1974). Influence of plant root on ion movement in the soil. In Carson, E. W. (Ed.), *The Plant Root and its Environment*, Charlottesville University Press, Charlottesville, Virginia, pp. 525–564.
Baroux, J. (1972). Toxicity of copper on the phenomena of ammonification in the soils of Bordeaux vineyards. *C. R. Hebd. Séances Acad. Sci. Ser. D. Sci. Nat.*, **275**, 499–502 (in French).
Barrett, G. W. (1968). The effects of an acute insecticide stress on a semi-enclosed grassland ecosystem. *Ecology*, **49**, 1019–1035.
Barrett, G. W., Van Dyne, G. M. and Odum, E. P. (1976). Stress ecology. *Bioscience*, **26**, 192–194.
Bartlett, L., Rabe, F. W. and Funk, W. H. (1974). Effects of copper, zinc and cadmium on *Selanastrum capricornutum*. *Water Res.*, **8**, 179–185.
Bartsch, A. F. (1948). Biological aspects of stream pollution. *Sewage Works J.*, **20**, 292–302.
Battaglia, B., Bisol, P. M., Fossato, V. U. and Rodino, E. (1980). Studies on the genetic effects of pollution in the sea. In McIntyre, A.D. and Pearce, J. B. (Eds.), *Biological Effects of Marine Pollution and the Problems of Monitoring*, Rapp. P.-v. Réun. Cons. Int. Explor. Mer, **179**, 267–274.
Bayne, B. L. (1975). Aspects of physiological condition in *Mytilus edulis* L., with special reference to the effects of oxygen tension and salinity. In Barnes, H. (Ed.) *Proceedings of the 9th European Marine Biology Symposium*, Aberdeen University Press, Aberdeen, pp. 213–238.
Bayne, B. L. (1978). Mussel watching. *Nature*, **275**, 87–88.

Bayne, B. L. (1980). Physiological measurements of stress. In McIntyre, A.D. and Pearce, J. B. (Eds.), *Biological Effects of Marine Pollution and the Problems of Monitoring*, Rapp. P.-v. Réun. Cons. Int. Explor. Mer, **179**, 56–61.

Bayne, B. L., Livingstone, D. R., Moore, M. N. and Widdows, J. (1976). A cytochemical and biochemical index of stress in *Mytilus edulis* L., *Mar. Pollut. Bull.*, **7**, 221–224.

Bayne, B. L., Moore, M. N., Widdows, J., Livingstone, D. R. and Salkeld, P. (1979). Measurement of the response of individuals to environmental stress and pollution: studies with bivalve molluscs. *Philos. Trans. R. Soc. Lond. B. Biol. Sci.*, **286**, 563–581.

Bayne, B. L., Anderson, J., Engel, D., Gilfillan, E., Hoss, D., Lloyd, R. and Thurberg, F. (1980). Physiological techniques for measuring the biological effects of pollution in the sea. In McIntyre, A. D. and Pearce, J.B. (Eds.), *Biological Effects of Marine Pollution and the Problems of Monitoring*, Rapp. P.-v. Réun. Cons. Int. Explor. Mer, **179**, 88–99.

Beamish, R. J. (1974). Loss of fish populations from unexploited remote lakes in Ontario, Canada as a consequence of atmospheric fallout of acid. *Water Res.*, **8**, 85–95.

Beamish, R. J. and Harvey, H. H. (1972). Acidification of the La Cloche Mountain Lakes, Ontario and resulting fish mortalities. *J. Fish. Res. Board Can.*, **29**, 1131–1143.

Beamish, R. J. and Van Loon, J. C. (1977). Precipitation loading of acid and heavy metals to a small acid lake near Sudbury, Ontario. *J. Fish. Res. Board Can.*, **34**, 649–658.

Beamish, R. J., Lockhart, W. L., Van Loon, J. C. and Harvey, H. H. (1975). Long-term acidification of a lake and resulting effects on fishes. *Ambio*, **4**, 98–102.

Beardmore, J. A., Barker, C. J., Battaglia, B., Berry, R. J., Longwell, A. C., Payne, J. F. and Rosenfield, A. (1980). The use of genetical approaches to monitoring biological effects of pollution. In McIntyre, A.D. and Pearce, J.B. (Eds.), *Biological Effects of Marine Pollution and the Problems of Monitoring*, Rapp. P.-v. Réun. Cons. Int. Explor. Mer, **179**, 299–305.

Bechtel, T. J. and Copeland, B. J. (1970). Fish species diversity indices as indicators of pollution in Galveston Bay, Texas. *Contrib. Mar. Sci.*, **15**, 103–132.

Beck, W. M. (1954). Studies in stream pollution biology I. A simplified ecological classification of organisms. *Quart. J. Fla. Acad. Sci*, **17**, 213–227.

Beckett, D. C. (1978). Ordination of macroinvertebrate communities in a multistressed river system. In Thorp, J. H. and Gibbons, J. W. (Eds.), *Energy and Environmental Stress in Aquatic Systems*, U.S. Department of Energy, CONF-771114, NTIS, U.S. Department of Commerce, Springfield, Virginia, pp. 748–770.

Beijer, K. and Jernelov, A. (1979). Sources, transport and transformation of metals in the environment. In Friberg, L., Nordberg, G. F. and Vouk, V. B. (Eds.), *Handbook on the Toxicology of Metals*, Elsevier/North Holland, Amsterdam.

Bengtson, C., Boström, C.-Å., Grennfelt, P., Skärby, L. and Troeng, E. (1980). Deposition of nitrogen oxides to Scots pine (*Pinus sylvestris* L.). In Drablφs, D. and Tollan, A. (Eds.) *Ecological Impact of Acid Precipitation*—Proceedings of an international conference, Sandefjord, Norway, The SNSF project, pp. 154–155.

Bengtsson, B. E. (1979). Biological variables, especially skeletal deformities in fish, for monitoring marine pollution. *Philos. Trans. R. Soc. Lond. B. Biol. Sci.* **286**, 457–464.

Bennett, J. H. and Hill, A. C. (1974). Acute inhibition of apparent photosynthesis by phytotoxic air pollutants. In Dugger, M. (Ed.), *Air Pollutant Effects on Plant Growth*, ACS Symposium Series 3, American Chemical Society, Washington, D.C., pp. 115–127.

Berger, W. H. and Parker, F. L. (1970). Diversity of planktonic foraminifera in deep-sea sediments. *Science*, **168**, 1345–1347.

Beyer, W. N. and Gish, C. D. (1980). Persistence in earthworms and potential hazards to birds of soil applied DDT, dieldrin and heptachlor. *J. Appl. Ecol.*, **17**, 295–307.

Bick, H. and Drews, E. F. (1973). Self purification and ciliate colonization in an acid environment (pilot experiment). *Hydrobiologia*, **42**, 393–402.

Bignoli, G. and Bertozzi, G. (1979). *Modelling of Artificial Radioactivity in the Environment: a Survey*, Commission of the European Communities, series on Nuclear Science and Technology, Brussels, 129 pages.

Bingham, F. T., Page, A. L. and Strong, J. E. (1980). Yield and cadmium content of rice grain in relation to addition rates of cadmium, copper, nickel, and zinc with sewage sludge and liming. *Soil Sci.*, **130**, 32–38.

Billard, R. (1978). Effect of heat pollution and organo-chlorinated pesticides on fish reproduction. In *Final Reports on Research Sponsored Under the First Environmental Research Programme*, Commission of the European Communities, Publ. by CEC Brussels, ISBN 99-825-0185-X, Brussels, Belgium, pp. 265-267.

Birge, W. J., Black, J. A., Westerman, A. G., Francis, P. C. and Hudson, J. E. (1977). *Embryopathic Effects of Waterborne and Sediment Accumulated Cadmium, Mercury and Zinc on Reproduction and Survival of Fish and Amphibian Populations in Kentucky*, University of Kentucky, Water Resources Research Institute, Research Report 100, Lexington, Kentucky, 28 pages.

Birge, W. J., Black, J. A. and Westerman, A. G. (1978). *Effects of Polychlorinated Biphenyl Compounds and Proposed PCB-Replacement Products on Embryo Larval Stages of Fish and Amphibians*, University of Kentucky, Water Resources Research Institute, Research Report 118, Lexington, Kentucky, 33 pages.

Birge, W. J., Black, J. A. and Bruser, D. M. (1979). *Toxicity of Organic Chemicals to Embryo-Larval Stages of Fish*. U.S. Environmental Protection Agency, EPA-560/11-79-007, Washington, D.C., 60 pages.

Birge, W. J., Black, J. A. and Kuehne, R. A. (1980). *Effects of Organic Compounds on Amphibian Reproduction*, University of Kentucky, Water Resources Research Institute, Research Report 121, Lexington, Kentucky, 39 pages.

Bisogni, J. J. and Lawrence, A. W. (1975). Kinetics of mercury methylation in aerobic and anaerobic aquatic environments. *J. Water Pollut. Control. Fed.*, **47**, 135–152.

Bjor, K. and Teigen, O. (1980). Effects of acid precipitation on soil and forest. 6. Lysimeter experiment in greenhouse. In Drabløs, D. and Tollan, A. (Eds.), *Ecological Impact of Acid Precipitation*—Proceedings of an international conference, Sandefjord, Norway, The SNSF project, pp. 200–201.

Blumer, M., Sanders, H. L., Grassle, J. F. and Hampson, G. R. (1971). A small oil spill. *Environment*, **13**, 3–12.

Boddington, M. J., deFreitas, A. S. W. and Miller, D. R. (1979). The effect of benthic invertebrates on the clearance of mercury from sediments. *Ecotoxicol. Environ. Safety*, **3**, 236–244.

Boesch, D. F. (1974). Diversity, stability and response to human disturbance in estaurine ecosystems. In *Proceedings of the First International Congress of Ecology*, Centre for Agricultural Publishing and Documentation, Wageningen, The Netherlands, pp. 109–114.

Boesch, D. F. (1977). *Application of Numerical Classification in Ecological Investigations of Water Pollution*. U.S. Environmental Protection Agency, Ecological Research Series, EPA-600/3-77-033, Washington, D.C., 115 pages.

Boesch, D. F. (1979). Ecosystem consequences of alterations of benthic community structure and function in the New York Bight region. Marine Ecosystems Analysis Program, *New York Bight Program Symposium, Ecological Effects of Environmental Stress*, June 1979, New York, New York.

Boling, R. H. Jr., Goodman, E. D., Van Sickle, J. A., Zimmer, J. O., Cummins, K. W., Peterson, R. C. and Reice, S. R. (1975). Towards a model of detritus processing in a Woodland stream. *Ecology*, **56**, 141–151.

Bolus, R. L., Fang, C. S. and Chia, S. N. (1973). The design of a thermal monitoring system. *Mar. Technol. Soc. J.*, **7**, 36–40.

Bond, H., Lighthart, B., Shimabuka, R. and Russell, L. (1976). Some effects of cadmium on coniferous forest soil and litter microcosms. *Soil Sci.*, **121**, 278–287.

Borgmann, U., Cove, R. and Loveridge, C. (1980). Effect of metals on the biomass production kinetics of freshwater copepods. *Can. J. Fish. Aquat. Sci.*, **37**, 567–575.

Botkin, D. B. (1980). A grandfather clock down the staircase: stability and disturbance in natural ecosystems. In Waring, R. H. (Ed.), *Forests: Fresh Perspectives from Ecosystem Analysis*, Oregon State University Press, Corvallis, Oregon, pp. 1–10.

Botkin, D. B. and Sobel, M. J. (1975). The complexity of ecosystem stability. In Levins, S. A. (Ed.), *Ecosystem Analysis and Prediction*, Society of Industrial and Applied Mathematics, Philadelphia, Pennsylvania, pp. 144–150.

Bott, T. L. and Rogenmuser, K. (1978). Effects of No. 2 fuel oil, Nigerian crude oil, and used crankcase oil on attached algal communities: acute and chronic toxicity of water-soluble constituents. *Appl. Environ. Microbiol.*, **36**, 673–682.

Bourdeau, P. and Treshow, M. (1978). Ecosystem response to pollution. In Butler, G.C. (Ed.), *Principles of Ecotoxicology*, SCOPE 12, John Wiley and Sons, Chichester, New York, Brisbane, Toronto, pp. 313–330.

Box, G. E. P. and Jenkins, G. M. (1970). *Time Series Analysis: Forecasting and Control*, Holden-Day, San Francisco, 575 pages.

Boyden, C. R. and Phillips, D. J. H. (1981). Seasonal variation and inherent variability of trace elements in oysters and their implications for indicator studies. *Mar. Ecol. Prog. Ser.*, **5**, 29–40.

Bradshaw, A. D. (1976). Pollution and evolution. In Mansfield, T. A. (Ed.), *Effects of Air Pollutants on Plants*, Cambridge University Press, Cambridge, London, New York, Melbourne, pp. 135–159.

Bradshaw, A. D., Humphries, R. N., Johnson, M. S. and Roberts, R.D. (1978). The restoration of vegetation on derelict land produced by industrial activity. In Holdgate, M. W. and Woodman, M. J. (Eds.), *The Breakdown and Restoration of Ecosystems*, Plenum Press, New York and London, pp. 249–274.

Bray, J. R. and Curtis, J. T. (1957). An ordination of the upland forest communities of southern Wisconsin. *Ecol. Monogr.*, **27**, 325–349.

Brinley, F. J. (1942). Biological studies, Ohio River pollution survey. I. Biological zones in a polluted stream. *Sewage Works J.*, **14**, 147–159.

Brock, D. A. (1977). Comparison of community similarity indexes. *J. Water Pollut. Control Fed.*, **49**, 2488–2494.

Brooks, J. A. (1981). Otolith abnormalities in *Limnodynastes tasmaniensis* tadpoles after embryonic exposure to the pesticide dieldrin. *Environ. Pollut. Ser. A. Ecol. Biol.*, **25**, 19–25.

Brown, A. W. A. (1971). Pest resistance to pesticides. In White-Stevens, R. (Ed.), *Pesticides in the Environment*, Vol. 1, Part II, Marcel-Dekker, New York, pp. 457–552.

Brown, A. W. A. (1978). *Ecology of Pesticides*, John Wiley and Sons, New York, Chichester, Brisbane, Toronto, 525 pages.

Brown, B. E. (1976). Observations on the tolerance of the isopod *Acellus meridianus* Rac. to copper and lead. *Water Res.*, **10**, 555–559.

Brown, D. A. and Parsons, T. R. (1978). Relationship between cytoplasmic distributions of mercury and toxic effects to zooplankton and Chum salmon (*Oncorhynchus keta*) exposed to mercury in a controlled ecosystem. *J. Fish. Res. Board Can.*, **35**, 880–884.

Brown, D. A., Bowden, C. A., Chatel, K. W. and Parsons, T. R. (1977). The wildlife community of Iona Island Jetty, Vancouver, B.C., and heavy-metal pollution effects. *Environ. Conserv.*, **4**, 213–216.

Brown, D. J. A. and Sadler, R. (1981). The chemistry and fisheries status of acid lakes in Norway and their relationship to European sulfur emissions. *J. Appl. Ecol.*, **18**, 433–441.

Brown, K. W. (1981). Pollutant monitoring in the Olympic National Park Biosphere Reserve. *Environ. Monitor. Assess.*, **1**, 37–48.

Brown, R. S. (1980). The value of the multidisciplinary approach to research marine pollution effects as evidenced in a three-year study to determine the etiology and pathogenesis of neoplasia in the softshell clam, *Mya arenaria*. In McIntyre, A. D. and Pearce, J. B. (Eds.), *Biological Effects of Marine Pollution and the Problems of Monitoring*, Rapp. P.-v. Réun. Cons. Int. Explor. Mer, **179**, 125–128.

Brunskill, G. J., Graham, B. W. and Rudd, J. W. M. (1980). Experimental studies on the effect of arsenic on microbial degradation of organic matter and algal growth. *Can. J. Fish. Aquat. Sci.*, **37**, 415–423.

Bryan, G. W. (1976). Some aspects of heavy metal tolerance in aquatic organisms. In Lockwood, A. P. M. (Ed.), *Effects of Pollutants on Aquatic Organisms*, Cambridge University Press, Cambridge, London, New York, Melbourne, pp. 7–34.

Bryan, G. W. (1979). Bioaccumulation of marine pollutants. *Philos. Trans. R. Soc. Lond. B. Biol. Sci.*, **286**, 483–505.

Bryant, R. D., Gordy, E. A. and Laishley, E. J. (1979). Effect of soil acidification on the soil microflora. *Water Air Soil Pollut.*, **11**, 437–445.

Bubel, A. (1976). Histological and electron microscopical observations on the effects of different salinities and heavy metal ions on the gills of *Jaera nordmanni* (Rathke) (Crustacea, Isopoda). *Cell Tissue Res.*, **167**, 65–95.

Buchanan, J. B., Sheader, M. and Kingston, P. F. (1978). Sources of variability in the benthic macrofauna of the south Northumberland coast, 1971–1976. *J. Mar. Biol. Ass. U.K.*, **58**, 191–209.

Buck, N. B. (1979). Animals as monitors of environmental quality. *Vet. Hum. Toxicol.*, **21**, 277–284.

Buikema, A. L. Jr., Niederlehner, B. R. and Cairns, J. Jr. (1980). Toxicant effects on reproduction and disruption of the egg-length relationship in grass shrimp. *Bull. Environ. Contam. Toxicol.*, **24**, 31–36.

Burlington, R. F. (1962). Quantitative biological assessment of pollution. *J. Water Pollut. Cont. Fed.* **34**, 179–183.

Burns, J. R. (1975). Error analysis of nonlinear simulations. *IEEE Trans. Sys. Man Cybernet.* **SMC-5**, 331–340.

Busby, D. G., Pearce, P. A. and Garrity, N. R. (1981). Brain Cholinesterase response in songbirds exposed to experimental fenitrothion spraying in New Brunswick, Canada. *Bull. Environ. Contam. Toxicol.*, **26**, 401–406.

Butcher, J. E., Boyer, M. G. and Fowle, C. D. (1975). Impact of Dursban and Abate on microbial numbers and some chemical properties of standing ponds. *Proc. 10th Canadian Symp., Water Pollution Research Canada*, pp. 33–41.

Butcher, J. E., Boyer, M. G. and Fowle, C. D. (1977). Some changes in pond chemistry and photosynthetic activity following treatment with increasing concentrations of Chlorpyrifos. *Bull. Environ. Contam. Toxicol.*, **17**, 752–758.

Butler, G. C. (Ed.) (1978). *Principles of Ecotoxicology*, SCOPE 12, John Wiley and Sons, Chichester, New York, Brisbane, Toronto, 350 pages.

Cairns, J. Jr. (1981). Biological monitoring. Part VI—future needs. *Water Res.*, **15**, 941–952.

Cairns, J. Jr. and Dickson, K. L. (1971). A simple method for the biological assessment of the effects of waste discharges on aquatic bottom-dwelling organisms. *J. Water Pollut. Control Fed.*, **43**, 755–772.

Cairns, J. Jr. and Dickson, K. L. (1977). Recovery of streams and spills of hazardous materials. In Cairns, J. Jr., Dickson, K. L. and Herricks, E. E. (Eds.), *Recovery and Restoration of Damaged Ecosystems*, University of Virginia Press, Charlotteville, pp. 24–44.

Cairns, J. Jr. and Kaesler, R. L. (1969). Cluster analysis of Potomac River survey stations based on protozoan presence-absence data. *Hydrobiologia*, **34**, 414–432.

Cairns, J. Jr. and van der Schalie, W. H. (1980). Biological monitoring part I—early warning systems. *Water Res.*, **14**, 1179–1196.

Cairns, J. Jr., Albaugh, D. W., Busey, F. and Chaney, M. D. (1968). The sequential comparison index—a simplified method for non-biologists to estimate relative differences in biological diversity in stream pollution studies. *J. Water Pollut. Control Fed.*, **40**, 1607–1613.

Cairns, J. Jr., Lanza, G. R. and Parker, B. C. (1972). Pollution related structural and functional changes in aquatic communities with emphasis on freshwater algae and protozoa. *Proc. Acad. Nat. Sci. Phila.*, **124**, 79–127.

Cairns, J. Jr., Patil, G. P. and Waters, W. E. (Eds.) (1979). *Environmental Biomonitoring, Assessment, Prediction, and Management—Certain Case Studies and Related Quantitative Issues*, Statistical Ecology Series, vol. 11, International Co-operative Publishing House, Fairland, Maryland, 438 pages.

Callahan, M. A., Slimak, M. W., Gabel, N. W., May, I. P., Fowler, C. F., Freed, J. R., Jennings, P., Durfee, R. L., Whitmore, F. C., Maestri, B., Mabey, W. R., Holt, B. R. and Gould, C. (1979). *Water-Related Environmental Fate of 129 Priority Pollutants*, Office of Water Planning and Standards, U.S. Environmental Protection Agency, EPA-440/4-79-029, 2 volumes.

Capuzzo, J. M. and Lancaster, B. A. (1981). Physiological effects of South Louisiana crude oil on larvae of the American lobster (*Homarus americanus*). In Vernberg, F. J., Calabrese, A., Thurberg, F. P. and Vernberg, W. B. (Eds.), *Biological Monitoring of Marine Pollutants*, Academic Press, New York, Toronto, London, Sydney, San Francisco, pp. 404–424.

Case, J. W. (1980). The influence of three sour gas processing plants on the ecological distribution of epiphytic lichens in the vicinity of Fox Creek and Whitecourt, Alberta, Canada. *Water Air Soil Pollut.*, **14**, 45–68.

CEC (Commission of the European Communities) (1979a). *Methodology for Evaluating the Radiological Consequences of Radioactive Effluents Released in Normal Operations*, Joint Report by NRPB, Harwell and CEA, Fontenay-aux-Roses, July, 1979.

CEC (Commission of the European Communities). (1979b). *Modelling of Artificial Radioactivity Migration in Environment: A survey*, Publication EUR 6179, Joint Research Centre, Ispra, Italy.

Cedeno-Maldonado, A., Swader, J. A. and Heath, R. L. (1972). The cupric ion as an inhibitor of photosynthetic electron transport in isolated chloroplasts. *Plant Physiol.*, **50**, 698–701.

Chan, E. I. (1977). Oil pollution and tropical littoral communities: biological effects of the 1975 Florida Keys oil spill. In *Proceedings of the 1977 Oil Spill Conference*, American Petroleum Institute, pp. 539–542.

Chandler, J. R. 1970. A biological approach to water quality management. *Water Pollut. Control*, **4**, 415–422.

Chaney, W. R., Kelly, J. M. and Strickland, R. C. (1978). Influence of cadmium and zinc on carbon dioxide evolution from litter and soil from a black oak forest. *J. Environ. Qual.*, **7**, 115–119.

Chau, Y. K., Wong, P. T. S., Kramar, O., Bengeret, G. A., Cruz, R. B., Kinrade, J. O., Lye, J. and Van Loon, J. C. (1980). Occurrence of tetraalkyllead compounds in the aquatic environment. *Bull. Environ. Contam. Toxicol.*, **24**, 265–269.

Chiou, C. T., Freed, H. V., Schmedding, D. W. and Kohnert, R. L. (1977). Partition coefficient and bioaccumulation of selected organic chemicals. *Environ. Sci. Technol.*, **11**, 475–478.

Chutter, F. M. (1972). An empirical biotic index of the quality of water in South African streams and rivers. *Water Res.*, **6**, 19–30.

Coello, W. F., Saleem, Z. A. and Khan, M. A. Q. (1974). Ecological effects of lead in auto-exhaust. In Khan, M. A. Q. and Bederka, J. P. Jr. (Eds.), *Survival in Toxic Environments*, Academic Press, Inc., New York, San Francisco, London, pp. 499–513.
Cole, H. A. (Ed.) (1979). The assessment of sublethal effects of pollutants in the sea. *Philos. Trans. R. Soc. Lond. B. Biol. Sci.*, **286**, 397–636.
Cole, T. J. (1978). Preliminary ecological-genetic comparison between unperturbed and oil-impacted *Urosalpinx cinerea* (Prosobranchia: Gastropoda) populations: Nobska Point (Woods Hole) and Wild Harbor (West Falmouth), Massachusetts. *J. Fish. Res. Board Can.*, **35**, 624–629.
Colebrook, J. M. (1978). Continuous plankton records: zooplankton and environment, North-east Atlantic and North Sea, 1948–1975. *Oceanol. Acta*, **1**, 9–22.
Colwell, A. E. and Schaefer, C. H. (1980). Diets of *Ictalurus nebulosus* and *Pomoxis nigromaculatus* altered by diflubenzuron. *Can. J. Fish Aquat. Sci.*, **37**, 632–639.
Colwell, R. R., Mills, A. L., Walker, J. D., Garcia-Tello, P. and Campos-P, V. (1978). Microbial ecology studies of the *Metula* spill in the Straits of Magellan. *J. Fish. Res. Board Can.*, **35**, 573–580.
Connell, W. E. and Patrick, W. H. Jr. (1968). Sulfate reduction in soil: effects of redox potential and pH. *Science*, **159**, 86–87.
Conrad, M. (1976). Patterns of biological control in ecosystems. In Patten, B. C. (Ed.), *Systems Analysis and Simulation in Ecology*, Volume IV, Academic Press, New York, San Francisco, London, pp. 431–456.
Constantinidou, H. A. and Kozlowski, T. T. (1979a). Effects of sulfur dioxide and ozone on *Ulmus americana* seedlings. I. Visible injury and growth. *Can. J. Bot.*, **57**, 170–175.
Constantinidou, H. A. and Kozlowski, T. T. (1979b). Effects of sulfur dioxide and ozone on *Ulmus americana* seedlings. II. Carbohydrates, proteins, and lipids. *Can. J. Bot.*, **57**, 176–184.
Cook, S. E. K. (1976). Quest for an index of community structure sensitive to water pollution. *Environ. Pollut.*, **11**, 269–288.
Cooke, A. S. (1970). The effect of pp'-DDT on tadpoles of the common frog (*Rana temporaria*). *Environ. Pollut.*, **1**, 57–71.
Cooke, A. S. (1971). Selective predation by newts on frog tadpoles treated with DDT. *Nature*, **229**, 275–276.
Cooke, A. S. (1972). The effects of DDT, dieldrin and 2,4-D on amphibian spawn and tadpoles. *Environ. Pollut.*, **3**, 51–68.
Cooke, A. S. (1973). Shell thinning in avian eggs by environmental pollutants. *Environ. Pollut.*, **4**, 85–152.
Cooke, A. S. (1981). Tadpoles as indicators of harmful levels of pollution in the field. *Environ. Pollut. Ser. A.*, **25**, 123–133.
Copeland, B. J. and Bechtel, T. J. (1971). Species diversity and water quality in Galveston Bay, Texas. *Water Air Soil Pollut.*, **1**, 89–105
Coughtrey, P. J., Jones, C. H., Martin, M. H. and Shales, S. W. (1979). Litter accumulation in woodlands contaminated by Pb, Zn, Cd and Cu. *Oecologia (Berl.)*, **39**, 51–60.
Cribben, L. D. and Scacchetti, D. D. (1977). Diversity in tree species in southeastern Ohio *Betula nigra* L. communities. *Water Air Soil Pollut.*, **8**, 47–55.
Crisman, T. L., Schulze, R. L., Brezonik, P. L. and Bloom, S. A. (1980). Acid precipitation: the biotic response in Florida lakes. In Drabløs, D. and Tollan, A. (Eds.), *Ecological Impact of Acid Precipitation*—Proceedings of an international conference, Sandefjord, Norway, The SNSF project, pp. 296–297.
Cronan, C. S. (1980). Solution chemistry of a New Hampshire subalpine ecosystem: a biogeochemical analysis. *Oikos*, **34**, 272–281.
Cronan, C. S. and Schofield, C. L. (1979). Aluminum leaching response to acid precipitation: effects on high-elevation watersheds in the northeast. *Science*, **204**, 304–306.

Cronan, C. S., Reiners, W. A., Reynolds, R. C. Jr. and Lang, G. E. (1978). Forest floor leaching: contributions from mineral, organic and carbonic acids in New Hampshire subalpine forests. *Science*, **200**, 309–311.

Cronan, C. S., Reiners, W. A. and Reynolds, R. C. Jr. (1980). The impact of acid precipitation on forest canopies and soils in the northeastern U.S. In Drabløs, D. and Tollan, A. (Eds.), *Ecological Impact of Acid Precipitation*—Proceedings of an international conference, Sandefjord, Norway, The SNSF project, pp. 158–159.

Cross, T. F., Southgate, T. and Myers, A. A. (1979). The initial pollution of shores in Bantry Bay, Ireland, by oil from the tanker *Betelgeuse*. *Mar. Pollut. Bull.*, **10**, 104–107.

Crossman, J. S., Kaesler, R. L. and Cairns, J. Jr. (1974). The use of cluster analysis in the assessment of spills of hazardous materials. *Am. Midl. Nat.*, **92**, 94–114.

Cummins, K. W. (1974). Structure and function of stream ecosystems. *Bioscience*, **24**, 631–641.

Cummins, K. W. and Klug, M. J. (1979). Feeding ecology of stream invertebrates. *Ann. Rev. Ecol. Syst.*, **10**, 147–172.

Cushing, D. H. (1979). The monitoring of biological effects: the separation of natural changes from those induced by pollution. *Philos. Trans. R. Soc. Lond. B. Biol. Sci.*, **286**, 597–609.

Custer, T. W. and Heinz, G. H. (1980). Reproductive success and nest attentiveness of mallard ducks fed Aroclor 1254. *Environ. Pollut. Ser. A.*, **21**, 313–318.

Czuba, M. and Ormrod, D. P. (1974). Effects of cadmium and zinc on ozone-induced phytotoxicity in cress and lettuce. *Can. J. Bot.*, **52**, 645–649.

Czuba, M. and Ormrod, D. P. (1981). Cadmium concentrations in cress shoots in relation to cadmium-enhanced ozone phytotoxicity. *Environ. Pollut. Ser. A.*, **25**, 67–76.

Davavin, I. A., Mironov, O. G. and Tsimbal, I. M. (1975). Influence of oil on nucleic acids of algae. *Mar. Pollut. Bull.*, **6**, 13–15.

Davenport, J. (1977). A study of the effects of copper applied continuously and discontinuously to specimens of *Mytilus edulis* (L.) exposed to steady and fluctuating salinity levels. *J. Mar. Biol. Ass. U.K.*, **57**, 63–74.

Davey, K. G., Saleuddin, A. S.M., Steel, C. G. H. and Webb, R. A. (1983). Methods for assessing the effects of chemicals on reproductive function in certain invertebrates: some principles and recommendations. In Vouk, V. B. and Sheehan, P. J. (Eds.), *Methods for Assessing the Effects of Chemicals on Reproductive Functions*, SCOPE 20, John Wiley and Sons, Chichester, New York, Brisbane, Toronto, pp. 483–497.

Dawson, J. L. and Nash, T. H. III (1980). Effects of air pollution from copper smelters on a desert grassland community. *Environ. Exp. Bot.*, **20**, 61–72.

Dean, J. M. and Burlington, R. F. (1963). A quantitative evaluation of pollution effects on stream communities. *Hydrobiologia*, **21**, 193–199.

DeAngelis, D. L. (1980). Energy flow, nutrient cycling and ecosystem resilience. *Ecology*, **61**, 764–771.

Denison, R., Caldwell, B., Bormann, B., Eldred, L., Swanberg, C. and Anderson, S. (1977). The effects of acid rain on nitrogen fixation in western Washington coniferous forests. *Water Air Soil Pollut.*, **8**, 21–34.

Dillon, P. J., Yan, N. D., Scheider, W. A. and Conroy, N. (1979). Acidic lakes in Ontario: characterization, extent, and responses to base and nutrient additions. *Arch. Hydrobiol. Beih. Ergebn. Limnol.*, **13**, 317–336.

Dillon, P. J., Jeffries, D. S., Scheider, W. A. and Yan, N. D. (1980). Some aspects of acidification in Southern Ontario. In Drabløs, D. and Tollan, A. (Eds.), *Ecological Impact of Acid Precipitation*—Proceedings of an international conference, Sandefjord, Norway, The SNSF project, pp. 212–213.

Dills, G. and Rogers, D. T. Jr. (1974). Macroinvertebrate community structure as an indicator of acid mine pollution. *Environ. Pollut.*, **6**, 239–262.

Dixon, D. R. (1983). Methods for assessing the effects of chemicals on reproductive function in marine molluscs. In Vouk, V. B. and Sheehan P. J. (Eds.), *Methods for Assessing the Effects of Chemicals on Reproductive Functions*, SCOPE 20, John Wiley and Sons, Chichester, New York, Brisbane, Toronto, pp. 439-457.

Dixon, K. R., Luxmoore, R. J. and Begovich, C. L. (1978). CERES—A model of forest stand biomass dynamics for predicting trace contaminant, nutrient and water effects. II. Model application. *Ecol. Modelling*, **5**, 93-114.

Doelman, P. and Haanstra, L. (1979a). Effect of lead on soil respiration and dehydrogenase activity. *Soil Biol. Biochem.*, **11**, 475-479.

Doelman, P., and Haanstra, L. (1979b). Effects of lead on the decomposition of organic matter. *Soil Biol. Biochem.*, **11**, 481-485.

Doetsch, R. N. and Cook, T. M. (1973). *Introduction to Bacteria and their Ecobiology*, University Park Press, Baltimore, Maryland, 371 pages.

Donaldson, E. M. and Dye, H. M. (1975). Corticosteroid concentrations in sockeye salmon (*Oncorhynchus nerka*) exposed to low concentrations of copper. *J. Fish. Res. Board Can.*, **32**, 533-539.

Donaldson, E. M. and Scherer, E. (1983). Methods to test and assess effects of chemicals on reproduction in fish. In Vouk, V. B. and Sheehan P. J. (Eds.), *Methods for Assessing the Effects of Chemicals on Reproductive Functions*, SCOPE 20, John Wiley and Sons, Chichester, New York, Brisbane, Toronto, pp. 365-404.

Drabløs, D. and Tollan, A. (Eds.) (1980). *Ecological Impact of Acid Precipitation*—Proceedings of an international conference, Sandefjord, Norway, March 11 – 14, 1980, the SNSF project, 383 pages.

Drummond, R. A., Spoor, W. A. and Olson, G. F. (1973). Some short-term indicators of sublethal effects of copper on brook trout, *Salvelinus fontinalis*. *J. Fish. Res. Board Can.*, **30**, 698-701.

Ducklow, H. W.and Mitchell, R. (1979). Bacterial populations and adaptations in the mucus layers on living corals. *Limnol. Oceanogr.*, **24**, 715-725.

Duddridge, J. E. and Wainwright, M. (1980). Heavy metal accumulation by aquatic fungi and reduction in viability of *Gammarus pulex* fed Cd^{2+} contaminated mycelium. *Water Res.*, **14**, 1605-1611.

Dutta, S. K. and Mohanty-Hejmadi, P. (1978). Life history and pesticide susceptible embryonic stages of the Indian bull frog *Rana tigerina*. *Indian J. Exp. Biol.*, **16**, 727-729.

Dwyer, R. L. and Perez, K. T. (1983). An experimental examination of ecosystem linearization. *Am. Nat.* **121**, 305-323.

Dye, H. M. and Donaldson, E. M. (1974). *A Preliminary Study of the Corticosteroid Stress Response in Sockeye Salmon to a Component of Kraft Pulp Mill Effluent Dehydroabietic Acid*, Dept. of the Environment, Canada, Fisheries and Marine Service Res. Dev. Tech. Rep. 461, 18 pages.

Eaton, J. S., Likens, G. E. and Bormann, F. H. (1973). Throughfall and stemflow chemistry in a northern hardwood forest. *J. Ecol.*, **61**, 495-508.

Eberhardt, L. L. (1978). Appraising variability in population studies. *J. Wildl. Manage.*, **42**, 207 – 238.

Ebregt, A. and Boldewijn, J. M. A. M. (1977). Influence of heavy metals in spruce forest soil on amylase activity, CO_2 evolution from starch and soil respiration. *Plant Soil*, **47**, 137-148.

Edwards, C. A. (1973). *Persistent Pesticides in the Environment*, 2nd ed., CRC Press, Cleveland, Ohio, 170 pages.

Edwards, C. A. and Thompson, A. R. (1973). Pesticides and the soil fauna. *Residue Rev.*, **45**, 1-79.

Edwards, R. W., Hughes, B. D. and Read, M. W. (1975). Biological survey in the detection and assessment of pollution. In Chadwick, M. J. and Goodman, G. T. (Eds.),

The Ecology of Resource Degradation and Renewal, The 15th Symposium of the British Ecological Society, Blackwell Scientific Publications, Oxford, London, Edinburgh, Melbourne, pp. 139–156.

Eisele, P. J. and Hartung, R. (1976). The effects of methoxychlor on riffle invertebrate populations and communities. *Trans. Am. Fish Soc.*, **105**, 628–633.

Eisler, R. (1979). Behavioural responses of marine poikilotherms to pollutants. *Philos. Trans. R. Soc. Lond. B. Biol. Sci.*, **286**, 507–521.

Ellgehausen, H., Guth, J. A. and Esser, H. O. (1980). Factors determining the bioaccumulation potential of pesticides in the individual compartments of aquatic food chains. *Ecotoxicol. Environ. Safety*, **4**, 134–157.

Elliott, J. M. (1977). *Some Methods for the Statistical Analysis of Samples of Benthic Invertebrates*, Freshwater Biological Association, Scientific Publication No. 25, 156 pages.

Emanuel, W. R., West, D. C. and Shugart, H. H. Jr. (1978a). Spectral analysis of forest model time series. *Ecol. Model.*, **4**, 313–326.

Emanuel, W. R., Shugart, H. H. Jr. and West, D. C. (1978b). Spectral analysis and forest dynamics: the effects of perturbations on long-term dynamics. In Shugart, H. H. Jr. (Ed.), *Time Series and Ecological Processes*, SIAM, Philadelphia, PA, pp. 193–207.

Engstrom, A. (Ed.) (1971). *Air Pollution Across National Boundaries*. The impact on the environment of sulfur in air and precipitation, Sweden's case study for the United Nations Conference on the Human Environment, Royal Ministry for Foreign Affairs and Royal Ministry of Agriculture, Sweden, 96 pages.

EPRI (Electric Power Research Institute). (1979). *Ecological Effects of Acid Precipitation.* EPRI EA-79-6-LD, Special Study Project SOA 77-403, Workshop Proceedings, Central Electricity Research Laboratories, Surrey, U.K.

Erickson, S. J. (1972). Toxicity of copper to *Thalassiosira pseudonana* in unenriched inhore seawater. *J. Phycol.*, **8**, 318–323.

Eriksson, M. O. G., Henrikson, L. Nilsson, B. -I., Nyman, G., Oscarson, H. G., Stenson, A. E. and Larsson, K. (1980). Predator-prey relations important for the biotic changes in acidified lakes. *Ambio*, **9**, 248–249.

Ernst, W. H. O. (1975). Physiology of heavy metal resistance in plants. In *International Conference on Heavy Metals in the Environment,* Toronto, Canada. Symposium Proceedings, vol. II, pp. 121–136, Institute for Environmental Studies, Toronto, Canada.

Evans, L. S., Conway, C. A. and Lewin, K. F. (1980). Yield responses of field-grown soybeans exposed to simulated acid rain. In Drabløs, D. and Tollan, A. (Eds.), *Ecological Impact of Acid Precipitation*–Proceedings of an international conference, Sandefjord, Norway, The SNSF project, pp. 162–163.

Fagerstrom, T. and Jernelov, A. (1971). Formation of methyl mercury from pure mercuric sulphide in aerobic organic sediment. *Water Res.*, **5**, 121–122.

FAO (Food and Agriculture Organization of the United Nations). (1977). *Impact of Oil on the Marine Environment*, Joint Group of Experts on the Scientific Aspects of Marine Pollution, Reports and Studies No. 6, 250 pages.

Farr, J. A. (1978). The effect of methyl parathion on predator choice of two estuarine prey species. *Trans. Am. Fish. Soc.*, **107**, 87–91.

Feder, W. A. (1973). Cumulative effects of chronic exposure of plants to low levels of air pollutants. In Naegele, J. A. (Ed.), *Air Pollution Damage to Vegetation*. Advances in Chemistry Series, vol. **122**, pp. 21–30, American Chemical Society, Washington, D.C.

Finn, J. T. (1976). Measures of ecosystem structure and function derived from analysis of flows. *J. Theor. Biol.*, **56**, 363–380.

Finn, J. T. (1980). Flow analysis of models of the Hubbard Brook ecosystem. *Ecology*, **61**, 562–571.

Fishelson, L. (1977). Stability and instability of marine ecosystems, illustrated by examples from the Red Sea. *Helgol. Wiss. Meeresunters.*, **30**, 18–29.
Fisher, N. S., Carpenter, E. J., Remsen, C. C. and Wurster, C. F. (1974). Effects of PCB on interspecific competition in natural and gnotobiotic phytoplankton communities in continuous and batch cultures. *Microb. Ecol.*, **1**, 39–50.
Fisher, R. A. Corbet, A. S. and Williams, C. B. (1943). The relation between the number of species and the number of individuals in a random sample of an animal population. *J. Anim. Ecol.*, **12**, 42–58.
Fisher, S. G. and Likens, G. E. (1973). Energy flow in Bear Brook, New Hampshire: an integrative approach to stream ecosystem metabolism. *Ecol. Monogr.*, **43**, 421–439.
Fjerdingstad, E. (1964). Pollution of streams estimated by benthal phytomicroorganisms. I. A saprobic system based on communities of organisms and ecological factors. *Int. Rev. Gesamtem. Hydrobiol.*, **49**, 63–131.
Forbes, S. A. and Richardson, R. E. (1913). Studies on the biology of the upper Illinois River. *Bull. Ill. Nat. History Survey*, **9**, 481–574.
Forbes, S. A. and Richardson, R. E. (1919). Some recent changes in Illinois River biology. *Bull. Ill. Nat. History Survey*, **13**, 139–156.
Fowler, C. W. and MacMahon, J. A. 1982). Selective extinction and speciation: their influence on the structure and functioning of communities and ecosystems. *Am. Nat.*, **119**, 480–498.
Fox, G. A., Gilman, A. P., Peakall, D. B. and Anderka, F. W. (1978). Behavioral abnormalities of nesting Lake Ontario herring gulls. *J. Wildl. Manage.*, **42**, 477–483.
Foy, C. D., Chaney, R. L. and White, M. C. (1978). The physiology of metal toxicity in plants. *Ann. Rev. Plant Physiol.*, **29**, 511–566.
Francis, A. J., Olson, D. and Bernatsky, R. (1980). Effect of acidity on microbial processes in a forest soil. In Drabløs, D. and Tollan, A. (Eds.), *Ecological Impact of Acid Precipitation*—Proceedings of an international conference, Sandefjord, Norway, The SNSF project, pp. 166–167.
Francis, A. J., Quinby, H. L. and Hendrey, G. R. (1983). Effect of lake pH on microbial decomposition of allochthonous litter. In Hendrey, G. R. and Teasley, J. (Eds.), *Acid Rain: Aquatic Effects*, ACS Acid Rain Symposium Proceedings, Ann Arbor Science Publishers, Ann Arbor, MI (in press).
Freedman, B. and Hutchinson, T. C. (1980a). Long-term effects of smelter pollution at Sudbury, Ontario, on forest community composition. *Can. J. Bot.*, **58**, 2123–2140.
Freedman, B. and Hutchinson, T. C. (1980b). Smelter pollution near Sudbury, Ontario, Canada, and effects on forest litter decomposition. In Hutchinson, T. C. and Havas, M. (Eds.), *Effects of Acid Precipitation on Terrestrial Ecosystems*, Plenum Press, New York and London, pp. 395–434.
Friberg, F., Otto, C. and Svensson, B. S. (1980). Effects of acidification on the dynamics of allochthonous leaf material and benthic invertebrate communities in running waters. In Drabløs, D. and Tollan, A. (Eds.), *Ecological Impact of Acid Precipitation*—Proceedings of an international conference. Sandefjord. Norway. The SNSF project, pp. 304–305.
Friberg, L., Nordberg, G. F. and Vouk, V. B. (1979). *Handbook on the Toxicology of Metals*, Elsevier/North Holland, Amsterdam, 709 pages.
Gächter, R. and Máreš, A. (1979). MELIMEX, an experimental heavy metal pollution study: effects of increased heavy metal loads on phytoplankton communities. *Schweiz. Z. Hydrol.* **41**, 228–246.
Gahnström, G., Andersson, G. and Fleischer, S. (1980). Decomposition and exchange processes in acidified lake sediment. In Drabløs, D. and Tollan, A. (Eds.), *Ecological Impact of Acid Precipitation*—Proceedings of an international conference, Sandefjord, Norway, The SNSF project, pp. 306–307.

Galloway, J. N., Likens, G. E. and Edgerton, E. S. (1976). Acid precipitation in the northeastern United States: pH and acidity. *Science*, **194**, 722–724.

Gamble, J. C., Davies, J. M. and Steele, J. H. (1977). Loch Ewe bag experiment, 1974. *Bull. Mar. Sci.*, **27**, 146–175.

Gardner, G. R. (1975). Chemically induced lesions in estuarine or marine teleosts. In Ribelin, W. E. and Magaki, G. (Eds.), *The Pathology of Fishes*, University of Wisconsin Press, Madison, Wisconsin, pp. 657–693.

Garrels, R. M., Mackenzie, F. T. and Hunt, C. (1975). *Chemical Cycles and the Global Environment*, William Kaufmann Publisher, Los Altos, California.

Gaufin, A. R. (1958). The effects of pollution on a mid-western stream. *Ohio J. Sci.*, **58**, 197–208.

Gaufin, A. R. and Tarzwell, C. M. (1952). Aquatic invertebrates as indicators of stream pollution. *Public Health Rep.*, **67**, 57–64.

Gaufin, R. F. and Gaufin, A. R. (1966). *Ecological Aspects of Organic Pollution in Streams*, World Health Organization, WHO/EBL/66.54.

Gaur, J. P. and Kumar, H. D. (1981). Growth response of four micro-algae to three crude oils and a furnace oil. *Environ. Pollut. Ser. A.*, **25**, 77–85.

Geckler, J. R., Horning, W. B., Neiheisel, T. M., Pickering, Q. H. and Robinson, E. L. (1976). *Validity of Laboratory Tests for Predicting Copper Toxicity in Streams*, Ecol. Res. Ser. Environ. Prot. Agency 600/3-76-116, Environ. Res. Lab., U.S. Environmental Protection Agency, Duluth, MN, 192 pages.

Giashuddin, M. and Cornfield, A. H. (1978). Incubation study on effects of adding varying levels of nickel (as sulphate) on nitrogen and carbon mineralization in soil. *Environ. Pollut.*, **15**, 231–234.

Giddings, J. M. and Eddlemon, G. K. (1978). Photosynthesis/respiration ratios in aquatic microcosms under arsenic stress. *Water Air Soil Pollut.*, **9**, 207–212.

Giere, O. (1979). The impact of oil pollution on intertidal meiofauna. Field studies after the La Coruna-spill, May (1976). *Cah. Biol. Mar.*, **20**, 231–251.

Giesy, J. P. Jr. (1978). Cadmium inhibition of leaf decomposition in an aquatic microcosm. *Chemosphere*, **6**, 467–475.

Gilboy, W. B., Mason, P. I. and Tout, R. E. (1979). Time variations in environmental pollution. *J. Radioanal. Chem.*, **48**, 327–335.

Gilfillan, E. S. (1980). The use of scope-for-growth measurements in monitoring petroleum pollution. In McIntyre, A. D. and Pearce, J. B. (Eds.), *Biological Effects of Marine Pollution and the Problems of Monitoring*, Rapp. P.-v. Réun. Cons. Int. Explor. Mer, **179**, 71–75.

Gilfillan, E. S. and Hanson, S. A. (1975). Effects of paralytic shellfish poisoning toxin on the behavior and physiology of marine invertebrates. In LoCicero, V. R. (Ed.), *Proceedings of the First International Conference on Toxic Dinoflagellate Blooms*, Boston, Massachusetts Science and Technology Foundation, Sea Grant Program, pp. 367–375.

Gilfillan, E. S. and Vandermeulen, J. H. (1978). Alterations in growth and physiology of soft-shell clams, *Mya arenaria*, chronically oiled with Bunker C from Chedabucto Bay, Nova Scotia, 1970–76. *J. Fish. Res. Board Can.*, **35**, 630–636.

Gilfillan, E. S., Mayo, D., Hanson, S., Donovan, D. and Jiang, L. C. (1976). Reduction in carbon flux in *Mya arenaria* caused by a spill of No. 6 fuel oil. *Mar. Biol.*, **37**, 115–123.

Gilfillan, E. S., Hanson, S. A., Page, D. S., Mayo, D., Cooley, J., Chalfant, J., Archanbeault, T., West, A. and Harshbarger, J. C. (1977). A chemical, biological and histopathological examination of the Searsport oil spill site, *Final Report to the State of Maine Department of Environmental Protection*, Augusta, Maine, 201 pages.

Glover, R. S. (1979). Natural fluctuations of populations. *Ecotoxicol. Environ. Safety*, **3**, 190–203.

Godfrey, P. J. (1978). Diversity as a measure of benthic macroinvertebrate community response to water pollution. *Hydrobiologia*, **57**, 111–122.

Goering, J. J., Boisseau, D. and Hattori, A. (1977). Effects of copper on silicic acid uptake by a marine phytoplankton population: controlled ecosystem pollution experiment. *Bull. Mar. Sci.*, **27**, 58–65.

Golley, F. B. (1973). Impact of small mammals on primary production. In Gessaman, J. A. (Ed.), *Ecological Energetics of Homeotherms*, Utah State University Press, Logan, Utah, pp. 142–147.

Golley, F. B., Ryszkowski, L. and Sokur, J. T. (1975). The role of small mammals in temperate forests, grasslands and cultivated fields. In Golley, F. B., Petrusewicz, K. and Ryszkowski, L. (Eds.), *Small Mammals: Their Productivity and Population Dynamics*, Cambridge University Press, Cambridge, London, New York, Melbourne, pp. 223–241.

Gonzalez, J. G., Everich, D., Hyland, J., Melzian, B. and Yevich, P. (1979). Effect of no. 2 heating oil on the filtration rate of the blue mussel, *Mytilus edulis*. In Jacoff, F. S. (Ed.), *Proc. Symposium on State of Marine Environmental Research*, U.S. Govt. Printing Office, Washington, D.C.

Goodall, D. W. (1972). Building and testing ecosystem models. In Jeffers, J. N. R. (Ed.), *Mathematical Models in Ecology*, The 12th Symposium of the British Ecological Society, March 1971, Blackwell Scientific Publications, Oxford, London, Edinburgh, Melbourne, pp. 173–194.

Goodman, D. (1975). The theory of diversity-stability relationships in ecology. *Quart. Rev. Biol.*, **50**, 237–266.

Goodman, G. T. (1974). How do chemical substances affect the environment? *Proc. R. Soc. Lond. B. Biol. Sci.*, **185**, 127–148.

Goodyear, C. P. (1972). A simple technique for detecting effects of toxicants or other stresses on a predator-prey interaction. *Trans. Am. Fish. Soc.*, **101**, 367–370.

Gordon, D. C. Jr., Dale, J. and Keizer, P. D. (1978). Importance of sediment working by the deposit-feeding polychaete *Arenicola marina* on the weathering rate of sediment-bound oil. *J. Fish. Res. Board Can.*, **35**, 591–603.

Gorham, E. and McFee, W. W. (1980). Effects of acid deposition upon outputs from terrestrial to aquatic ecosystems. In Hutchinson, T. C. and Havas, M. (Eds.), *Effects of Acid Precipitation on Terrestrial Ecosystems*, Plenum Press, New York and London, pp. 465–480.

Gorham, E., Vitousek, P. M. and Reiners, W. A. (1979). The regulation of chemical budgets over the course of terrestrial ecosystem succession. *Ann. Rev. Ecol. Syst.*, **10**, 53–84.

Gosz, J. R., Likens, G. E. and Bormann, F. H. (1973). Nutrient release from decomposing leaf and branch litter in the Hubbard Brook Forest, New Hampshire. *Ecol. Monogr.*, **43**, 173–191.

Gowdy, C. M., Mulholland, R. J., and Emanuel, W. R. (1975). Modelling the global carbon cycle. *Int. J. Systems Sci.*, **6**, 965–976.

Grahn, O. (1977). Macrophyte succession in Swedish lakes caused by deposition of airborne acid substances. *Water Air Soil Pollut.*, **7**, 295–305.

Grahn, O. and Hultberg, H. (1974). *Effect of Acidification on the Ecosystem of Oligotrophic Lakes: Integrated Changes in Species Composition and Dynamics*, Swedish Water and Air Pollution Research Laboratory, Gothenburg, Meddelande nr 2, 12 pages (in Swedish, abstract in English).

Grahn, O., Hultberg, H. and Landner, L. (1974). Oligotrophication—a self-accelerating process in lakes subjected to excessive supply of acid substances. *Ambio*, **3**, 93–94.

Grant, I. F., Bancroft, K. and Alexander, M. (1979). Effect of sulfur dioxide and bisulfite on heterotrophic activity in an acid soil. *Appl. Environ. Microbiol.*, **38**, 78–83.

Grassle, J. F., and Grassle, J. P. 1974. Opportunistic life histories and genetic systems in marine benthic polychaetes. *J. Mar. Res.*, **32**, 253-284.

Gray, J. S. (1976). The fauna of the polluted river Tees estuary. *Estuarine Coastal Mar. Sci.*, **4**, 653-676.

Gray, J. S. (1979). Pollution-induced changes in populations. *Philos. Trans. R. Soc. Lond. B. Biol. Sci.*, **286**, 545-561.

Gray, J. S. (1980). The measurement of effects of pollutants on benthic communities. In McIntyre, A.D. and Pearce, J. B. (Eds.), *Biological Effects of Marine Pollution and the Problems of Monitoring*, Rapp. P.-v. Réun. Cons. Int. Explor. Mer, **179**, 188-193.

Gray, J. S. and Mirza, F. B. (1979). A possible method for the detection of pollution-induced disturbance on marine benthic communities. *Mar. Pollut. Bull.*, **10**, 142-146.

Gray, J. S., Boesch, D., Heip, C., Jones, A. M., Lassig, J., Vanderhorst, R. and Wolfe, D. (1980). The role of ecology in marine pollution monitoring. In McIntyre, A. D. and Pearce, J. B. (Eds.), *Biological Effects of Marine Pollution and the Problems of Monitoring*, Rapp. P.-v. Réun. Cons. Int. Explor. Mer, **179**, 237-252.

Green, R. H. (1979). *Sampling Design and Statistical Methods for Environmental Biologists*, John Wiley and Sons, New York, 257 pages.

Green, R. H. and Vascotto, G. L. (1978). A method for the analysis of environmental factors controlling patterns of species composition in aquatic communities. *Water Res.*, **12**, 583-590.

Grimwood, P. and Webb, G. (1977). Can nuclear wastes be buried at sea? *New Sci.*, **73**, 709-711.

Grollé, T. and Kuiper, J. (1980). Development of marine periphyton under mercury stress in a controlled ecosystem experiment. *Bull. Environ. Contam. Toxicol.*, **24**, 858-865.

Grzywacz, A. and Ważny, J. (1973). The impact of industrial air pollutants on the occurrence of several important pathogenic fungi of forest trees in Poland. *Eur. J. For. Pathol.*, **3**, 129-141.

Guderian, R. and Kueppers, K. (1980). Response of plant communities to air pollution. In Proceedings of Symposium on *Effects of Air Pollutants on Mediterranean and Temperate Forest Ecosystems*, United States Department of Agriculture, Forest Service, Pacific Southwest Forest and Range Experimental Station, General Technical Report PSW-43, Berkeley, California, pp. 187-199.

Haedrich, R. L. (1975). Diversity and overlap as measures of environmental quality. *Water Res.*, **9**, 945-952.

Haegele, M. A. and Hudson, R. H. (1977). Reduction of courtship behaviour induced by DDE in male ringed turtle doves. *Wilson Bull.*, **89**, 593-601.

Hågvar, S. (1980). Effects of acid precipitation on soil and forest. 7. Soil animals. In Drabløs, D. and Tollan, A. (Eds.), *Ecological Impact of Acid Precipitation*— Proceedings of an international conference, Sandefjord, Norway, The SNSF project, pp. 202-203.

Hagvar, S. and Amundsen, T. (1981). Effects of liming and artificial acid rain on the mite (Acari) fauna in coniferous forest. *Oikos*, **37**, 7-20.

Haines, T. A. (1981). Acidic precipitation and its consequences for aquatic ecosystems: A review. *Trans. Am. Fish. Soc.* **110**, 669-707.

Hakanson, L. (1980). An ecological risk index for aquatic pollution control. A sedimentological approach. *Water Res.*, **14**, 975-1001.

Hales, D. C. 1962. Stream bottom sampling as a research tool. *Utah Academy Proc.*, **39**, 84-91.

Halfon, E. (1976). Relative stability of ecosystem linear models. *Ecol. Model.*, **2**, 279-296.

Hall, R. J. and Likens, G. E. (1980). Ecological effects of experimental acidification of stream ecosystems. In Drabløs, D. and Tollan, A. (Eds.), *Ecological Impact of Acid*

Precipitation—Proceedings of an international conference, Sandefjord, Norway, The SNSF project, pp. 375–376.
Hall, R. J., Likens, G. E., Fiance, S. B. and Hendrey, G. R. (1980). Experimental acidification of a stream in the Hubbard Brook Experimental Forest, New Hampshire. *Ecology*, **61**, 976–989.
Hällgren, J. E. and Huss, K. (1975). Effects of SO_2 on photosynthesis and nitrogen fixation. *Physiol. Plant.*, **34**, 171–176.
Hammons, A. S. (Ed.) (1980). *Methods for Ecological Toxicology. A Critical Review of Laboratory Multi-species Tests*, U.S. Environmental Protection Agency, EPA 560/11-80-026, Oak Ridge National Laboratories, Oak Ridge, Tennessee.
Hampson, G. R. and Moul, E. T. (1978). No. 2 fuel oil spill in Bourne, Massachusetts: immediate assessment of the effects on marine invertebrates and a 3-year study of growth and recovery of a salt marsh. *J. Fish. Res. Board Can.*, **35**, 731–744.
Hanna, B. M., Hellebust, J. A. and Hutchinson, T. C. (1975). Field studies on the phytotoxicity of crude oil to subarctic aquatic vegetation. *Verh. Internat. Verein. Limnol.*, **19**, 2165–2171.
Harrison, H. L., Loucks, O. L., Mitchell, J. W., Parkhurst, D. F., Tracy, C. R., Watts, D. G. and Yannacone, V. J. Jr. (1970). Systems studies of DDT transport. *Science*, **170**, 503–508.
Harrison, W. G., Eppley, R. W. and Renger, E. H. (1977). Phytoplankton nitrogen metabolism, nitrogen budgets and observations on copper toxicity: controlled ecosystem pollution experiment. *Bull. Mar. Sci.*, **27**, 44–57.
Harriss, R. C. (1976). *Suggestions for the Development of a Hazard Evaluation Procedure for Potentially Toxic Chemicals*, Report Number 3, MARC (Monitoring and Assessment Research Centre), Chelsea College, University of London, 18 pages.
Harvey, H. H. (1980). Widespread and diverse changes in biota of North American lakes and rivers coincident with acidification. In Drabløs, D. and Tollan, A. (Eds.), *Ecological Impact of Acid Precipitation*—Proceedings of an international conference, Sandefjord, Norway, The SNSF project, pp. 93–98.
Harvey, H. H. and Lee, C. (1980). *Fishes of the La Cloche Mountain Lakes of Ontario, 1965–1980*. Report to the Ontario Ministry of Natural Resources, Canada, 89 pages.
Hatakeyama, S. and Yasuno, M. (1981a). Effects of cadmium on the periodicity of parturition and brood size of *Moina macrocopa* (Cladocera). *Environ. Pollut. Ser. A. Ecol. Biol.*, **26**, 111–120.
Hatakeyama, S. and Yasuno, M. (1981b). The effects of cadmium-accumulated chlorella on the reproduction of *Moina macrocopa* (Cladocera). *Ecotoxicol. Environ. Safety*, **5**, 341–350.
Hawksworth, D. L. and Rose, F. (1970). Qualitative scales for estimating sulphur dioxide air pollution in England and Wales using epiphytic lichens. *Nature*, **227**: 145–148.
Heagle, A. S. (1973). Interactions between air pollutants and plant parasites. *Annu. Rev. Phytopathol.*, **11**, 363–388.
Heal, O. W. and MacLean, S. F. Jr. (1975). Comparative productivity in ecosystems—secondary productivity. In van Dobben, W. H. and Lowe-McConnell, R. H. (Eds.), *Unifying Concepts in Ecology*, Dr. Junk B. V. Publishers, The Hague, pp. 89–108.
Heath, R. L. (1975). Ozone. In Mudd, J. B. and Kozlowski, T. T. (Eds.), *Response of Plants to Air Pollution*, Academic Press, New York, London, San Francisco, pp. 23–55.
Heath, R. T. (1979). Holistic study of an aquatic microcosm: theoretical and practical implications. *Int. J. Environ. Stud.*, **13**, 87–93.
Heath, R. G., Spann, J. W. and Kreitzer, J. F. (1969). Marked DDE impairment of mallard reproduction in controlled studies. *Nature*, **224**, 47–48.

Heinz, G. H. (1976). Behavior of mallard ducklings from parents fed 3 ppm DDE. *Bull. Environ. Contam. Toxicol.*, **16**, 640–645.

Heinz, G. H. (1979). Methylmercury: Reproductive and behavioral effects on three generations of mallard ducks. *J. Wildl. Manage.*, **43**, 394–401.

Heip, C. (1974). A new index measuring evenness. *J. Mar. Biol. Ass. U.K.*, **54**, 555–557.

Heip, C. (1980). Meiobenthos as a tool in the assessment of marine environmental quality. In McIntyre, A. D. and Pearce, J. B. (Eds.), *Biological Effects of Marine Pollution and the Problems of Monitoring*, Rapp. P.-v. Réun. Cons. Int. Explor. Mer, **179**, 182–187.

Hellawell, J. M. (1977). Change in natural and managed ecosystems: detection, measurement and assessment. *Proc. R. Soc. Lond. B. Biol. Sci.* **197**, 31–56.

Hellawell, J. M. (1978). *Biological Surveillance of Rivers: A Biological Monitoring Handbook*, Water Research Centre, Medmenham, Stevenage, England, 331 pages.

Helle, E., Olsson, M. and Jensen, S. (1976a). DDT and PCB levels and reproduction in ringed seal from the Bothnian Bay. *Ambio*, **5**, 188–189.

Helle, E., Olsson, M. and Jensen, S. (1976b). PCB levels correlated with pathological changes in seal uteri. *Ambio* **5**, 261–263.

Hendrey, G. R. and Vertucci, F. A. (1980). Benthic plant communities in acidic Lake Colden, New York. In Drabløs, D. and Tollan, A. (Eds.), *Ecological Impact of Acid Precipitation*—Proceedings of an international conference, Sandefjord, Norway, The SNSF project, pp. 314–315.

Hendrey, G. R., Baalsrud, K., Traaen, T. S., Laake, M. and Raddum, G. (1976). Acid precipitation: Some hydrobiological changes. *Ambio*, **5**, 224–227.

Henrikson, L., Oscarson, H. G. and Stenson, J. A. E. (1980). Does the change of predator system contribute to the biotic development of acidified lakes? In Drabløs, D. and Tollan, A. (Eds.), *Ecological Impact of Acid Precipitation*—Proceedings of an international conference, Sandefjord, Norway, The SNSF project, p. 316.

Heral, M., Berthome, J. P., Polanco Torres, E., Alzieu, C., Deslous-Paoli, J. M., Razet, D. and Garnier, J. (1981). Growth anomalies of the shell of *Crassostrea gigas* in the Marennes-Oléron. Account of three years observation. Conseil International pour l'Exploration de la Mer, Comité des mollusques et crustacés, France (mimeo, in French).

Herricks, E. E. and Cairns, J. Jr. (1982). Biological monitoring part III—receiving system methodology based on community structure. *Water Res.*, **16**, 141–153.

Hershner, C. and Lake, J. (1980). Effects of chronic oil pollution on a salt-marsh grass community. *Mar. Biol.*, **56**, 163–173.

Hetrick, F. M., Knittel, M. D. and Fryer, J. L. (1979). Increased susceptibility of rainbow trout to infectious hematopoietic necrosis virus after exposure to copper. *Appl. Environ. Microbiol.*, **37**, 198–201.

Hill, M. O. (1973). Diversity and evenness: a unifying notion and its consequences. *Ecology*, **54**, 427–432.

Hodson, R. E., Azam, F. and Lee, R. F. (1977). Effects of four oils on marine bacteria populations: controlled ecosystem pollution experiment. *Bull. Mar. Sci.*, **27**, 119–126.

Hodson, P. V., Blunt, B. R. and Spry, D. J. (1978). pH-induced changes in blood lead of lead-exposed rainbow trout (*Salmo gairdneri*). *J. Fish. Res. Board Can.*, **35**, 437–445.

Holcombe, G. W., Benoit, D. A., Leonard, E. N. and McKim, J. M. (1976). Long-term effects of lead exposure on three generations of brook trout (*Salvelinus fontinalis*). *J. Fish. Res. Board Can.*, **33**, 1731–1741.

Holdgate, M. W. (1979). *A Perspective of Environmental Pollution*, Cambridge University Press, Cambridge, London, New York, Melbourne, 278 pages.

Holdgate, M. W. and White, G. F. (Eds.) (1977). *Environmental Issues*, SCOPE 10, John Wiley and Sons, Chichester, New York, Brisbane, Toronto, 224 pages.

Holling, C. S. 1973. Resilience and stability of ecological systems. *Annu. Rev. Ecol. Syst.*, **4**, 1–24.

Horne, A. J. and Goldman, C. R. (1974). Supression of nitrogen fixation by blue-green algae in a eutrophic lake with trace additions of copper. *Science*, **183**, 409–411.

Houba, C. and Remacle, J. (1980). Composition of the saprophytic bacterial communities in freshwater systems contaminated by heavy metals. *Microb. Ecol.*, **6**, 55–69.

Houston, D. B. and Dochinger, L. S. (1977). Effects of ambient air pollution on cone, seed, and pollen characteristics in eatern white and red pines. *Environ. Pollut.*, **12**, 1–5.

Hovland, J. and Abrahamsen, G. (1976). Acidification experiments in conifer forest. 1. Studies on decomposition of cellulose and wood material. SNSF Project, Oslo, Norway.

Hovland, J. and Ishac, Y. Z. (1975). Effects of simulated acid precipitation and liming on nitrification in forest soil. SNSF Project IR/14, Oslo, Norway, 15 pages.

Hughes, D. N., Boyer, M. G., Papst, M. H., Fowle, C. D., Rees, G. A. V. and Baulu, P. (1980a). Persistence of three organophosphorous insecticides in artificial ponds and some biological implications. *Arch. Environ. Contam. Toxicol.* **9**, 269–279.

Hughes, M. K., Lepp, N. W. and Phipps, D. A. (1980b). Aerial heavy metal pollution and terrestrial ecosystems. In Macfadyen, A. (Ed.), *Advances in Ecological Research*, vol. 11, Academic Press, London, New York, Toronto, Sydney, San Francisco, pp. 218–307.

Huisingh, D. (1974). Heavy metals: implications for agriculture. *Annu. Rev. Phytopathol.*, **12**, 375–388.

Hultberg, H. (1977). Thermally stratified acid water in late winter—a key factor inducing self-accelerating processes which increase acidification. *Water Air Soil Pollut.*, **7**, 279–294.

Hummon, W. D., Evans, W. A., Hummon, M. R., Doherty, F. G., Wainberg, R. H. and Stanley, W. S. (1978). Meiofaunal abundance in sandbars of acid mine polluted, reclaimed, and unpolluted streams in southeastern Ohio. In Thorp, J. H. and Gibbons, J. W. (Eds.), *Energy and Environmental Stress in Aquatic Systems*, US Department of Energy, CONF-771114, NTIS, US Department of Commerce, Springfield, Virginia, pp. 188–203.

Hunding, C. and Lange, R. (1978). Ecotoxicology of aquatic plant communities. In Butler, G. C. (Ed.), *Principles of Ecotoxicology*, SCOPE 12, John Wiley and Sons, Chichester, New York, Brisbane, Toronto, pp. 239–255.

Hurlbert, S. H. (1971). The nonconcept of species diversity: a critique and alternative parameters. *Ecology*, **52**, 577–586.

Hurlbert, S. H. 1975. Secondary effects of pesticides on aquatic ecosystems. *Residue Rev.*, **57**, 81–148.

Hurlbert, S. H., Mulla, M. S. and Willson, H. R. (1972). Effects of an organophosphorous insecticide on the phytoplankton, zooplankton and insect populations of fresh-water ponds. *Ecol. Monogr.*, **42**, 269–299.

Hutchinson, G. E. (1944). Limnological studies in Connecticut: VII. A critical examination of the supposed relationship between phytoplankton periodicity and chemical changes in lake water. *Ecology*, **25**, 3–26.

Hutchinson, G. E. (1973). Eutrophication—the scientific background of a contemporary practical problem. *Am. Sci.*, **61**, 269–279.

Hutchinson, T. C. (1980). Effects of acid leaching and cation loss from soil. In Hutchinson, T. C. and Havas, M. (Eds.), *Effects of Acid Precipitation on Terrestrial Ecosystems*, Plenum Press, New York and London, pp. 481–497.

Hutchinson, T. C. and Havas, M. (Eds.) (1980). *Effects of Acid Rain on Terrestrial Ecosystems*, Proceedings of a Conference held in Toronto May 21–27, 1978, NATO Conference Series 1:4, Plenum Press, New York and London, 654 pages.

Hutchinson, T. C. and Whitby, L. M. (1974). Heavy metal pollution in the Sudbury mining and smelting region of Canada. I. Soil and vegetation contamination by nickel, copper and other metals. *Environ. Conserv.*, **1**, 123–132.

Hutchinson, T. C. and Whitby, L. M. (1977). The effects of acid rainfall and heavy metal particulates on a boreal forest ecosystem near the Sudbury smelting region of Canada. *Water Air Soil Pollut.*, **7**, 421–438.

Hutchinson, T. C., Fedorenko, A., Fitchko, J., Kuja, A., Van Loon, J. and Lichwa, J. (1975). Movement and compartmentation of nickel and copper in an aquatic ecosystem. In Nriagu, J. O. (Ed.), *Environmental Biogeochemistry*, Vol. 2, *Metals Transfer and Ecological Mass Balance*, Proc. 2nd Intern. Symp. Environ. Biogeochem., Ann Arbor Science Publishers, Ann Arbor, Michigan, pp. 565–585.

Hutter, K. J. and Oldiges, H. (1980). Alterations of proliferating microorganisms by flow cytometric measurements after heavy metal intoxication. *Ecotoxicol. Environ. Safety.*, **4**, 57–76.

Hutton, M. (1980). Metal contamination of feral pigeons *Columba livia* from the London area: Part 2—biological effects of lead exposure. *Environ. Pollut. Ser. A. Ecol. Biol.*, **22**, 281–293.

Hutton, M. (1981). Accumulation of heavy metals and selenium in three seabird species from the United Kingdom. *Environ. Pollut. Ser. A. Ecol. Biol*, **26**, 129–145.

Inman, J. C. and Parker, G. R. (1978). Decomposition and heavy metal dynamics of forest litter in northwestern Indiana. *Environ. Pollut.*, **17**, 39–51.

IRPTC (International Register of Potentially Toxic Chemicals) (1978). *Data Profiles for Chemicals for the Evaluation of their Hazards to the Environment of the Mediterranean Sea*, United Nations Environment, Program, Geneva.

Irving, P. A. and Miller, J. E. (1980). Response of field-grown soybeans to acid precipitation alone and in combination with sulfur dioxide. In Drablφs D. and Tollan, A. (Eds.), *Ecological Impact of Acid Precipitation*—Proceedings of an international conference, Sandefjord, Norway, The SNSF project, pp. 170–171.

Jaccard, P. (1912). The distribution of flora in the alpine zone. *New Phytol.*, **11**, 37–50.

Jackim, E. (1973). Influence of lead and other metals on fish δ-aminolevulinate dehydrase activity. *J. Fish. Res. Board Can.*, **30**, 560–562.

Jackson, D. R. and Watson, A. P. (1977). Disruption of nutrient pools and transport of heavy metals in a forested watershed near a lead smelter. *J. Environ. Qual.*, **6**, 331–338.

Jackson, D. R., Selvidge, W. J. and Ausmus, B. S. (1978). Behavior of heavy metals in forest microcosms. II. Effects on nutrient cycling processes. *Water Air Soil Pollut.*, **10**, 13–18.

Jackson, D. R., Ausmus, B. S. and Levin, M. (1979). Effects of arsenic on the nutrient dynamics of grassland microcosms and field plots. *Water Air Soil Pollut.*, **11**, 13–21.

Jacobs, J. (1975). Diversity, stability and maturity in ecosystems influenced by human activities. In van Dobben, W. H. and Lowe-McConnell, R. H. (Eds.), *Unifying Concepts in Ecology*. First International Congress of Ecology, Dr. W. Junk B. V. Publishers, The Hague, pp. 187–207.

Jacobs, R. P. W. M. 1980. Effects of the 'Amoco Cadiz' oil spill on the seagrass community at Roscoff with special reference to the benthic infauna. *Mar. Ecol. Prog. Ser.*, **2**, 207–212.

Jeffree, R. A. and Williams, N. J. (1980). Mining pollution and the diet of the purple-striped gudgeon *Mogurnda mogurnda* Richardson (Eleotridae) in the Finniss River, Northern Territory, Australia. *Ecol. Monogr.*, **5**, 457–485.

Jensen, A. (1980). The use of phytoplankton cage culture for *in situ* monitoring of marine pollution. In McIntyre, A. D. and Pearce, J. B. (Eds.), *Biological Effects of Marine Pollution and the Problems of Monitoring*, Rapp. P.-v. Réun. Cons. Int. Explor. Mer, **179**, 306–309.

Jensen, A. (1983). Methods for assessing the effects of chemicals on algal reproduction. In Vouk, V. B. and Sheehan, P. J. (Eds.), *Methods for Assessing the Effects of Chemicals on*

Reproductive Functions, SCOPE 20, John Wiley and Sons, Chichester, New York, Brisbane, Toronto, pp. 515–523.

Jensen, S. and Jernelov, A. (1969). Biological methylation of mercury in aquatic organisms. *Nature*, **223**, 753–754.

Jensen, V. (1977). Effects of lead on biodegradation of hydrocarbons in soil. *Oikos*, **28**, 220–224.

Jernelov, A. and Rosenberg, R. (1976). Stress tolerance of ecosystems. *Environ. Conserv.*, **3**, 43–46.

Johannes, R. E. (1975). Pollution and degradation of coral reef communities. In Ferguson Wood, E. J. and Johannes, R. E. (Eds.), *Tropical Marine Pollution*, Elsevier Scientific Publishing Company, Amsterdam, Oxford, New York, pp. 13–51.

Johansen, C. A. (1977). Pesticides and pollinators. *Annu. Rev. Entomol.*, **22**, 177–192.

Johansson, S., Larsson, U. and Boehm, P. (1980). The Tsesis oil spill—Impact on the pelagic ecosystem. *Mar. Pollut. Bull.*, **11**, 284–293.

Johnson, D. W. and Shriner, D. S. (1980). Effects of acidification on soil microbes, including decomposition processes. In Hutchinson, T. C. and Havas, M. (Eds.), *Effects of Acid Precipitation on Terrestrial Ecosystems*, Plenum Press, New York and London, pp. 601–608.

Johnson, N. M., Driscoll, C. T., Eaton, J. S., Likens, G. E. and McDowell, W. H. (1981). Acid rain, dissolved aluminum and chemical weathering at the Hubbard Brook Experimental Forest, New Hampshire. *Geochim. Cosmochim. Acta*, **45**, 1421–1438.

Jones, A. M. (1979). Structure and growth of a high-level population of Cerastroderma edule (Lamellibranchiata). *J. Mar. Biol. Ass. U.K.*, **59**, 277–287.

Jones, C. W. (1978). The ranking of hazardous materials by means of hazard indices. *J. Hazard. Mater.*, **2**, 363–389.

Jones, M. B. (1975). Synergistic effects of salinity, temperature and heavy metals on mortality and osmoregulation in marine and estuarine isopods (Crustacea). *Mar. Biol.*, **30**, 13–20.

Jordan, M. J. (1975). Effects of zinc smelter emissions and fire on a chestnut-oak woodland. *Ecology*, **56**, 78–91.

Jordan, M. J. and Lechevalier, M. P. (1975). Effects of zinc-smelter emissions on forest soil microflora. *Can. J. Microbiol.*, **21**, 1855–1865.

Jordan, M., and Likens, G.E. (1975). An organic carbon budget for an oligotrophic lake in New Hampshire, USA. *Verh. Internat. Verein. Limnol.*, **19**, 994–1003.

Jørgensen, S. E. and Mejer, H. (1977). Ecological buffer capacity. *Ecol. Model.*, **3**, 39–61.

Jørgensen, S. E., and Mejer, H. (1979). A holistic approach to ecological modelling. *Ecol. Model.*, **7**, 169–189.

JFRBC (1978). Proceedings of the Symposium on Recovery Potential of Oiled Marine Northern Environments. *J. Fish. Res. Board Can.*, **35**, 499–795.

Kaiser, K. L. E. (1977). Organic contaminant residues in fishes from Nipigon Bay, Lake Superior. *J. Fish. Res. Board Can.*, **34**, 850–855.

Kania, H. J. and O'Hara, J. O. (1974). Behavioral alterations in a simple predator–prey system due to sublethal exposure to mercury. *Trans. Am. Fish. Soc.*, **103**, 134–136.

Keefe, T.J. and Bergersen, E. P. (1977). A simple diversity index based on the theory of runs. *Water Res.*, **11**, 689–691.

Keller, Th. (1976). Wintertime atmospheric pollutants—do they affect the performance of deciduous trees in the ensuing growing season? *Environ. Pollut.*, **16**, 243–247.

Kelso, J. R. M. (1977). Density, distribution and movement of Nipigon Bay fishes in relation to a pulp and paper mill effluent. *J. Fish. Res. Board Can.*, **34**, 879–885.

Kercher, J. R., Axelrod, M. C. and Bingham, G. E. (1980). Forecasting effects of SO_2 pollution on growth and succession in a western conifer forest. In Proceedings of

Symposium on *Effects of Air Pollutants on Mediterranean and Temperate Forest Ecosystems*, United States Department of Agriculture, Forest Service, Pacific Southwest Forest and Range Experimental Station, General Technical Report PSW-43, Berkeley, California, pp. 200–202.

Kevan, P. G. (1975). Forest application of the insecticide fenitrothion and its effect on wild bee pollinators (Hymenoptera: Apoidea) of low bush blueberries (*Vaccinium* Spp.) in southern New Brunswick, Canada. *Biol. Conserv.*, **7**, 301–309.

Killham, K. and Wainwright, M. (1981). Deciduous leaf litter and cellulose decomposition in soil exposed to heavy atmospheric pollution. *Environ. Pollut. Ser. A. Ecol. Biol.*, **26**, 79–85.

Kinako, P. D. S. (1981). Short-term effects of oil pollution on species numbers and productivity of a simple terrestrial ecosystem. *Environ. Pollut. Ser. A. Ecol. Biol.*, **26**, 87–91.

Kitchell, J. F., O'Neill, R. V., Webb, D., Gallepp, G. W., Bartell, S. M., Koonce, J. F. and Ausmus, B. S. (1979). Consumer regulation of nutrient cycling. *Bioscience*, **29**, 28–34.

Knabe, W. (1976). Effects of sulfur dioxide on terrestrial vegetation. *Ambio*, **5**, 213–218.

Koenig, C. C., Livingston, R. J. and Cripe, C. R. (1976). Blue crab mortality: interaction of temperature and DDT residues. *Arch. Environ. Contam. Toxicol.*, **4**, 119–128.

Koster, A. S. J. and Van den Biggelaar, J. A. M. (1980). Abnormal development of *Dentalium* due to the Amoco Cadiz oil spill. *Mar. Pollut. Bull.*, **11**, 166–169.

Kothé, P. (1962). The 'number of missing species', a simple, good criterion and its application to the biological examination of outfalls. *Dt. Gewässerkundl. Mitt*, **6**, 60–65 (in German).

Krause, G. H. M. and Kaiser, H. (1977). Plant response to heavy metals and sulfur dioxide. *Environ. Pollut.*, **12**, 63–71.

Krebs, C. T. and Burns, K. A. (1977). Long-term effects of an oil spill on populations of the salt-marsh crab *Uca pugnax*. *Science*, **197**, 484–487.

Krenkel, P. A. (1979). Problems in the establishment of water quality criteria. *J. Water Pollut. Control Fed.*, **51**, 2168–2188.

Krieger, R. I., Gee, S. J. and Lim, L. O. (1981). Marine bivalves particularly mussels, *Mytilus*, sp., for assessment of environmental quality. *Ecotoxicol. Environ. Safety*, **5**, 72–86.

Kynard, B. (1974). Avoidance behavior of insecticide susceptible and resistant populations of mosquito fish to four insecticides. *Trans. Am. Fish. Soc.*, **103**, 557–561.

La Du, B. N., Mandel, H. G. and Way, E. L. (1971). *Fundamentals of Drug Metabolism and Drug Disposition*, Williams and Wilkins, Baltimore.

Lakhani, K. H. and Miller, H. G. (1980). Assessing the contribution of crown leaching to the element content of rainwater beneath trees. In Hutchinson, T. C. and Havas, M. (Eds.), *Effects of Acid Precipitation on Terrestrial Ecosystems*, Plenum Press, New York and London, pp. 161–172.

Landa, V., Bennettová, B., Gelbič, I., Matolín, S. and Soldán, T.(1983). Methods for the assessment of the effects of chemicals on the reproductive function of insects. In Vouk, V. B. and Sheehan P. J. (Eds.), *Methods for Assessing the Effects of Chemicals on Reproductive Functions*, SCOPE 20, John Wiley and Sons, Chichester, New York, Brisbane, Toronto, pp. 415–438.

La Salle, J. and Lefschetz, S. (1961). *Stability by Liapunov's Direct Method With Applications*, Academic Press, New York, London, 134 pages.

Last, F. T., Likens, G. E., Ulrich, B. and Wallφe, L. (1980). Acid precipitation—progress and problems. Conference summary. In Drablφs, D. and Tollan, A. (Eds.), *Ecological Impact of Acid Precipitation*—Proceedings of an international conference, Sandefjord, Norway, The SNSF project, pp. 10–12.

Lee, R., Davies, J. M., Freeman, H. C., Ivanovici, A., Moore, M. N., Stegeman, J. and Uthe, J. F. (1980). Biochemical techniques for monitoring biological effects of pollution in the sea. In McIntyre, A. D. and Pearce, J. B. (Eds.), *Biological Effects of Marine Pollution and the Problems of Monitoring*, Rapp. P.-v. Réun. Cons. Int. Explor. Mer, **179**, 48–55.

Legge, A. H. (1980). Primary productivity, sulfur dioxide, and the forest ecosystem: an overview of a case study. In Proceedings of Symposium on *Effects of Air Pollutants on Mediterranean and Temperate Forest Ecosystems*, United States Department of Agriculture, Forest Service, Pacific Southwest Forest and Range Experimental Station, General Technical Report PSW-43, Berkeley, California, pp. 51–62.

Legge, A. H. (1982). Sulphur gas emissions in the boreal forest: the West Whitecourt case study—II: research plan and background. *Water Air Soil Pollut.*, **17**, 379–398.

Legge, A. H., Jaques, D. R., Harvey, G. W., Krouse, H. R., Brown, H. M., Rhodes, E. C., Nosal, M., Schellhase, H. U., Mayo, J., Hartgerink, A. P., Lester, P. F., Amundson, R. G. and Walker, R. B. (1981). Sulphur gas emissions in the boreal forest: the West Whitecourt case study—I: executive summary. *Water Air Soil Pollut.*, **15**, 77–85.

Leivestad, H., Hendrey, G., Muniz, I. P. and Snekvik, E. (1976). Effects of acid precipitation on freshwater organisms. In Braekke, F. H. (Ed.), *Impact of Acid Precipitation on Forest and Freshwater Ecosystems in Norway*, Research Report 6/76, The SNSF project, pp. 87–111.

Lepp, N. W. (Ed.) (1981). *Effect of Heavy Metal Pollution on Plants*, Vol. 2, *Metals in the Environment*, Applied Science Publishers, London and New Jersey.

Leppäkoski, E. J. (1975). Assessment of degree of pollution on the basis of macrozoobenthos in marine and brackish-water environments. *Acta Acad. Abo. Ser. B*, **35**, 1–90.

Leppäkoski, E. J. and Lindström, L. S. (1978). Recovery of benthic macrofauna from chronic pollution in the sea area off a refinery plant, southwest Finland. *J. Fish. Res. Board Can.*, **35**, 766–775.

Letterman, R. D. and Mitsch, W. J. (1978). Impact of mine drainage on a mountain stream in Pennsylvania. *Environ. Pollut.*, **17**, 53–73.

Levins, S. A. (Ed.) (1975). *Ecosystem Analysis and Prediction*, SIAM Press, Philadelphia, PA, 337 pages.

Levy, E. M. (1980). Oil pollution and seabirds: Atlantic Canada 1976–77 and some implications for northern environments. *Mar. Pollut. Bull.*, **11**, 51–56.

Lewis, F. G. III and Livingston, R. J. (1977). Avoidance of bleached kraft pulpmill effluent by pinfish (*Lagodon rhomboides*) and gulf killifish (*Fundulus grandis*). *J. Fish. Res. Board Can.*, **34**, 568–570.

Lewis, J. R. (1976). Long-term ecological surveillance: practical realities in the rocky littoral. *Oceanogr. Mar. Biol. Annu. Rev.*, **14**, 371–390.

Lewis, J.R. (1978). The implications of community structure for benthic monitoring studies. *Mar. Pollut. Bull.*, **9**, 64–67.

Liang, C. N. and Tabatabai, M. A. (1978). Effects of trace elements on nitrification in soils. *J. Environ. Qual.*, **7**, 291–293.

Lighthart, B. (1980). Effects of certain cadmium species on pure and litter populations of microorganisms. *Antonie van Leeuwenhoek*, **46**, 161–167.

Likens, G. E. (1975). Nutrient flux and cycling in freshwater ecosystems. In Howell, F. G., Gentry, J. B. and Smith, M. H. (Eds.), *Mineral Cycling in Southeastern Ecosystems*, U.S. Energy Research and Development Administration, CONF-740513, U.S. Department of Commerce, Springfield, Virginia, pp. 314–348.

Likens, G. E, and Bormann, F. H. (1975). An experimental approach in New England landscapes. In Hasler, A. D. (Ed.), *Coupling of Land and Water Systems*, Ecological Studies, vol. 10, Springer-Verlag, New York, Heidelberg, Berlin, pp. 7–29.

Likens, G. E., Bormann, F. H., Johnson, N. M., Fisher, D. W. and Pierce, R. S. (1970). Effects of forest cutting and herbicide treatment on nutrient budgets in the Hubbard Brook watershed-ecosystem. *Ecol. Monogr.*, **40**, 23–47.

Likens, G. E., Bormann, F. H., Pierce, R. S. and Reiners, W. A. (1978). Recovery of a deforested ecosystem. *Science*, **199**, 492–496.

Lincer, J. L. (1975). DDE-induced eggshell-thinning in the American kestrel: a comparison of the field situation and laboratory results. *J. Appl. Ecol.*, **12**, 781–793.

Lindén, O., Elmgren, R. and Boehm, P. (1979). The Tsesis oil spill: its impact on the coastal ecosystem of the Baltic Sea. *Ambio*, **8**, 244–253.

Livingston, R. J. (1977). Review of current literature concerning the acute and chronic effects of pesticides on aquatic organisms. *CRC Crit. Rev. Environ. Control*, **7**, 325–351.

Livingston, R. J. (1979). Multiple factor interactions and stress in coastal systems: a review of experimental approaches and field implications. In Vernberg, W. B., Calabrese, A., Thurberg, F. P. and Vernberg, F. J. (Eds.), *Marine Pollution: Functional Responses*, Academic Press, New York, San Francisco, London, pp. 389–414.

Lloyd, M. and Ghelardi, R. J. (1964). A table for calculating the equitability component of species diversity. *J. Anim. Ecol.*, **33**, 217–225.

Lockwood, A. P. M. (Ed.) (1976). *Effects of Pollutants on Aquatic Organisms*, Cambridge University Press, Cambridge, London, New York, Melbourne, 180 pages.

Lohm, U. (1980). Effects of experimental acidification on soil organism populations and decomposition. In Drabløs, D. and Tollan, A. (Eds.), *Ecological Impact of Acid Precipitation*—Proceedings of an international conference, Sandefjord, Norway, The SNSF project, pp. 178–179.

Lohm. U., and Persson, T. (Eds.) (1977). *Soil Organisms as Components of Ecosystems*, Proc. 6th Int. Coll. Soil Zool., Ecological Bulletin (Stockholm), Vol. **25**, 614 pages.

Longhurst, A., Colebrook, M. Gulland, J., Le Brasseur, R., Lorenzen, C. and Smith, P. (1972). The instability of ocean populations. *New Sci.*, **54**, 500–502.

Longwell, A. C. (1976). *Chromosome Mutagenesis in Developing Mackerel Eggs Sampled from the New York Bight*, NOAA Technical Memorandum ERL-MESA-7, U.S. Dept. Commerce, NOAA, ERL.

Longwell, A. C. (1977). A genetic look at fish eggs and oil. *Oceanus*, **20**, 46–58.

Longwell, A. C. and Hughes, J. B. (1980). Cytologic, cytogenic, and developmental state of Atlantic mackerel eggs from sea surface waters of the New York Bight, and prospects for biological effects monitoring with ichthyoplankton. In McIntyre, A. D. and Pearce, J. B. (Eds.), *Biological Effects of Marine Pollution and the Problems of Monitoring*, Rapp. P.-v. Réun. Cons. Int. Explor. Mer, **179**, 275–291.

Lorz, H. W. and McPherson, B. P. (1976). Effects of copper or zinc in fresh water on the adaptation to sea water and ATPase activity, and the effects of copper on migratory disposition of Coho salmon (*Oncorhynchus kisutch*). *J. Fish. Res. Board Can.*, **33**, 2023–2030.

Loya, Y. (1975). Possible effects of water pollution on the community structure of Red Sea corals. *Mar. Biol.*, **29**, 177–185.

Loya, Y. (1976). Recolonization of Red Sea corals affected by natural catastrophies and man-made perturbations. *Ecology*, **57**, 278–289.

Loya, Y. and Rinkevich, B. (1980). Effects of oil pollution on coral reef communities. *Mar. Ecol. Prog. Ser.*, **3**, 167–180.

Ludke, J. L., Hill, E. F. and Dieter, M. P. (1975). Cholinesterase (ChE) response and related mortality among birds fed ChE inhibitors. *Arch. Environ. Contam. Toxicol.*, **3**, 1–21.

Lue, K. Y. and de la Cruz, A. A. (1978). Mirex incorporation in the environment: toxicity in two soil macroarthropods and effects on soil community respiration. *Water Air Soil Pollut.*, **9**, 177–191.

Lulman, P. D., Fessenden, R. J. and McKinnon, S. A. (1980). Lichens as air quality monitors. In Proceedings of Symposium on *Effects of Air Pollutants on Mediterranean and Temperate Forest Ecosystems*, United States Department of Agriculture, Forest Service, Pacific Southwest Forest and Range Experimental Station, General Technical Report PSW-43, Berkeley, California, p. 241.

Lundkvist, H. (1977). Effects of artificial acidification on the abundance of Enchytraeidae in a Scots pine forest in northern Sweden. In Lohm, U. and Persson, T. (Eds.), *Soil Organisms as Components of Ecosystems*, Ecological Bulletin (Stockholm), Vol. **25**, pp. 570–573.

Luoma, S. N. (1977). Detection of trace contaminant effects in aquatic ecosystems. *J. Fish. Res. Board Can.*, **34**, 436–439.

MAB (Program on Man and the Biosphere) (1974). *Final Report, Task Force on Criteria and Guidelines for the Choice and Establishment of Biosphere Reserves*, MAB Report No. 22, UNESCO, Paris.

MacMahon, J. A. (1980). Ecosystems over time: Succession and other types of change. In Waring, R. H. (Ed.), *Forests: Fresh Perspectives from Ecosystem Analysis*, Oregon State University Press, Corvallis, Oregon, pp. 27–58.

Majkowski, J., Ridgeway, J. M. and Miller, D. R. (1981). Multiplicative sensitivity analysis and its role in development of simulation models. *Ecol. Model.*, **12**, 191–208.

Maki, A. W. and Johnson, H. E. (1976). Evaluation of a toxicant on the metabolism of model stream communities. *J. Fish. Res. Board Can.*, **33**, 2740–2746.

Malmer, N. (1976). Acid precipitation: chemical changes in the soil. *Ambio*, **5**, 231–234.

Malouf, R. E. and Breeze, W. P. (1978). Intensive culture of the Pacific oyster *Crassostrea gigas* (Thunberg), in heated effluents. *Agricultural Experimental Station Bulletin 627*, ORESU-T-78-003, Oregon State University, Corvallis.

Mann, R. (1979). Some biochemical and physiological aspects of growth and gametogenesis in *Crassostrea gigas* and *Ostrea edulis* grown at sustained elevated temperatures. *J. Mar. Biol. Assoc. U.K.*, **59**, 95–110.

Mansfield, T. A. (Ed.) (1976). *Effects of Air Pollutants on Plants*, Cambridge University Press, Cambridge, London, New York, Melbourne, 209 pages.

Margalef, R. (1951). Species diversity in natural communities. *Publnes, Inst. Biol. Appl. Barcelona*, **6**, 59–72 (in Spanish).

Marshall, J. S. and Mellinger, D. L. (1980). Dynamics of cadmium-stressed plankton communities. *Can. J. Fish. Aquat. Sci.*, **37**, 403–414.

Marshall, P. E. and Furnier, G. R. (1981). Growth responses of *Ailanthus altissama* seedlings to SO_2. *Environ. Pollut. Ser. A. Ecol. Biol.*, **25**, 149–153.

Martin, A. (1983). Assessment of the effects of chemicals on the reproductive functions of reptiles and amphibians. In Vouk, V. B. and Sheehan, P. J. (Eds.), *Methods for Assessing the Effects of Chemicals on Reproductive Functions*, SCOPE 20, John Wiley and Sons, Chichester, New York, Brisbane, Toronto, pp. 405–413.

Martin, J. H., Elliot, P. D., Anderlini, V. C., Girvin, D., Jacobs, S. A., Risebrough, R. W., Delong, R. L. and Gilmartin, N. G. (1976). Mercury-selenium-bromine imbalance in premature parturient California sea lions, *Mar. Biol.*, **35**, 91–104.

Matthews, R. A., Buikema, A. L. Jr., Cairns, J. Jr. and Rodgers, J. H. Jr. (1982). Biological monitoring part IIA—receiving system functional methods, relationships and indices. *Water Res.*, **16**, 129–139.

Mattson, W. J. and Addy, N. D. (1975). Phytophagous insects as regulators of forest primary production. *Science*, **190**, 515–522.

Maugh, T. H. II (1978). Chemicals: how many are there? *Science*, **199**, 162.

May, R. M. (1973). *Stability and Complexity in Model Ecosystems*, Princeton University Press, Princeton, New Jersey, 235 pages.

May, R. M. (1975). Stability in ecosystems: some comments. In van Dobben, W. H. and Lowe-McConnell, R. H. (Eds.), *Unifying Concepts in Ecology*, Dr. W. Junk B. V. Publishers, The Hague, pp. 161–168.

May, R. M. (1977). Thresholds and breakpoints in ecosystems with a multiplicity of stable points. *Nature*, **269**, 471–477.

May, R. M. (1981). A cycling index for ecosystems. *Nature*, **292**, 105.

McCain, B. B., Pierce, K. V., Willings, S. R. and Miller, B. S. (1977). Hepatomas in marine fish from an urban estuary *Bull. Environ. Contam. Toxicol.*, **18**, 1–2.

McCain, B. B., Hodgins, H. O., Gronlund, W. D., Hawkes, J. W., Brown, D. W., Myers, M. S. and Vandermeulen, J. H. (1978). Bioavailability of crude oil from experimentally oiled sediments to English sole (*Parophrys vetulus*) and pathological consequences. *J. Fish. Res. Board Can.*, **35**, 657–664.

McClenahen, J. R. (1978). Community changes in a deciduous forest exposed to air pollution. *Can. J. For. Res.*, **8**, 432–438.

McGreer, E. R. (1979). Sublethal effects of heavy metal contaminated sediments on the bivalve *Macoma balthica* (L.). *Mar. Pollut. Bull.*, **10**, 259–262.

McIntosh, R. P. (1967). An index of diversity and the relation of certain concepts to diversity. *Ecology*, **48**, 392–404.

McIntyre, A. D. and Pearce, J. B. (Eds.) (1980). *Biological Effects of Marine Pollution and the Problems of Monitoring*, Rapp. P.-v. Réun. Cons. Int. Explor. Mer, **179**, 341 pages.

McKim, J. M. (1977). Evaluation of tests with early life cycle stages of fish for predicting long-term toxicity. *J. Fish. Res. Board Can.*, **34**, 1148–1154.

McKim, J. M. and Benoit, D. A. (1971). Effects of long-term exposures to copper on survival, growth and reproduction of brook trout (*Salvelinus fontinalis*) *J. Fish. Res. Board Can.*, **28**, 655–662.

McKim, J. M., Olson, G. F., Holcombe, G. W. and Hunt, E. P. (1976). Long-term effects of methylmercuric chloride on three generations of brook trout (*Salvelinus fontinalis*): toxicity, accumulation, distribution, and elimination. *J. Fish. Res. Board Can.*, **33**, 2726–2739.

McNaughton, S. J. (1968). Structure and function in California grasslands. *Ecology*, **49**, 962–972.

Miles, L. J. and Parker, G. R. (1980). Effects of cadmium and a one-time drought stress on survival, growth, and yield of native plant species. *J. Environ. Qual.*, **9**, 278–282.

Miller, D. R. (1974). Sensitivity analysis and validation of simulation models. *J. Theor. Biol.*, **48**, 345–360.

Miller, D. R. (Ed.) (1977). *Ottawa River Project, Final Report*, National Research Council of Canada, Ottawa, Canada, October 1977, 2 volumes.

Miller, D. R. and Akagi, H. (1979). pH affects mercury distribution, not methylation. *Ecotoxicol. Environ. Safety*, **3**, 36–38.

Miller, D. R. and Buchanan, J. M. (1979). *Atmospheric Transport of Mercury: Exposure Commitment and Uncertainty Calculations*, Report Number 14, MARC (Monitoring and Assessment Research Centre), Chelsea College, University of London, 75 pages.

Miller, D. R., Butler, G. and Bramall, L. (1976). Validation of ecological system models. *J. Environ. Manage.*, **4**, 383–401.

Miller, J. E., Hassett, J. J. and Koeppe, D. E. (1977). Interactions of lead and cadmium on metal uptake and growth of plant species. *J. Environ. Qual.*, **6**, 18–20.

Minshall, G. W. and Minshall, J. N. (1978). Further evidence on the role of chemical factors in determining the distribution of benthic invertebrates in the River Duddon. *Arch. Hydrobiol.*, **83**, 324–355.

Mitchell, R. and Chet, I. (1975). Bacterial attack of corals in polluted sea water. *Microb. Ecol.*, **2**, 227–233.

Mitchell, R., Fogel, S. and Chet, I. (1972). Bacterial chemoreception: an important ecological phenomenon inhibited by hydrocarbons. *Water Res.*, **6**, 1137–1140.

MITRE Corporation (1976). *Preliminary Scoring of Organic Air Pollutants*, U.S. Environmental Protection Agency, EPA-450-77-008a.

Mohanty-Hejmadi, P. and Dutta, S. K. (1981). Effects of some pesticides on the development of the Indian bull frog *Rana tigerina*. *Environ. Pollut. Ser. A. Ecol. Biol.*, **24**, 145–161.

Mood, A. M. (1940). The distribution theory of runs. *Ann. Math. Statist.*, **11**, 367–392.

Moon, T. C. and Lucostic, C. M. (1979). Effects of acid mine drainage on a southwestern Pennsylvania stream. *Water Air Soil Pollut.*, **11**, 377–390.

Moore, J. W. (1979). Diversity and indicator species as measures of water pollution in a subarctic lake. *Hydrobiologia*, **66**, 73–80.

Moore, J. W. (1980). Seasonal and species-dependent variability in the biological impact of mine wastes in an alpine river. *Bull. Environ. Contam. Toxicol.*, **25**, 524–529.

Moore, M. N. (1980). Cytochemical determination of cellular responses to environmental stressors in marine organisms. In McIntyre, A. D., and Pearce, J. B. (Eds.) *Biological Effects of Marine Pollution and the Problems of Monitoring*, Rapp. P.-v. Réun. Cons. Int. Explor. Mer, **179**, 7–15.

Moore, M. N. and Stebbing, A. R. D. (1976). The quantitative cytochemical effects of three metal ions on a lysosomal hydrolase of a hydroid. *J. Mar. Biol. Ass. U.K.*, **56**, 995–1005.

Moraitou-Apostolopoulou, M. and Verriopoulos. G. (1979). Some effects of sublethal concentrations of copper on a marine copepod. *Mar. Pollut. Bull.*, **10**, 88–92.

Moriarty, F. (1978). Terrestrial animals. In Butler, G. C. (Ed.), *Principles of Ecotoxicology*, SCOPE 12, John Wiley and Sons, Chichester, New York, Brisbane, Toronto, pp. 169–186.

Mossberg, P. and Nyberg, P. (1979). Bottom fauna in small acid lakes. *Institute for Freshwater Research, Drottningholm*, Report 58, pp. 77–87.

Mosser, J. L., Fisher, N. S. and Wurster, C. F. (1972). Polychlorinated biphenyls and DDT alter species composition in mixed cultures of algae. *Science*, **176**, 533–535.

Mouw, D., Kalitis, K., Anver, M., Schwartz, J., Constan, A., Hartung, R., Cohen, B. and Ringler, D. (1975). Lead-possible toxicity in urban versus rural rats. *Arch. Environ. Health*, **30**, 276–280.

Mudd, J. B., and Kozlowski, T. T. (Eds.) (1975). *Responses of Plants to Air Pollution*, Academic Press, New York, San Francisco, London, 383 pages.

Müller, P. (1980). Effects of artificial acidification on the growth of periphyton. *Can. J. Fish. Aquat. Sci.*, **37**, 355–363.

Muniz, I. P. and Leivestad, H. (1980). Acidification-effects on freshwater fish. In Drabløs, D., and Tollan, A. (Eds.) *Ecological Impact of Acid Precipitation*. Proceedings of an international conference, Sandefjord, Norway, The SNSF project, pp. 84–92.

Munn, R. E. and Bolin, B. (1971). Global air pollution—meteorological aspects. *Atmos. Environ.*, **5**, 363–402.

Munn, R. E. (1978). Physical and chemical changes in the environment with indirect biological effects. In Butler, G. C. (Ed.) *Principles of Ecotoxicology, SCOPE 12*, John Wiley and Sons, Chichester, New York, Brisbane, Toronto, pp. 295–312.

Myers, A. A., Southgate, T. and Cross, T. F. (1980). Distinguishing the effects of oil pollution from natural cyclical phenomena on the biota of Bantry Bay, Ireland. *Mar. Pollut. Bull.*, **11**, 204–207.

Naiman, R. J. and Sibert, J. (1977). Annual and diel variations in a small marine bay: interpretation of monitoring data. *J. Exp. Mar. Biol. Ecol.*, **26**, 27–40.

NAS (National Academy of Sciences) (1981). *Testing for Effects of Chemicals on*

Ecosystems, A Report by the Committee to Review Methods for Ecotoxicology, Commission of Natural Resources, National Research Council, National Academy Press, Washington, D.C., 98 pages.

NASA (National Aeronautics and Space Administration) (1977). *Chlorofluoromethanes and the Stratosphere*, NASA Report Ref. Publication 1010.

Nash, T. H. (1973). Sensitivity of lichens to sulphur dioxide. *Bryologist*, **76**, 333–339.

Natusch, D. F. S., Wallace, J. R. and Evans, C. A. (1974). Toxic trace elements: preferential concentrations in respirable particles. *Science*, **183**, 202–204.

Needham, P. R. and Usinger, R. L. (1956). Variability in the macrofauna of a single riffle in Prosser Creek, California, as indicated by the Surber Sampler. *Hilgardia*, **24**, 383–409.

Neely, W. G., Bronson, D. R. and Blau, G. E. (1974). The use of the partition coefficient to measure the bioconcentration potential of organic chemicals in fish. *Environ. Sci. Technol.*, **8**, 1113–1115.

Nehring, R. B. (1976). Aquatic insects as biological monitors of heavy metal pollution. *Environ. Contam. Toxicol.*, **19**, 147–154.

Nehring, R. B., Nisson, R. and Minasian, G. (1979). Reliability of aquatic insects versus water samples as measures of lead pollution. *Bull. Environ. Contam. Toxicol.*, **22**, 103–108.

Nesbit, I. C. T. and Sarofim, A. F. (1972). Rates and routes of transport of PCB's in the environment. *Environ. Health Perspect.*, **1**, 21–38.

Newbold, J. D., Elwood, J. W., O'Neill, R. V., and Van Winkle, W. (1981). Measuring nutrient spiralling in streams. *Can. J. Fish. Aquat. Sci.*, **38**, 860–863.

Nicol, J. A. C., Donahue, W. H., Wang, R. T. and Winters, K. (1977). Chemical composition and effects of water extracts of petroleum on eggs of the sand-dollar (*Melitta quinquiesperforata*). *Mar. Biol.*, **40**, 309–316.

Nieboer, E., Richardson, D. H. S., Puckett, K. J. and Tomassini, F. D. (1976). The phytotoxicity of sulphur dioxide in relation to measurable responses of lichens. In Mansfield, T. A. (Ed.), *Effects of Air Pollutants on Plants*, Cambridge University Press, Cambridge, London, New York, Melbourne, pp. 61–85.

Norton, S. A., Hanson, D. W. and Compana, R. J. 1980. The impact of acid precipitation and heavy metals on soils in relation to forest ecosystems. In Proceedings of Symposium on *Effects of Air Pollutants on Mediterranean and Temperate Forest Ecosystems*, United States Department of Agriculture, Forest Service, Pacific Southwest Forest and Range Experimental Station, General Technical Report PSW-43, Berkeley, California, pp. 152–164.

Notini, M. 1978. Long-term effects of an oil spill on *Fucus* macrofauna in a small Baltic bay. *J. Fish. Res. Board Can.*, **35**, 745–753.

Nounou, P. (1980). The oil spill age—Fate and effects of oil in the marine environment. *Ambio*, **9**, 297–302.

NRCC (National Research Council of Canada) (1978). *Phenoxy Herbicides—Their Effects on Environmental Quality with Accompanying Scientific Criteria for 2, 3, 7, 8-Tetrachlorodibenzo-p-dioxin (TCDD)*, NRCC Associate Committee on Scientific Criteria for Environmental Quality, National Research Council, NRCC 16075, Ottawa, Canada, 440 pages.

NRCC (National Research Council of Canada) (1979). *Effects of Mercury in the Canadian Environment*, NRCC Associate Committee on Scientific Criteria for Environmental Quality, National Research Council, NRCC 16739, Ottawa, Canada, 290 pages.

NRCC (National Research Council of Canada) (1981a). *Acidification in the Canadian Aquatic Environment*, NRCC Associate Committee on Scientific Criteria for Environmental Quality, National Research Council, NRCC 18475, Ottawa, Canada, 369 pages.

NRCC (National Research Council of Canada) (1981b). *Pesticide-Pollinator Interactions*, NRCC Associate Committee on Scientific Criteria for Environmental Quality, National Research Council, NRCC 18471, Ottawa, Canada, 190 pages.

Nriagu, J. O. (Ed.) (1979). *The Biogeochemistry of Mercury in the Environment*, Elsevier/North Holland, Amsterdam, 696 pages.

Occhiogrosso, T. J., Waller, W. T. and Lauer, G. J. (1979). Effects of heavy metals on benthic macroinvertebrate densities in Foundry Cove on the Hudson River. *Bull. Environ. Contam. Toxicol.*, 22, 230–237.

Odum, E. P. (1969). The strategy of ecosystem development. *Science*, 164, 262–270.

Odum, E. P. 1977. The emergence of ecology as a new integrative discipline. *Science*, 195, 1289–1293.

Odum, H. T. (1956). Primary production in flowing waters. *Limnol. Oceanogr.*, 1, 102–117.

Odum, H. T. (1976). *The Effects of Herbicides in South Vietnam. Part B: Working Papers, Models of Herbicides, Mangroves, and War in Vietnam*, National Academy of Sciences, National Research Council, Washington, D.C.

Odum, H. T. and Pinkerton, R. C. (1955). Time's speed regulator, the optimum efficiency for maximum power output in physical and biological systems. *Am. Sci.*, 43, 331–343.

Odum, H. T., Kemp, W., Sell, M., Boynton, W. and Lehman, M. (1977). Energy analysis and the coupling of man and estuaries. *Environ. Manag.*, 1, 297–315.

Odum, W. E. and Johannes, R. E. (1975). The response of mangroves to man-induced environmental stress. In Ferguson Wood, E. J. and Johannes, R. E. (Eds.), *Tropical Marine Pollution*, Elsevier, Amsterdam, Oxford, New York, pp. 52–62.

Økland, J. and Økland, K. (1980). pH level and food organisms for fish: studies of 1000 lakes in Norway. In Drabløs, D. and Tollan, A. (Eds.), *Ecological Impact of Acid Precipitation*—Proceedings of an international conference, Sandefjord, Norway, The SNSF project, pp. 322–323.

Oladimeji, A. A., Qadri, S. U., Tam, G. K. H. and deFreitas, A. S. W. (1979). Metabolism of inorganic arsenic to organoarsenicals in rainbow trout (*Salmo gairdneri*). *Ecotox. Environ. Safety*, 3, 394–400.

Olla, B. L., Pearson, W. H. and Studholme, A. L. (1980). Applicability of behavioral measures in environmental stress assessment. In McIntyre, A. D. and Pearce, J. B. (Eds.), *Biological Effects of Marine Pollution and the Problems of Monitoring*, Rapp. P.-v. Réun. Cons. Int. Explor. Mer, 179, 162–173.

Olson, B. H. and Cooper, R. C. (1976). Comparison of aerobic and anaerobic methylation of mercuric chloride by San Francisco Bay sediments. *Water Res.*, 10, 113–116.

OME (Ontario Ministry of the Environment) (1979). *Acid Precipitation in Ontario*, Ontario Ministry of the Environment, Toronto, Canada, 30 pages.

O'Neill, R. V. (1976). Ecosystem persistence and heterotrophic regulation. *Ecology*, 57, 1244–1253.

O'Neill, R. V. and Giddings, J. M. (1979). Population interactions and ecosystem function. In Innis, G. S. and O'Neill, R. V. (Eds.), *Systems Analysis of Ecosystems*, International Cooperative Publishing House, Fairland, Maryland, pp. 103–123.

O'Neill, R. V. and Reichle, D. E. (1980). Dimensions of ecosystem theory. In Waring, R. H. (Ed.), *Forests: Fresh Perspectives for Ecosystem Analysis*, Oregon State University Press, Corvallis, Oregon, pp. 11–26.

O'Neill, R. V. and Waide, J. B. (1982). Ecosystem theory and the unexpected: implications for environmental toxicology. In Cornaby, B. (Ed.), *Management of Toxic Substances in Our Ecosystems: Taming the Medusa*, Ann Arbor Science, Ann Arbor, Michigan, pp. 43–73.

O'Neill, R. V., Harris, W. F., Ausmus, B. S. and Reichle, D. E. (1975). A theoretical basis for ecosystem analysis with particular reference to element cycling. In Howell, F. G.,

Gentry, J. B. and Smith, M. H. (Eds.), *Mineral Cycling in Southeastern Ecosystems*, U.S. Energy Research and Development Administration, CONF-740513, U.S. Department of Commerce, Springfield, Virginia, pp. 28–40.

O'Neill, R. V., Ausmus, B. S., Jackson, D. R., Van Hook, R. I., Van Vorris, P., Washburne, C. and Watson, A. P. (1977). Monitoring terrestrial ecosystems by analysis of nutrient export. *Water Air Soil Pollut.*, **8**, 271–277.

O'Neill, R. V., Suter, G. W. II and Giddings, J. M. (1981). Measures of ecosystem function. Ecotoxicology Workshop, Cornell University, Ecosystems Research Center, Ithaca, New York, mimeo, 21 pages.

Opler, P. A., Baker, H. G. and Frankie, G. W. (1977). Recovery of tropical lowland forest ecosystems. In Cairns, J. Jr., Dickson, K. L. and Herricks, E. E. (Eds.), *Recovery and Restoration of Damaged Ecosystems*, University of Virginia Press, Charlottesville, Virginia, pp. 379–421.

Orians, G. H. (1975). Diversity, stability and maturity in natural ecosystems. In van Dobben, W. H. and Lowe-McConnell, R. H. (Eds.), *Unifying Concepts in Ecology*, First International Congress of Ecology, Dr. W. Junk B. V. Publishers, The Hague, pp. 139–150.

Ormrod, D. P. 1977. Cadmium and nickel effects on growth and ozone sensitivity of pea. *Water Air Soil Pollut.*, **8**, 263–270.

Osborne, L. L., Davis, R. W. and Linton, K. J. (1979). Effects of limestone strip mining on benthic macroinvertebrate communities. *Water Res.*, **13**, 1285–1290.

Ottawa River Project Group (1979). Mercury in the Ottawa River. *Environ. Res.*, **19**, 231–243.

Overrein, L. N. (1972). Sulphur pollution—patterns observed; leaching of calcium in forest soils determined. *Ambio*, **1**: 145–147.

Owen, D. F. and Wiegert, R. G. (1976). Do consumers maximize plant fitness? *Oikos*, **27**, 488–492.

Paine, R. T. (1969). A note on trophic complexity and community stability. *Am. Nat.*, **103**, 91–93.

Paine, R. T. (1974). Intertidal community structure. Experimental studies on the relationship between a dominant competitor and its principal predator. *Oecologia (Berl.)*, **15**, 93–120.

Pasquill, F. (1974). *Atmospheric Diffusion*, 2nd ed., Ellis Horwood Limited, Chichester, and John Wiley and Sons, New York, 429 pages.

Patil, G. P. and Rosenzweig, M. L. (Eds.) (1979). *Contemporary Quantitative Ecology and Related Ecometrics*, Statistical Ecology Series, vol. 12, International Co-operative Publishing House, Fairland, Maryland, 725 pages (approx.)

Patrick, R. (1949). A proposed biological measure of stream conditions, based on a survey of the Conestoga Basin, Lancaster County, Pennsylvania. *Proc. Acad. Nat. Sci. Phila.*, **101**, 277–341.

Patrick, R. (1954). Diatoms as an indication of river change. In *Proceedings of the 9th Industrial Waste Conference*, Purdue University Engineering Extension Service, Report 87, pp. 325–330.

Patrick, R. (1968). The structure of diatom communities in similar ecological conditions. *Am. Nat.*, **102**, 173–183.

Patrick, R., Roberts, N. A. and Davis, B. (1968). The effect of changes in pH on the structure of diatom communities. *Notulae Naturae, Acad. Nat. Sci. Phila. No. 416*, 16 pages.

Peakall, D. B. (1983). Methods for assessment of the effects of pollutants on avian reproduction. In Vouk. V. B. and Sheehan, P. J. (Eds.), *Methods for Assessing the Effects of Chemicals on Reproductive Functions*, SCOPE 20, John Wiley and Sons, Chichester, New York, Brisbane, Toronto, pp. 345–363.

Peakall, D. B., Tremblay, J., Kinter, W. B. and Miller, D. S. (1981). Endocrine dysfunction in seabirds caused by ingested oil. *Environ. Res.*, **24**, 6–14.
Pearson, T. H. (1975). The benthic ecology of Loch Linnhe and Loch Eil, and a sea-loch system on the west coast of Scotland. IV. Changes in the fauna attributable to organic enrichment. *J. Exp. Mar. Biol. Ecol.*, **20**: 1–41.
Peet, R. K. (1974). The measurement of species diversity. *Annu. Rev. Ecol. Syst.*, **5**, 285–307.
Pelz, E. (1963). Studies in the seed production of spruce stands damaged by fumes. *Arch. Forstw.*, **12**: 1066–1077 (in German).
Percy, J. A. (1978). Effect of chronic exposure to petroleum upon the growth and molting of juveniles of the arctic marine isopod crustacean *Mesidotea entomon*. *J. Fish. Res. Board Can.*, **35**, 650–656.
Perfect, T. J. (1979). The environmental impact of DDT in a tropical agro-ecosystem. *Ambio*, **9**, 16–21.
Perfect, T. J., Cook, A. G., Critchley, B. R., Critchley, U., Davis, A. L., Swift, M. J., Russell-Smith, A. and Yeadon, R. (1979). The effect of DDT contamination on the productivity of a cultivated forest soil in the sub-humid tropics. *J. Appl. Ecol.*, **16**, 705–719.
Persoone, G. and De Pauw, N. (1979). Systems of biological indicators for water quality assessment. In Ravera, O. (Ed.), *Biological Aspects of Freshwater Pollution*, published for the Commission of the European Communities, Directorate General Scientific and Technical Information and Information Management, Luxembourg, Pergamon Press, New York, pp. 39–76.
Pfister, R. M. (1972). Interactions of halogenated pesticides and microorganisms: a review. *CRC Crit. Rev. Microbiol.*, **2**, 1–33.
Phelps, D. K., Galloway, W., Thurberg, F. P., Gould, E. and Dawson, M. A. (1981). Comparison of several physiological monitoring techniques as applied to the blue mussel, *Mytilus edulis*, along a gradient of pollutant stress in Narragansett Bay, Rhode Island. In Vernberg, F. J., Calabrese, A., Thurberg, F. P. and Vernberg, W. B. (Eds.), *Biological Monitoring of Marine Pollutants*, Academic Press, New York, Toronto, London, Sydney, San Francisco, pp. 335–355.
Phillips, D. J. (1977). The use of biological indicator organisms to monitor trace metal pollution in marine and estuarine environments—a review. *Environ. Pollut.*, **13**, 281–317.
Phillips, D. J. H. (1980). *Quantitative Biological Indicators*, Applied Science Publishers Ltd., London, 488 pages.
Pielou, E. C. (1966). The measurement of diversity in different types of biological collections. *J. Theor. Biol.*, **13**, 131–144.
Pielou, E. C. (1969). *An Introduction to Mathematical Ecology*, Wiley-Interscience, New York, 286 pages.
Pimentel, D. (1971). *Ecological Effect of Pesticides on Non-target Species*, U.S. Government Printing Office, Washington, D.C. 220 pages.
Pimentel, D. (1972). Ecological impact of pesticides. *Environ. Biol.*, **72**, 1–27.
Platt, T. (1975). Analysis of importance of spatial and temporal heterogeneity in the estimation of annual production by phytoplankton in a small, enriched, marine basin. *J. Exp. Mar. Biol. Ecol.*, **18**, 99–109.
Platt, T., Dickie, L. M. and Trites, R. W. (1970). Spatial heterogeneity of phytoplankton in a near-shore environment. *J. Fish. Res. Board Can.*, **27**, 1453–1473.
Plowright, R. C. and Rodd, F. H. (1980). The effect of aerial insecticide spraying on hymenopterous pollinators in New Brunswick. *Can. Entomol.*, **112**, 259–269.
Poole, R. W. (1978). The statistical prediction of population fluctuations. *Ann. Rev. Ecol. Syst.*, **9**, 427–448.

Pough, F. H. (1976). Acid precipitation and embryonic mortality of spotted salamander, *Ambystoma maculatum. Science*, **192**, 68–70.
Price, D. R. H. (1978). Fish as indicators of water quality. *Water Pollut. Control*, **77**, 285–296.
Price, D. R. H. (1979). Fish as indicators of water quality. In James, A. and Evinson, L. (Eds.), *Biological Indicators of Water Quality*, John Wiley and Sons, Chichester, New York, Brisbane, Toronto, pp. 8.1–8.23.
Price, R. K. J. and Uglow, R. F. (1979). Some effects of certain metals on development and mortality within moult cycle of *Crangon crangon* (L.). *Mar. Environ. Res.*, **2**, 287–299.
Puckett, K. J., Richardson, D. H. S., Flora, W. P., and Niebour, E. 1974. Photosynthetic ^{14}C fixation by the lichen *Umbilicaria muhlenbergii* (Ach.) Tuck. following short exposures to aqueous sulphur dioxide. *New Phytol.*, **73**, 1183–1192.
Raabe, E. W. (1952). Use of the 'affinity value' in plant sociology. *Vegetatio*, **4**, 53–68.
Raddum, G. G. (1980). Comparison of benthic invertebrates in lakes with different acidity. In Drabløs, D. and Tollan, A. (Eds.), *Ecological Impact of Acid Precipitation*—Proceedings of an international conference, Sandefjord, Norway, The SNSF project, pp. 330–331.
Rafes, P. M. (1970). Estimation of the effects of phytophagous insects on forest production. In Reichle, D. E. (Ed.), *Analysis of Temperate Forest Ecosystems*, Springer-Verlag, Berlin, Heidelberg, New York, pp. 100–106.
Ragsdale, H. L. and Rhoads, W. A. (1974). Four-year post-exposure assay of vegetation surrounding Project Pinstripe: demonstration of the utility of delayed damage appraisals. *Radiat. Bot.*, **14**, 229–236.
Ratcliffe, D. A. (1973). Studies of the recent breeding success of the peregrine, *Falco peregrinus. J. Reprod. Fertil. Suppl.*, **19**, 377–389.
Read, P. A., Renshaw, T. and Anderson, K. J. (1978). Pollution effects on intertidal macrobenthic communities. *J. Appl. Ecol.*, **15**, 15–31.
Reeve, M. R., Gamble, J. C. and Walter, M. A. (1977a). Experimental observations on the effects of copper on copepods and other zooplankton: controlled ecosystem pollution experiment. *Bull. Mar. Sci.*, **27**, 92–104.
Reeve, M. R., Walter, M. A., Darcy, K. and Ikeda, T. (1977b). Evaluation of potential indicators of sub-lethal toxic stress on marine zooplankton (feeding, fecundity, respiration, and excretion): controlled ecosystem pollution experiment. *Bull. Mar. Sci.*, **27**, 105–113.
Regier, H. A. and Cowell, E. B. (1972). Applications of ecosystem theory, succession, diversity, stability, stress and conservation. *Biol. Conserv.*, **4**, 83–88.
Reich, P. B., Amundson, R. G. and Lassoie, J. P. (1982). Reduction in soybean yield after exposure to ozone and sulfur dioxide using a linear gradient exposure technique. *Water Air Soil Pollut.*, **17**, 29–36.
Reichle, D. E. (1975). Advances in ecosystem analysis. *Bioscience*, **25**, 257–264.
Reichle, D. E., Dinger, B. E., Edwards, N. T., Harris, W. F. and Sollins, P. (1973a). Carbon flow and storage in a forest ecosystem. In Woodwell, G. M. and Pecan, E. V. (Eds.), *Carbon in the Biosphere*, AEC Symposium Series, vol. 30, pp. 345–365, NTIS, U.S. Dept. Commerce, Springfield, Virginia.
Reichle, D. E., Goldstein, R. A., Van Hook, R. I. Jr. and Dodson, G. J. (1973b). Analysis of insect consumption in a forest canopy. *Ecology*, **54**, 1076–1084.
Reichle, D. E., O'Neill, R. V., Kaye, S. V., Sollins, P. and Booth, R. S. (1973c). Systems analysis as applied to modelling ecological processes. *Oikos*, **24**, 337–343.
Reichle, D. E., O'Neill, R. V. and Harris, W. F. (1975). Principles of energy and material exchange in ecosystems. In van Dobben, W. H. and Lowe-McConnell, R. H. (Eds.),

Unifying Concepts in Ecology, First International Congress of Ecology, Dr. W. Junk B. V. Publishers, The Hague, pp. 27–43.

Reijnders, P. J. H. (1980). Organochlorine and heavy metal residues in harbour seals from the Wadden Sea and their possible effects on reproduction. *Neth. J. Sea Res.*, **14**, 30–65.

Reimer, A. A. (1975). Effects of crude oil on corals. *Mar. Pollut. Bull.*, **6**, 39–43.

Reinert, R. A. and Spurr, H. W. Jr. (1972). Differential effects of fungicides on ozone injury and brown spot disease of tobacco. *J. Environ. Qual.*, **1**, 450–452.

Remacle, J. (1977). The role of heterotrophic nitrification in acid forest soils—preliminary results. In Lohm, U. and Persson, T. (Eds.), *Soil Organisms as Components of Ecosystems*, Proc. 6th Int. Coll. Soil Zool., Ecological Bulletin (Stockholm), vol. 25, pp. 560–561.

Renfro, J. L., Schmidt-Nielsen, B., Miller, D., Benos, D. and Allen, J. (1974). Methyl mercury and inorganic mercury: uptake, distribution, and effect on osmoregulatory mechanisms in fishes. In Vernberg, F. J. and Vernberg, W. B. (Eds.), *Pollution and Physiology of Marine Organisms*, Academic Press, New York, San Francisco, London, pp. 101–122.

Resh, V. H. (1979). Sampling variability and life history features: basic considerations in the design of aquatic insect studies. *J. Fish. Res. Board Can.*, **36**, 290–311.

Resh, V. H. and Unzicker, J. D. (1975). Water quality monitoring and aquatic organisms: the importance of species identification. *J. Water Pollut. Control Fed.*, **47**, 9–19.

Rice, W. A., Penney, D. C. and Nyborg, M. (1977). Effects of soil acidity on rhizobia numbers, nodulation, and nitrogen fixation by alfalfa and red clover. *Can. J. Soil Sci.*, **57**, 197–204.

Ricklefs, R. E. (1973). *Ecology*, Thomas Nelson and Sons Ltd, London, 861 pages.

Rigler, F. H. (1975). The concept of energy flow and nutrient flow between trophic levels. In van Dobben, W. H. and Lowe-McConnell, R. H. (Eds.), *Unifying Concepts in Ecology*, First International Congress of Ecology, Dr. W. Junk B. V. Publishers, The Hague, pp. 15–26.

Roberts, J. R., Rodgers, D. W., Bailey, J. R. and Rorke, M. A. (1978). *Polychlorinated Biphenyls: Biological Criteria for an Assessment of Their Effects on Environmental Quality*, NRCC Associate Committee on Scientific Criteria for Environmental Quality, National Research Council Canada, NRCC 16077, Ottawa, Canada, 172 pages.

Roberts, J. R., Mitchell, M. F., Boddington, M. J., Ridgeway, J. M., McGarrity, J. T. and Marshall, W. K. (1981). *A Screen for the Relative Persistence of Lipophilic Organic Chemicals in Aquatic Ecosystems—An Analysis of the Role of a Simple Computer Model in Screening*, NRCC Associate Committee on Scientific Criteria for Environmental Quality, National Research Council, NRCC 18570, Ottawa, Canada, 300 pages.

Roberts, T. M., Clarke, T. A., Ineson, P. and Gray, T. R. (1980). Effects of sulphur deposition on litter decomposition and nutrient leaching in coniferous forest soils. In Hutchinson, T. C. and Havas, M. (Eds.), *Effects of Acid Precipitation on Terrestrial Ecosystems*, Plenum Press, New York and London, pp. 381–393.

Rorison, I. H. (1980). The effects of soil acidity on nutrient availability and plant response. In Hutchinson, T. C. and Havas, M. (Eds.), *Effects of Acid Precipitation on Terrestrial Ecosystems*, Plenum Press, New York and London, pp. 283–304.

Rosenberg, R. (1976). Benthic faunal dynamics during succession following pollution abatement in a Swedish estuary, *Oikos* **27**, 414–427.

Rosenberg, R. (1977). Benthic macrofaunal dynamics, production, and dispersion in an oxygen-deficient estuary of west Sweden. *J. Exp. Mar. Biol. Ecol.*, **26**, 107–133.

Rosenberg, R. and Costlow, J. D. Jr. (1976). Synergistic effects of cadmium and salinity combined with constant and cycling temperatures on the larval development of two estuarine crab species. *Mar. Biol.*, **38**, 291–303.

Rosenthal, H. and Alderdice, D. F. (1976). Sublethal effects of environmental stressors, natural and pollutional, on marine fish eggs and larvae. *J. Fish. Res. Board Can.*, **33**, 2047–2065.

Rotty, R. M. (1979). Atmospheric CO_2 consequences of heavy dependence on coal. *Environ. Health Perspect.*, **33**, 273–283.

Rühling, Å. and Tyler, G. (1973). Heavy metal pollution and decomposition of spruce needle litter. *Oikos*, **24**, 402–416.

Ruschmeyer, O. R. and Schmidt, E. L. (1958). Cellulose decomposition in soil burial beds. II. Cellulolytic activity as influenced by alteration of soil properties. *Appl. Microbiol.*, **6**, 115–120.

Saila, S. B., Pikanowski, R. A. and Vaughan, D. S. (1976). Optimum allocation strategies for sampling benthos in the New York Bight. *Estuarine and Coastal Mar. Sci.*, **4**, 119–128.

Saliba, L. J. and Vella, M. G. (1977). Effects of mercury on the behavior and oxygen consumption of *Monodonta articulata*. *Mar. Biol.*, **43**, 277–282.

Sanders, H. L. (1960). Benthic studies in Buzzards Bay III. The structure of the soft-bottom community. *Limnol. Oceanogr.* **5**, 138–153.

Sanders, H. L. (1968). Marine benthic diversity: a comparative study. *Am. Nat.*, **102**, 243–282.

Sanders, H. L. (1978). Florida oil spill impact on the Buzzards Bay benthic fauna: West Falmouth. *J. Fish. Res. Board Can.*, **35**, 717–730.

Sanders, H. O. and Cope, O. B. (1968). The relative toxicities of several pesticides to naiads of three species of stoneflies. *Limnol. Oceanogr.*, **13**, 112–117.

Sastry, A. N. and Miller, D. C. (1981). Application of biochemical and physiological responses to water quality monitoring. In Vernberg, F. J., Calabrese, A., Thurberg, F. P. and Vernberg, W. B. (Eds.), *Biological Monitoring of Marine Pollutants*, Academic Press, New York, Toronto, London, Sydney, San Francisco, pp. 265–294.

Satchell, J. E. (1967). Lumbricidae. In Burges, A. and Raw, F. (Eds.), *Soil Biology*, Academic Press, London, New York, pp. 259–322.

Saunders, G. W. (1976). Decomposition in freshwater. In Anderson, J. M. and Macfadyen, A. (Eds.), *The Role of Terrestrial and Aquatic Organisms in Decomposition Processes*, Blackwell Scientific Publications, Oxford, London, Edinburgh, Melbourne, pp. 341–373.

Saward, D., Stirling, A. and Topping, G. (1975). Experimental studies on the effects of copper on a marine food chain. *Mar. Biol.*, **29**, 351–361.

Scale, P. R. (1980). Changes in plant communities with distances from SO_2 sources. In Proceedings of Symposium on *Effects of Air Pollutants on Mediterranean and Temperate Forest Ecosystems*, United States Department of Agriculture, Forest Service, Pacific Southwest Forest and Range Experimental Station, General Technical Report PSW-43, Berkeley, California, p. 248.

Scale, P. R. (1982). *The Effects of Emissions from an Iron-Sintering Plant in Wawa, Canada on Forest Communities*. MSc. Thesis, Institute for Environmental Studies, University of Toronto, Toronto, Canada, 141 pages.

Scheider, W., Adamski, J. and Paylor, M. (1975). *Reclamation of Acidified Lakes Near Sudbury Ontario*. Ontario Ministry of the Environment, Rexdale, Ontario, 129 pages.

Schindler, D. W. (1980). Experimental acidification of a whole lake: a test of the oligotrophication hypothesis. In Drabløs, D. and Tollan, A. (Eds.), *Ecological Impact of Acid Precipitation*—Proceedings of an international conference, Sandefjord, Norway, The SNSF project, pp. 370–374.

Schindler, D. W., Wagemann, R., Cook, R. B., Ruszczynski, T. and Prokopowich, J.

(1980a). Experimental acidification of Lake 223, Experimental Lakes Area: background data and the first three years of acidification. *Can. J. Fish. Aquat. Sci.*, **37**, 342–354.

Schindler, J. E., Waide, J. B., Waldron, M. C., Hains, J. J., Schreiner, S. P., Freeman, M. L., Benz, S. L., Pettigrew, D. P., Schissel, L. A. and Clarke, P. J. (1980b). A microcosm approach to the study of biogeochemical systems. I. Theoretical rationale. In Giesy, J. P. (Ed.), *Microcosms in Ecological Research*, U.S. Department of Energy, CONF-781101, NTIS, U.S. Department of Commerce, Springfield, Virginia, pp. 192–203.

Schneider, M. J., Barraclough, S. A., Genoway, R. G. and Wolford, M. L. (1980). Effects of phenol on predation of juvenile rainbow trout, *Salmo gairdneri. Environ. Pollut. Ser. A. Ecol. Biol.*, **23**, 121–130.

Schnitzer, M. and Khan, S. U. (1972). *Humic Substances in the Environment*, Marcel Dekker, New York, 327 pages.

Schofield, C. L. (1976). Acid precipitation effects on fish. *Ambio*, **5**, 228–230.

Shannon, C. E. (1948). A mathematical theory of communication. *Bell Systems. Tech. J.*, **27**, 379–423; 623–656.

Shannon, C. E. and Weaver, W. (1963). *The Mathematical Theory of Communications*, University of Illinois Press, Chicago, Illinois, 117 pages.

Sheehan, P. J. (1980). *The Ecotoxicology of Copper and Zinc: Studies on a Stream Macroinvertebrate Community*, Ph.D. Dissertation, University of California, Davis, California, 262 pages.

Sheehan, P. J., Klein, W., Bourdeau, Ph. and Korte, F. (Eds.) (1984). *Appraisal of Tests to Predict the Environmental Behaviour of Chemicals*, John Wiley and Sons, Chichester, New York, Brisbane, Toronto, (in press).

Sheldon, A. L. (1969). Equitability indices: dependence on species counts. *Ecology*, **50**, 466–467.

Sheldon, A. L. (1977). Colonization curves: application to stream insects on semi-natural substrates. *Oikos*, **28**, 256–261.

Shindler, D. B., Scott, B. F. and Carlisle, D. B. (1975). Effect of crude oil on populations of bacteria and algae in artificial ponds subject to winter weather and ice formation. *Verh. Internat. Verein. Limnol.*, **19**, 2138–2144.

Shkarlet, O. D. (1972). Influence of industrial pollution of the atmosphere and soil on the size of pollen grain of the Scots pine. *Ekologiya*, **3**, 53–57 (in Russian).

Shriner, D. S. (1976). Effects of simulated rain acidified with sulfuric acid on host-parasite interactions. In Dochinger, L. S. and Seliga, T. A. (Eds.), *Proceedings of the First International Symposium on Acid Precipitation and the Forest Ecosystem*, U.S.D.A. Forest Service, Northeastern Forest Experiment Station, General Technical Report NE-23, Upper Darby, Pennsylvania, pp. 919–925.

Shure, D. J. (1971). Insecticide effects on early succession in an old field ecosystem. *Ecology*, **52**, 271–279.

Sick, L. V. and Windom, H. L. (1975). Effects of environmental levels of mercury and cadmium on the rates of metal uptake and growth physiology of selected genera of marine phytoplankton. In Howell, F. G., Gentry, J. B. and Smith, M. H. (Eds.), *Mineral Cycling in Southeastern Ecosystems*, U. S. Energy Research and Development Administration, CONF-740513, U.S. Department of Commerce, Springfield, Virginia, pp. 257–267.

Sigal, L. L. and Nash, T. H. III (1980). Lichens as ecological indicators of photochemical oxidant air pollution. In Proceedings of Symposium on *Effects of Air Pollutants on Mediterranean and Temperate Forest Ecosystems*, United States Department of

Agriculture, Forest Service, Pacific Southwest Forest and Range Experimental Station, General Technical Report PSW-43, Berkeley, California, p. 249.

Simmons, G. M. Jr. (1972). A preliminary report on the use of the sequential comparison index to evaluate acid mine drainage on the macrobenthos in a pre-impoundment basin. *Trans. Am. Fish. Soc.*, **101**, 701–713.

Simpson, E. H. (1949). Measurement of diversity. *Nature*, **163**, 688.

Sindermann, C. J. (1979). Pollution associated disease and abnormalities of fish and shell fish: a review. *Fish. Bull.*, **76**, 717–748.

Sindermann, C. J. (1980). The use of pathological effects of pollutants in marine environmental monitoring programs. In McIntyre, A. D. and Pearce, J. B. (Eds.), *Biological Effects of Marine Pollution and the Problems of Monitoring*, Rapp. P.-v. Réun. Cons. Int. Explor. Mer, **179**, 129–134.

Sindermann, C. J., Bang, F. B., Christensen, N. D., Dethlefsen, V., Harshbarger, J. C., Mitchell, J. R. and Mulcahy, M. F. (1980). The role and value of pathobiology in pollution effects monitoring programs. In McIntyre, A. D. and Pearce, J. B. (Eds.), *Biological Effects of Marine Pollution and the Problems of Monitoring*, Rapp. P.-v. Réun. Cons. Int. Explor. Mer, **179**, 135–151.

Skidmore, J. F. (1964). Toxicity of zinc compounds to aquatic animals, with special reference to fish. *Quart. Rev. Biol.*, **39**, 227–248.

Skye, E. (1979). Lichens as biological indicators of air pollution. *Annu. Rev. Phytopathol.*, **17**, 325–341.

Slater, H. L. (1983). Methods for assessing the effects of chemicals on the reproductive functions of microorganisms. In Vouk, V. B., and Sheehan, P. J. (Eds.), *Methods for Assessing the Effects of Chemicals on Reproductive Functions*, SCOPE 20, John Wiley and Sons, Chichester, New York, Brisbane, Toronto, pp. 525–535.

Smith, W. H. (1974). Air pollution—effects on the structure and function of the temperate forest ecosystem. *Environ. Pollut.*, **6**, 111–129.

Snyder, N. F. R., Snyder, H. A., Lincer, J. L. and Reynolds, R. T. (1973). Organochlorines, heavy metals, and the biology of North American accipiters. *Bioscience*, **23**, 300–305.

Södergren, S. (1976). Ecological effects of heavy metal discharge in a salmon river. *Institute of Freshwater Research, Report No. 55*, Drottingholm, Sweden, pp. 91–131.

Solbé, J. F. de L. G. (1977). Water quality, fish and invertebrates in a zinc polluted stream. In Alabaster, J. S. (Ed.), *Biological Monitoring of Inland Fisheries*, Applied Science Publishers, London, pp. 97–105.

Sørensen, T. (1948). A method of establishing groups of equal amplitude in plant sociology based on similarity of species content and its application to analysis of the vegetation on Danish Commons. *Biol. S.Kr.* (*K. Danske Vidensk. Selsk. N. S.*), **5**, 1–34.

Soule, D. F. and Soule, J. D. (1979). Bryozoa: (Ectoprocta). In Hart, C. W. Jr. and Fuller, S. L. H. (Eds.), *Pollution Ecology of Estuarine Invertebrates*, Academic Press, New York, London, Toronto, Sydney, San Francisco, pp. 35–76.

Southward, A. J. and Southward, E. C. (1978). Recolonization of rocky shores in Cornwall after use of toxic dispersants to clean up the Torrey Canyon spill. *J. Fish. Res. Board Can.*, **35**, 682–706.

Spalding, B. (1979). Effects of divalent metal chlorides on respiration and extractable enzymatic activities of Douglas-fir needle litter. *J. Environ. Qual.*, **8**, 105–109.

Spangler, W. J., Spigarelli, J. L., Rose, J. M. and Miller, H. M. (1973). Methylmercury: bacterial degradation in lake sediments. *Science*, **180**, 192–193.

Sprague, J. B. (1971). Measurement of pollutant toxicity to fish. III. Sublethal effects and safe concentrations. *Water Res.*, **5**, 245–266.

Sprague, J. B., Elson, P. F. and Saunders, R. L. (1965). Sublethal copper-zinc pollution in a salmon river—a field and laboratory study. In Jaag, O. (Ed.), *Advances in Water Pollution Research*, Pergamon Press, New York and London, pp. 65–82.

Sprules, W. G. (1975). Midsummer crustacean zooplankton communities in acid-stressed lakes. *J. Fish. Res. Board Can.*, **32**, 389–395.

SRI (Stanford Research Institute). *Environmental Pathways of Selected Chemicals in Freshwater Systems*, U.S. Environmental Protection Agency, EPA-600/7-77-113, October 1977.

Stainken, D. M. (1978). Effects of uptake and discharge of petroleum hydrocarbons on the respiration of the soft-shell clam, *Mya arenaria*. *J. Fish. Res. Board Can.*, **35**, 637–642.

Staveland, J. T. (1979). Effects on hatching in *Littorina littorea* after an oil spill. *Mar. Pollut. Bull.*, **10**, 255–258.

Steemann Nielsen, E. and Bruun Laursen, H. (1976). Effect of $CuSO_4$ on the photosynthetic rate of phytoplankton in four Danish lakes. *Oikos*, **27**, 239–242.

Stegeman, J. J. (1977). Fate and effects of oil in marine animals. *Oceanus*, **20**, 59–66.

Stegeman, J. J. (1978). Influence of environmental contamination on cytochrome P-450 mixed-function oxygenases in fish: implications for recovery in the Wild Harbor Marsh. *J. Fish. Res. Board Can.*, **35**, 668–674.

Stegeman, J. J. (1980). Mixed-function oxygenase studies in monitoring for effects of organic pollution. In McIntyre, A. D. and Pearce, J. B. (Eds.), *Biological Effects of Marine Pollution and the Problems of Monitoring*, Rapp. P.-v. Réun. Cons. Int. Explor. Mer, **179**, 33–38.

Stein, W. D. (1967). *The Movement of Molecules Across Cell Membranes*, Academic Press, New York, 369 pages.

Stenseth, N. C. (1980). Spatial heterogeneity and population stability: some evolutionary consequences. *Oikos*, **35**, 165–184.

Stich, H. F., Acton, A. B. and Forrester, C. R. (1976). Fish tumors and sublethal effects of pollutants. *J. Fish. Res. Board Can.*, **33**, 1993–2001.

Stickel, L. F. (1973). Pesticide residues in birds and mammals. In Edwards, C. A. (Ed.), *Environmental Pollution by Pesticides*, Plenum Press, London and New York, pp. 254–312.

Stickel, W. H. (1975). Some effects of pollutants in terrestrial ecosystems. In McIntyre, A. D. and Mills, C. F. (Eds.), *Ecological Toxicology Research—Effects of Heavy Metals and Organohalogen Compounds*, Plenum Press, New York and London, pp. 25–74.

Stirling, E. A. (1975). Some effects of pollutants on the behavior of the bivalve *Tellina tenuis*. *Mar. Pollut. Bull.*, **6**, 122–124.

Stoner, A. W. and Livingston, R. J. (1978). Respiration, growth and food conversion efficiency of pinfish (*Lagodon rhomboides*) exposed to sublethal concentrations of bleached Kraft mill effluent. *Environ. Pollut.*, **17**, 207–217.

Stout, J. D. and Heal, O. W. (1967). Protozoa. In Burges, A. and Raw, F. (Eds.), *Soil Biology*, Academic Press, London and New York, pp. 149–195.

Strojan, C. L. (1978a). Forest leaf litter decomposition in the vicinity of a zinc smelter. *Oecologia (Berl.)*, **32**, 203–212.

Strojan, C. L. (1978b). The impact of zinc smelter emissions on forest litter arthropods. *Oikos*, **31**, 41–46.

Stumm, W. and Morgan, J. J. (1981). *Aquatic Chemistry. An Introduction Emphasizing Chemical Equilibria in Natural Waters*, John Wiley and Sons, New York, Chichester, Brisbane, Toronto, 748 pages.

Sullivan, J. F., Atchison, G. J., Kolar, D. J. and McIntosh, A. W. (1978). Changes in the predator–prey behavior of fathead minnows (*Pimephales promelas*) and largemouth

bass (*Micropterus salmoides*) caused by cadmium. *J. Fish. Res. Board Can.*, **35**, 446–451.
Summers, A. O. and Silver, S. 1978. Microbial transformations of metals. *Ann. Rev. Microbiol.*, **32**, 637–672.
Sundström, K.-R. and Hällgren, J.-E. (1973). Using lichens as physiological indicators of sulfurous pollutants. *Ambio*, **2**, 13–21.
Swift, M. J., Heal, O. W. and Anderson, J. M. (1979). *Decomposition in Terrestrial Ecosystems, Studies in Ecology*, vol. 5, University of California Press, Berkeley and Los Angeles, 372 pages.
Tagatz, M. E. (1976). Effect of mirex on predator–prey interaction in an experimental estuarine ecosystem. *Trans. Am. Fish. Soc.*, **105**, 546–549.
Talmage, S. S. and Coutant, C. C. (1980). Thermal effects. In Literature Review Issue, *J. Water Pollut. Control Fed.*, **52**, 1575–1616.
Tam, T. Y. and Trevors, J. T. (1981). Effects of pentachlorophenol on asymbiotic nitrogen fixation in soils. *Water Air Soil Pollut.*, **16**, 409–414.
Tamm, C. O. (1976). Acid precipitation: Biological effects in soil and forest vegetation. *Ambio*, **5**, 235–238.
Tamm, C. O., Wiklander, G. and Popović, B. (1977). Effects of application of sulphuric acid to poor pine forests. *Water Air Soil Pollut.*, **8**, 75–87.
Teal, J. M. (1957). Community metabolism in a temperate cold spring. *Ecol. Monogr.*, **27**, 283–302.
Teal, J. M., Burns, K. and Farrington, J. (1978). Analysis of aromatic hydrocarbons in intertidal sediments resulting from two spills of No. 2 fuel oil in Buzzards Bay, Massachusetts. *J. Fish. Res. Board Can.*, **35**, 510–520.
Teigen, O. (1975). *Experiments with Germination and Establishment of Spruce and Pine in Artificially Acidified Mineral Soils*, SNSF Project, IR.10/75, Norway, 36 pages.
Thomas, M. L. H. (1978). Comparison of oiled and unoiled intertidal communities in Chedabucto Bay, Nova Scotia. *J. Fish. Res. Board Can.*, **35**, 707–716.
Thomas, W. H. and Seibert, D. L. R. (1977). Effects of copper on the dominance and diversity of algae: controlled ecosystem pollution experiment. *Bull. Mar. Sci.*, **27**, 23–33.
Thomas, W. H., Holm-Hansen, O. Seibert, D. L. R., Azam, F., Hodson, R. and Takahashi, M. (1977). Effects of copper on phytoplankton standing crop and productivity: controlled ecosystem pollution experiment. *Bull. Mar. Sci.*, **27**, 34–43.
Thurberg, F. P., Dawson, M. A. and Collier, R. S. (1973). Effects of copper and cadmium on osmoregulation and oxygen consumption in two species of estuarine crabs. *Mar. Biol.*, **23**, 171–175.
Timoney, J. F., Port, J., Giles, J. and Spanier, J. (1978). Heavy metal and antibiotic resistance in the bacterial flora of sediments of the New York Bight. *Appl. Environ. Microbiol.*, **36**, 465–472.
Tingey, D. T., Reinert, R. A., Dunning, J. A. and Heck, W. W. (1971). Vegetation injury from the interaction of nitrogen dioxide and sulfur dioxide. *Phytopathology*, **61**, 1506–1511.
Tingey, D. T., Reinert, R. A., Dunning, J. A. and Heck, W. W. (1973). Foliar injury responses of eleven plant species to ozone/sulfur dioxide mixtures. *Atmos. Environ.*, **7**, 201–208.
Tomassini, F. D., Lavoie, P., Puckett, K. J., Nieboer, E. and Richardson, D. H. S. (1977). The effect of time of exposure to sulphur dioxide on potassium loss and photosynthesis in the lichen, *Cladina rangiferina* (L.) Harm. *New Phytol.*, **79**, 147–155.
Tont, S. and Platt, T. (1979). Fluctuations in the abundance of phytoplankton on the California coast. In Naylor, E. and Hartnoll, R. G. (Eds.), *Cyclic Phenomena in Marine Plants and Animals*, Proc. 13th European Marine Biology Symposium, Pergamon Press, Oxford, New York, Toronto, Sydney, Paris, Frankfurt, pp. 11–18.

REFERENCE LIST FOR CHAPTERS 2 TO 6

Traaen, T. (1974). Effect of pH on microbial decomposition of organic matter. In Braekke, F. H. (Ed.), *Hydrokjemiske og Hydrobiologiske Rapporter fra NIVA*, SNSF Project, IR. 3/74, Norway, pp. 15-22.

Traaen, T. S. (1980). Effects of acidity on decomposition of organic matter in aquatic environments. In Drabløs, D. and Tollan, A. (Eds.), *Ecological Impact of Acid Precipitation*—Proceedings of an international conference, Sandefjord, Norway, The SNSF project, pp. 340-341.

Transport Canada, Transport of Dangerous Goods Branch (1979). *Emergency Response Guide for Dangerous Goods*, Copp Clark Pitman, Toronto, and Canadian Government Publishing Centre, Supply and Services Canada, Ottawa, Canada.

Treshow, M. (1975). Interaction of air pollutants and plant disease. In Mudd, J. B. and Kozlowski, T. T. (Eds.), *Responses of Plants to Air Pollutants*, Academic Press, New York, San Francisco, London, pp. 307-334.

Treshow, M. (1978). Terrestrial plants and plant communities. In Butler, G. C. (Ed.), *Principles of Ecotoxicology*, SCOPE 12, John Wiley and Sons, Chichester, New York, Brisbane, Toronto, pp. 223-237.

Treshow, M. and Allan, J. (1979). Annual variation in the dynamics of a woodland plant community. *Environ. Conserv.*, **6**, 231-236.

Tu, C. M. and Miles, J. R. W. (1976). Interactions between insecticides and soil microbes. *Residue Rev.*, **64**, 17-65.

Tukey, H. B. and Morgan, J. V. (1963). Injury to foliage and its effect upon the leaching of nutrients from above-ground plant parts. *Physiol. Plant.*, **16**, 557—564.

Türk, R., Wirth, V. and Lange, O. L. (1974). CO_2 exchange measurements for determination of SO_2 resistance of lichens. *Oecologia (Berl.)*, **15**, 33-64. (in German).

Turnpenny, A. W. H. and Williams, R. (1981). Factors affecting the recovery of fish populations in an industrial river. *Environ. Pollut. Ser. A. Ecol. Biol.*, **26**, 39-58.

Turoboyski, L. (1973). The indicator organisms and their ecological variability. *Acta Hydrobiol.*, **15**, 259-274.

Tveite, B. (1980). Effects of acid precipitation on soil and forest. 9.Tree growth in field experiments. In Drabløs, D. and Tollan, A. (Eds.), *Ecological Impact of Acid Precipitation*—Proceedings of an international conference, Sandefjord, Norway, The SNSF project, pp. 206-207.

Tveite, B. and Abrahamsen, G. (1980). Effects of artificial acid rain on the growth and nutrient status of trees. In Hutchinson, T. C. and Havas, M. (Eds.), *Effects of Acid Precipitation on Terrestrial Ecosystems*, Plenum Press, New York and London, pp. 305-318.

Tyler, G. (1972). Heavy metals pollute nature, may reduce productivity. *Ambio*, **1**, 52-59.

Tyler, G. (1974). Heavy metal pollution and soil enzymatic activity. *Plant Soil*, **41**, 303-311.

Tyler, G. (1976). Heavy metal pollution, phosphatase activity, and mineralization of organic phosphorus in forest soils. *Soil Biol. Biochem.*, **8**, 327-332.

UNSCEAR (United Nations Scientific Committee on the Effects of Atomic Radiation) (1972). *Ionizing Radiation: Levels and Effects*, Report to the General Assembly, United Nations Publication E. 72. IX. 17, 2 volumes.

Uthe, J. F., Freeman, H. C., Mounib, S. and Lockhart, W. L. (1980). Selection of biochemical techniques for detection of environmentally induced sublethal effects in organisms. In McIntyre, A. D. and Pearce, J. B. (Eds.), *Biological Effects of Marine Pollution and the Problems of Monitoring*, Rapp. P.-v. Réun. Cons. Int. Explor. Mer, **179**, 39-47.

Vaccaro, R. F., Farooq, A. and Hodson, R. E. (1977). Response of natural marine bacterial populations to copper: controlled ecosystem pollution experiments. *Bull. Mar. Sci.*, **27**, 17-22.

Vale, T. R. (1977). Fire and man in Sequoia National Park. *Ann. Assoc. Am. Geogr.*, **67**, 28–45.

Van den Bosch, R. (1969). The toxicity problem—comments by an applied insect ecologist. In Miller, M. W. and Berg, G. G. (Eds.), *Chemical Fallout: Current Research on Persistent Pesticides*, Charles C. Thomas Publisher, Springfield, Illinois, pp. 97–112.

Vandermeulen, J. H., Keizer, P. D. and Penrose, W. R. (1977). Persistence of non-alkane components of Bunker C oil in beach sediments of Chedabucto Bay, and the lack of their metabolism by molluscs. *Proceedings 1977 Oil Spill Conference: (Prevention, Behavior, Control, Cleanup)*, Sponsored by the American Petroleum Institute, United States Environmental Protection Agency and United States Coast Guard, pp. 469–473.

Vander Wal, J. (1977). Relations between Nipigon Bay benthic macroinvertebrates and selected aspects of their habitat. *J. Fish. Res. Board Can.*, **34**, 824–829.

Van Voris, P., O'Neill, R. V., Emanuel, W. R. and Shugart, H. H. (1980). Functional complexity and ecosystem stability. *Ecology*, **61**, 1352–1360.

Varshney, S. R. K. and Varshney, C. K. (1981). Effects of sulfur dioxide on pollen germination and pollen tube growth. *Environ. Pollut. Ser. A. Ecol. Biol.*, **24**, 87–92.

Veith, G. D., Defoe, D. L. and Bergstedt, B. V. (1979). Measuring and estimating the bioconcentration factors of chemicals in fish. *J. Fish. Res. Board Can.*, **36**, 1040–1048.

Verma, S. R., Tonk, I. P., Gupta, A. K. and Dalela, R. C. (1981). *In vivo* enzymatic alterations in certain tissues of *Saccobranchus fossilis* following exposure to four toxic substances. *Environ. Pollut. Ser. A. Ecol. Biol.*, **26**, 121–127.

Vernberg, F. J. (1978). Multiple-factor and synergistic stresses in aquatic systems. In Thorp, J. H. and Gibbons, J. W. (Eds.), *Energy and Environmental Stress in Aquatic Systems*, U.S. Department of Energy, CONF-771114, NTIS, U.S. Department of Commerce, Springfield, Virginia, pp. 726–745.

Vernberg, F. J. and Vernberg, W. B. (Eds.) (1974). *Pollution and Physiology of Marine Organisms*, Academic Press, New York, London, San Francisco, 492 pages.

Vernberg, F. J., Calabrese, A., Thurberg, F. P. and Vernberg, W. B. (Eds.) (1981). *Biological Monitoring of Marine Pollutants*, Academic Press, New York, Toronto, London, Sydney, San Francisco, 559 pages.

Vernberg, W. B., Calabrese, A., Thurberg, F. P. and Vernberg, F. J. (Eds.) (1979). *Marine Pollution: Functional Responses*, Academic Press, New York, San Francisco, London, 454 pages.

Verschueren, K. (1977). *Handbook of Environmental Data on Organic Chemicals*, Van Nostrand Reinhold, New York, 659 pages.

Vitousek, P. M. and Reiners, W. A. (1975). Ecosystem succession and nutrient retention: a hypothesis. *Bioscience*, **25**, 376–381.

Voigt, G. K. (1980). Acid precipitation and soil buffering capacity. In Drabløs, D. and Tollan, A. (Eds.), *Ecological Impact of Acid Precipitation*—Proceedings of an international conference, Sandefjord, Norway, The SNSF project, pp. 53–57.

Vouk, V. B. and Sheehan, P. J. (Eds.) (1983). *Methods for Assessing the Effects of Chemicals on Reproductive Functions*. SCOPE 20, John Wiley and Sons, Chichester, New York, Brisbane, Toronto, 541 pages.

Waide, J. B., Krebs, J. E., Clarkson, S. P. and Setzler, E. M. (1974). A linear systems analysis of the calcium cycle in a forested watershed ecosystem. *Prog. Theor. Biol.*, **3**, 261–345.

Waide, J. B., Schindler, J. E., Waldron, M. C., Hains, J. J., Schreiner, S. P., Freedman, M. L., Benz, S. L., Pettigrew, D. P., Schissel, L. A. and Clark. P. J. (1980). A microcosm approach to the study of biogeochemical systems. 2. Responses of aquatic laboratory microcosms to physical, chemical and biological perturbations. In Geisy, J. P. (Ed.), *Microcosms in Ecological Research*, U.S. Department of Energy, CONF-781101, U.S. Department of Commerce, Springfield, Virginia, pp. 204–223.

Waldichuk, M. (1979). Review of the problems. *Philos. Trans. R. Soc. Lond. B. Biol. Sci.*, **286**, 399–424.

Walker, S. D., Seesman, P. A. and Colwell, R. R. (1975). Effects of South Louisiana crude oil and No. 2 fuel oil on growth of heterotrophic microorganisms including proteolytic, lipolytic, chitinolytic, and cellulolytic bacteria. *Environ. Pollut.*, **9**, 13–33.

Wallace, J. B., Webster, J. R. and Woodall, W. R. (1977). The role of filter feeders in flowing waters. *Arch. Hydrobiol.*, **79**, 506–532.

Ward, D. V. and Busch, D. A. (1976). Effects of Temefos, an organophosphorus insecticide, on survival and escape behavior of the marsh fiddler crab *Uca pugnax*. *Oikos*, **27**, 331–335.

Ward, D. V., Howes, B. L. and Ludwig, D. F. (1976). Interactive effects of predation pressure and insecticide (Temefos) toxicity on populations of the marsh fiddler crab *Uca pugnax*. *Mar. Biol.*, **35**, 119–126.

Warren, C. E. (1971). *Biology and Water Pollution Control*, W. B. Saunders Company, Philadelphia, Pennsylvania, 434 pages.

Waters, T. F. (1977). Secondary production in inland waters. *Adv. Ecol. Res.*, **10**, 91–150.

Watson, A. P., Van Hook, R. I., Jackson, D. R. and Reichle, D. E. (1976). *Impact of a Lead Mining-Smelting Complex on the Forest Floor Litter Arthropod Fauna in the New Lead Belt Region of Southwest Missouri*, ORNL/NSF/EATC-30, Oak Ridge National Laboratories, Oak Ridge, Tennessee.

Watson, T. A. and McKeown, B. A. (1976). The activity of $\Delta 5$-3β hydroxy steroid dehydrogenase enzyme in the interrenal tissue of rainbow trout (*Salmo gairdneri* Richardson) exposed to sublethal concentrations of zinc. *Bull. Environ. Contam. Toxicol.*, **16**, 173–181.

Webster, J. R. and Patten, B. C. (1979). Effects of watershed perturbation on stream potassium and calcium dynamics. *Ecol. Monogr.*, **49**, 51–72.

Webster, J. R., Waide, J. B. and Patten, B. C. (1975). Nutrient recycling and the stability of ecosystems. In Howell, F. G. Gentry, J. B. and Smith, M. H. (Eds.), *Mineral Cycling in Southeastern Ecosystems*, U.S. Energy Research and Development Administration, CONF-740513, U.S. Department of Commerce, Springfield, Virginia, pp. 1–27.

Weiss, H. V., Koede, M. and Goldberg, E. D. (1971). Mercury in a Greenland ice sheet: evidence of recent input by man. *Science*, **174**, 692–694.

Weiss, P. A. (1971). The basic concept of hierarchic systems. In Weiss, P. A. (Ed.), *Hierarchically Organized Systems in Theory and Practice*, Hafner Publishing Company, New York, pp. 1–44.

Wellings, S. R., Alpers, C. E., McCain, B. B. and Miller, B. S. (1976). Fin erosion disease of the starry flounder (*Platichthys stellatus*) in the estuary of the Duwamish River, Seattle, Washington. *J. Fish. Res. Board Can.*, **33**, 2577–2586.

Wentsel, R., McIntosh, A., McCafferty, W. P., Atchison, G. and Anderson, V. (1977). Avoidance response of midge larvae (*Chironomus tentans*) to sediments containing heavy metals. *Hydrobiologia*, **52**, 171–175.

Westman, W. E. (1978). Measuring the inertia and resilience of ecosystems. *Bioscience*, **28**, 705–710.

Wetzel, R. G. (1975). *Limnology*, W. B. Saunders Company, Philadelphia, London, Toronto, 743 pages.

Whelpdale, D. M. (1978). *Atmospheric Pathways of Sulphur Compounds*, Report Number 7, MARC (Monitoring and Assessment Research Centre), Chelsea College, University of London, 39 pages.

Whitby, L. M. and Hutchinson, T. C. (1974). Heavy-metal pollution in the Sudbury mining and smelting region of Canada. II. Soil toxicity tests. *Environ. Conserv.*, **1**, 191–200.

White, M. C., Chaney, R. L. and Decker, A. M. (1974). Differential varietal tolerance in soybean to toxic levels of zinc in sassafras sandy loam. *Agron. Abstr.*, **1974**, 144-145.

Whittaker, R. H. (1972). Evolution and measurement of species diversity. *Taxon*, **21**, 213-251.

Whittaker, R. H. and Fairbanks, C. W. (1958). A study of plankton copepod communities in the Columbia Basin, southeastern Washington. *Ecology*, **39**, 46-65.

Whittaker, R. H. and Woodwell, G. M. (1971). Evolution of natural communities. In Wiens, J. A. (Ed.), *Ecosystem Structure and Function*, Oregon State University Press, Corvallis, Oregon, pp. 137-159.

Whittaker, R. H. and Woodwell, G. M. (1973). Retrogression and coenocline distance. In Whittaker, R. H. (Ed.), *Handbook of Vegetation Science*, part V. *Ordination and Classification of Communities*, Dr. W. Junk B. V. Publishers, The Hague, pp. 53-73.

Whitworth, W. R. and Lane, T. H. (1969). Effects of toxicants on community metabolism in pools. *Limnol. Oceanogr.*, **14**, 53-58.

Wiederholm, T. (1980). Use of benthos in lake monitoring. *J. Water Pollut. Control Fed.*, **52**, 537-547.

Wiegert, R. G. and Owens, D. F. (1971). Trophic structure, available resources and population density in terrestrial vs. aquatic ecosystems. *J. Theor. Biol.*, **30**, 69-81.

Wiklander, L. (1975). The role of neutral salts in the ion exchange between acid precipitation and soil. *Geoderma*, **14**, 93-105.

Wiklander, L. (1980). Interaction between cations and anions influencing adsorption and leaching. In Hutchinson, T. C. and Havas, M. (Eds.) *Effects of Acid Precipitation on Terrestrial Ecosystems*, Plenum Press, New York and London, pp. 239-254.

Wildish, D. J., Akagi, H. and Poole, N. J. (1977). Avoidance by herring of dissolved components in pulp mill effluents. *Bull. Environ. Contam. Toxicol.*, **18**, 521-525.

Wilhm, J. L. (1967). Comparison of some diversity indices applied to populations of benthic macroinvertebrates in a stream receiving organic wastes. *J. Water Pollut. Control Fed.*, **39**, 1674-1683.

Wilhm, J. L. (1968). Biomass units versus numbers of individuals in species diversity indices. *Ecology*, **49**, 153-159.

Wilhm, J. L. (1970). Range of diversity index in benthic macroinvertebrate populations. *J. Water Pollut. Control Fed.*, **42**, R221-R224.

Wilhm, J. L. (1972). Graphic and mathematical analysis of biotic communities in polluted streams. *Annu. Rev. Entomol.*, **17**, 223-252.

Wilhm, J. L. and Dorris, T. C. (1968). Biological parameters of water quality. *Bioscience*, **18**, 447-481.

Williams, S. T., McNeilly, T. and Wellington, E. M. H. (1977). The decomposition of vegetation growing on metal mine waste. *Soil Biol. Biochem.*, **9**, 271-275.

Williams, W. T. (1971). Principles of clustering. *Annu. Rev. Ecol. Syst.*, **2**, 303-326.

Williamson, P. and Evans, P. R. (1972). Lead levels in roadside invertebrates and small mammals. *Bull. Environ. Contam. Toxicol.*, **8**, 280-288.

Williamson, P. and Evans, P. R. (1973). A preliminary study of the effects of high levels of inorganic lead on soil fauna. *Pedobiologia*, **13**, 16-21.

Willis, A. J. (1972). Long-term ecological changes in sward composition following application of maleic hydrazide and 2,4-D. In *Proceedings of the 11th British Weed Control Conference*, British Crop Protection Council, pp. 360-367.

Wilson, D. O. (1977). Nitrification in three soils amended with zinc sulfate. *Soil Biol. Biochem.*, **9**, 277-280.

Wilson, E. O. (1969). The species equilibrium. *Brookhaven Symp. Biol.*, **22**, 38-47.

Winner, R. W., Van Dyke, J. S., Caris, N. and Farrell, M. P. (1975). Response of the macroinvertebrate fauna to a copper gradient in an experimentally-polluted stream. *Verh. Internat. Verein. Limnol.*, **19**, 2121-2127.

Winner, R. W., Keeling, T., Yeager, R. and Farrell, M. P. (1977). Effects of food type on the acute and chronic toxicity of copper to *Daphnia magna*. *Freshwater Biol.*, **7**, 343–349.

Winner, R. W., Boesel, M. W. and Farrell, M. P. (1980). Insect community structure as an index of heavy-metal pollution in lotic ecosystems. *Can. J. Fish. Aquat. Sci.*, **37**, 647–655.

Winteringham, F. P. W. (1977). Comparative ecotoxicology of halogenated hydrocarbon residues. *Ecotoxicol. Environ. Safety*, **1**, 407–425.

Winteringham, F. P. W. (1980). Food and agriculture in relation to energy, environment and resources. *Atomic Energy Rev.*, **18**, 223–245.

Wissmar, R. C. (1972). *Some Effects of Mine Drainage in Coeur d'Alene River and Lake, Idaho*. Ph.D. dissertation, University of Idaho, Moscow, Idaho.

Witkamp, M. (1966). Decomposition of leaf litter in relation to environment, microflora, and microbial respiration. *Ecology*, **47**, 197–201.

Witkamp, M. and Ausmus, B. S. (1976). Processes in decomposition and nutrient transfer in forest systems. In Anderson, J. M. and Macfadyen, A. (Eds.), *The Role of Terrestrial and Aquatic Organisms in Decomposition Processes*, Blackwell Scientific Publications, Oxford, London, Edinburgh, Melbourne, pp. 375–396.

Witkamp, M. and Frank, M. L. (1970). Effects of temperature, rainfall, and fauna on transfer of ^{137}Cs, K, Mg, and mass in consumer-decomposer microcosms. *Ecology*, **51**, 465–474.

Wodzinski, R. S., Labeda, D. P. and Alexander, M. (1977). Toxicity of SO_2 and NO: selective inhibition of blue-green algae by bisulfite and nitrite. *J. Air Pollut. Control Assoc.*, **27**, 891–893.

Wood, T. G. (1974). The distribution of earthworms (Megascolecidae) in relation to soils, vegetation and altitude on the slopes of Mt. Kosciusko, Australia. *J. Anim. Ecol.*, **43**, 87–106.

Wood, T. and Bormann, F. H. (1975). Increases in foliar leaching caused by acidification of an artificial mist. *Ambio*, **4**, 169–171.

Woodall, W. R. Jr. and Wallace, J. B. (1975). Mineral pathways in small Appalachian streams. In Howell, F. G., Gentry, J. B. and Smith, M. H. (Eds.), *Mineral Cycling in Southeastern Ecosystems*, U.S. Energy Research and Development Administration, CONF-740513, U.S. Department of Commerce, Springfield, Virginia, pp. 408–422.

Woodiwiss, F. S. (1964). A biological system of stream classification used by the Trent River Board. *Chem. Ind. Lond.*, **11**, 443–447.

Woodwell, G. M. (1967). Radiation and the patterns of nature. *Science*, **156**, 461–470.

Woodwell, G. M. (1970). Effects of pollution on the structure and physiology of ecosystems. *Science*, **168**, 429–433.

Woodwell, G. M., Ballard, J. T. and Pecan, E. V. (1975). Ecological succession and ionic leakage in terrestrial systems. In *International Conference on Heavy Metals in the Environment*, Toronto, Canada. Symposium Proceedings, Vol. II, pp. 189–198, Institute for Environmental Studies, Toronto, Canada.

Wormald, A. P. (1976). Effects of a spill of marine diesel oil on the meiofauna of a sandy beach at Picnic Bay, Hong Kong. *Environ. Pollut.*, **11**, 117–130.

Wright, R. F. and Gjessing, E. T. (1976). Acid precipitation: changes in the chemical composition of lakes. *Ambio*, **5**, 219–223.

Wright, R. F. and Henriksen, A. (1978). Chemistry of small Norwegian lakes, with special reference to acid precipitation. *Limnol. Oceanogr.*, **23**, 487–498.

Wright, R. F. and Johannessen, M. (1980). Input-output budgets of major ions at gauged catchments in Norway. In Drablos, D. and Tollan, A. (Eds.), *Ecological Impact of Acid Precipitation*—Proceedings of an international conference, Sandefjord, Norway. The SNSF project, pp. 250–251.

Wright, R. and Snekvik, E. (1978). Acid precipitation: Chemistry and fish populations in 700 lakes in southernmost Norway. *Verh. Internat. Verein. Limnol.*, **20**, 765-775.

Wu, L. and Antonovics, J. (1978). Zinc and copper tolerance of *Agrostis stolonifera* L. in tissue culture. *Am. J. Bot.*, **65**, 268-271.

Wu, R. S. S. and Levings, C. D. (1980). Mortality, growth and fecundity of transplanted mussel and barnacle populations near a pulp mill outfall. *Mar. Pollut. Bull.*, **11**, 11-15.

Wurster, C. F. Jr. (1968). DDT reduces photosynthesis by marine phytoplankton. *Science*, **159**, 1474-1475.

Yan, N. (1979). Phytoplankton community of an acidified, heavy metal-contaminated lake near Sudbury, Ontario: 1973-1977. *Water Air Soil Pollut.*, **11**, 43-55.

Zeiman, J. C. (1975). Tropical sea grass ecosystems and pollution. In Ferguson Wood, E. J. and Johannes, R. E. (Eds.), *Tropical Marine Pollution*, Elsevier Scientific Publishing Company, Amsterdam, Oxford, New York, pp. 63-74.

Zelinka, M. and Marvan, P. (1966). Comments on a new method for judgement of the sapro-biological character of water. *Verh. Internat. Verein. Limnol*, **16**, 817-822 (in German).

Ziegler, I. (1975). The effect of SO_2 pollution on plant metabolism. *Residue Rev.*, **56**, 79-105.

Part II
Case Studies

Part II
Case Studies

Effects of Pollutants at the Ecosystem Level
Edited by P. J. Sheehan, D. R. Miller, G. C. Butler and Ph. Bourdeau
© 1984 SCOPE. Published by John Wiley & Sons Ltd

CHAPTER 7
Introduction to Case Studies

These studies are presented to compliment the survey of pollutant effects on populations, communities and ecosystems proffered in Part I. The case studies illustrate the variety of approaches used and problems encountered in assessing ecosystem-level changes. Contributors were asked to prepare case studies of stressed ecosystems which had been examined in detail and to synthesize general principles from their experiences. The studies were to be limited to those in which the ecosystem itself had been deleteriously affected. Cases in which only distribution and cycling of pollutants and not their toxic effects were studied were specifically excluded, as were those in which possible danger was identified for only a single 'target' species.

The authors were encouraged to follow a common format and to address, whenever applicable, the following points: pollutant input, ecosystem description, pollutant behaviour, effects noted on individual species, on communities, on structural and functional properties of the ecosystem (species diversity, total biomass, productivity, material balance and cycling), and recovery (characteristics and time scale, new stable configuration achieved).

The cases presented cover a variety of current or recent chemical or physical disturbances to freshwater, marine and terrestrial environments. The 'Thames Estuary Study' (Case 7.1) contains a brief historical account of the problems caused by release of sewage and industrial effluents into the Thames River, and then provides a quantitative description of ecosystem recovery following abatement measures. This contribution exemplifies nicely the usefulness of a number of analysis techniques and indices in assessing community changes related to reduction in pollution stress.

The 'Clearwater Lake Study' (Case 7.2) examines the effects of atmospheric inputs of H^+ and metals from diffuse sources on a poorly buffered lake system. A number of changes in phytoplankton, zooplankton and macrophyte communities are related to acidification. The importance of community interactions and of shifts in primary production to ecosystem function are discussed.

'Copper Gradient Studies' (Case 7.3) compares the responses of a variety of structural indices to levels of heavy metal pollution in three contaminated streams. The similarities in stressed macroinvertebrate communities between a stream receiving a controlled dose of copper and others receiving seasonally and industrially influenced inputs are discussed in terms of ecosystem tests. The dominance of chironomids as an index of metal stress is proposed.

The 'Ecological Effects of Hydrodevelopment' (Case 7.4) illustrates the use of ecosystem assessment techniques for the analysis of a physical perturbation. The mitigation of disturbances in major projects such as hydrodevelopment are also discussed, in relation to the history of the disrupted system.

The influence of oil spills on marine biota, particularly commercially exploitable species, is the primary emphasis of Case 7.5. This study examines some of the critical life processes (e.g. reproduction, recruitment) of marine fishes and macroinvertebrates which were seriously impaired as a result of the Amoco Cadiz and Gino spills.

The three concluding case studies exemplify the effects of toxic substances on terrestrial biota and ecosystem processes. The 'Impact of Airborne Metal Contamination' study (Case 7.6) underscores the gradient approach to assessing the response of a forest system exposed to a mixture of toxic heavy metals. The influence of metals on the rate of organic decomposition and the consequent disruption of nutrient cycles is examined, as a case of impaired ecosystem function.

Several of the problems encountered in the analysis of secondary effects of pesticides are exposed in the 'Fenitrothion' study (Case 7.7). The need for coordinated exposure and response assessment programmes is highlighted. The problem of identifying sensitive species with functionally important roles is examined in detail.

The reestablishment of an ecosystem on a virtually sterile land base (mine tailings) is addressed as a problem facing mining industries (Case 7.8). Despite the fact that the reseeding project was not primarily undertaken as a study of ecosystem rehabilitation, it was hoped that such a report would identify those ecosystem processes which require the greatest amount of assistance in order to become successfully reestablished (e.g. N-recycling).

Conclusions from these case studies reinforce those drawn from the literature survey and lead to the generalizations and recommendations developed in Chapter 8.

Effects of Pollutants at the Ecosystem Level
Edited by P. J. Sheehan, D. R. Miller, G. C. Butler and Ph. Bourdeau
© 1984 SCOPE. Published by John Wiley & Sons Ltd

CASE 7.1
Thames Estuary: Pollution and Recovery

M. J. ANDREWS
Thames Water Authority
Metropolitan Pollution Control
Northumberland House
Mogden S. T. Works, Isleworth
Middlesex TW7 7LP, England

7.1.1	Introduction	195
7.1.2	Pollution Input	197
	7.1.2.1 Sewage Effluents	198
	7.1.2.2 Industrial Effluents	201
	7.1.2.3 Synthetic Detergents	203
	7.1.2.4 Thermal Pollution	204
7.1.3	Pollution Effects	205
7.1.4	Recovery of the Macrofauna	207
	7.1.4.1 Number of Individuals (N)	209
	7.1.4.2 Number of Species (S)	211
	7.1.4.3 Species Diversity (H')	212
	7.1.4.4 Evenness Function (J')	212
	7.1.4.5 Species Richness (D)	213
	7.1.4.6 Relative Dominance	213
	7.1.4.7 Log-normal Distribution	213
7.1.5	Fish Communities and Water Quality	217
	7.1.5.1 Cluster Analysis of Biological Data	219
	7.1.5.2 Principal Component Analysis of Physico-chemical Data	219
7.1.6	Conclusions	224
7.1.7	References	225

7.1.1 INTRODUCTION

A change in the condition of the Thames estuary (Figure 7.1.1) from an 'open-sewer' to a waterway through which salmon may freely pass occurred over the 15 years between the early 1960s and late 1970s.

The opportunity for observing the development of the ecosystem as water quality improved was recognized by researchers from the University of London King's College Pollution Unit, who undertook quantitative ecological studies between 1967 and 1973. Huddart (1971) and Sedgwick (1979) provide a valuable

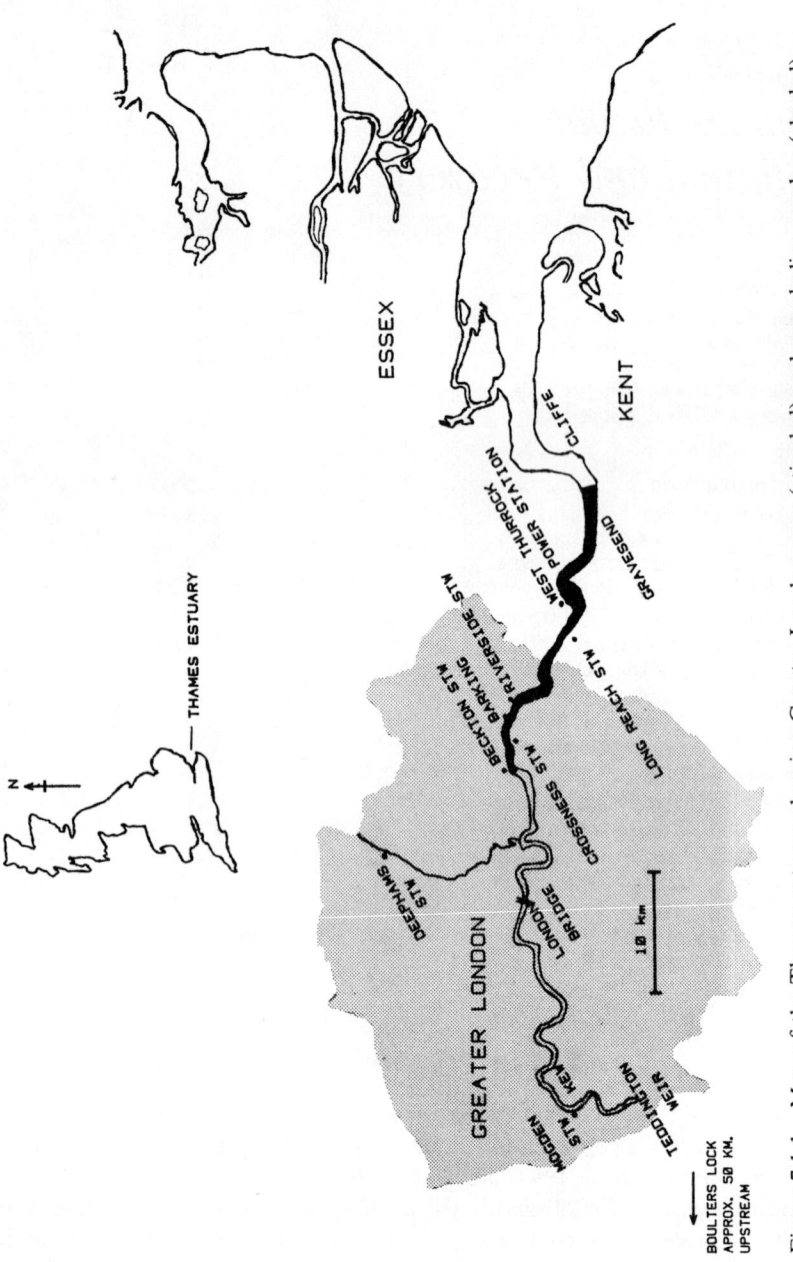

Figure 7.1.1 Map of the Thames estuary showing Greater London area (stippled) and mesohaline reaches (shaded)

record of the status of fishes, shrimp and other macroinvertebrates in the estuary at that time, while Birtwell (1972) examined the ecophysiology of tubificid worms and, in particular, the estuarine *Tubifex costatus*. The return of water fowl was also observed and reported by Harrison and Grant (1976).

Since 1974 the fishes and macroinvertebrate communities in the estuary have been studied further by staff of the Thames Water Authority, partly in recognition of a statutory requirement to have regard to the desirability of conserving fauna (section 22 of the Water Act, 1973), but also to provide information on the water quality needed to allow aquatic animals to thrive. The information was required by the Authority whose functions included the exploitation of resources and use of the estuary for effluent disposal, which essentially competed with the need for conserving fish stocks and other aquatic life.

In the Thames situation it is difficult to determine the environmental 'masterfactors' which influence the distribution of biota. Consideration has to be given to the effects of strong tidal currents, fluctuating salinity, high silt load, presence of sewage effluents and associated reduced levels of dissolved oxygen, and thermal pollution, which in combination provide for a community of animals where the diversity of species is low compared with neighbouring freshwater and open sea habitats. Multivariate analysis has helped to differentiate environmental changes mediated by natural and human causes, but these changes can only be interpreted in ecological terms with the provision of reliable background data which describes seasonal fluctuations of migrants and local movements of resident species.

The objective of this chapter is to give an historical perspective of pollution in the estuary and to describe some of the ecological changes that occurred between 1975 and 1980 when macrofaunal populations stabilized. The reported results expand existing knowledge by providing information from field observations on the level of dissolved oxygen required to sustain a 'normal' estuarine fish community.

7.1.2 POLLUTION INPUT

In terms of the oxygen-demanding load entering the estuary, the relative magnitudes of various sources of pollution, expressed as percentages of total load, were as follows:

1976 values

Sewage effluents	74%
Direct industrial discharges	9%
Rivers	6.5%
Upper freshwater Thames	7.5%
Storm water	3.0%

7.1.2.1 Sewage Effluents

The importance of sewage effluents as pollutants of the estuary was reported by the Water Pollution Research Laboratory (Department of Scientific and Industrial Research, 1964).

Historically, sewers were constructed in the early 19th century as the area of the metropolis was covered with buildings. Originally they were intended to carry away the rainfall from the fields, roads and roofs of houses, but following the introduction of water-closets from about 1810 onwards drains were constructed from the overflowing cesspools, which could not cope with the additional load, to connect with the street sewers. An act passed in 1847 made it compulsory to drain houses directly into the sewers with the result that, within a period of about six years, 30 000 cesspools were abolished, and all house and street refuse was turned into the Thames (Fitzmaurice, 1912). Drinking water was still abstracted from the tidal Thames in London at that time and the severe pollution of the water was responsible for the deaths of 14 000 people during the London cholera outbreak in 1849. The stench in London arising from the anaerobic river was described by Dr William Budd, quoted by Gray (1940):

> ...for many weeks the atmosphere of Parliamentary Committee rooms at Westminster was only rendered barely tolerable by the suspension before every window, of blinds saturated with chloride of lime and by the lavish use of this and other disinfectants.

In 1856 the newly appointed Metropolitan Board of Works instructed their chief engineer to report as to the plans necessary for the complete interception of the sewage of the metropolis, so as to discharge it into the river *below* London, instead of within the boundaries of the metropolis. The main feature of the scheme which was adopted in 1856, and carried out between that time and 1874, was the construction of intercepting sewers parallel to the river, connecting with the main sewers that previously delivered the sewage into the river. The intercepting sewers were carried to Beckton on the north side of the river, and to Crossness on the south side of the river (see Figure 7.1.1). Large reservoirs were made at both these locations, which were sufficient to contain all sewage coming down the outfall sewers during a period of about 6 hours. The sewage was only discharged into the Thames on the ebb tide, in the belief that this would prevent sewage being swept back within the boundaries of the metropolis on a flooding tide.

The overall result was a transfer of the pollution from central London eastwards to the Barking area, where sludge banks built up in the estuary. To overcome this particular problem works were constructed at Beckton and Crossness for the precipitation or chemical clarification of the sewage. Both sets of works were completed by 1891, and with the removal to sea of the precipitated sewage solids in a fleet of specially designed ships, an improvement in the estuary

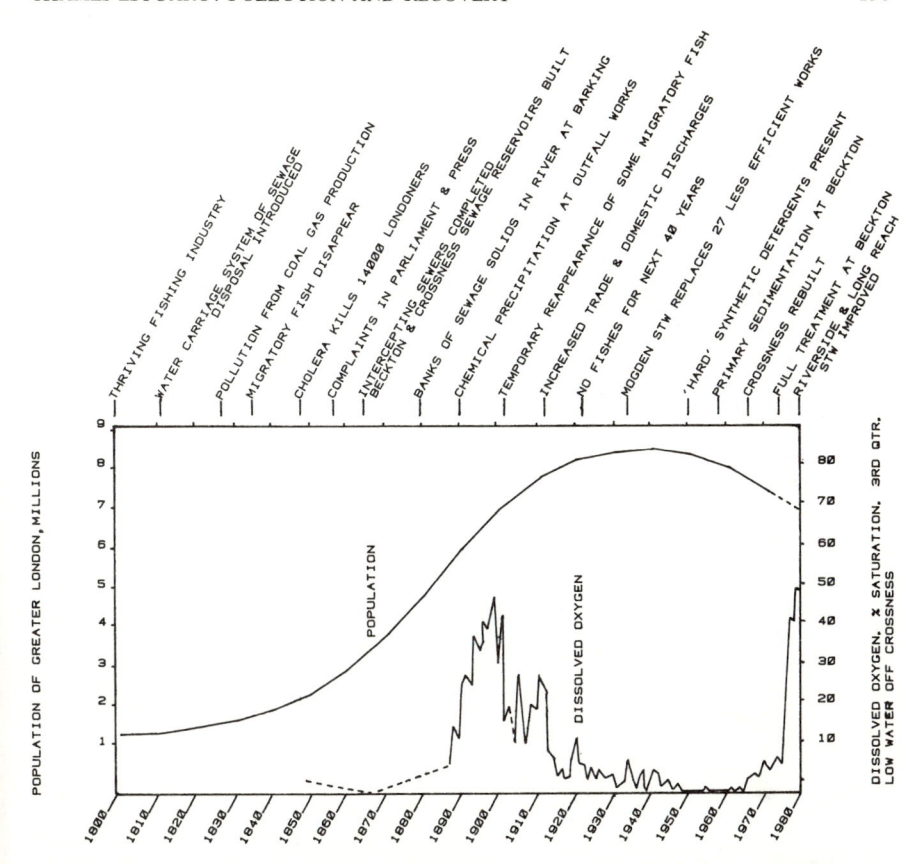

Figure 7.1.2 Population of Greater London and average concentration of dissolved oxygen in the estuary off Crossness (third quarter). Comments relate to events referred to in the text

was observed as an increase in dissolved oxygen levels (Figure 7.1.2). The state of the river declined again after the first World War when London's population increased further. Numerous small and inefficient sewage works were built to cope with the increased flow of sewage and by 1930 there was on average one sewage works to every 25.2 km² around the centre of London (Wood, 1980). Gradually, over the next three decades, an attempt at regionalization was made, and the number of sewage treatment works in Greater London was reduced from 198 to only 10.

New and efficient primary sedimentation plant was built at Beckton in 1954, and in 1959 a modern diffused air activated sludge plant brought the fraction of flow receiving secondary treatment to 50 per cent. In 1964 Crossness works was completely rebuilt with the largest mechanized aeration plant in the United Kingdom, capable of producing a nitrified effluent from an incoming flow of

500 000 m³ d⁻¹. From the time that the plant was commissioned in 1964 the estuary ceased to be anaerobic (Wood, 1980). Major extensions were made at other treatment works; Deephams sewage treatment works was built in 1962 to replace 14 smaller obsolete works and discharged 180 000 m³ d⁻¹ of effluent of very high quality to the River Lee, a major tributary of the estuary in the metropolitan area. Extensions were made to Riverside Works (87 000 m³ d⁻¹) discharging to the tideway and to other smaller works using as receiving streams the smaller metropolitan tributaries.

A most significant change in river water quality resulted from the completion in 1976 of extensions to Beckton sewage treatment works. This works treats an average of 818 000 m³ d⁻¹ and incorporates the latest design of diffused air activated sludge tanks. It has consistently produced a better quality effluent than originally anticipated, and the effect on the river was dramatic, both in terms of dissolved oxygen level and in the recovery of the ecology of the region (Cockburn, 1979).

The completion of these and other works has resulted in a reduction in the pollution load discharged to the estuary in recent years of nearly 80 per cent as shown in Figure 7.1.3.

A remaining problem occurs when sewage enters the estuary during long-

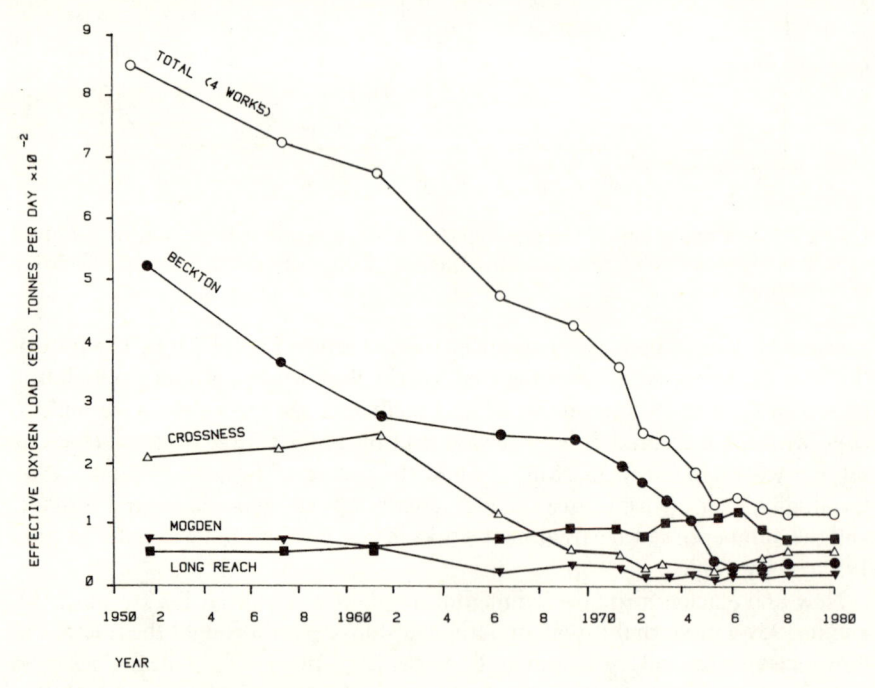

Figure 7.1.3 Effective oxygen load (tonnes d⁻¹) from the four major sewage treatment works on the tideway

duration summer storms of moderate intensity, which can bring about a marked reduction in the dissolved oxygen content of the river. The original sewerage scheme can cope with a quantity of storm sewage approximately two and a half times the flow of sewage provided for, but beyond that level sewage is transferred to the river along with storm water.

Recent investigations have been aimed at preventing fish mortalities arising from dissolved oxygen depletions caused by storm sewage (Wood *et al.*, 1980), and for technical and financial reasons a compensation method has been chosen which supplies 'pure' oxygen direct to the river. Using a pressure swing adsorption unit applied from a British Oxygen Company 'Vitox' system mounted on a barge, the minimum of the oxygen sag curve is followed as it moves down river, and oxygen is supplied sufficient to raise the level in the river to that acceptable for preservation of the fish.

7.1.2.2 Industrial Effluents

In the nineteenth century the types of trade effluents entering the estuary were very varied, including for example, those from soap manufacturers, slaughter houses, shot factories, coal-wharfs, cow-houses, tanneries, fish markets and gut spinners. From the 1820s serious pollution of the estuary was caused by effluents from the production of coal gas, and from the processing of the by-products, carried out by a large number of small works.

The gas-works near Barking were started in 1868, and soon grew to become the largest works in Europe. Wheeler (1979) describes the experiments of Gunther in 1882, who noted that an eel placed into discharge from the works became covered in white mucus and was dead in five minutes. Almost a century later, material in the waste tip from the gas-works was regarded as so toxic that an embargo was placed on the carting away of the material from the site to a tip elsewhere (Longlands and Townsend, 1980). The liquid fraction carried high concentrations of phenols, tar oils, naphthalenes and pitch, all of which would have drained into the Thames during the early years of gas production.

Serious attempts at controlling industrial effluents in London date from the Third Report of the Royal Commission of Sewage Disposal in 1903. The Commissioners recommended that laws should be introduced so as to make if the duty of the local authorities (then responsible for sewage disposal) to provide sewers and treatment works for the conveyance and treatment of both domestic and trade effluents. Discharge of trade effluents to sewers was seen as preferable since treatment could then be supervised by experts to meet the standards necessary for discharge to the river.

Under current UK law responsibility for sewage disposal has shifted to the Water Authorities. They have a duty to control trade effluents under the consent procedure which (1) requires any trader wishing to discharge effluent to a sewer or watercourse to give details of this intention in advance of doing so and (2) allows the Water Authorities to prohibit the discharge of certain substances, to limit others—and the quantity and state of the effluent as a whole, and to make a charge for reception, conveyance and treatment.

A report by the Royal Commission on Environmental Pollution (1972) attempted to reconcile the extreme view of the pure environmentalists, who considered that no pollution should be permitted to enter an estuary, with those of the industrialists, who thought that such a policy would throw a very heavy burden on industry in waste disposal costs. The Commission considered that while it was permissible to exploit an estuary for waste disposal up to a level which does not endanger aquatic life or transgress the standards of amenity which the public needs and for which they are prepared to pay, toxic and nonbiodegradable substances such as heavy metals and persistent organic residues should be excluded from discharges. More stringent trade effluent control since the 1972 report has reduced these substances to the levels shown in Table 7.1.1 (Wood, 1980).

One example of effective pollution control has been the reduction of mercury input to the outer estuary. In 1971 approximately 18 kg d^{-1} of mercury entered the waters of the estuary in association with London's sewage sludge, which was being dumped there. The major sources of the mercury were traced to trade premises in London, and were successfully controlled, so that by 1976 the level of

Table 7.1.1 Concentrations of toxic non-biodegradable substances in the Thames estuary (Taken from Wood, 1980)

Organohalogen pesticides and PCBs	
α-BHC (α-HCH)	$1-15 \text{ ng l}^{-1}$
γ-BHC (HCH)	$1-15 \text{ ng l}^{-1}$
Aldrin	$1-7 \text{ ng l}^{-1}$
Dieldrin	$1-13 \text{ ng l}^{-1}$
DDT	$1-40 \text{ ng l}^{-1}$
Endrin	1 ng l^{-1}
PCB as Arochlor 1254	$5-50 \text{ ng l}^{-1}$
Metals	
(unfiltered samples)	
Lead	$5-100 \text{ μg l}^{-1}$
Cadmium	$1-4 \text{ μg l}^{-1}$
Mercury	$0.01-0.6 \text{ μg l}^{-1}$*
Copper	$2-80 \text{ μg l}^{-1}$
Nickel	$10-70 \text{ μg l}^{-1}$
Zinc	$10-200 \text{ μg l}^{-1}$
Detergents	
Nonionic (as Lissapol NX)	$30-60 \text{ μg l}^{-1}$
Anionic (as Manoxol OT)	$20-200 \text{ μg l}^{-1}$

*High values of mercury associated with high levels of suspended solids

THAMES ESTUARY: POLLUTION AND RECOVERY 203

mercury put out with the shipped sludge had fallen to less than $4\,\mathrm{kg\,d^{-1}}$. Monitoring work in the estuary showed a reduction in the average concentration of the metal in muscle tissue of Thames flounder *Platichthys flesus* from $0.74\,\mathrm{mg\,kg^{-1}}$ wet weight to $0.59\,\mathrm{mg\,kg^{-1}}$ within that period (MAFF, 1971, 1973). The average mercury concentration in flounder tissue has continued to fall, to $0.38\,\mathrm{mg\,kg^{-1}}$ in 1977, $0.35\,\mathrm{mg\,kg^{-1}}$ in 1978 and $0.26\,\mathrm{mg\,kg^{-1}}$ in 1979. This last figure is below the 'safe' value of $0.3\,\mathrm{mg\,kg^{-1}}$ which has been included in a proposal for a Council Directive by the European Economic Community as a quality objective for the aquatic environment into which mercury is discharged by the chlor-alkali electrolysis industry (Commission of the European Communities, 1979).

7.1.2.3 Synthetic Detergents

The introduction of synthetic detergents in the post-war period, and the rapid increase in the quantities used (Table 7.1.2), caused a marked deterioration of the already polluted estuary (Wood, 1980).

These surfactants, based on the active principle tetrapropylene benzene sulphonate, were not readily biodegraded at sewage treatment works (approximately 30 per cent surfactant was removed), but passed in concentrations of over $10\,\mathrm{mg\,l^{-1}}$ in effluents into the river. It was estimated that the synthetic detergents reduced the exchange coefficient for surface aeration of the river by up to 19 per cent and its ability to oxidise polluting loads correspondingly. The problem was eventually resolved by the setting up in 1957 of the Standing Technical Committee on Synthetic Detergents composed of scientists from central government, pollution control authorities and the detergent industry, and, by voluntary agreement, the nonbiodegradable surfactant was replaced by

Table 7.1.2 Consumption of surface active materials in the United Kingdom, 1949–1959 (thousands of tonnes)

Year	Domestic	Industrial	Total
1949	10.7	2.5	13.2
1950	12.7	2.7	15.4
1951	14.2	3.9	18.1
1952	21.3	5.3	26.6
1953	29.5	6.1	35.6
1954	33.5	6.1	39.6
1955	34.5	6.6	41.1
1956	35.0	6.6	41.6
1957	34.7	7.0	41.7
1958	35.6	7.4	43.0
1959	38.1	7.5	45.6

alkylbenzene sulphonate surfactants with linear side chains of which 96 per cent degraded during sewage treatment. Further reformulations, using alcohols, either as sulphates or as ethoxylates or their sulphates, has even further reduced the chance of foaming and residue problems in many cases (Taylor, 1980).

7.1.2.4 Thermal Pollution

A mathematical model of the Thames estuary developed by the Water Pollution Research Laboratory can be used as a predictive tool (Department of Scientific and Industrial Research, 1964; Barrett *et al.*, 1978; Gilligan 1972) and has been used in assessing the effect that directly cooled riverside power stations would have on the temperature of the estuary (Casapieri and Owers, 1979). It has been found that the development of power stations since 1920 has raised the temperature of the estuary by about 3°C in some reaches (Figure 7.1.4) and by much higher levels in limited local areas. An increase in temperature of 3°C above the original value of 15°C represents a decrease in solubility of oxygen of 6 per cent and hence a reduction in the dissolved oxygen resources of the river.

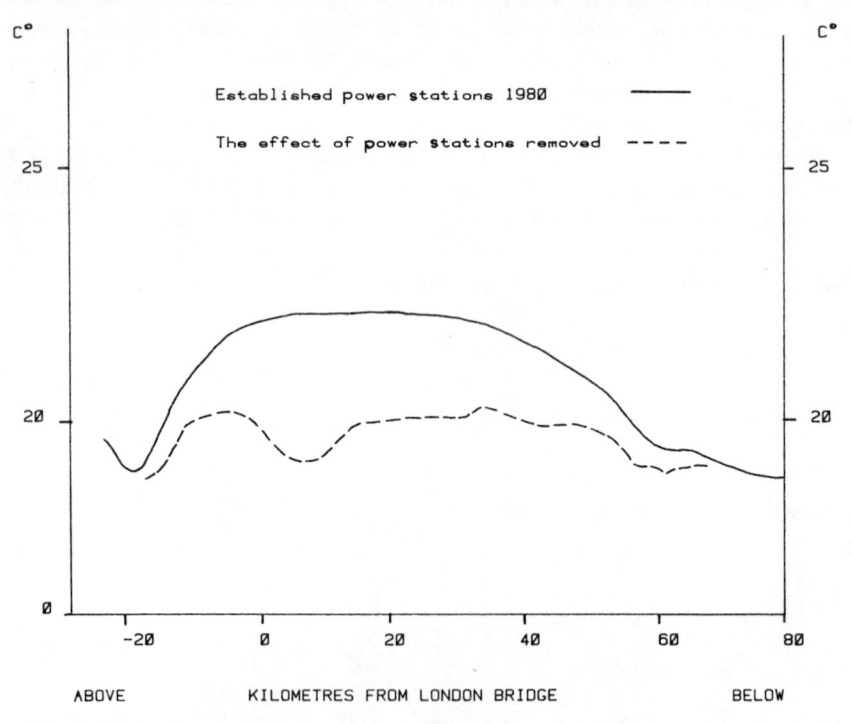

Figure 7.1.4 Temperature profile of Thames estuary, showing the effect of established directly cooled power stations under third quarter low flow conditions, 1980

7.1.3 POLLUTION EFFECTS

One sign that the estuary was being polluted in the late 18th century and early 19th century was the decline in estuarine migratory fish stocks, with the loss of commercially sought species causing most comment.

There is little objective evidence of the former abundance of salmon *Salmo*

Table 7.1.3 Catches of salmon at Boulter's Lock, Taplow (River Thames), 1794–1821

Year	Number of salmon	Total weight lbs	Average weight lbs	5-Year running average number of salmon
1794	15	248	16.5	
1795	19	168	8.8	
1796	18	328	18.2	
1797	37	670	18.1	
1798	16	317	19.8	21
1799	36	507	14.1	25
1800	29	388	13.4	27
1801	66	1124	17.0	37
1802	18	297	16.5	33
1803	20	374	18.7	34
1804	62	943	15.2	39
1805	7	116	16.5	35
1806	12	245	20.4	24
1807	16	253	15.8	23
1808	5	88	17.5	20
1809	8	116	14.5	10
1810	4	70	17.5	9
1811	16	182	11.4	10
1812	18	224	12.4	10
1813	14	220	15.7	12
1814	13	98	7.5	13
1815	4	52	13.0	13
1816	14	179	12.8	13
1817	5	76	15.3	10
1818	4	49	12.1	8
1819	5	84	16.8	6
1820	0	0	0	5
1821	2	31	15.5	3

salar and sea trout *Salmo trutta* in the Thames, mainly because stocks were extinct before reliable fishery statistics were regularly collected. One contemporary author writing in the seventeeth century regarded the Thames as 'of principal note as a salmon fishery' and suggested that Thames salmon were the best in Europe (Chetham, 1688). However, Wheeler (1979), with reference mainly to nineteenth century accounts, concludes that it is unlikely that the species was especially abundant. The decline in the stock took place in the first two decades of the nineteenth century, and by the 1830s the salmon was a rare fish in the Thames. The record of catches from 1794 to 1821 by the Lovegrove family at Boulter's Lock, which is the nearest approach to a statistical record we have, illustrates this decline (Table 7.1.3). Between 1794 and 1804, catches fluctuated from 15 to 66 fish per season and the annual total weight from 168 to 1124 lbs. In the succeeding 15 years (omitting 1820 and 1821) the number of fish varied markedly from year to year and such variations in annual catch are frequently observed in declining fish populations (Thames Water Authority, 1977).

The causes for the decline of the stock were probably several, and all were related to the activities of man. Pollution, both from untreated sewage and industrial effluents in the metropolitan reaches of the river and some tributaries, was undoubtedly a contributing factor. Coinciding with the increase in pollution was the development of navigation for large boats in the reaches above London which involved the erection of the extensive series of pound locks and weirs under

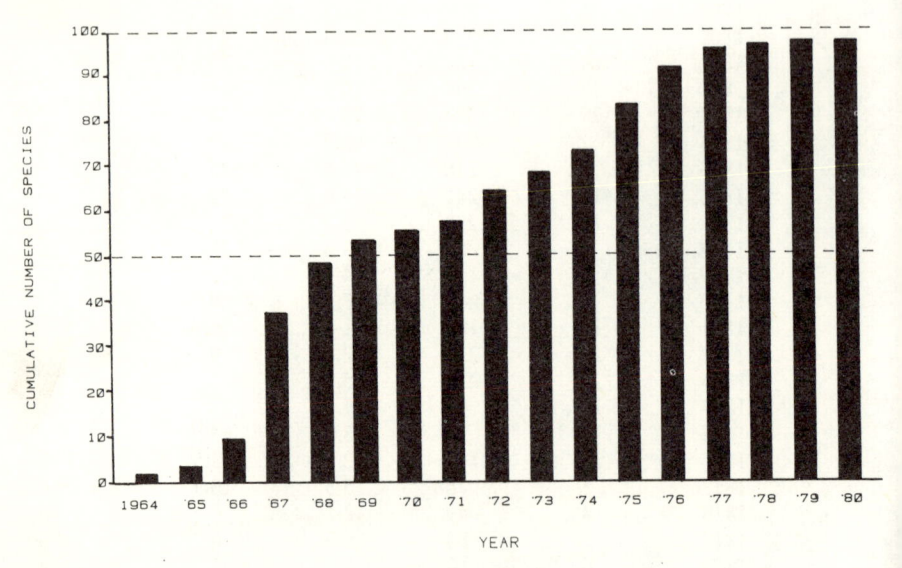

Figure 7.1.5 The cumulative number of fish species recorded in the Thames between Kew and Gravesend from 1964 to 1980

the Thames Navigation Acts of 1788, 1795 and 1812. Not only were these a considerable impediment to the returning adult salmon and sea trout, but they altered the flow of the river over or through the spawning beds, which were in some cases also dredged out in the interests of navigation (Thames Water Authority, 1977).

Other species sought commercially in the estuary were smelt *Osmerus eperlanus*, lampern *Lampetra fluviatilis*, shads *Alosa fallax* and *A. alosa*, eel *Anguilla anguilla* and flounder *Platichthys flesus*. Wheeler (1969) describes how stocks of these fish in the estuary below London had also been affected by pollution by the second or third decades of the nineteenth century, and how the number of fish and fish species in the Thames declined thereafter. A revival towards the end of the century is suggested by the records of lampern and smelt in the reaches upstream of London, possibly correlated with the improvement of the sewage effluent discharges due to methods adopted in 1889–91, which temporarily improved oxygen levels in the estuary. The improvement was shortlived (Figure 7.1.2) and, with the exception of eels, no fish seem to have been present in the lower Thames from around Richmond, 20 km above London Bridge, to Gravesend, 40 km below London Bridge, from about 1920 through to about 1960 (Wheeler, 1969). It was not until 1964, following the rebuilding of Crossness sewage treatment works, that marine fish were reported again above Gravesend. Since then there has been a steady recolonization as indicated in Figure 7.1.5.

7.1.4 RECOVERY OF MACROFAUNA

Andrews and Rickard (1980) considered that recovery of the Thames estuary ecosystem could be separated into two phases, the first following the rebuilding of Crossness sewage treatment works, in 1964, and the second following the commissioning of extensions to Beckton works in 1976. The first phase of recovery involved the return of many fish species between 1964 and 1976, although over that period an equilibrium community was not achieved. The concept of the equilibrium community is described by Gray and Mirza (1979), and occurs where emigrations of species have stabilized and the proportion of individuals per species remains fairly constant.

The invertebrate community was unbalanced during the initial recovery period, especially in the region of Crossness and Beckton outfalls. There, the macroinvertebrates were represented by a super-abundance of one species of tubificid worm, with densities, on occasion, exceeding $600 \times 10^3 \, \mathrm{m}^{-1}$ (Birtwell, 1972), but with no established populations of any other species (Huddart, 1971). The food web shown in Figure 7.1.6 developed only after 1976, following the full commissioning of extensions to Beckton sewage treatment works. Occupying central positions in the web are the ragworm *Nereis* spp., sand hopper *Corophium* spp. and the 'shrimp' group *Crangon* sp., *Gammarus* spp. and mysids

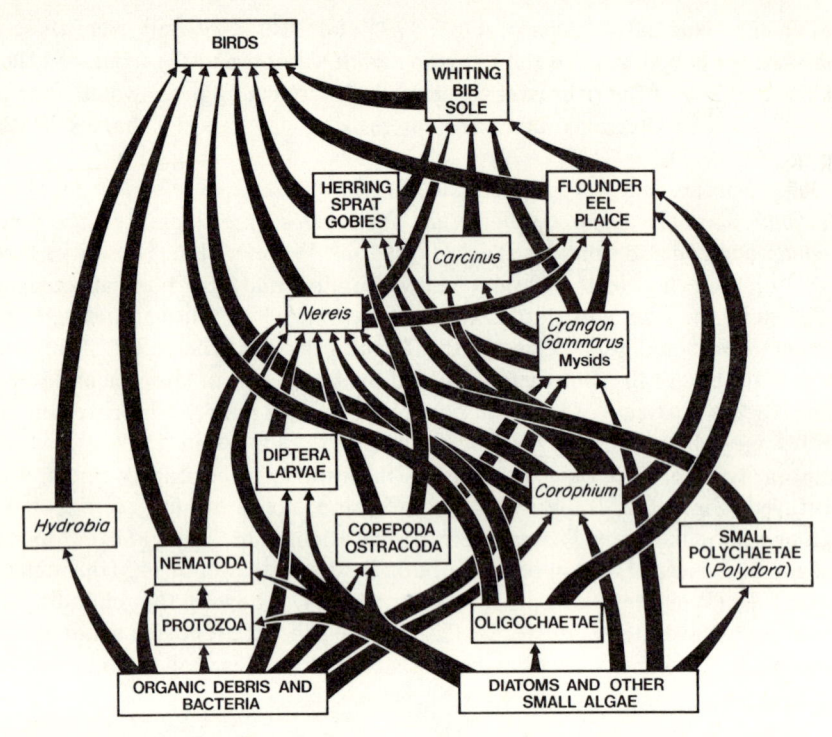

Figure 7.1.6 Food web occurring in the brackish Thames after 1976–77. Before then oligochaete worms occupied a central position, providing the major diet item for both waterfowl and fishes

Praunus sp. and *Neomysis* sp. The food-web interconnections have increased and oligochaete worms no longer occupy a central position, thus improving the buffering capacity of the system, making it less prone to 'crashes and booms'. The formerly super-abundant tubificid *Limnodrilus hoffmeisteri* was replaced in the river a few kilometres above the outfalls by another species *Tubifex costatus*. Interestingly, this change in dominance could be predicted from laboratory studies performed by Birtwell and Arthur (1980) which indicated that *T. costatus* would extend its distributive range following a local improvement in levels of dissolved oxygen.

The improved oxygen level in the water also supported the increased penetration upriver of common shrimp *Crangon crangon*. Laboratory studies reported by Huddart and Arthur (1971) indicated that shrimp would not penetrate above West Thurrock in 1970 because the low oxygen levels at that time bordered on the lethal limit for the species. By 1976 the situation had improved to the degree that shrimp could be trawled throughout the brackish reaches, and penetrated above London to locations more than 50 km beyond West Thurrock.

Fishes returned as water quality improved. The number of species recorded in the metropolitan reaches increased from 3 to 98 between 1964 and 1980 (Figure 7.1.5). Huddart (1971), Sedgwick (1979) and Wheeler (1979) have detailed the partial recovery of fish populations that occurred between 1967 and 1973, and Andrews and Rickard (1980) provided further information for the period 1974–79 which showed that fish communities attained stability in about 1977, when populations indicated the existence of a stress-free environment (see also subsection 7.1.4.7, p. 213).

Livingston (1976) recommended that for comprehensive analysis of community structure for estuarine fishes the following parameters should be included: number of individuals (N), numbers of species (S), species richness (D), diversity index (H') and relative dominance. These measures, along with evenness index (J') and examination for log-normal distribution, were applied to species-abundance data for fish entrapments at West Thurrock Power Station where a total of 67 species representing 32 families and comprising 219 147 individuals was collected in samples taken over the 6-year period 1975–80. (For species-abundance information refer to Andrews and co-workers, 1982.)

Results are described in the following seven subsections.

7.1.4.1 Number of Individuals (N)

Catch rates of fishes at West Thurrock Power Station are standardized with respect to throughput of cooling water and expressed in terms of catch per 0.455×10^6 m^3 (100 M gal) circulating water drawn from the river. This volume of water represents 4 to 6 hours sampling, depending on the number of pumps in operation during a sampling period. Spawning activity of adult fish and the seasonal migration of juveniles to and from the study zone produce cyclical fluctuations in the number of individuals present in the estuary (Figure 7.1.7). Numbers at West Thurrock are lowest in the spring to early summer period when approximately 100–200 fishes are impinged on the power station screens each hour. Numbers steadily increase through the summer and early autumn months to reach a peak late in the year, when on average more than 1000 fishes per hour appear on the screens.

Over the 6-year period there was a slight, but noticeable, improvement in summer catches. Winter catches in the first 2 years of the study were dominated by exceptionally large numbers of whiting and sprat and the abundance of these fish in 1975 and 1976 compared with the period 1977–80 resulted in a slight reduction in total winter peak catches over successive winter periods. However, the trend line using a least squares regression fit does show a slight upward trend when catch data for the whole study period are included.

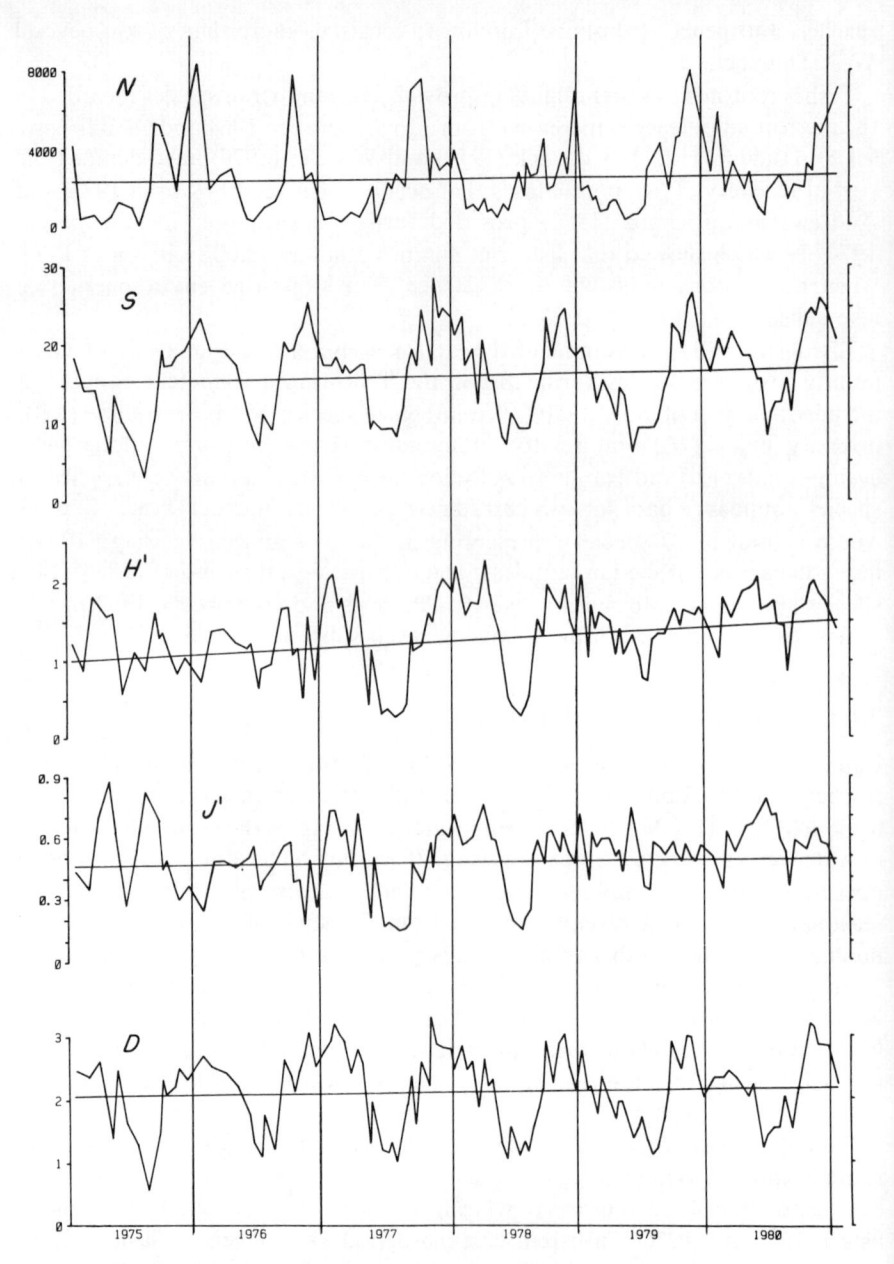

Figure 7.1.7 Results of fish entrapment studies at West Thurrock Power Station, 1975–80, N: number of individuals taken; S: number of species; H' species diversity: J' evenness function; D: species richness

7.1.4.2 Number of Species (S)

Figure 7.1.5 gives a general indication of improvement through the number of fish species present in the middle estuary. The limitations of this method of presentation are twofold: firstly, seasonal improvements are not discernible and, secondly, because species availability is finite, and all the commonly occurring and several of the rarer English East Coast fishes had been recorded in the middle estuary by 1976, any improvements after that date would not be very apparent. Increases in number of species taken at West Thurrock, including invertebrates, during the winter period, when numbers are at a maximum, and during the summer when numbers are low, are shown in Table 7.1.4. Summer catches at the Power Station in 1968–69 were usually nil or one species, whereas a decade later 10 to 15 species were normally present. The number of species recorded in the winter almost trebled over the same period.

In Figure 7.1.7 can be seen the marked seasonal cycle that exists for number of fish species present which is characterized by autumn/winter abundance and spring/summer paucity. The cycle is related to the migration of nonresident species, including bass *Dicentrarchus labrax*, dab *Limanda limanda*, plaice *Pleuronectes platessa* and members of the cod family. Trend analysis for 1975–80 results shows a slight improvement over the period which is more marked than the numerical abundance trend shown previously.

Table 7.1.4 Number of fish, crustacean and jelly fish species at West Thurrock Power Station, sampled quantitatively, in summer and winter periods between 1968 and 1980 (where information is available). Pre-1974 information from Huddart (1971), and Sedwick (1979)

	Summer (May–August)		Winter (October–February)	
	Number of species per sample		Number of species per sample	
	Average	(range)	Average	(range)
1968	0	(0–1)	2	(0–12)
1969	0	(0–0)	5	(0–11)
1970	–		4	(0–13)
1971	2	(0–4)	10	(0–19)
1972	6	(4–10)	–	
1973	–		–	
1974	–		–	
1975	12	(8–15)	24	(20–27)
1976	18	(15–21)	27	(22–31)
1977	18	(15–26)	29	(22–34)
1978	17	(13–21)	26	(18–36)
1979	18	(12–23)	28	(21–33)
1980	21	(16–27)	27	(19–32)

7.1.4.3 Species Diversity (H')

The diversity of each sample was calculated using the Shannon–Weaver Index (Shannon and Weaver, 1963):

$$H' = -\sum_{i}^{s} p_i \log p_i$$

where p_i is the proportion of individuals in the ith species. This index increases as both the numbers of species and equitability of species abundance increase, and thus reflects community structure. It is relatively insensitive to simple species-number effects that may bias a species richness index. Figure 7.1.7 includes plots for H' over the period 1975–80. There is marked fluctuation of H' values in the first 2 years of the study, but in subsequent years a pattern of low summer values and high winter values is generally established, with oscillation becoming less pronounced in the last two years. Summer H' is influenced by the large numbers in summer dominant species, especially juvenile flounder and smelt which congregate in the brackish reaches. Quite pronounced troughs on H' plots can occur in the winter when values are otherwise high, and these are linked with sudden increases in the number of whiting *Merlangius merlangus*, sand goby *Pomatoschistus minutus*, herring *Clupea harengus* or sprat *Sprattus sprattus* entering the estuary. Time trend analysis of H' indicates that diversity increased during the study period, and probably reflects gains in the population of the dominant estuarine species, smelt *Osmerus eperlanus*, and flounder *Platichthys flesus*, which offset the tendency for H' to be reduced with the large influx of winter migrants. The time trend towards increased species diversity is the most pronounced of the five community structure parameters plotted in Figure 7.1.7.

7.1.4.4 Evenness Function (J')

Under certain circumstances the previous index H' may be 'seasonally compensating', and may be insensitive to seasonal fluctuations (Dahlberg and Odum, 1970). The use of the evenness index J' resolves seasonal differences since it relates the observed density to its potential maximum (Pielou, 1966):

$$J' = H'/H' \max = H'/\log S$$

in which log S is the maximum possible value of H'.

The 5-year plots of J' from 1976–80 (Figure 7.1.7) generally provide a similar pattern to those for H'. The pattern for 1975, however, is somewhat different, with peaks in J' not coincident with peaks in H'. J' peaks occur in April and August 1975, at times when total numbers of species are very low, and with dominant species (eel/sprat and eel/flounder, respectively) present in equal numbers. H' values for April and August were not as high as would have been expected from the apparent even distribution of species, because of the low total number of species present on those two occasions. The fitted regression line to all the J' data including 1975, shows only a slight upward trend over the period which is not nearly as marked as in H'.

The decreasing amplitude of fluctuations in J' values for 1979 and 1980 compared with previous years is interpreted as being a measure of gain in diversity—increase in numbers of species—and decreased dominance resulting from pollution amelioration.

7.1.4.5 Species Richness (D)

To measure species richness in sample populations the Margalef index (Margalef, 1969) was employed:

$$D = (S - 1)/\log N$$

The index weights addition or deletion of species more than the numerical changes and, therefore, reflects seasonal migrational fluxes more than the Shannon-Weaver index. The pattern for D over the period 1975-80 (Figure 7.1.7) is one of summer troughs and winter peaks, and is similar to that observed for S. Trend analysis shows an increased function with time, and is confirmation that an increase in diversity has occurred. This has been caused by a true gain of species in some cases, and a shift in population structure in others.

7.1.4.6 Relative Dominance

The proportion of individuals in the dominant and subdominant fish species calculated for each sample is represented as histograms of dominance (Figure 7.1.8).

Dominance was high during all seasons, with the collections usually represented by associations of only a few dominant species plus a considerable abundance of rarer species (Table 7.1.5). A pattern of dominant species replacement did not become established until 1976. In general, winter dominance is relatively low with marine whiting, herring, sand goby or euryhaline smelt being abundant. There follows a change to flounder abundance in the summer, although since 1978 there does seem to have been a trend towards reduced dominance of this particular species, which is also seen as a more evenly balanced fish assemblage as shown by J' values (Figure 7.1.7). The increased incidence of occurrence of the estuarine smelt is apparent in the dominance histograms for 1978 onwards.

7.1.4.7 Log-normal Distribution

Departure from a log-normal distribution of individuals among species offers a sensitive and objective method of assessing perturbation effects on communities (Gray, 1979). In recognition of the fact that large sample sizes are required which

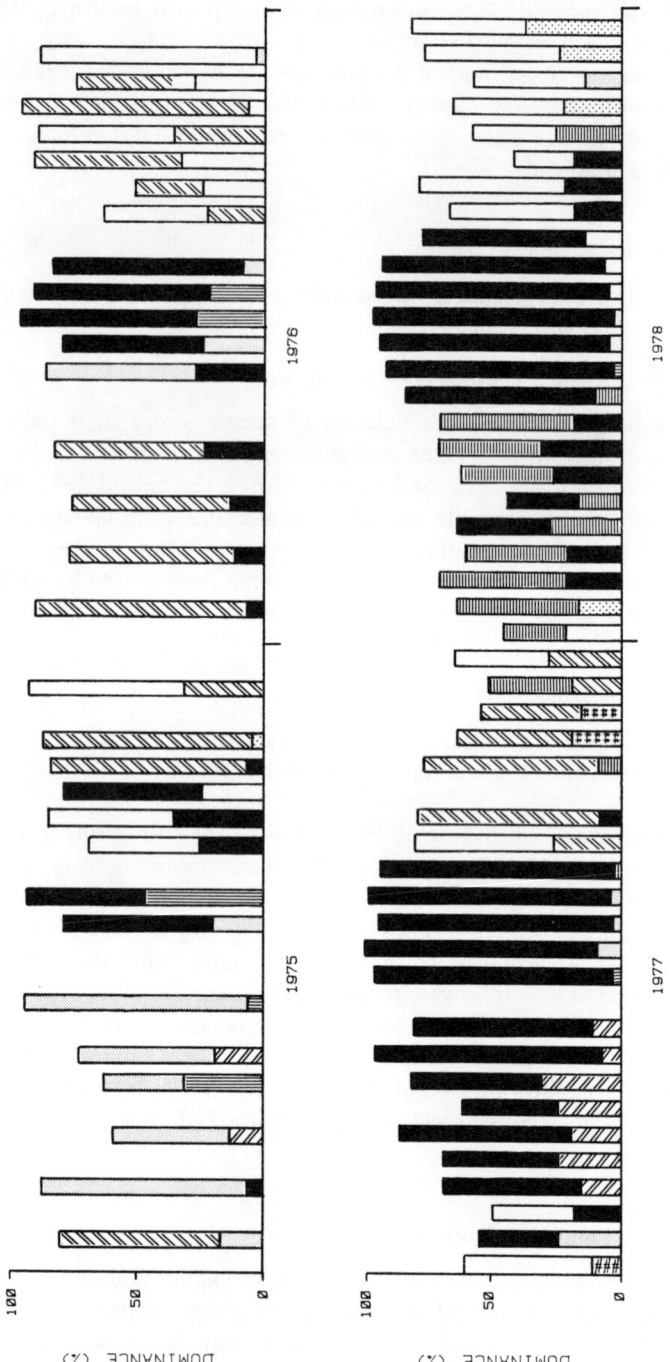

214 EFFECTS OF POLLUTANTS AT THE ECOSYSTEM LEVEL

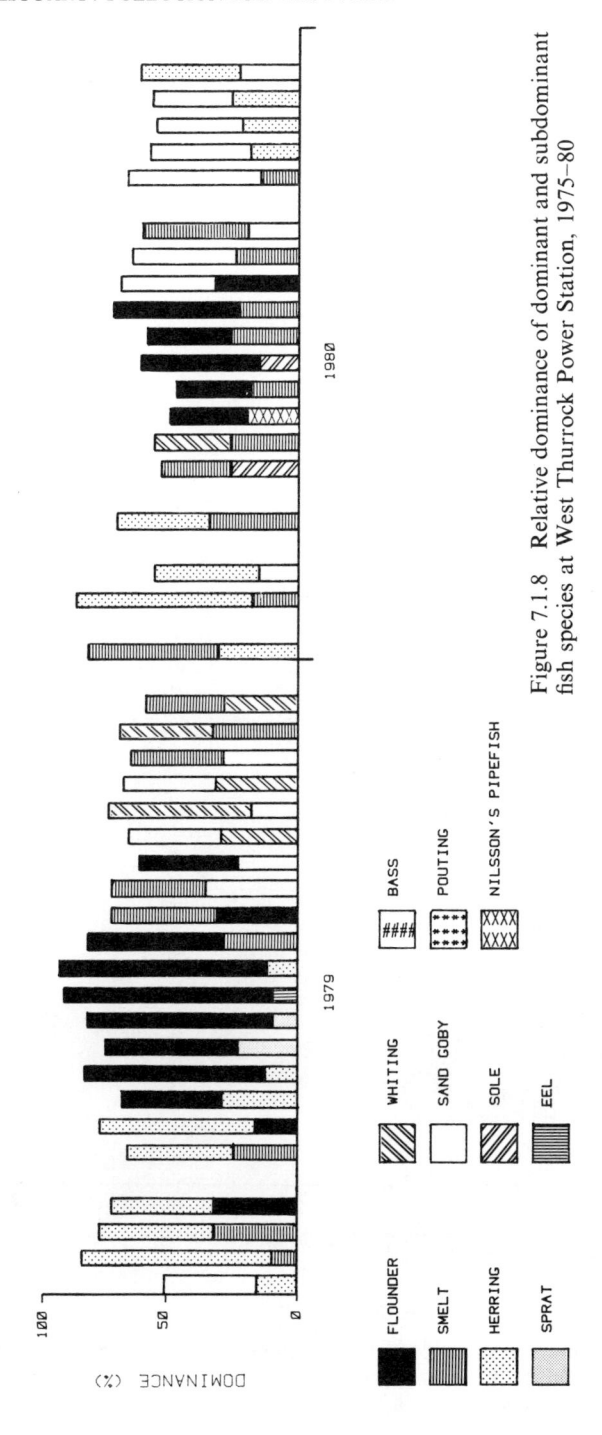

Figure 7.1.8 Relative dominance of dominant and subdominant fish species at West Thurrock Power Station, 1975–80

Table 7.1.5 The 15 numerically most abundant fish species at West Thurrock Power Station over the period 1975–79, as a cumulative percentage of the total power station catch

	Species	Cumulative percentage
Whiting	*Merlangius merlangus*	23.3
Sand Goby	*Pomatoschistus minutus*	42.2
Flounder	*Platichthys flesus*	61.3
Smelt	*Osmerus eperlanus*	75.3
Herring	*Clupea harengus*	85.4
Sprat	*Sprattus sprattus*	89.5
Sole	*Solea solea*	92.2
Pouting	*Trisopterus luscus*	94.4
Eel	*Anguilla anguilla*	95.1
Bass	*Dicentrarchus labrax*	95.8
Poor Cod	*Trisopterus minutus*	96.5
Dab	*Limanda limanda*	97.1
Nilssons Pipefish	*Syngnathus rostellatus*	97.5
Sea Snail	*Liparis liparis*	97.9
Hooknose	*Agonus cataphractus*	98.3

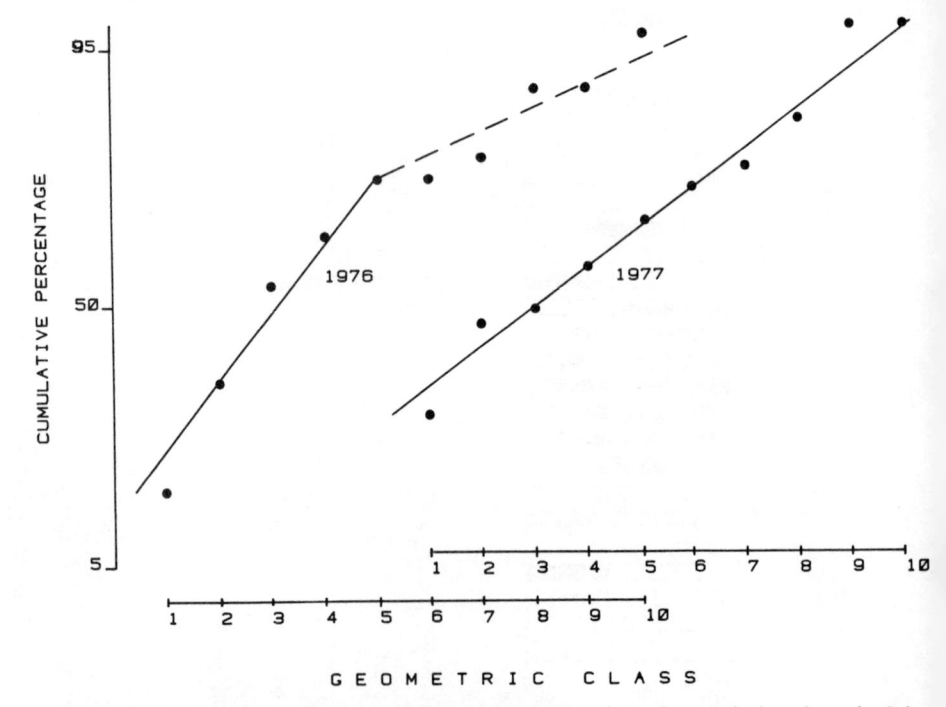

Figure 7.1.9 Log-normal plots of fish community data for periods of peak fish abundance at West Thurrock Power Station, 1976 and 1977, which indicate a change from a polluted to a nonpolluted regime in the Thames occurring between these years

represent a heterogeneous species assemblage (Gray, 1979; Gray and Mirza, 1979) the Thames data chosen for plotting were those for fish captures at West Thurrock during each winter period of peak numerical abundance.

Examples of results obtained for plots of cumulative percentage species on probability paper against species grouped in geometric classes are given in Figure 7.1.9. The straight line obtained for 1977 results represents a log-normal distribution of individuals among species and contrasts with the distinctly broken line obtained for 1976. The break-point observed for 1976 data indicates a skewed distribution in which the common species (geometric classes 8–10) are more abundant than in 1977 and the rarer species (geometric classes 1–4) are less abundant.

These plots may be interpreted as indicating the removal of pollution stress, in terms of its effect on fish communities, from the critical brackish reaches of the Thames occurring between 1976 and 1977 (Andrews and Rickard, 1980).

7.1.5 FISH COMMUNITIES AND WATER QUALITY

The Marine Pollution Subcommittee of the British National Committee on Oceanic Research identified as being urgent the need to separate the effect of pollution-induced and natural changes on estuarine populations or ecosystems, based on long-period study (The Royal Society, 1979). The Thames situation offerred a rare opportunity to study the process of change in an ecosystem, and is unique in that major population changes occurred over a relatively very short time period.

In attempting to establish some cause–effect relationship between environment and population success, Green and Vascotto (1978) point out the inappropriateness of any statistical model which assumes a linear relationship between species success (abundance, presence, frequency, diversity or any other measure) and environmental variables. Such a relationship is assumed by simple linear or multiple regression of species on environmental variables, but it is also assumed by multivariate methods such as canonical correlation, principal components or factor analysis when both biotic and environmental variables are included in the same analysis.

In this study the fish data is reduced by classification (cluster) analysis of the species abundance information to provide a relatively small number of groups, or clusters, which are characterized by homogeneous species assemblages. The environmental data, on the other hand, is structured by reducing the numerous test variables to a similar number of variables (components) which account for much of the variation in the original data, using the method of principal component analysis (PCA). This is a very useful technique where variables are highly intercorrelated which has been shown to be the case in the Thames estuary (Department of Scientific and Industrial Research, 1964). Results of classification and principal component analysis are compared in an attempt to match discrete clusters derived from biological information with discrete groupings

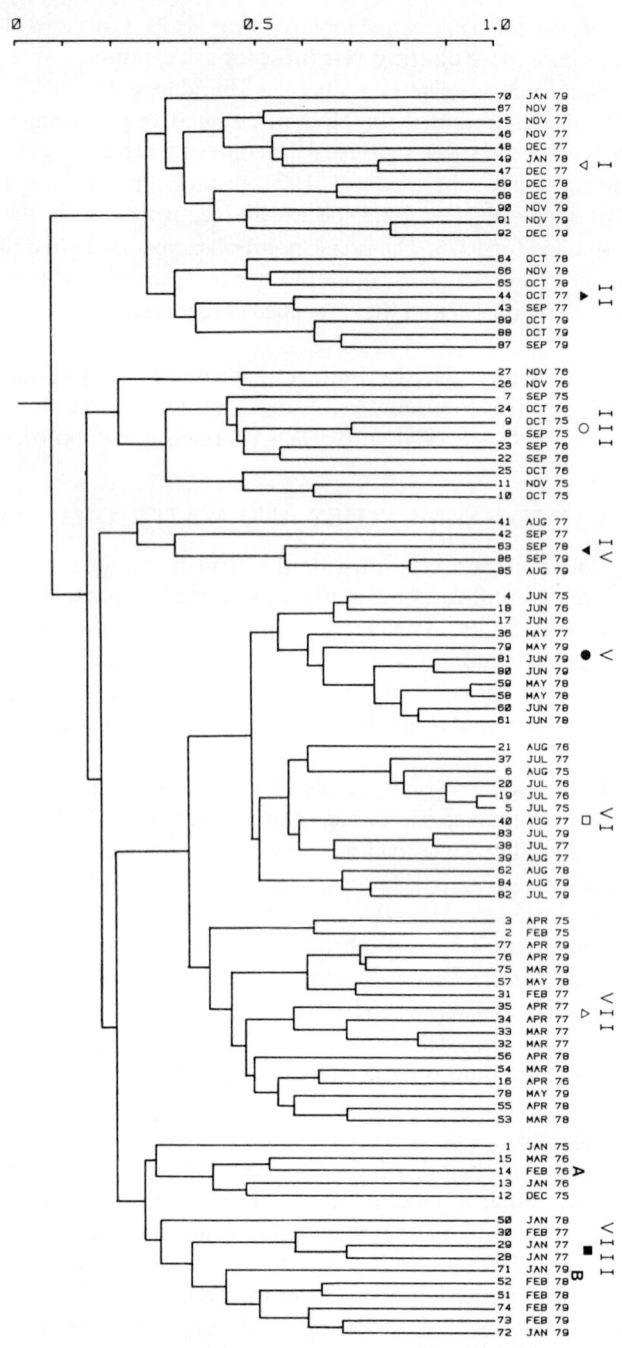

Figure 7.1.10 Cluster analysis plot of 92 fish assemblages sampled quantitatively between 1975 and 1979 at West Thurrock Power Station

provided by PCA of environmental data, with the objectives of linking fish assemblages with specific water states.

7.1.5.1 Cluster Analysis of Biological Data

The species abundance information used in this analysis is that obtained for catches of the 15 most commonly occurring fish at West Thurrock Power Station (Table 7.1.5) on the 92 sampling occasions between 1975 and 1979. Analysis for discrete communities or 'recurrent groupings' is performed using a measure of biotic similarity, B, between samples (Pinkham and Pearson, 1976), where:

$$B = \frac{1}{k} \sum_{i=1}^{k} \frac{\text{Min}(x_{ia}, x_{ib})}{\text{Max}(x_{ia}, x_{ib})}$$

with

k = number of different taxa in all samples compared.
x_{ia} = number of individuals in taxon i of sample a.
x_{ib} = number of individuals in taxon i of sample b.

This similarity coefficient is incorporated in a computer programme which displays a number of useful options (Pinkham et al., 1975).

For the analysis the options chosen are:

1. Clustering of the matrix by sampling dates, according to the degree of similarity of fish assemblages on each date.
2. Mutual absence matches (i.e. 0/0) considered as significant.
3. The size of the group of individuals of a species considered as important.

The programme generates a matrix of measures of biotic similarity between all possible pairs of original samples, and then analyses the matrix for underlying relationships between the samples.

The output from the computer programme is displayed as a dendrogram shown in Figure 7.1.10. Eight discrete clusters can be distinguished, which in general are related to season. Estuarine fish populations in other estuaries have shown similar marked seasonality (see McErlean et al., 1973; Livingston, 1976; Hoff and Ibara, 1977; van den Broek, 1979). However, an interesting separation of groups identified as II and III in Figure 7.1.10 occurs which is not explainable in terms of season, since both clusters represent mainly September/October communities. The significance of this group separation is discussed in the next section.

7.1.5.2 Principal Component Analysis of Physico-chemical Data

Physico-chemical information on effluent and estuary water quality was obtained for each of the 92 occasions that fish samples were collected from West Thurrock Power Station. The variables input to factor analysis are as follows:

T: water temperature (°C)
Cl: chlorinity (mg l^{-1} Cl^{-})
$D.O.$ dissolved oxygen (% air saturation value)
$D.O.'$: dissolved oxygen (mg l^{-1})
$\Sigma.O.$: calculated total oxygen reserves in the brackish region of estuary between London Bridge and Gravesend (tonnes)
$F(t)$: average river flow gauged at Teddington Weir over the previous week (m^{3}s^{-1})
$F(b)$: average flow measured at Beckton sewage treatment works over the previous week (m^{3}s^{-1})
$Poll\ (B)$: pollution load from Beckton sewage treatment works for the previous week, measured as effective oxygen load (EOL^1, tonnes d^{-1})
$Poll\ (\Sigma)$: total pollution load from Beckton, Crossness and Long Reach sewage treatment works for the previous week, (as EOL, tonnes d^{-1})
$Tide$: difference in height between high and low water (m)
$\Sigma O(-1)$: calculated total oxygen (see $\Sigma.O.$) previous week (tonnes)
$\Sigma NH(-1)$: calculated total ammonia (see $\Sigma.O.$) previous week (tonnes)
$\Sigma NO(-1)$: calculated total oxidised nitrogen (see $\Sigma.O.$) previous week (tonnes)
$\Sigma Cl(-1)$: calculated total chloride (see $\Sigma.O.$) previous week (tonnes)
$\Sigma BOD(-1)$: calculated total oxygen demand (see $\Sigma.O.$) previous week (tonnes).

In PCA the interpretation is usually done by looking at the loadings (eigenvectors) which show the extent of the correlation of the original variables with the principal components. The components are then named accordingly. The eigenvectors for the analysis of physico-chemical data are given in Table 7.1.6. PC 1 gives high positive weightings to temperature (T) and high negative weightings to oxygen reserves in the water ($D.O.'$, $\Sigma O(-1)$, $\Sigma.O.$) and to river flow ($F(t)$). Thus PC 1 may be interpreted as an index of season. It tells us that in summer when water temperatures are high, rainfall and hence upriver flow may be expected to be low and oxygen reserves in the estuary are also low (upland flow is a major source of dissolved oxygen (Department of Industrial and Scientific Research, 1964). The opposite applies in winter. PC 2 gives high positive weightings to pollution load from sewage works ($Poll\ (B)$. $Poll\ (\Sigma)$) and high negative weightings to dissolved oxygen ($D.O.$) in the water. This is thus an index of pollution in the estuary caused by sewage effluents, which, when they are of poor quality reduce the percentage saturation of oxygen in the receiving water.

The interpretation of PC 3 is difficult. The highest vector weighting is given to chlorinity ($\Sigma Cl(-1)$) in the week prior to sampling. There is also a negative correlation between present pollution load ($Poll.\ (\Sigma)$) and oxygen demand of the river water of the previous week ($BOD(-1)$). This might be explained by

[1] Effective oxygen load $(EOL) = F \times 3/2\ (B+3N) \times 10^{-6}$ tonnes d^{-1}; where F = flow, m^{3}d^{-1}; B = biochemical oxygen demand, mg l^{-1} (five day, nitrification inhibited); N = oxidizable nitrogen, mg l^{-1}.

Table 7.1.6 Eigenvectors and eigenvalues for principal components of environmental variables

Variable	Eigenvector for component					
	1	2	3	4	5	6
T	0.333	0.027	−0.153	0.132	−0.089	0.014
Cl	0.288	−0.265	0.297	−0.064	0.067	−0.009
D.O.	−0.250	−0.403	0.218	−0.120	−0.199	0.218
D.O.'	−0.311	−0.309	0.148	−0.108	−0.138	0.075
Σ.O.	−0.330	−0.263	0.111	−0.108	−0.109	0.098
F(t)	−0.318	0.242	−0.142	0.114	−0.139	0.061
F(b)	−0.239	0.090	−0.134	0.397	−0.509	−0.149
Poll (B)	−0.163	0.491	0.350	0.090	−0.019	0.127
Poll (Σ)	−0.198	0.431	0.394	−0.005	0.144	0.168
Tide	−0.046	−0.231	−0.178	0.714	0.368	0.463
ΣO(−1)	−0.330	−0.157	0.065	−0.018	0.126	0.028
ΣNH(−1)	−0.282	−0.012	−0.038	0.218	0.068	−0.390
ΣNO(−1)	−0.236	0.037	−0.049	−0.178	0.653	−0.066
ΣCl(−1)	0.107	−0.168	0.553	0.403	0.103	−0.539
ΣBOD(−1)	−0.240	−0.036	−0.383	−0.098	0.158	−0.454
Eigenvalue	7.111	2.136	1.376	1.044	0.957	0.685
Cumulative % of variability	47.4	61.7	70.8	77.8	84.2	88.7

sudden changes in weather sequence affecting water quality and occurring on or just before the sampling date, e.g. summer storms.

PC 4 is also difficult to name. The large weighting given to 'Tide' indicates a possible relationship with the neap/spring tide cycle. A simple but arbitrary rule when deciding on the number of components which have practical significance is to consider only those components which have eigenvalues of 1.00 or greater (Jeffers, 1967). In this study the first four components, accounting for about 78 per cent of the total variability (Table 7.1.6) could be regarded as of practical significance. Because of difficulties in interpreting many of the components, only values of PC 1 and PC 2 are plotted. Graphical representation highlights the association of data points having similar characteristics with respect to PC 1 and PC 2. However, the objective was to link physico-chemical results with biological information contained in cluster analysis, and hence the positions of coordinates for weeks grouped according to similarity of fish populations (Figure 7.1.10) were indicated on the PC graphs using appropriate symbols (Figure 7.1.11). To avoid data overlap it was found to be necessary to produce three PC 1 verses PC 2 plots. Figure 7.1.11a shows the distribution on PC axes 1 and 2, of weeks grouped in clusters VII and IV. The latter is a small group and clusters fairly well with more separation on PC 2 than 1.

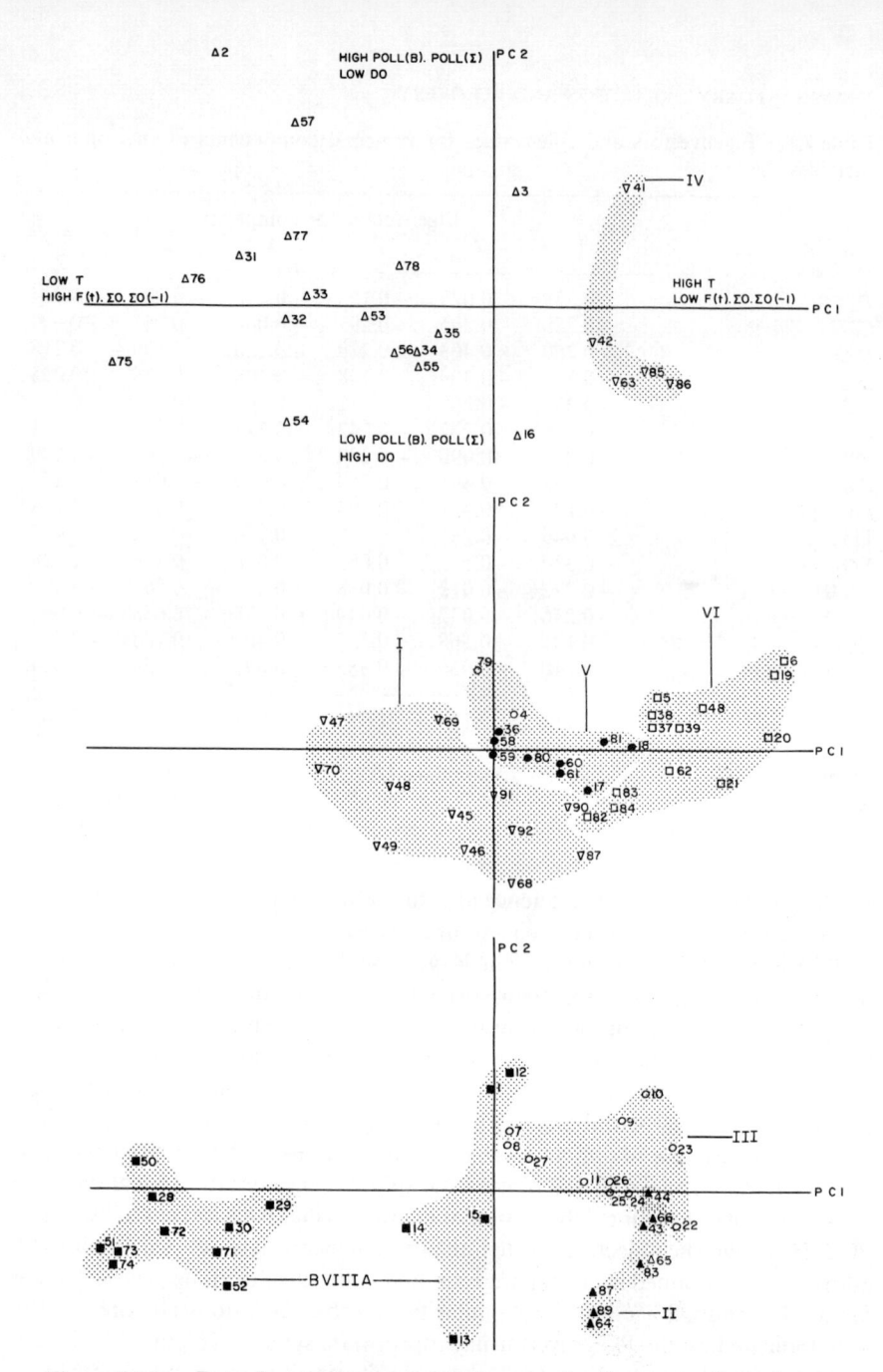

Figure 7.1.11 Data for environmental factors on the first two principal components. The position of groups clustered according to similarity of fish populations (Figure 7.1.10) are indicated by the appropriate symbols: (a) groups IV and VII; (b) groups I, V and VI; (c) groups II, III and VIIIA/B

Group VII is larger and ill-defined, the points being scattered in an apparently random manner along the two axes. It represents unstable or changing environmental conditions in the estuary which occur during the spring, and may account for the finding that fish populations themselves are relatively unstructured at that time of year. The numbers of species are at a minimum and the timing of the change from 'winter' to 'summer' assemblages is unpredictable. Often the influx of summer migrants begins as early as March, but is then halted and the community regains a winter-type fish distribution for several more weeks.

Figure 7.1.11b shows the separation of weeks grouped in clusters I, V and VI when plotted on PC 1 and PC 2. These clusters represent groupings that recur in November/December, May/June and July/August, respectively. As could be predicted, major separation on the principal components graphs is along the 'seasonal' axis, PC 1, although there is considerable overlap along this axis of a few I and V components. This overlap can be explained in terms of river flows which are abnormal for the season, e.g. after high rainfall in summer or dry periods in winter, the flow being a significant contributory factor to variance along PC 1.

Figure 7.1.11c separates groups II, III and VIII on PC 1 and PC 2. The latter group has been subdivided according to an indicated intragroup separation in the cluster analysis. Subgroup VIIIA includes winter/early spring data for 1975/76, while VIIIB covers a similar season but over the period 1977/79. The separation which is observed along PC 1 of data from these two subgroups is not in this case so much a seasonal separation, but is linked more with river flow. The winters of 1975 and 1976 were exceptionally dry, both falling in a drought period, whereas in 1977/79 rainfall was near normal.

The unexplained separation of cluster groups II and III has been mentioned already. When physico-chemical components for each of these groups are separated on PC 1/PC 2 axes there is shown to be little separation along the seasonal axis, PC 1, since both relate to similar September/October periods. More obvious separation occurs along the 'pollution' axis PC 2, with results for 1975/76 tending towards 'high pollution' and those for 1977/79 occupying 'low pollution' positions. Bearing in mind the change that occurred in community structure plots over the same period (Figure 7.1.9), it may be surmised that physico-chemical conditions incorporated in group III on the PCA plots provide an environment such that the development of fish communities in still limited by that environment, while those conditions represented by II on PCA allow for a normal untressed fish assemblage.

Further investigation shows that while there is not a significant difference between mean values of 'seasonal' factors incorporated in II and III, the mean value of $D.O.$ in group II is 51.4 per cent ASV which significantly differs from a mean of 42.5 per cent for group III ($P<0.025$). Other 'pollution' factors that contribute to variance along PC 2 are $Poll(B)$ with a significant difference ($P<0.01$) between mean values of group II (19.8 tonnes d^{-1}) and group III (35.1

tonnes d^{-1}) and *Poll* (Σ) where means are also significantly different ($P < 0.005$) between group II values (118.3 tonnes d^{-1}) and those of group III (168.4 tonnes d^{-1}). Using these figures it is possible to predict the level and kind of pollution which will cause only minor environmental change. This is important because it would be a great economy of resources to discharge as much waste as possible to the estuary to the point where only minor perturbations of the local ecosystems are caused. The view has been taken that minor perturbations are those which do not upset the equilibrium of the fish community. The considerable respiratory and cardiovascular adaptive responses of fish to change in, for example, *D.O.*, are not necessarily indicative of ecologically important functions, and probably should not be used as a basis for judgements on *D.O.* requirements (Alabaster, 1973).

7.1.6 CONCLUSIONS

Species abundance information for fish from the Thames estuary was treated in several ways in an attempt to quantify the changes that occurred as water quality improved. Number of individuals (N), number of species (S), species diversity (H'), evenness function (J') and species richness (D) all proved to be sensitive to seasonal change. The most marked responses in trend analysis of each characteristic over 6 years were observed with indices of species diversity (H') and species richness (D). Changing dominance patterns over the 6 years were detectable using the log-normal model of community structure.

The difficulty of quantifying the relative effects of natural and pollution derived influences was resolved by using the data reduction methods of cluster analysis for biological information, and principal components analysis of physico-chemical information, and then comparing results obtained. It was found that to achieve a 'normal' community, as measured by a log-normal distribution of individuals among species, a level of dissolved oxygen near to or slightly in excess of 50 per cent ASV is required. The principal components analysis also suggests ways in which this may be achieved through the control of the effective oxygen load imparted to the estuary by way of sewage effluents.

The level of 50 per cent ASV compares well with the tentative minimum *D.O.* of 5 mg l^{-1} which Alabaster (1973) suggests would be a satisfactory limit for most of the processes required for a successful fish life-cycle (5 mg l^{-1} *D.O.* represents an ASV of 50 per cent at the average autumn temperature in the Thames of 15.5°C).

The successful clean up of the Thames shows that it is possible to redeem even an extremely polluted aquatic ecosystem. Estuaries were identified by the Royal Commission on Environmental Pollution (1972) as areas requiring priority attention since they remain more vulnerable to pollution than any other part of the British environment. The Thames is of paramount importance, providing a sink for approximately one third of the sewage of Great Britain. In dry weather the flow of sewage effluents entering the estuary is four times as great as the freshwater flow over Teddington weir.

The major aim provided by the Royal Commission on Environmental Pollution (1972), which was to exploit the estuary for waste disposal up to a level which does not endanger aquatic life, has been achieved. Not only has this involved an enormous amount of foresight and scientific investigation, but also the expenditure of large sums of money. At least £100M has been spent over the last 20 years or so to transform the estuary from an anaerobic tidal channel, through Greater London and further downstream, to an aerobic tideway supporting normal fisheries, and able to support the passage of migratory salmon. Apart from the desirability of providing a healthy environment, one important issue should not pass without comment. In the drought of 1976, virtually all freshwater flow of the Thames was taken for public supply, giving an additional yield of some 772 tcmd (179 mgd). If the estuary had not been cleaned up none of this additional yield could have been taken without enveloping riverside London in a noxious and poisonous atmosphere of hydrogen sulphide of unprecedented and totally unacceptable proportions. In the event, the estuary, with no significant freshwater input over many months, reached its highest degree of cleanliness for almost a century, demonstrating beyond any doubt that the estuary clean-up had given, and can continue to give when necessary, an additional massive yield of freshwater resources for public supply. Thus, by this incontrovertible fact, the improvement in estuary quality can more than pay for itself as a water conservation measure (Fish, 1977).

ACKNOWLEDGEMENTS

The author thanks Dr M. C. Dart, Director of Scientific Services, Thames Water Authority, for permission to prepare this case history for publication, and for the great assistance given by colleagues in the biology and data handling sections of the Directorate. Thanks are also due to the Central Electricity Generating Board, and especially to the manager and chemists at West Thurrock Power Station for allowing access to the screens in order to collect specimens.

7.1.7. REFERENCES

Alabaster, J. S. (1973). Oxygen in estuaries, requirements for fisheries. In *Mathematical and hydraulic modelling of estuarine pollution*. Wat. Pollut. Res. Tech. Pap., Lond.

Andrews, M. J. and Rickard, D. G. (1980). Rehabilitation of the inner Thames estuary, *Mar. Pollut. Bull.*, **11**, 327–332.

Andrews, M. J., Aston, K. F. A., Rickard, D. G. and Steel, J. E. C. (1982). Macrofauna of the Thames estuary, *Lond. Nat.*, **61**, 30–62.

Barrett, M. J., Mallowney, B. M. and Casapieri, P. (1978). The Thames model: an assessment. *Prog. Wat. Tech.*, **10**, 409–416.

Birtwell, I. K. (1972). *Ecophysiological aspects of tubificids in the Thames estuary*. Ph.D. thesis, University of London.

Birtwell, I. K. and Arthur, D. R. (1980). The ecology of tubificids in the Thames estuary with particular reference to *Tubifex costatus* (Claparède). In R. O. Brinkhurst and D. G. Cook (Eds.), *Aquatic Oligochaete Biology*, Plenum Press, New York and London, pp. 331–381.

Casapieri, P. and Owers, P. J. (1979). Modelling the Thames, a management use of a mathematical model. Water Research Centre conference, Keble College, Oxford, April 1979.

Chetham, J. (1688). *The Anglers Vade Mecum, etc.* p. 111.

Cockburn, A. G. (1979). Beckton's part in the great Thames clean-up. *Surveyor*, 15 March, 15–18.

Commission of the European Communities (1979). Proposal for a Council Directive on the quality objectives for the aquatic environment into which mercury is discharged by the chlor-alkali electrolysis industry. *Official Journal of the European Communities*, No. C169, 6 July 1979, 2–10.

Dahlberg, M. P. and Odum, E. P. (1970). Annual cycles of species occurrence, abundance and diversity in Georgia estuarine populations. *American Midland Naturalist*, **83**, 382–392.

Department of Scientific and Industrial Research (1964). Effects of polluting discharges on the Thames estuary. *Wat. Poll. Res. Tech. Pap. 11*, HMSO, London, 609pp.

Fish, H. (1977). Rivers and river quality. At symposium on *Water, the Environment and the Future*, 30 November, The Institution of Municipal Engineers.

Fitzmaurice, Sir M. (1912). Main drainage of London. Descriptive account, with accompanying plans and illustrations, London County Council, 19pp. + plans.

Gilligan, R. M. (1972). Forecasting the effects of polluting discharges on estuaries. Parts 1, 2 and 3. *Chem. and Industry*, **22**, 865–874; **23**, 909–916; **24**, 950–958.

Gray, H. F. (1940). Sewerage in ancient and medievael times. *Sewage Works Journal*, September, 939–946.

Gray, J. S. (1979). Pollution-induced changes in populations. *Phil. Trans. R. Soc. Lond. B.*, **286**, 545–561.

Gray, J. S. and Mirza, F. B. (1979). A possible method for the detection of pollution-induced disturbance on marine benthic communities. *Mar. Pollut. Bull.*, **10**, 142–146.

Green, R. H. and Vascotto, G. L. (1978). A method for the analysis of environmental factors controlling patterns of species composition in aquatic communities. *Water Research*, **12**, 583–590.

Harrison, J. and Grant, P. (1976). *The Thames Transformed: London's River and its Waterfowl*, André Deutsch, London, 240pp.

Hoff, J. G. and Ibara, R. M. (1977). Factors affecting the seasonal abundance, composition and diversity of fishes in a southeastern New England estuary. *Estuarine and Coastal Marine Science*, **5**, 665–678.

Huddart, R. (1971). *Some Aspects of the Ecology of the Thames Estuary in Relation to Pollution.* Ph.D. thesis, University of London.

Huddart, R. and Arthur, D. R. (1971). Shrimps in relation to oxygen depletion and its ecological significance in a polluted estuary. *Envir. Pollut.*, **2**, 13–35.

Jeffers, J. N. R (1967). Two case studies in the application of principal component analysis. *Applied Statistics*, **163**, 225–236.

Livingston, R. J. (1976). Diurnal and seasonal fluctuations of organisms in a north Florida estuary. *Estuarine and Coastal Marine Science*, **4**, 373–400.

Longlands, H. G. and Townsend, G. H. (1980). Sewerage: with particular reference to the Beckton development in East London. In *Reclamation of Contaminated Land*, The Society of Chemical Industry, London, contribution E2.

MAFF (1971). *Survey of mercury in food.* Working party on monitoring of foodstuffs for mercury and other heavy metals. First report. HMSO, London, 33 pages.

MAFF (1973). *Survey of mercury in food*. Working party on monitoring of foodstuffs for heavy metals. A supplementary report: mercury in food HMSO, London, 23 pages.

Margalef, R. (1969). *Perspectives in ecological theory*. University of Chicago Press, Chicago, 111 pages.

McErlean, A. J., O'Connor, S. G., Mihursky, J. A. and Gibson, C. I. (1973). Abundance, diversity and seasonal patterns of estuarine fish populations. *Estuarine and Coastal Marine Science*, **1**, 19–36.

Pielou, E. C. (1966). The measurement of diversity in different types of biological collections. *J. Theoret. Biol.*, **13**, 131–144.

Pinkham, C. F. A., Clontz, W. L. and Asaki, A. E. (1975). A computer programme for the calculation of measures of biotic similarity between samples and the plotting of the relationship between the measures. *Edgewood Arsenal Tech. Rept. EB-TR 75013*, Aberdeen proving ground, Md.

Pinkham, C. F. A. and Pearson, J. G. (1976). Applications of a new coefficient of similarity to pollution surveys. *Journal W.P.C.F.*, **48**, 717–723.

Royal Commission on Environmental Pollution (1972). *Third report*, HMSO, London.

Sedgwick, R. W. (1979). *Some Aspects of the Ecology and Physiology of Nekton in the Thames Estuary, with Special Reference to the Shrimp Crangon vulgaris*. Ph.D. thesis, University of London.

Shannon, E. C. and Weaver, W. (1963). *The Mathematical Theory of Communication*, University of Illinois Press, Urbana.

Taylor, A. (1980). Soaps and detergents and the environment. *Journal of American Oil Chemists Society*, **57**, no. 11, 859A–861A.

Taylor J. R., Humphreys, G. W. and Peirson, F. T. (1935). *Greater London Drainage* (report on), HMSO, London.

Thames Water Authority, (1977). Thames Migratory Fish Committee Report, Thames Water, London, 40 pages.

The Royal Society (1979). The effects of marine pollution: some research needs. A memorandum prepared by the Marine Pollution Sub-committee of the British National Committee on Oceanic Research. The Royal Society, London, 78 pages.

Van den Broek, W. L. F. (1979). A seasonal survey of fish populations in the lower Medway estuary, Kent, based on power station screen samples. *Estuarine and Coastal Marine Science*, **9**, 1–15.

Water Act (1973). HMSO, London, 120 pages.

Wheeler, A. (1969). Fish-life and pollution in the lower Thames: a review and preliminary report. *Biological Conservation*, **2**, 25–30.

Wheeler, A. (1969). The tidal Thames. *The History of River and Its Fishes*, Routledge and Kegan Paul, London, 228 pages.

Wood, L. B. (1980). The rehabilitation of the tidal river thames. *The Public Health Engineer*, **8**, 112–120.

Wood, L. B., Borrows, P. F. and Whiteland, M. R. (1980). Scheme for remedying the effects of storm sewage overflows to the tidal river Thames, *Prog. Wat. Tech.*, **12**, 93–107.

Effects of Pollutants at the Ecosystem Level
Edited by P. J. Sheehan, D. R. Miller, G. C. Butler and Ph. Bourdeau
© 1984 SCOPE. Published by John Wiley & Sons Ltd

CASE 7.2
Clearwater Lake: Study of an Acidified Lake Ecosystem

PAMELA STOKES
Institute for Environmental Studies
University of Toronto
Toronto, Ontario, Canada M5S 1A4

7.2.1	Background and Rationale	229
7.2.2	Type of Ecosystem	230
7.2.3	Pollutant Input	231
7.2.4	Pollutant Behaviour	233
7.2.5	Effects on Species and Communities	236
	7.2.5.1 Phytoplankton	236
	7.2.5.2 Zooplankton	238
	7.2.5.3 Macrophytes	239
	7.2.5.4 Bacteria	240
	7.2.5.5 Benthos	240
	7.2.5.6 Fish	240
7.2.6	Ecosystem Function	241
	7.2.6.1 Shifts in Primary Production	241
	7.2.6.2 Productivity	241
	7.2.6.3 Community Interactions	241
	7.2.6.4 Reduction and Nutrient Cycling	243
7.2.7	Ecosystem Recovery	243
	7.2.7.1 Effect of Base Additions	243
	7.2.7.2 Effects of Phosphorus Additions	244
	7.2.7.3 Fish Stocking	244
7.2.8	Stability and Resilience of the Clearwater Lake Ecosystem	245
7.2.9	Conclusions	250
7.2.10	References	251

7.2.1 BACKGROUND AND RATIONALE

The lake and drainage basin which this study addresses is a small, oligotrophic water body, currently acidic and metal polluted. Since 1973, the lake has been part of the Sudbury Environmental Study by the Ontario Ministry of the Environment. The work of scientists in the MOE group provides a rather comprehensive data base of physical, chemical and biological information. It is assumed that the present condition of the lake ecosystem has resulted from atmospheric inputs of H^+, Cu^{2+}, and Ni^{2+}, both as a direct response to the pollutants as well as higher order effects. Experimental chemical manipulations have been made on nearby lakes, providing some information on the potential for reversibility of the effects of metals and acid.

Lake sediment chemistry combined with dating techniques is available to provide information on historical loading of substances to the lake. A study of diatom frustules in recent sediment provides some evidence for changes in species composition and biomass of diatoms over the past 200 years, which more than covers the period since smelting began in the early 1900's.

Over several years of study on calibrated watersheds, chemical and hydrological data on all inputs to and outputs from the lake and watersheds and mass balances were measured for the lake.

An opportunity therefore exists to present this as an ecotoxicological study at the ecosystem level, indeed, as one of the most complete of its kind. At the same time it has to be recognized that the objectives of certain study components were distinct from those of the present report.

7.2.2 TYPE OF ECOSYSTEM

Clearwater is an oligotrophic lake situated 13 km south-southwest of the INCO smelter near Sudbury, Ontario, on precambrian bedrock. Table 7.2.1 summarizes the physical characteristics. There are several cottages on the shoreline, and there is some recreational exploitation of the area. The major human impact has resulted from deposition of airborne pollutants on the lake and its watershed, and the major source of this pollution is assumed to be the Sudbury smelting complex. The chemical characteristics are presented in a later section, since the chemistry has been extremely modified by the inputs of pollutants.

Table 7.2.1 Clearwater Lake: physical characteristics. (Sources: Dillon et al., 1979; Scheider and Dillon, 1976; MOE, 1981)

Lake area A_0 (ha)	76.5	Bedrock geology: gneiss, migmatite, quartzite	
Lake volume $V(10^5 m^3)$	64.2	Shoreline length L(km)	4.97
Drainage area A_d(ha)	34.0	Shoreline development [1]D_L	1.6
Mean depth $\bar{Z}$(m)	8.3	Flushing rate $\rho(y^{-1})$	0.4
Maximum depth Z_m(m)	21.5	Coordinates 46° 22' latitude 81° 3' longitude	

[1]$D_L = \dfrac{L}{2\sqrt{\pi A_0}}$

7.2.3 POLLUTANT INPUT

The major toxic pollutants are H^+, Cu^{2+}, and Ni^{2+}. For any air pollution scenario, impacts of a local source have to be superimposed upon the regional patterns of air pollutants resulting from long-range transport. For the current status of local versus regional contributions of air pollutants to Clearwater Lake, information is available from a recent study made during an 8-month shutdown of the INCO operation (Scheider et al., 1981). Bulk deposition of hydrogen ions, sulphate and a number of metals was compared prior to and during the shutdown which occurred from September 1978 to June 1979. The results indicated that total copper, nickel and sulphate, over a 12 km radius, decreased by more than 90 per cent during shutdown, supporting the hypothesis that smelting and related operations in the Sudbury area were contributing significantly to the materials deposited on the lakes in the Sudbury area. In contrast, hydrogen ion deposition did not decrease significantly during the same period. The interpretation of this statistic is discussed (MOE, 1980) and comparisons made with deposition measurements for Muskoka-Haliburton, 225 km away. In this latter region, H^+ deposition *increased* during the Sudbury shutdown period, suggesting that the apparent lack of H^+ decrease in Sudbury during shutdown was a result of increased input from long-range transport. This assumes that Sudbury and Muskoka-Haliburton are affected by the same pattern of long range transport, from the same regional air masses. If this assumption is correct, and if there were no local sources of H^+, then the Sudbury region should have shown a comparable increase over the same period. Since there was no such increase, the authors concluded that there was in fact a local *decrease* in H^+ during the shutdown period; from this it follows that there is normally a local source of acid deposition (Dillon, 1981).

Table 7.2.2 Annual bulk deposition of metals, nitrate and sulphates in Sudbury and Muskoka-Haliburton areas. (Sources: NRCC, 1981; Jeffries and Snyder, 1981)

Element/ion	Annual average deposition (mg m^{-2} yr^{-1}) Sudbury S[1]	South-Central Ontario[2]
NO_3^- N	553	350–470
SO_4^{2-}	5300	3000
Cu	136	2.12
Ni	84.0	1.6
Zn	68.7	15.7
Al	53.9	58.8
Mn	5.79	7.56
Fe	2.12	84.7
pH	3.8–4.5	4.03–4.38

[1] January 1976–June 1978
[2] October 1977–September 1978

Information of the type discussed above emphasizes the fact that the contemporary situation for pollution of Clearwater Lake is complex, resulting as it does from a combination of local and regional air pollution. The 'experiment' of the 8-month shutdown permits some clarification of this. However, the authors point out that it does not clarify the situation concerning historical causes of acidification of lakes and watersheds in the Sudbury area.

Rates of deposition for sulphate and metals in precipitation in the Clearwater Lake area are given in Table 7.2.2. These values are for bulk (i.e. wet and dry) deposition. Values for Muskoka-Haliburton (South-Central Ontario) are provided for comparison. The inputs of copper, nickel and sulphate for the Sudbury location are extremely high. Mass balance studies for the calibrated watershed indicated that SO_4^{2-} output was substantially greater than the input measured as bulk deposition, leading to the tentative conclusion that an additional input directly onto the lake surface was occurring. The most reasonable explanation with the current state of knowledge is that there are direct inputs from SO_2 in the Sudbury area (Dillon, 1981). In support of this, the fraction of SO_4^{2-} unaccounted for in the mass balance decreased with distance from Sudbury.

Table 7.2.2 indicates that the pH of the bulk precipitation was low, but no lower than that of precipitation in other parts of Ontario including the South-Central region.

In summary, the major inputs of pollutants to the lake are:

Hydrogen ions: local source and long-range transport distinguished by comparisons between operational and shutdown periods
Sulphate ions: predominantly local source, but some long range transport
SO_2: local source, assumed from the mass balance studies
Cu^{2+} and Ni^{2+}: local source

Each of these components currently impinge directly onto the lake surface, and most of them also come from the watershed. A large pulse of metals and acid occurs with the spring run off. In any terrestrial ecosystem, the chemical composition of rainfall is modified by the tree canopy and by interaction with soils. The watershed of Clearwater Lake is itself disturbed and lacks the forest which is otherwise typical of this region. Without the tree canopy, and with soils themselves polluted, the chemistry of water entering the lake is different from rainwater. In particular, calcium and magnesium are leached from soils by acidic water, and Clearwater Lake has higher levels of these cations than the more remote lakes (Table 7.2.3).

The major nutrients, phosphorus and nitrogen, are of considerable significance to the biota in lakes. The levels of total phosphorus in Clearwater Lake in the 1970s varied from 2.4 to 5.9 μg l^{-1} (MOE, 1981), which is not atypical of softwater Precambrian Shield lakes (Armstrong and Schindler, 1971). Total nitrogen to total phosphorus ratios ranged from 25:1 to 80:1, indicating that P was in shortest supply (Schindler, 1977).

Table 7.2.3 Clearwater Lake, major ions 1973–79, whole lake composite samples. (Source: Dillon et al., 1979)

	1973	1975	1977	1979	ELA[2]	M-H[3]
conductivity, μmhos cm^{-1}, 25°C	89	85	89	85	19	33
[1]H$^+$	51	47	79	39	1	1
$\sum$ cations (H$^+$, Ca^{2+}, Mg^{2+}, Na$^+$, K$^+$)	616	504	567	561	203	271
SO$_4^{2-}$	579	519	550	458	63	170
HCO$_3^-$	0	0	0	0	62	110
$\sum$ anions (SO$_4^{2-}$, Cl$^-$, HNO$_3^-$ – N, HCO$_3^-$)	—	566	616	565	164	298

[1] All ions in μeq l^{-1}.
[2] Values for mean of 40 lakes in the Experimental Lakes Area (Armstrong and Schindler, 1971).
[3] Muskoka-Haliburton, South-Central Ontario.

Inputs of nitrate from the atmosphere in precipitation comprise a significant component of the nitrogen budget in any ecosystem, and normally almost 100 per cent of the incoming nitrate is retained in a vegetated watershed. Levels of nitrate were high in Clearwater Lake, as they are in most severely acidified lakes in the area most affected by the Sudbury operation. The major reason for this is believed to be the lack of vegetation in the disturbed watersheds and associated losses of nitrate to the lake (MOE, 1981).

7.2.4 POLLUTANT BEHAVIOUR

For an aquatic ecosystem, the chemical composition of the water and sediments at any point in time is related to the past history of inputs, outputs and transformations of chemical pollutants, as well as the 'natural' geochemistry of the region. Since there is a flushing of water in a lake, water chemistry reflects inputs and outputs of materials for a lake over a short time scale of years or even months. Tables 7.2.3 and 7.2.4 show the major ions and some of the metal ions measured in Clearwater Lake over the past 8 years. Values for other areas are included for comparison. As might be anticipated from the chemistry of precipitation, H$^+$, sulphate and metal ions in the water are exceptionally high. Bicarbonate is absent and is replaced by sulphate as the major anion, a feature which is characteristic of lakes acidified by acid sulphate deposition. High aluminum concentrations, typically associated with acidification, result from dissolution of Al from soils and bedrock, i.e. aluminum can be perceived as an internally generated contaminant.

The form or species of a metal in water has an effect on its toxicity to biota, mainly by determining its availability. In general, as conditions become more acidic, more of the metal exists in the 'free' or ionic form which, for most

Table 7.2.4 Clearwater Lake, total metal concentrations 1973–79, whole lake composite samples. (Source: Dillon et al., 1979)

	1973[2]	1975	1977	1979	ELA[1]	M-H[4]
[3]Copper	98.3	100	81.3	59.8		<2
Nickel	283	272	278	220		<2
Zinc	47.9	49.0	39.2	31.4		5
Iron	114	132	88.2	55.0		98
Aluminum	420	447	381	272		42
Manganese	328	289	290	282		33
pH	4.3	4.3	4.2	4.4	5.6–6.7	5.6–6.4

[1] Experimental Lakes Area, Patalas, 1971.
[2] Sample was 1 m from bottom in 1973.
[3] All in mg m^{-3} (ppb).
[4] Muskoka-Haliburton, South-Central Ontario.

elements, is the form toxic to biota. Analysis of copper and zinc by anodic stripping voltammetry (Chau and Lum-Sheu Chan, 1974) indicated that 50–80 per cent of these metals in Clearwater Lake were in a free or weakly bound form (MOE, 1981).

In contrast to water chemistry, which reflects only recent years or decades of inputs and outputs, lake sediment chemistry can often be used to provide historical information on a lake and its drainage area over its entire life span.

Sediment cores collected from Clearwater Lake were dated by ^{210}Pb. Sedimentation rates increased at a point about 50 years ($\pm$ 5) before present, but there are uncertainties about recent years because of a mixing of the top 3 cm of the sediment profile. Nevertheless, there is evidence from the sedimentation rates to support the idea that disturbances in the land have occurred over the last 50 years, which coincides with changes in vegetation cover resulting from smelter fume and related damage.

Metal concentrations in recently deposited sediments of Clearwater Lake are shown in Table 7.2.5, with ratios of peak to background levels. The values indicate surface enrichment of Cu, Ni, Zn, Pb, Cd and Fe, and a decrease of Al. This decrease in Al results from acid leaching of Al from sediments.

Profiles of metal concentrations provide a chronology for the deposition of metals, and it is concluded that:

1. The elevation of Cu, Ni, Zn, Fe and Cd, and the decrease in Al occurred 50 years ($\pm$ 5) before present.
2. The elevation of Pb began considerably earlier than 50 years ago, but continued to increase so that the highest levels of Pb are found in the last 3–4 decades (MOE, 1981).

Sediments are generally regarded as sinks for metals. In a series of experimental aquaria, release of copper and nickel across the acidic sediment water interface

Table 7.2.5 Clearwater Lake, metals in sediments. Peak metal concentrations in recently deposited sediments, background levels and ratio of peak to background concentrations for Clearwater Lake sediment cores. Bracketed Al concentrations correspond to recent low values. (Source: MOE, 1981)

	Cu $\mu g\,g^{-1}$	Ni $\mu g\,g^{-1}$	Zn $\mu g\,g^{-1}$	Pb $\mu g\,g^{-1}$	Cd $\mu g\,g^{-1}$	Al $mg\,g^{-1}$	Fe $mg\,g^{-1}$
Core #1 Peak	630	1300	96	89	0.90	(12)	83
Bkgd	37	56	59	4	0.28	27	30
Ratio	17	23	1.6	22	3.2	–	3
#2 Peak	1700	1600	350	150	6.70	(20)	91
Bkgd	63	76	145	3	0.96	28	32
Ratio	27	21	2.4	50	7.0	–	3
#3 Peak	1500	1600	290	160	5.10	(19)	79
Bkgd	60	60	143	3	0.82	25	27
Ratio	25	27	2.0	53	6.2	–	3
#4 Peak	2000	1800	390	150	6.20	(15)	34
Bkgd	83	40	93	3	0.63	21	14
Ratio	24	45	4.2	50	10	–	2
#5 Peak	620	1300	86	120	0.60	(14)	110
Bkgd	38	40	108	4	0.73	24	21
Ratio	16	33	<1.0	30	<1.0	–	5
Average ratio	22	30	2.6	41	6.6	–	3

and into the overlying water was demonstrated for Clearwater Lake sediments which were oxygenated and contained living organisms (Stokes and Szokalo, 1977). If this process is important in the field, the long-term fate of the lake could be affected, even in the event that inputs of metal were controlled or prevented.

An analysis of diatoms in sediment cores was made by Walsh (1977). Her study showed that surface sediments were of lower pH than deeper sediments, and that acid-tolerant diatoms predominated in the top 6–10 cm of each core. Diatoms of alkaline preference were more common below 10 cm and pH-indifferent species were present at all depths but were more common below 4 cm. Cores were dated by ^{210}Pb and by the *Ambrosia* rise, and sedimentation rates were estimated at 1 cm per 10 years. Thus the major shift towards acidic diatoms occurred over the last 60–100 years, which coincides with the development of smelting in the Sudbury area over the past 70 years. Other lakes in the area produced similar results, providing evidence linking acidification with the smelting activities.

By their ecological preferences, diatoms are believed to be reliable indicators of pH (Nygaard, 1956; Patrick *et al.*, 1968) but their preferences for metals are less well known and they may be less useful as indicators of historical metal contamination.

At this point it is appropriate to raise the problem of distinguishing between the effects of acidity and metal contamination in a lake which has both. Clearwater is acidic and polluted by nickel and copper, and the chemical composition of the water is atypical (e.g. sulphate replaces bicarbonate). The effects of lake acidification *per se* are complex, involving direct effects of H^+ on biota, community interactions resulting from changes in community structure, release of metals or change in speciation of existing metals and the response of biota to the metals. Clearly, without experimental evidence, it may not be possible to attribute the biotic changes which have been observed to one specific toxic material. It will become apparent, however, that by comparison with other systems, rather specific interpretations can be made of some of the data in terms of cause and effect.

7.2.5 EFFECTS ON SPECIES AND COMMUNITIES

7.2.5.1 Phytoplankton

Clearwater Lake is extremely transparent and one would expect this clarity to have physical and biological effects. The epilimnion of Clearwater Lake was usually 10–12 m thick compared with the more normal 6–7 m in comparable sized nonacidic shield lakes (MOE, 1981). NRCC (1981) consider that the clarity of the acidic lakes is due to precipitation of organic matter by aluminum and not to reduced algal biomass. Schindler (1980) similarly found increased clarity but no decrease in chlorophyll in an artificially acidified lake. The total annual biomass of phytoplankton in Clearwater varied from 0.3 to 0.7 mg l^{-1}, which is not significantly lower than lakes in Muskoka-Haliburton (Yan, 1979).

Extinction coefficients for Clearwater, measured in 1978, were 0.13 microeinsteins m^{-2} s^{-1} compared with 0.56 for Harp Lake and 0.41 for Red Chalk Lake, both oligotrophic lakes in South-Central Ontario (NRCC, 1981). Because of the unusual light regime, vertical distribution of phytoplankton may be altered in acid lakes, and increased cell concentration may occur at greater depths than in neutral lakes. In fact, vertical chlorophyll profiles indicated that phytoplankton did accumulate in the deeper strata of Clearwater Lake.

Biomass of phytoplankton was originally considered to be decreased in acid lakes. Hendrey *et al.* (1976) suggested this but other data do not consistently support their conclusions. The discrepancies may well be the result of technical inconsistencies. It is of great importance to determine biomass from a sufficient number of samples over the season and from sufficient depths. In Clearwater Lake, there was considerable seasonable variability in biomass. Very significant factors in sampling are the clarity of the water, the depth of the epilimnion and the vertical distribution of phytoplankton in acid lakes. Samples taken from the surface of the epilimnion are, therefore, unlikely to be representative of the phytoplankton community and this may have led to premature conclusions in early studies.

In comparison with softwater shield lakes in the ELA and Muskoka-Haliburton, the number of genera in the phytoplankton of Clearwater Lake was low, and the flora was deficient in the number of genera of Chrysophyceae, Cyanophyceae and Cryptophyceae (Table 7.2.6). The most common genera of algae in Clearwater Lake in a 1973–77 survey were *Peridinium* (Dinophyceae), *Cryptomonas* (Cryptophyceae), *Dinobryon* (Chrysophyceae), *Chlamydomonas* and *Oocystis* (Chlorophyta). Numbers of genera of diatoms and desmids were comparable to those in the less acid lakes, which is not surprising since acid tolerant diatoms and desmids are known to occur (Yan, 1979).

Metal tolerance of algae taxa is less well known, but Yan (1979) suggested that algae present in metal-contaminated water were metal tolerant, pointing out that for other lakes in the Sudbury area, metal tolerance of algae has been demonstrated (Stokes *et al.*, 1973). However, the presence of the algae in metal-polluted waters is only circumstantial evidence that they are metal tolerant, and in some instances the algae when tested under laboratory conditions were less tolerant than would have been anticipated from the total metal levels in the lake (Stokes *et al.*, 1973; Stokes, unpublished). Factors which determine effective metal concentrations include organic content and pH; in other words, not all of the measured metal is 'seen' by the algae. This would be one explanation why tolerance is lower than expected based on total levels. No metal tolerance tests have been made on algae isolated from Clearwater.

In summary, the phytoplankton biomass of Clearwater Lake was not significantly lower than that of neutral softwater lakes, although its vertical distribution was affected by acidification. The species composition and community structure, however, were markedly altered, with decrease in richness and a marked predominance (30–35 per cent of biomass) of dinoflagellates, especially *Peridinium inconspicuum*. This species has been described as dominating the

Table 7.2.6 Clearwater Lake, phytoplankton genera represented. (Source: Yan, 1979)

	Number of genera		
	Clearwater	ELA	Blue Chalk (M–H)
Chlorophyta (excluding Desmidiaceae)	17	32	25
Chrysophyceae (including Prymnesiophyceae)	10	22	15
Cyanophyta	6	18	11
Bacillariophyceae	12	7	11
Cryptophyceae	3	6	6
Desmidiceae	7	5	4
Dinophyceae	4	5	3
Xanthophyceae	0	4	1
Euglenophyta	1	4	0
Total genera	60	103	76

phytoplankton biomass in a number of lakes which were acidic but not nickel and copper contaminated (e.g. Hörnström et al., 1973; Yan and Stokes, 1976). This species is thus considered indicative of acid lakes and suggests that the low pH of Clearwater Lake may influence the phytoplankton composition more than do the metals.

Any phytoplankton which survives at low pH must have the capacity to use CO_2 as its carbon source. Furthermore, experimental work has shown that there is a change in the permeability to ions of the cell membrane at low pH (Mierle and Stokes, 1976; Mierle, personal communication) so that, depending upon the transport mechanisms for nutrient uptake, there may be profound effects on nutrition at low pH.

7.2.5.2 Zooplankton

In Clearwater Lake, zooplankton showed low species richness compared with less acid lakes—an average of 3.7 species of crustacea per sample compared with 11.9 for lakes in Muskoka-Haliburton—and a predominance of one species, *Bosminia longirostris* (Table 7.2.7). *Cyclops vernalis* was often codominant in early spring. *Daphnia* species, important in nonacid lakes, were completely absent from Clearwater. Although biomass of zooplankton in Clearwater was significantly lower than that of the nonacidic lakes, correlation analysis revealed no significant relationship between zooplankton biomass and pH, copper or nickel in a set of acidic lakes which included Clearwater (Yan and Strus, 1980). The lack of such relationships was unexpected, but may have been related to the rather narrow range of pH included in the data set. There were some similarities between the zooplankton species composition of Clearwater and other acidic non-metal-contaminated lakes but one notable difference was in the dominance of *Diaptomus minutus* in acidic lakes (Sprules, 1975; Haydu, personal communication) while *Diaptomus* was very infrequent in Clearwater Lake. Thus there is fairly good evidence that metals influenced the crustacean zooplankton species composition. This is supported by the fact that copper levels alone in

Table 7.2.7 Clearwater Lake, crustacean zooplankton. (Source: Yan and Strus, 1980.)

	Clearwater Lake 1976–78 (Sudbury)	Blue Chalk Lake 1977 (Muskoka-Haliburton)
Average no. species per collection	3.7	11.9
No. animals per litre	14–24	43
Biomass of *Bosminia longirostris* (% of crustacea)	79–93	0.3
Biomass of copepods (% of crustacea)	6–20	48.4

Clearwater exceeded the 16 per cent reproductive impairment concentrations for *Daphnia magna* reported by Beisinger and Christensen (1972).

Changes also occurred in the rotifer community. The rotifers contributed 16–30 per cent of the total zooplankton biomass in Clearwater, compared with approximately 1 per cent in South-Central Ontario lakes. Furthermore the dominant species *Keratella taurocephala* formed up to 98 per cent of the rotifer biomass in Clearwater Lake; this species was also dominant in the acidic lakes of the La Cloche Mountains (Roff and Kwiatkowski, 1977) although rotifers as a group were not abundant in the latter study.

7.2.5.3 Macrophytes

Softwater lakes have typically isoetid-dominated macrophyte flora. In Clearwater Lake *Ericaulon septangulare* was the dominant macrophyte. In this respect it resembled the nonacid lakes of Muskoka-Haliburton. Species richness was poor, with a total of 7 compared with 11 in the nonacid lakes (Table 7.2.8). Macrophyte growth extended to 8 m in Clearwater Lake, compared with 3–4 m in the nonacidic lakes. These values approximated the secchi disc readings (Wile *et al.*, 1981).

The average macrophyte biomass was high in Clearwater Lake (Table 7.2.8)

Table 7.2.8 Clearwater Lake, macrophytes. A: abundant, C: common, O: occasional, R: rare. (Source: Wile *et al.*, 1981)

Species	Clearwater Lake Sudbury	Red Chalk Lake Muskoka-Haliburton
Vascular plants		
Eleocharis acicularis R. & S.	O	R
Eriocaulon septangulare With.	A	A
Isoetes sp.	–	O
Juncus pelocarpus Mey.	O	O
Lobelia dortmanna L.	–	C
Lycopus sp.	R	–
Myriophyllum tenellum Bigel.	O	O
Nuphar variegatum Engelm.	–	O
Sparganium spp.	–	O
Utricularia purpurea Walt.	–	R
Utricularia vulgaris L.	–	R
Mosses and liverworts		
Cladopodiella fluitans (Nees) Buch.	O	O
Drepanocladus exannulatus (BSG) Warnst.	C	C
No. of macrophyte taxa	7	12
Average macrophyte biomass ($g\,m^{-2}$ dry weight)	510 ± 98	71.6 ± 11.2

compared with the nonacidic lakes, whose macrophyte biomass was within the ranges reported in the literature. The high biomass in Clearwater Lake is indicative of the tolerance of the angiosperm *E. septangulare* and the bryophyte *Drepanocladus exannulatus* to low pH and elevated metals; biomass values are nevertheless surprisingly high (Wile *et al.*, 1981) and the authors have not proposed any other explanation.

The simplification which apparently occurs in communities of plants and animals during the acidification process is likely to change many interactions such as competition, grazing and predation, and, in general, it is not surprising to find an increase in the biomass of certain constituents of the food chain. No specific mechanism for the observed macrophyte biomass is suggested.

7.2.5.4 Bacteria

Very little data are available on the reducer communities in the lake. A comparison between the acidic Clearwater and comparable nonacidic lakes has been made by Thompson and Wilson (1973) who considered that the composition and population levels for yeasts and moulds in Clearwater was within the normal range for oligotrophic lakes. Ammonium oxidizing bacteria were not found, indicating limited nitrogen cycling.

Indirect evidence for decreased heterotrophic aerobes in Clearwater Lake comes from comparisons with the neutralized Lohi Lake in which the heterotrophic plate count increased significantly after liming. Concerning specific groups, the sulphate reducers in sediments were relatively low in Clearwater Lake, compared with eutrophic or neutralized oligotrophic lakes even though the necessary substrate was abundant in Clearwater (Scheider *et al.*, 1975). There were relatively low heterotrophic: aciduric ratios—0.176 (1 m below surface) and 0.276 (1 m above bottom) (Scheider and Dillon, 1976).

Functional studies of the heterotrophic microbiota are lacking.

7.2.5.5 Benthos

The benthic macroinvertebrate community in Clearwater Lake was dominated by chironomids; certain taxonomic groups, the molluscs and ephemeroptera, were absent. Actual numbers of benthic invertebrates ranged from 654 to 1172 m^{-2}, which is similar to the values in ELA lakes of 518 to 1642 m^{-2} (MOE, 1981).

7.2.5.6 Fish

No fish were caught in extensive trapping in 1972, no fish were recorded in the lake during the study period 1973, and no reliable historical records are available. Local residents recall seeing fish in a number of the lakes around Sudbury, which suggests that 40–50 years ago the lake did support fishery, but there was no consensus on the species present (Scheider *et al.*, 1975). Numerous studies have documented the disappearance of fish species with acidification (e.g. Beamish and Harvey, 1972) and it is not surprising that Clearwater at pH 4.3 had no fishery. However, the relative importance of metals and hydrogen ions and the time trends of the loss of fish remain unknown.

7.2.6 ECOSYSTEM FUNCTION

7.2.6.1 Shifts in Primary Production

Subjective observations have indicated that with lake acidification, primary production shifts from the pelagic to the littoral and benthic. Many acid lakes have prominent benthic mats of algae and mosses (e.g. Grahn *et al.*, 1974; Hendrey and Vertucci, 1980; Stokes, 1981), and this benthic production is apparently facilitated by the clarity of acid lakes. In Clearwater Lake there was a very high macrophyte production in the littoral zone with a resulting increase in the macrophyte: phytoplankton biomass ratio. To date, no prominent algal or moss mats have been recorded in Clearwater, although the nearby Swan Lake, whose chemistry is very similar, has dense mats of *Pleurodiscus* from the littoral to the mid-lake benthic region.

7.2.6.2 Productivity

Limited data on primary productivity for Clearwater Lake suggest, by comparison with the Muskoka-Haliburton lakes, that there was no significant decrease (Dillon *et al.*, 1979). However, the authors stress that differences in light regimes were not taken into account and that sample numbers were small. Using Schindler and Nighswander's (1970) production efficiency calculation, MOE (1981) showed that efficiencies of conversion of light energy to carbohydrate were similar for Clearwater Lake (Sudbury) and Blue Chalk Lake (Muskoka-Haliburton), but they caution that conclusions are tentative. Nevertheless, this finding would be consistent with the other limited data in the literature on acidic lake productivity (Schindler, 1980). It is at first surprising that the metal burden does not appear to affect productivity in Clearwater Lake, since copper is known to be a potent inhibitor of photosynthesis (Steemann Nielsen *et al.*, 1969). However, if the organisms are metal tolerant then this would be a reasonable explanation for the lack of effect on productivity.

7.2.6.3 Community Interactions

It has been shown (Section 7.2.5) that species composition and community structure of phytoplankton and zooplankton in Clearwater Lake are different from those in circumneutral lakes, and three hypotheses have been made concerning their interactions (Yan and Strus, 1980).

1. Only 50 per cent of the phytoplankton biomass in Clearwater is available as food for the zooplankton.
2. Changes in the structure of the phytoplankton in Clearwater are independent of zooplankton grazing.
3. The contamination of Clearwater results in reduced efficiency of energy transfer from phytoplankton to herbivorous zooplankton.

The dinoflagellate, *Peridinium inconspicuum*, which formed on average 45 per cent of the phytoplankton biomass, is a large organism (average diameter 14 μm) and was never detected in the gut contents of *Bosminia longirostris*, the dominant zooplankton. The balance of the phytoplankton present in appreciable quantity were of a size range considered 'edible'. However, the fact that zooplankton biomass in Clearwater Lake was substantially reduced while phytoplankton biomass and renewal rates were not reduced, makes hypothesis 1 unlikely. The capacity of the zooplankton to exert pressure on the phytoplankton, is also likely to be negligible in comparison with nonacidified lakes, so hypothesis 2 is supported.

Some studies have been made of rates of filtering by zooplankton and values by Haney (1973) for *B. longirostris* were used by Yan and Strus (1980) to estimate filtering rates. They considered that herbivore grazing would probably exert little effect on the phytoplankton, and concluded that for phytoplankton, pH *per se* controlled the community composition in acid lakes. This is supported by other data including Yan and Stokes' (1978) study on experimentally acidified columns in soft water lakes which were not metal polluted.

There is some evidence that the energy transfer from primary to secondary trophic levels is affected by acidification (Smith and Frey, 1971). Filtering rates estimated in Clearwater Lake were low (Yan and Strus, 1980). Recent studies on lakes of a wider range of pH, from pH 7.0 to 4.5 (Haydu, personal communication) have not demonstrated pH related in decreases in filtering rates. However, Haydu's lakes were not metal contaminated. With the extremely low pH and additional contamination by Ni and Cu in Clearwater, it is still likely that energy transfer is less efficient.

It has been suggested recently that the role of invertebrate predators may be a major determining factor for zooplankton in acid lakes (Eriksson *et al.*, 1980). Reference is made in Yan and Strus (1980) to the predator *Cyclops vernalis* which appeared to control the *B. longirostris* population over each season but the long-term control by predators is not discussed further. In some acid lakes, predatory cyclopoids are reduced or absent (Haydu, personal communication). This aspect has not been evaluated in detail for Clearwater Lake, but remains as one of potential importance for an understanding of community dynamics.

7.2.6.4 Reduction and Nutrient Cycling

As a general observation, accumulation of detritus and decreased heterotrophic bacterial decomposition is said to accompany acidification in lakes. This has the potential ultimately to affect nutrient cycling, which would have a feedback effect on all communities. As far as is known, Clearwater Lake is no more nutrient limited than other nonacidic softwater lakes even though P and inorganic carbon are very low. However, more work is required on the effect of H^+, Ni^{2+} and Cu^{2+} on the availability of these nutrients, as well as the effects of the pollutants on bacterial decomposition.

7.2.7 ECOSYSTEM RECOVERY

The chemical status of Clearwater Lake at the present time means that pH is rather stable, being buffered by systems other than the bicarbonate system, and it is likely to remain acidic for a long period of time, even if inputs of H^+ are decreased. For the metals, data during the 8-month shutdown (MOE, 1981) suggest that decreased inputs may be reflected by lower concentrations in the water. However, these data have not yet been reviewed as part of a long-term trend, and until data are available for the years after 1979 such a review cannot be done. It is also important to consider the role which sediments may play in cycling of metals through the water column.

Speculation can be made on the prospects for recovery resulting from intervention in the form of chemical treatments, since several lakes in the vicinity of Clearwater have been the subject of experimental manipulation. 'Recovery' implies a return to the original state and it is perhaps more accurate to use the term 'rehabilitation'. The major objective from a practical standpoint of manipulation of these lakes is to produce a viable fishery. From the scientific standpoint, however, well-designed experimental manipulations can also assist in elucidating the responses of biota to chemical perturbation.

7.2.7.1 Effect of Base Additions

In 1973, Lohi Lake, which is downstream of Clearwater, was neutralized by additions of Ca(OH)$_2$ (Dillon et al., 1979). The pH rose to 7.0, but decreased rapidly and further additions of Ca(OH)$_2$ were made in 1974. Metal levels were substantially reduced by the base additions. Other chemical data are summarized in Table 7.2.9. There were no changes in total P and no hypolimnetic oxygen deficit as a result of the treatment.

After liming, secchi disc transparency decreased, bacteria increased and the biomass of phytoplankton decreased but recovered within a few months. Species composition of phytoplankton changed with a decrease in dinoflagellates and cryptophytes and an increase in chrysophytes, i.e. there appeared to be a reversal

Table 7.2.9 Lohi Lake, effects of base additions. (Source: MOE, 1981)

Parameter	1973	1974
pH	4.4	6.1
Alkalinity mg l^{-1} CaCO$_3$	<0	0.7
Ca^{2+} mg l^{-1}	6.2	7.9
Total Cu mg m^{-3}	84	44
Total Ni mg m^{-3}	254	195
Total Zn mg m^{-3}	48	47

of the acidification process in the phytoplankton response. Zooplankton standing stocks were reduced and did not recover to resemble the communities of nonacidified lakes. The zooplankton biomass eventually recovered after 3 years, but a typical non-acidic community was not established. This is explained by the fact that the need for recruitment of certain zooplankton species may be limiting the process of recovery (Dillon et al., 1979).

7.2.7.2 Effects of Phosphorus Additions

Lohi Lake was not fertilized but data are available for Middle Lake, another of the Sudbury study set. Phosphorus was added to raise total P from 2–5, up to 8 μg l^{-1} in 1975 and 12 μg l^{-1} in 1976. This lake had already been neutralized. Significant increases in phytoplankton biomass and impressive changes in species composition resulted from the phosphorus additions. Most notable was the increase in the biomass of blue-green algae (Cyanophyta), particularly *Mastigocladus*.

The phosphorus had less effect on the zooplankton; Middle Lake resembled Lohi in its zooplankton biomass as well as species composition. Although the increased phytoplankton biomass might lead to the expectation of an increase in zooplankton, the authors (Dillon et al., 1979) point out that *Mastigocladus* is not likely to be a suitable food source for zooplankton.

7.2.7.3 Fish Stocking

As part of the liming study, fish were introduced into the study lakes in Sudbury (Gunn, personal communication; MOE, 1981). No fish had been caught in Clearwater, Lohi or Middle Lake in a 1972 survey. In 1976, after liming, Middle Lake had a pH of 6.2 and metal levels were 60, 470 and 35 μg l^{-1} for copper, nickel and zinc, respectively. Smallmouth bass (*Micropterus dolomieui*), Iowa darters (*Etheostoma exile*) and brook stickleback (*Culae inconstans*) were introduced and in 1977 no fish were captured by traps, gillnets or trap nets. In 1977, brook trout (*Salvelinus fontinalis*) were introduced into Lohi Lake, and

Table 7.2.10 Fish mortality in neutralized Sudbury lakes. (Source: MOE, 1981)

Lake	n	pH	Cu	Ni (mg m^{-3})	Zn	GMST (days)
Middle	5	6.4	53	406	27	3.6(2.1–5.7)[2]
Lohi	3	5.3	43	227	31	2.3(1.3–3.5)[3]
Panache	1[1]	6.5	6	28	6	34.0

[1] In this experiment fish were placed in transportation enclosures and driven over local roads for 2 hours prior to being placed into enclosures. In a later experiment fish were taken from the holding pool on the shores of Lake Panache and placed directly into the enclosure. No mortalities were observed for 50 days in this experiment, indicating that fish survival in the enclosures was possible given suitable water quality.
[2] Range for 5 experiments in Middle Lake.
[3] Range for 3 experiments in Lohi Lake.

again no fish were recovered. The cause of death was concluded to be copper toxicity (Powell, 1977).

In order to test this assumption, experimental enclosures containing rainbow trout were introduced into Lohi and Middle lakes and also into Panache Lake to act as a control for the handling techniques. Table 7.2.10 shows the results. It was concluded that since low pH was not sufficient to have caused the observed mortalities, then metal toxicity was the cause of death. Metal concentration in the gills and liver of the Lohi and Middle Lake fish were higher than those in Panache. Although the liming had decreased the metal concentrations, the combined metal levels were still high enough to have accounted for the death of the fish (MOE, 1981).

7.2.8 Stability and Resilience of the Clearwater Lake Ecosystems

The concept of ecosystem stability is not far removed from the physicist's concept of stability: resistance to perturbation. The complexity of the ecosystem and the time scales of fluctuation may limit the usefulness of physical models in ecology, but conceptually such models have great potential to direct and organize the collection of ecological data, as well as to aid in their interpretation. According to ecological theory, communities and ecosystems have mechanisms which maintain stability. In this sense, stability refers to both structure and function. Components which are believed to affect stability include species diversity, spatial heterogeneity and population interactions such as competition, herbivore–food and predator–prey interactions. Not all ecologists agree on the details of the respective roles of these components and their interactions, but there is little dispute as to the importance of homeostatic mechanisms in the maintenance of an ecosystem's viability. The stable state of a system most certainly is not a static state. Change is normal, which is why the ecosystem is often referred to as a dynamic system.

Changes result in variations of ecosystem properties such as population sizes and species composition which, when they occur in a stable or steady state ecosystem, fluctuate or oscillate around an average or 'normal' condition. Stability is also used in the context of the implicit or explicit assumption that response to a disturbance (a perturbation) is temporary, i.e. that upon removal of the disturbing factor the system will revert to the 'normal' condition. Holling (1978) has referred to this as 'a view of an infinitely forgiving Mother Nature'. This is not the only model of Mother Nature, however. Holling also illustrates by models of population responses to disturbance that there are alternatives to the above mentioned 'Beneficient Nature'. These include 'Ephemeral Nature', in which a disturbance results in instability and the fate is extinction of populations, and 'Mischievous Nature', in which a new domain of stability is eventually reached. For a more detailed discussion of the theoretical and practical aspects of these models, the reader is referred to Holling (1973, 1978).

Resilience is persistence under stress and often is a result of stress. To illustrate this by a simple example we can consider intertidal communities which are regularly exposed and covered by tidal movements, and contrast these with deepwater communities whose environment is much more stable in terms of water level. The regularly stressed intertidal communities are more able to tolerate or adapt to stress. In Holling's words, 'the continual "testing" of these systems gives them the resilience they have'. He juxtaposes the concepts of resilience and stability and considers that the resilience of a system is a more important characteristic of its viability than any fluctuation in the numbers of its components. In fact, 'protection' from influences which cause fluctuation may ultimately *decrease* resilience. Put very simply, the resilient properties of the components and structure of an ecosystem allow the system to persist.

In the case of Clearwater Lake, three questions are of interest concerning the state of the ecosystem:

1. Has the disturbance (60–80 years of metal and H^+ inputs) been so great that the system will not revert to its original state?
2. Is the present state a point on a trajectory towards extinction of whole functional groups, and continuing degradation of the system?
3. Is the system at, or directed towards, a 'new' stable state which is different from the original?

The greatest challenge still lies in knowing what to measure in order to answer these questions. No one has yet produced a satisfactory methodology to monitor an intact ecosystem, even though many pay lip service to the ecosystem or holistic approach. In a practical sense, we are still reduced to monitoring or experimenting with a small component of the system; given the 'right' set of parameters, this can

CLEARWATER LAKE STUDY OF AN ACIDIFIED LAKE ECOSYSTEM   247

yield some very respectable results as the preceding chapters of this book have discussed. The alternative strategy is to 'measure everything' which is certainly unnecessary, but at present there is rarely an adequate theoretical basis for selection of the key or 'rate limiting' components.

In a consideration of the three questions posed above, information has been utilized from the large scale data gathering approach as well as from specific so-called 'reductionist' approaches.

With regard to the first question, which addresses reversibility of the change,

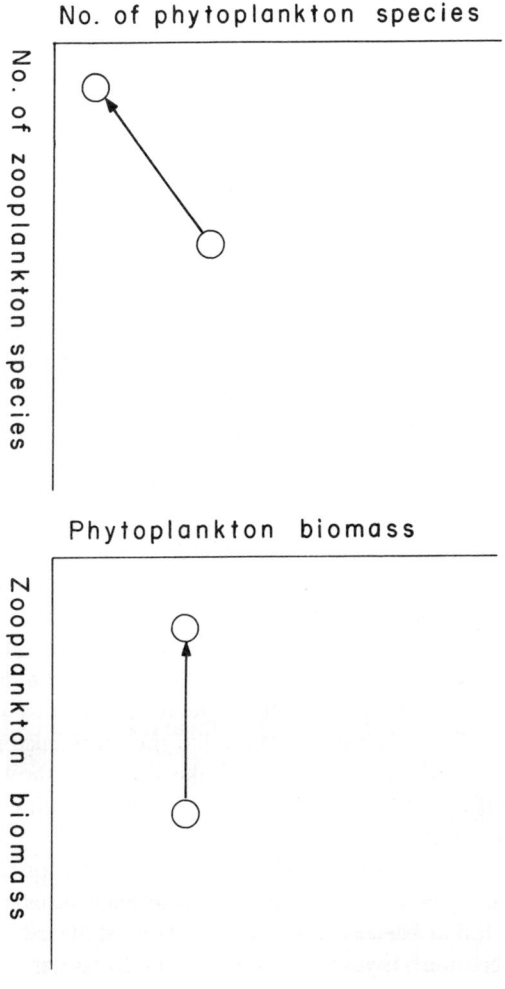

Figure 7.2.1 Schematics for the changes in numbers of species or in biomass for two interacting communities

the extinction of fish populations alone would indicate that a return to the original state will not occur. Furthermore, massive changes in the soils and vegetation of the watershed, and residues of metals in sediments, make it unlikely that recovery will occur, at least over decades or even centuries. In view of this, it is irrelevant to the whole system to consider whether communities of lower organisms, plants, invertebrate animals and microorganisms, would 'recover', although, as the neutralization experiments suggested, the answer to this would be positive at least for the phytoplankton.

In answer to the second question, predictions on the future of the lake chemistry can be made with some confidence, since we can anticipate that the aluminium buffering system will operate to control the pH above 4.0. The present evidence suggests that since a sufficient number of species populations are functioning and interacting in the lake, continued degradation with further loss of functional groups, leading to collapse of the system, is unlikely in the near future. This assumption is based in part on our knowledge of the adaptation of organisms to low pH and metal stress. The literature illustrates many examples of species which are inherently tolerant or which have produced tolerant ecotypes for a wide range of chemical stresses. Confirmation of this relatively constant composition over recent decades could be sought for the diatoms at least, in a rather fine evaluation of community structure in very recent sediments.

Holling's (1973) phase portraits representing models of ecosystem change can be applied to the Clearwater Lake situation. The phase plane axes could be zooplankton and phytoplankton, respectively, since we have data for both. Very simply, the system is perceived as having changed from its original equilibrium to a new state. In Holling's words, the system has 'flipped from one domain of stability to another'. This would imply a positive response to the third question and a negative response to the first and second. To exemplify this, two cases are illustrated schematically in Figure 7.2.1.

This 'new' system is simplified in terms of the richness or diversity of populations, as well as in the number of trophic levels. In this state, it is pertinent to ask whether individual communities show less short-term temporal stability than comparable communities in circumneutral (i.e. less disturbed) lakes. Recently, Marmorek (1982) has suggested that the zooplankton communities in two acid lakes, one of which is Clearwater Lake, do in fact show more short-term fluctuation than those in a set of less acidic lakes, including some situated in South and Central Ontario.

Marmorek computed annual coefficients of variation for total zooplankton biomass for 33 lake-years of data, including four enclosures in a circumneutral lake, two of which had been experimentally fertilized and acidified, and two of which had been fertilized. Figure 7.2.2 supports his hypothesis that the more acid lakes show greater within-season fluctuations, in that the coefficients of variation are much higher for lakes with lower pH. The data set is limited, but the hypothesis (Marmorek, personal communication) has a sound theoretical basis:

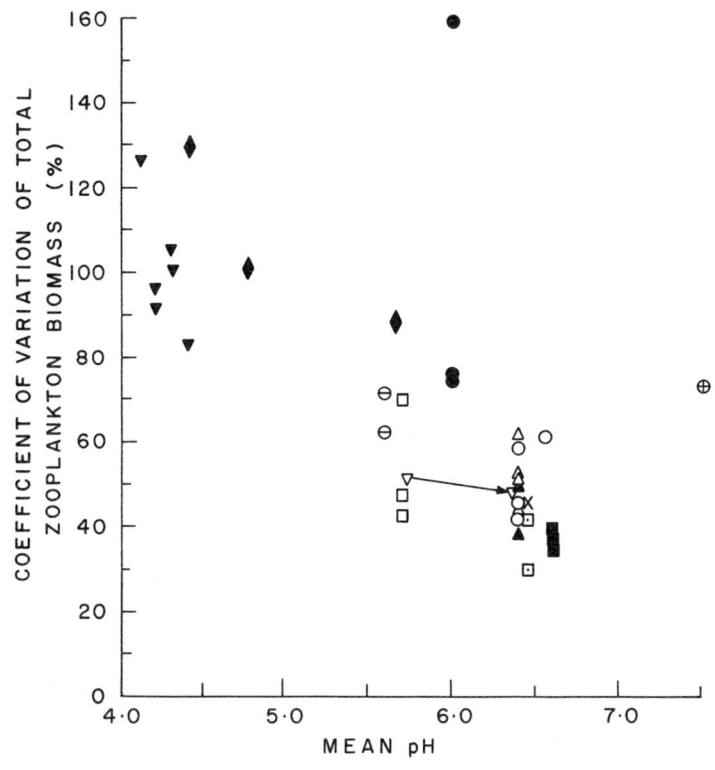

CLEARWATER LAKE STUDY OF AN ACIDIFIED LAKE ECOSYSTEM 249

Symbol	Lake	# Samples	Source	Symbol	Lake	# Samples	Source
▼	Clearwater	9-24	(1)	▲	Red Chalk	20-24	(1)
□	Chubb	23-24	(1)	♦	Cheat	13-14	(2,3)
▽	Nelson	8-9	(1)	⊕	Maggiore	9	(4)
●	Dickie	22-27	(1)	⊖	AF Cylinders	11	
○	Harp	21-25	(1)	⊡	F Cylinders	11	this study
△	Jerry	16-24	(1)	X	Eunice	11	

Figure 7.2.2 Coefficients of variation for May to November, of total zooplankton biomass for 33 lake-years of data (after Marmorek, 1982)

'The coefficients of variation of total zooplankton biomass could vary inversely with pH because of 'holes' in the temporal organization of the communities, with no acid-tolerant species available with the appropriate life history and temperature response physiology to fill them.'

Interestingly, a preliminary analysis of the phytoplankton biomass data for a limited data set including Clearwater does not follow the pattern of pH-related coefficient of variation which was shown for the zooplankton.

Whether or not the idea illustrated in Figure 7.2.2 can be substantiated, it certainly merits further investigation. At present, the need for an understanding of temporal as well as spatial variation over the short term is emphasized. The most productive approach to such questions would be through a series of well-designed experiments, rather than more extensive monitoring. If such decreases in short-term stability turn out to be consistent with simplified community structure, then it would be necessary to determine whether they indicate mere 'noise' around the 'normal' condition, or whether they signal a trend towards extinction.

7.2.9 CONCLUSIONS

1. The acidified metal polluted lake shows major chemical and biological differences and some physical differences when compared with circumneutral lakes of comparable geological and morphometric characteristics. The introduction of the major pollutants H^+, Ni^{2+} and Cu^{2+} and the biological and chemical changes have occurred over the last 60 years, which is a relatively short time span for the life of a lake. For the plants and invertebrate biota however, particularly algae which have short generation times, 60 years is quite long enough for adaptations to have occurred.

2. As a system, the lake appears to be dynamic and functional, with viable communities at producer and consumer levels, and cycling of essential nutrients. Over the recent decade, during which time intensive studies have been made, no major trends can be discerned. It is therefore tempting to speculate that the system as it is now is relatively stable. One significant factor in this stability is the existence of buffering systems which would maintain the pH in the mid-4 range and thus prevent major shifts in pH.

3. In terms of the specific pollutants which are related to the changes observed, pH seems to be the major controlling variable for the phytoplankton community and probably also for the benthos and microorganisms. For the zooplankton, pH is important but the metals are clearly additional controlling factors. The situation for fish is more difficult to assess; however, while the low pH *per se* in Clearwater is sufficient to explain the absence of fish, the metals were toxic even when the pH was raised by liming in Middle Lake, so that both types of stress are likely to be important. Experimental work, e.g. the type described for fish in enclosures, would be necessary to delimit precisely the role of H^+, Cu^{2+} and Ni^{2+} for all of the communities.

4. Recovery to the original condition, in the absence of intervention, cannot be anticipated. One obvious reason for this is the massive damage and major changes sustained by the watershed of the lake. This observation emphasizes the importance of considering a lake ecosystem as the basin plus its surrounding watersheds, and also emphasizes the potential shortcomings of experimental manipulation such as lake or enclosure (limno-corral) experiments to simulate

pollution 'in the field'. Such experiments have considerable value, but caution is needed in the interpretation, especially for long-term responses.

5. Recovery, in the sense of rehabilitation with the objective of restoring a viable fishery, also has to be addressed with guarded optimism. The experimental liming procedure may have been too 'coarse' but it achieved the desired objective in terms of chemical properties. However, the only community which 'recovered' rapidly in response to liming was the phytoplankton. Since other biota may require more time to adjust to the 'shock' of neutralization, it is important to monitor the long-term effects. This particular type of chemical treatment is, naturally, of great interest in view of the large number of lakes whose biota are endangered by acid stress.

6. While much remains to be explained, the approach which was used for the Sudbury Study, i.e. a comprehensive collection of data over an extended period of time, with reference or control sites included, is clearly essential for a study of a system under stress. Parallel experimental enclosures or microcosms should be considered as a valuable adjunct to the descriptive field studies.

7. In the future, it would be efficient and economical for studies on polluted systems if some general indicators or 'vital signs' could be identified in order to avoid the expensive and detailed collection of data. Alternatively, the dynamics of certain indicator species which have been shown to represent rate limiting steps in the over-all system may provide sufficient information to assess the condition of an ecosystem. The older concept of an indicator species, meaning one most sensitive to change, is in itself inadequate in the assessment process, unless there is compelling evidence that the species is also playing a key role in the system.

8. At the present time, on the basis of the Clearwater Lake study, it is not possible to recommend useful generalized vital signs or indicators. The concepts of instability and of high coefficients of variation in simplified systems merit further evaluation.

7.2.10 REFERENCES

Armstrong, F. A. J. and Schindler, D. W. (1971). Preliminary chemical characterisation of waters in the Experimental Lakes Area, northwestern Ontario. *J. Fish. Res. Bd. Canada*, **28**, 171–187.

Beamish, R. J. and Harvey, H. H. (1972). Acidification of the La Cloche Mountain Lakes, Ontario and resulting fish mortalities. *J. Fish. Res. Bd. Canada*, **29**, 1131–1143.

Beisinger, K. E. and Christensen, G. M. (1972). Effects of various metals on survival, growth, reproduction and metabolism of *Daphnia magna*. *J. Fish. Res. Bd. Canada*, **29**, 1691–1700.

Chau, Y. K. and Lum-Shue Chan, K. (1974). Determination of labile and strongly bound metals in lake water. *Water Res.*, **8**, 383–388.

Dillon, P. J. (1981). Calibrated lakes and watersheds in southern Ontario. AMS/CMOS Conference on Long-Range Transport of Airborne Pollutants, Albany, N. Y., April 27–30, 1981.

Dillon, P. J., Yan, N. D., Schlider, W. A. and Conroy, N. (1979). Acidic lakes in Ontario, Canada: characterisation, extent and responses to base and nutrient additions. *Arch. Hydrobiol. Beih. Ergebn. Limnol.*, **13**, 317–336.
Eriksson, M. O. G., Henrikson, L., Nilsson, B. I., Nyman, G., Oscarson, H. G. and Stenson, A. E. (1980). Predator-prey relations important for the biotic changes in acidic lakes. *Ambio*, **9**, 248–249.
Grahn, O., Hultberg, H. and Linder, L. (1974). Oligotrophication—A self-accelerating process in lakes subjected to excessive supply of acid substances. *Ambio*, **3**, 93–94.
Gunn, J. M. Ministry of Natural Resources, Ontario, personal communication.
Haney, J. F. (1973). An *in situ* examination of the grazing activities of natural zooplankton. *Arch. Hydrobiol.*, **72**, 87–132.
Haydu, A. S. Botany Department, University of Toronto, personal communication.
Hendrey, G. R., Baalsrud, K., Tragen, T. S., Laake, M. and Raddum, G. (1976). Acid precipitation: some hydrobiological changes. *Ambio*, **5**, 224–227.
Hendrey, G. R. and Vertucci, F. A. (1980). Benthic plant communities in acidic Lake Colden, New York: Sphagnum and the algal mat. In Drablφs, D. and Tollan, A. (Eds.), *Ecological Impact of Acid Precipitation*, SNSF Project. pp. 314–315.
Holling, C. S. (Ed.) (1978). *Adaptive Environmental Assessment and Management*, John Wiley and Sons, New York, 377 pages.
Holling, C. S. (1973). Resilience and stability of ecological systems. *Ann. Rev. Ecol. and Systematics.*, **4**, 1–23.
Hörnström, E., Ekström, C., Miller, U. and Dickson, W. (1973). Sotvattans-Laboratoriet, Drottingholm NR. 4, 81 pages.
Jeffries, D. S. and Snyder, W. R. (1981). Atmospheric deposition of heavy metals in central Ontario, Canada. *Water Air Soil Pollut.*, **15**, 127–152.
Marmorek, D. R. (1982). *The Effects of Lake Acidification on Zooplankton Community Structure: An Experimental Approach*. M.Sc. thesis, University of British Columbia.
Mierle, G. Botany Department, University of Toronto, personal communication.
Mierle, G. M. and Stokes, P. M. (1976). Heavy metal tolerance and metal accumulation by planktonic algae. In Hemphill, D. D. (Ed.), *Trace Substances in Environmental Health*, vol. x, pp. 113–121.
MOE (1980). Bulk deposition in the Sudbury and Muskoka-Haliburton areas of Ontario during the shutdown in INCO Ltd in Sudbury. Ministry of the Environment, Province of Ontario.
MOE (1981). Sudbury Lakes Report (in press). Ministry of the Environment, Province of Ontario.
NRCC (1981). Acidification in the Canadian aquatic environment: scientific criteria for assessing the effects of acidic deposition on aquatic ecosystems. NRCC No. 18475.
Nygaard, G. (1956). The ancient and recent flora of diatoms and chrysophyceae in Lake Gribsφ. In Berg, K. and Peterson, I. C. (Eds.), *Studies on Humic Acid, Lake Gribsφ*, vol. 8, Folia Limnolog. Scandinavia, Copenhagen.
Patalas, K. (1971). Crustacean plankton communities in forty-five lakes in the Experimental Lakes Area, northwestern Ontario. *J. Fish. Res. Bd. Canada.*, **28**, 231–244.
Patrick, R., Roberts, N. A. and Davis, B. (1968). The effect of changes in pH on the structure of diatom communities. *Notulae Naturae*, **416**, 1–16.
Powell, M. J. (1977). An assessment of brooktrout planting in a neutralised lake as compared to four other Sudbury area lakes. Ontario Ministry of Natural Resources, Fisheries and Wildlife Branch. Unpublished report, 32 pages.
Roff, J. C. and Kwiatkowski, R. E. (1977). Zooplankton and zoobenthos of selected northern Ontario lakes of different acidities. *Can. J. Zool.*, **55**, 899–911.

Scheider, W. A., Adamski, J. and Payler, M. (1975). Reclamation of acidified lakes near Sudbury, Ontario. Ministry of the Environment, Province of Ontario, 129 pages.
Scheider, W. and Dillon, P. J. (1976). Neutralisation and fertilisation of acidified lakes near Sudbury, Ontario. *Water Pollut. Res. Can.*, **11**, 93–100.
Scheider, W. A., Jeffries, D. S. and Dillon, P. J. (1981). Bulk deposition in the Sudbury and Muskoka-Haliburton areas of Ontario during the shutdown of INCO Ltd in Sudbury. *Atmos. Environ.*, **15**, 945–956.
Schindler, D. W. (1977). Evolution of phosphorus limitation in lakes—natural mechanisms compensate for deficiencies of nitrogen and carbon in eutrophied lakes. *Science*, **195**, 260–262.
Schindler, D. W. (1980). Experimental acidification of a whole lake: a test of the oligotrophication hypothesis. In Drabløs, D. and Tollan, A. (Eds.), *Ecological Impact of Acid Precipitation*, SNSF Project, pp. 370–374.
Schindler, D. W. and Nighswander, J. E. (1970). Nutrient supply and primary production in Clear Lake, eastern Ontario. *J. Fish. Res. Bd. Canada*, **27**, 2009–2036.
Smith, R. W. and Frey, D. G. (1971). Acid mine pollution effects on lake biology. U.S. EPA Water Pollut. Cont. Res. Ser. 18050 EEC 12/17, 129 pages.
Sprules, W. G. (1975). Midsummer crustacean zooplankton communities in acid-stressed lakes. *J. Fish. Res. Bd. Canada*, **32**, 389–395.
Steemann Nielsen, E., Kamp-Nielsen, L. and Wium-Andersen, S. (1969). The effect of deleterious concentrations of copper on the photosynthesis of *Chlorella pyrenordosa*. *Physiol. Planatarium*, **22**, 1121–1133.
Stokes, P. M., Hutchinson, T. C. and Krauter, K. (1973). Heavy metal tolerance in algae isolated from contaminated lakes near Sudbury, Ontario. *Can. J. Bot.*, **51**, 2155–2168.
Stokes, P. M. and Szokalo, A. M. (1977). Sediment-water interchange of copper and nickel in experimental aquaria. *Water Pollut. Res. Can.*, **12**, 157–177.
Stokes, P. M. (1981). Benthic algal communities in acidic lakes. In Singer, R. (Ed.), *Effects of Acid Rain on Benthos*, The North American Benthological Society, pp. 119–138.
Thompson, M. and Wilson, D. (1973). 1973 microbiology report on the Sudbury Environmental Study, part A, Lake Reclamation Study. Ministry of the Environment, Province of Ontario. 38 pages.
Walsh, M. (1977). *Effects of Acidification on Diatom Occurrence in Recent Sediments of Four Sudbury Area Lakes*. B.Sc. (Hons). thesis, Trent University.
Wile, I., Miller, G. E., Hitchin, G. G. and Beggs, G. (1981). Species composition, structure and biomass of the macrophyte vegetation in one acidified and two acid sensitive lakes in Ontario. Unpublished MS.
Yan, N. D. (1979). Phytoplankton community of an acidified, heavy metal-contaminated lake near Sudbury, Ontario: 1973–1977. *Water Air Soil Pollut.*, **11**, 43–55.
Yan, N. D. and Stokes, P. M. (1976). The effects of pH on lake water chemistry and phytoplankton in a La Cloche Mountain Lake. *Water Pollut. Res. Can.*, **11**, 127–137.
Yan, N. D. and Stokes, P. M. (1978). Phytoplankton of an acidic lake, and its response to experimental alterations of pH. *Environ. Conserv.*, **5**, 93–100.
Yan, N. D. and Strus, R. (1980). Crustacean zooplankton communities of acidic, metal-contaminated lakes near Sudbury, Ontario. *Can. J. Fish. Aquat. Sci.*, **37**, 2282–2293.

Effects of Pollutants at the Ecosystem Level
Edited by P. J. Sheehan, D. R. Miller, G. C. Butler and Ph. Bourdeau
© 1984 SCOPE. Published by John Wiley & Sons Ltd

CASE 7.3
Comparison of Gradient Studies in Heavy-Metal-Polluted Streams

PATRICK J. SHEEHAN[1], ROBERT W. WINNER[2]

[1] Division of Biological Sciences
National Research Council of Canada
Ottawa, Ontario, Canada K1A0R6

[2] Department of Zoology
Miami University
Oxford, Ohio 45056, U.S.A.

7.3.1 Introduction 255
7.3.2 The Stream Ecosystems 257
7.3.3 Heavy-Metal Inputs 257
7.3.4 Community Analysis 261
    7.3.4.1 Density 261
    7.3.4.2 Species Richness and Diversity 264
    7.3.4.3 Community Composition 266
    7.3.4.4 Coefficient of Variation 267
7.3.5 Notions of Ecosystem Stability Related to Pollution Stress ... 268
7.3.6 Conclusions 270
7.3.7 References 270

7.3.1 INTRODUCTION

The impact of heavy-metal pollution in streams on macroinvertebrates was reported several decades prior to the more recent concern over the potential hazards of synthetic organic compounds (e.g. Jones, 1940). However, there have been few critical evaluations of the specific changes in macroinvertebrate community structure which can be consistently attributed to heavy-metal stress.

The gradient approach, that is, the analysis of macroinvertebrate community response to a decline in pollutant concentrations common with downstream distance from a point-source input, has provided a means of assessing changes in structural indices in relation to pollutant levels. This approach has several notable advantages, provided that care is taken in the choice of appropriate sampling sites to assure environmental homogeneity in such factors as benthic substrate, discharge rate, temperature, turbidity and ionic concentrations. If this is done, general environmental conditions will in all probability be similar throughout the study sites with the exception of the variable of interest, pollutant concentrations. Comparisons of changes can then be made within the same system and can be evaluated in terms of a reference (upstream from the outfall), also within the system. Such factors as dilution, sedimentation and adsorption provide a gradient of exposures which can be related to community and ecosystem changes. Gradient studies are a form of 'natural bioassay' of concentration and effect, incorporating seasonal and other environmental factors.

This case study compares data from three stream-gradient studies in which copper was the principal contaminant. The purposes of this exercise are to examine which structural characteristics of the marcoinvertebrate community exhibit a predictable graded response to heavy-metal pollution and to elucidate how seasonal and other cyclic factors influence the recognition of pollution-induced changes in structural patterns.

Criteria for evaluating indices of response to the impact of heavy metals on a community include:

1. The strength (statistical significance) of the relationship between the index and the metal concentrations.
2. The sensitivity of the index to changes in concentration along the gradient.
3. The influence of spatial and temporal variability on the index.
4. The degree of difficulty in collecting and analysing data from which to estimate the response.
5. The ecological relevance of the index.

Changes in communities are generally assessed in terms of the presence or absence of species or the more basic parameters of population density and biomass. The latter indices are relatively easy to measure and require little taxonomic skill but they also provide the least ecological information. In comparing mieobenthos samples from a brackish pond, Heip (1980) noted that most of the natural variance in density and biomass was introduced by long-term biological periodicity, making these parameters potentially valuable in monitoring change. On the other hand, the problems of quantitatively estimating density and biomass in the heterogeneous habitat of a stream riffle are more difficult. Specifically, considerably more variability is expected in most estimates of stream macroinvertebrate density or biomass than in taxonomic parameters (see Chapter 3). Indeed, Heip (1980) found a much larger coefficient of variation for density than for diversity measures for the more homogeneous lentic benthos.

Although diversity and evenness indices are generally highly correlated, a number of researchers have shown a preference for using the number of species or Shannon's index (Heip and Engels, 1974; Rosenberg, 1975; Marshall and Mellinger, 1980) and have been inclined to avoid evenness indices in monitoring pollution effects (Heip, 1980). Both indices have desirable statistical properties, as their temporal behaviour responds primarily to events with long periodicities. Shannon's diversity index measures something different from the number of species. The loss or the addition of a species is a more dramatic ecological event than those subtle interactions which affect relative abundance and thereby influence the diversity index. The variability in all of the community characteristics discussed above is expected to be of much less magnitude than that of the numbers of individuals in various populations.

In order to evaluate the sensitivity and utility of the various structural indices applied in this study, analysis of variance techniques and Duncan's New Multiple Range Test (Steel and Torrie, 1980) were employed to facilitate multiple comparisons along specific heavy-metal concentration gradients.

7.3.2 THE STREAM ECOSYSTEMS

Shayler Run and Elam's Run are small, second-order Southwestern Ohio streams and Little Grizzly Creek is a second-order stream in the Sierra Nevada Mountains of Northern California. Although Little Grizzly Creek is located at nearly 1500 m greater elevation, the three streams are similar in size, substrate and macrobenthic fauna. The Ohio streams are subject to occasional flooding, and the Sierra stream to spring snow melt, events which increase their volumes substantially. Sampling stations were located in rock-rubble riffle habitats where aquatic insects were the predominant macroinvertebrates.

7.3.3 HEAVY-METAL INPUTS

The studies chosen for comparison, Winner *et al.* (1975, 1980) and Sheehan (1980), present some interesting contrasts in terms of the source, delivery and concentration of heavy metals. Shayler Run was experimentally dosed with copper as part of a United States Environmental Protection Agency study (Geckler *et al.*, 1976). It had been exposed to a low level of copper stress (Table 7.3.1) for 2.5 years prior to evaluation of biotic changes (Winner *et al.*, 1975). The copper was input at a level which allowed metal concentrations to decrease sufficiently downstream to a point where the biota were virtually unaffected by the chemical stress. This study represents one of the few in which the effects of a toxic effluent were tested in a natural ecosystem.

The other Ohio stream, Elam's Run, had been subjected to much higher and more highly variable concentrations of three metals (Cu, Cr and Zn) from the effluent of a metal-plating industry (Table 7.3.2) for 8 years prior to the assessment of impacts (Winner *et al.*, 1980). Concentrations of each metal fluctuated widely and were frequently in the $mg\,l^{-1}$ range, and slugs of cyanide were occasionally detected in the effluent and along the stream. However, there were also periods when the plant was not operational (nights and weekends), and during these periods there was virtually no introduction of pollutants into Elam's

Table 7.3.1 Copper concentrations and structural characteristics of macroinvertebrate communities in Shayler Run, 1972

Station	Cu(μg l^{-1})	Insects m^{-2}	% Chironomids/ sample	% Caddis-flies/ sample	% Mayflies/ sample
1	9a	2448	5e	11fg	42.0h
2	120	462c	75	6f	2.0 j
3	66	876cd	58	11f	0.3 j
4	52 b	805cd	41	22 g	10.0 j
5	38 b	1056cd	8e	40	12.0 j
6	23a	1652 d	3e	21 g	36.0h

Means followed by a common letter are not significantly different ($p \leq 0.05$).

Run. This stream merged with a much larger stream about 3.5 km below the effluent outfall and at this point the fauna still exhibited signs of being significantly stressed. Due to adverse publicity during the course of the Winner et al. (1980) study, the metal-plating industry initiated improvements in its waste-water treatment facility which resulted in a 4- to 25-fold decrease in average concentrations of the three metals in its effluent between 1978 and 1981 (Table 7.3.2).

Because of the periodic release of highly concentrated chemical slugs into Elam's Run, variability within water samples was so high that mean concentrations did not differ significantly between stations for any of the three metals. There was, however, a difference in the frequency with which the stations received these high concentrations as shown for Cu in Figure 7.3.1. Metal concentrations were also measured in the sediments from soft-bottomed pools adjacent to the riffle stations in Elam's Run. Metal concentrations were more stable in the sediments and were significantly different for each of the three metals among each of the five stations (Table 7.3.3). Sediment metal concentrations confirmed that there was a significant decrease in metal pollution between 1978 and 1981. These data also reflect the relationship between metal concentrations

Table 7.3.2 Metal concentrations (μg l^{-1}) in water of Elam's Run

Station	Cu 1978	Cu 1981	Cr 1978	Cr 1981	Zn 1978	Zn 1981
Effluent	1250	191	910	36	508	126
1	336	$-^1$	70	$-^1$	101	$-^1$
2	237	$-^1$	54	$-^1$	55	$-^1$
3	221	58	58	11	51	25
4	87	$-^1$	15	$-^1$	24	$-^1$
5	74	13	2	4	24	5

[1] Samples not collected.

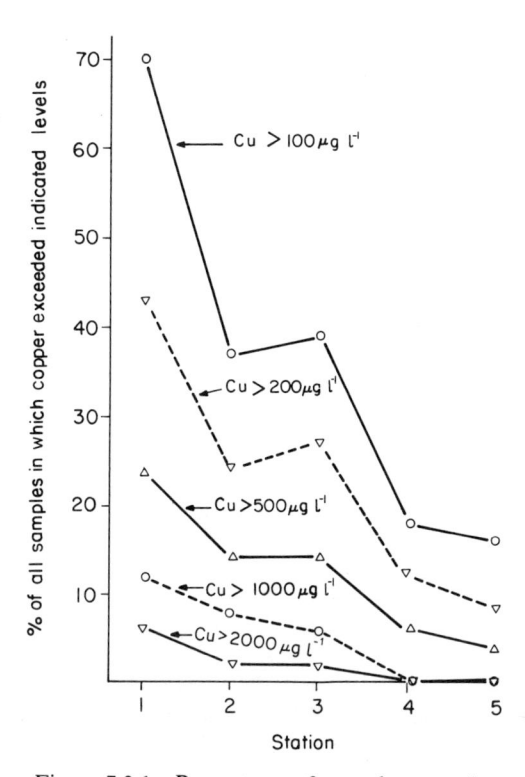

COMPARISON OF GRADIENT STUDIES IN HEAVY-METAL-POLLUTED STREAMS 259

Figure 7.3.1 Percentage of samples containing selected copper concentrations at five stations on Elam's Run, 1978. (From Winner et al., 1980. Reproduced by permission of the *Canadian Journal of Fisheries and Aquatic Sciences*)

Table 7.3.3 Mean metal concentrations (μgg^{-1}) in sediments of Elam's Run, 1978 and 1981

Station	% Organic content 1978	1981	Copper 1978	1981	Zinc 1978	1981	Chromium 1978	1981
1	2.8	$-^1$	239	$-^1$	259	$-^1$	121	$-^1$
2	5.5	3.6	669 *	29	515 *	17	508 *	17
3	4.5	4.8	431 *	60	488 *	37	213 *	43
4	3.6	3.2	157 *	17	195 *	10	63 *	12
5	2.5	2.8	24 *	6	70 *	4	3	4

Corresponding 1978, 1981 means marked with asterisks are significantly different ($p \leq 0.05$).
[1] Samples not collected.

Table 7.3.4 Copper concentrations and structural characteristics of macroinvertebrate communities in Little Grizzly Creek, July through October, 1975 and May through October, 1976.

Station	Cu(μg l^{-1}) 1975 max	1975 min	1976 max	1976 min	Insects m^{-2} 1075	1976	% Chironomids 1975	1976	% Caddis-flies 1975	1976	% Mayflies 1975	1976	% Stoneflies 1975	1976	% Beetles 1975	1976
7 Reference	<4		<4		7150	7815b	14	29e	19g	12h	28	24k	10m	9	22p	25
8	630	200	150	80	24a	1345 c	—¹	95 f	—¹	1h	—¹	0k	—¹	0n	—¹	1r
9	390	130	120	30	118a	1830 c	65d	79 f	13g	6h	0j	7k	0m	0n	14pq	0r
12	320	50	60	20	463a	4562bc	54d	46ef	31g	7h	2j	38k	2m	2n	4 q	0r

For structural characteristics, means followed by a common letter are not significantly different ($p \leq 0.05$).
¹ Samples not collected.

and organic content of the sediment. In 1978, the highest organic content and metal concentrations were at station 2. In 1981 the highest organic content and metal concentrations had shifted to station 3.

Little Grizzly Creek has been exposed to heavy-metal drainage from tunnels of a nonoperational copper mine for nearly 40 years. Although the effluent has traces of several potentially toxic metals (see Sheehan, 1980), only the levels of copper are high enough to be toxic during most of the year (Table 7.3.4). The input of copper is greatly increased during the snow-melt period and is highly dependent on the amount of snow accumulated during the winter. In 1975, a normal snow-fall year, Cu levels as high as $1700\,\mu g\,l^{-1}$ were recorded entering Little Grizzly Creek during snow-melt. This elevated input was noticeable even in the site 12 levels ($320\,\mu g\,l^{-1}$), at greater than 15 km downstream. At the entry site in 1976 (a drought year), only $830\,\mu g\,l^{-1}$ Cu were measured during the high-flow period (Sheehan, 1980). Copper levels throughout the summer period, accompanying minimal flow, were always significantly less than those measured during the high-flow spring season. The contrast between soluble copper levels in a normal year and a drought year in California can be seen in the range of values recorded from water samples taken from Little Grizzly Creek during the summers of 1975 and 1976, respectively (Table 7.3.4).

Other investigations at this creek have shown that Cu^{2+} made up a large portion of the soluble copper pool at all contaminated sites; however, the greatest portion of copper in the water column was adsorbed on suspended particulates (Sheehan, 1980). As reported for Elam's Run, metal concentrations in the sediments of Little Grizzly Creek were more stable than soluble levels, although there was still significant seasonal variation.

7.3.4 COMMUNITY ANALYSIS

7.3.4.1 Density

The densities of insect communities were affected below the point of metal introduction in all streams, although there were some obvious differences in response. In Little Grizzly Creek, macroinvertebrate densities were most severely impacted during the 1975 increases in copper exposure. At the sampling site closest to the outfall, density was found to be only 1 percent of that measured for the reference community (station 7). Even at a distance greater than 15 km from the pollutant source, the density was only 10 per cent of the reference value. In both 1975 and 1976, significant reductions in densities at all contaminated sites were observed, with the exception of site 12 for the summer of 1976 (Table 7.3.4.).

A similar pattern is evident in the density data from Shayler Run, although the uncontaminated riffle in this stream contains only about one third as many macroinvertebrates as are found at the comparable station in Little Grizzly Creek (Tables 7.3.1 and 7.3.4). Mean density was reduced by 81 per cent just below the point of discharge, and although it had increased fourfold at a point 2.6 km downstream, it was still significantly lower than at station 1, upstream from the outfall. The numbers of individuals in both the Little Grizzly Creek and the Shayler Run samples were inversely correlated ($p < 0.01$) with soluble copper concentrations.

Although macroinvertebrate densities appeared to be depressed in Elam's Run below the point of metal introduction, the variablity in the numbers of individuals made it impossible to distinguish significant differences among the samples at varying distances from the outfall. Insect densities were generally higher in Elam's Run than in Shayler Run (Tables 7.3.1 and 7.3.5). It would seem that the situation should be reversed since the Elam's Run fauna were exposed to much higher concentrations of three metals. The difference may be related to the continuous versus intermittent nature of the stress in the two streams. In Shayler Run, the copper dosage was continuous throughout the diel period for each day of the week. In Elam's Run, the composition of the effluent was extremely variable from hour to hour and from day to day and was virtually zero between the hours of 21.30 and 07.30 and on holidays and weekends. There is evidence which suggests that some components of the fauna can avoid peak concentrations of toxic chemicals by burrowing into the stream bed. For example, at 10.45 on 15 September 1978, a slug of 4 mg Cu l^{-1} and 3.8 mg CN $^{-}l^{-1}$ was detected at station 5 in Elam's Run. Concentrations of both chemicals remained in the mg l^{-1} range throughout the remainder of the day. On 16 September, two bottom samples were taken from the riffle at this station and *no* macroinvertebrates were found. However, another set of samples was taken on September 28 and the macroinvertebrate density was 880 m $^{-2}$, with 75 per cent of the total being late-instar caddis-flies. The only reasonable explanation for this apparent repopulation is that the animals had avoided the chemical slug by burrowing into the substrate. This is probably a manifestation of a more general avoidance response which has evolved in the fauna of small streams to promote survival during periods of drought. Such a strategy was not possible in Shayler Run because of the continuous addition of copper to the stream, or in Little Grizzly Creek, where shallow sediments in combination with the high influx of heavy metals during snow-melt and normal bottom scouring severely stress the macrobenthic community on a regular cyclic basis.

Macroinvertebrate density data from Shayler Run and Little Grizzly Creek showed a strong inverse correlation with Cu levels and the values were relatively easy to measure. However, there was a considerable degree of variability even among replicate samples, and any combination of samples representing a sampling period of several months will certainly have as much variability, if not more. This situation corroborates the assumption that estimates based on a small number of replicate stream samples (two in Shayler Run and three in Little Grizzly Creek) can provide only a crude approximation of the actual density. Biomass of macroinvertebrates in Little Grizzly Creek also fluctuated greatly. Because biomass and

Table 7.3.5 Structural characteristics of macroinvertebrate communities in Elam's Run, June 1977–1978 and 1981

Station	Insects m^{-2} 1978	Insects m^{-2} 1981	%Chironomids/sample 1978	%Chironomids/sample 1981	%Caddis-flies/sample 1978	%Caddis-flies/sample 1981	%Mayflies/sample 1978	%Mayflies/sample 1981
1	1299a	2787b	86c	99f	2h	<1	0j	0k
2	1526a	2953b	77cd	98f	2h	<1	0j	0k
3	1301a	593b	66cde	89f	14h	6	0j	1k
4	2498a	1667b	60 de	22 g	20h	61	0j	14 m
5	2101a	1880b	48 e	39 g	45	38	0.1j	16 m

Means followed by a common letter are not significantly different.

density measures illustrated a similar pattern of change along the gradient, Sheehan (1980) suggested that it is unnecessary to assess both indices, with biomass being the less useful if secondary productivity estimates are not being made.

Due to the inherent variability in macroinvertebrate density data and the lack of ecological information transmitted by this parameter, density estimates by themselves do not provide a satisfactory description of quantitative changes. As pointed out in Chapter 3, their use should be encouraged only as a first approximation of the degree of impact or as part of a set of indices including taxonomic information.

7.3.4.2. Species Richness and Diversity

In both Shayler Run and Little Grizzly Creek species richness (number of species per sample) was more strongly and significantly correlated with copper level than

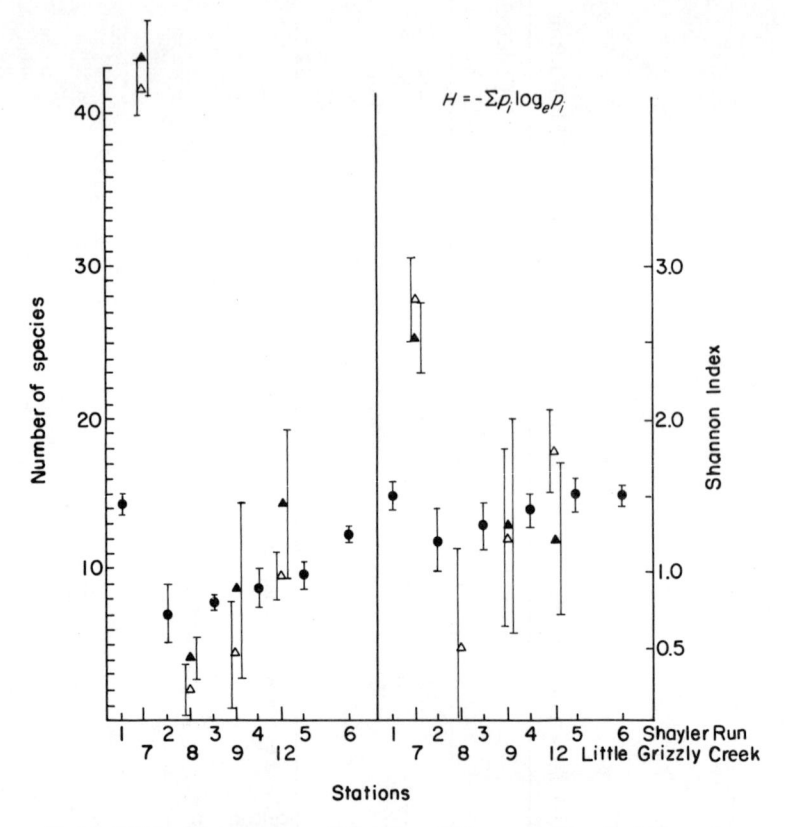

Figure 7.3.2 The numbers of species and Shannon diversity index values as means ± SDs of macroinvertebrate fauna along copper gradients in Shayler Run, 1972 (●), and in Little Grizzly Creek, 1975 (Δ), 1976 (▲)

was the Shannon diversity index. In fact, the 1976 macroinvertebrate diversity estimates for the Little Grizzly Creek stations were not significantly correlated with Cu levels ($p > 0.05$), while for the same period, the number of species (excluding chironomids) was significantly related to Cu in an inverse fashion ($p < 0.01$).

Although both taxonomic indices displayed reduced diversity with increased copper level (Figure 7.3.2), the number of species was the more sensitive parameter. In Shayler Run, using species richness as an index, stations 1 and 6 (the reference and the treated station having the lowest copper concentration, respectively) were significantly different from each other and from all the other stations; in addition, station 2 had significantly fewer species than station 5. Station 1 was significantly different from all other stations when evaluated with the Shannon index but there were no significant differences among any of the other stations.

Similar patterns emerged from the Little Grizzly Creek data although summer comparisons in 1976 showed less distinctive differences. In 1975, the numbers of species at stations 7 and 12 were significantly different from each other and from the more contaminated sites, while the Shannon index distinguished differences only between the reference and the stressed communities.

The ability of an index to distinguish significant differences within the intermediate range of pollutant concentrations is essential to a more fine-tuned appraisal of community response (Gray, 1976). The fact that the numbers of species provided greater resolution than the Shannon index in comparisons along the Cu gradient, in both Shayler Run and Little Grizzly Creek, is noteworthy because the latter index require considerable effort in calculation. This result also contradicts earlier assumptions that diversity indices based on information theory were superior to more direct indices, such as species richness, in detecting the severity of stress on macroinvertebrate communities (Wilhm and Dorris, 1968).

The Little Grizzly Creek data, from the two summers (Figure 7.3.2), clearly

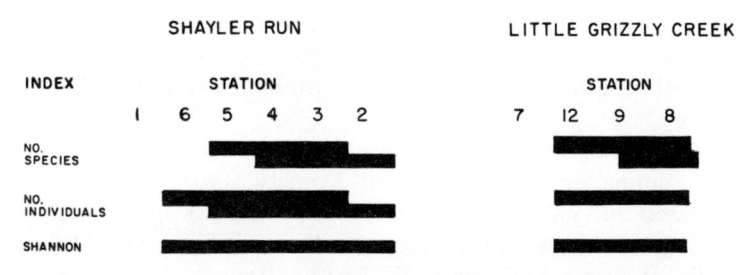

Figure 7.3.3 The effect of downstream location on measures of macroinvertebrate community response as determined by Duncan's New Multiple Range Test. Any two stations not underscored by the same bar are significantly different ($p < 0.05$)

indicate that the response of both taxonomic measures is most distinct under conditions of high copper stress. As taxonomic diversity increased in the stressed communities during the reduced effluent inputs of 1976, changes in the dominance component of the Shannon index masked any indication of the communities' recoveries despite an approximate doubling of the numbers of species found at polluted stations. This problem of a misleading interpretation, due to the influence of the evenness component when the Shannon index is applied within a community in which the total population density is low, evokes perhaps the most strenuous objection to its use (Godfrey, 1978; Gray, 1979). Equitability dramatically increases the importance of rare species in the index value under these conditions.

A comparison of density and diversity indices (Figure 7.3.3) showed that the two less complex measures were more useful than Shannon's index in distinguishing the degree of community response.

7.3.4.3 Community Composition

Although chironomids comprise a very small fraction of the macroinvertebrate riffle communities of unpolluted streams in North America and Europe (Hynes, 1961; Morgan and Egglishaw, 1965; Cummins *et al.*, 1966; Wynes, 1979), in all three heavy-metal contaminated streams chironomid larvae dominated the benthic communities at the most severely polluted riffles. In Shayler Run, chironomids comprised 40–75 per cent of the fauna collected at the most contaminated sites (Table 7.3.1), while in Elam's Run and Little Grizzly Creek this group made up 75–95 per cent of the organisms collected at stations near the outfall (Tables 7.3.4 and 7.3.5). There are some indications that this pattern may be common in heavy-metal contaminated streams. Butcher (1946) found that chironomid larvae were among the first macroinvertebrates to recolonize an English stream grossly polluted with industrial copper waste. Chironomid larvae, specifically *Cricotopus bicinctus*, have been shown to be particularly resistant to electroplating wastes containing high levels of chromium, cyanide and copper (Surber, 1959). In Elam's Run *C. bicinctus* and *C. infuscatus* were the dominant species at stations 1, 2, 3 and 4, comprising 92 per cent, 77 per cent, 77 per cent and 41 per cent, respectively, of the chironomids collected (Winner *et al.*, 1980).

While the numerical importance of chironomids, as a group, increased with the degree of pollution, species richness and diversity as measured by Shannon's index decreased with increasing pollution (Table 7.3.6). A downstream increase in the diversity of chironomid species generally results from an addition, not a substitution, of species (Winner *et al.*, 1980). Although the same number of species was collected from station 4 as from station 5 in Elam's Run, Shannon's index was somewhat higher at station 5 due to a more even distribution of individuals among species.

Table 7.3.6 Changes in the structure of chironomid communities along a heavy metal gradient in Elam's Run, 1977–1978

Station	Total species collected	Shannon's Index (H') Adults	Larvae
1	15	1.10	0.80
2	28	1.10	1.04
3	24	1.07	1.35
4	39	1.64	1.55
5	39	2.67	1.79

Although essentially eliminated from the most grossly polluted riffles, caddis-flies were co-dominant with chironomids at moderately polluted stations in all streams (Tables 7.3.1, 7.3.4 and 7.3.5). Elmid beetles also comprised a significant fraction of the fauna at moderately polluted sites in Little Grizzly Creek during the summer of 1975, following high copper inputs. Mayflies were a significant fraction of the community only at the unpolluted station and at the station farthest downstream from the point of copper introduction in both Shayler Run and Little Grizzly Creek. Prior to the improvement in wastewater treatment in Elam's Run, mayflies were virtually absent, even at the farthest downstream site. Subsequent to the improvement, mayfly nymphs appeared in samples from the two stations farthest downstream (Table 7.3.5). Concurrent with this appearance, there was a reduction in the chironomid fraction of the communities at these two stations. Stoneflies made up 10 per cent of the macroinvertebrate community at station 7 in Little Grizzly Creek but were not found at all in samples from stations 8 and 9. This order was not common in the Ohio streams (Winner et al., 1980).

7.3.4.4 Coefficient of Variation

The proportions of midges, caddis-flies and mayflies in samples collected from riffle communities, therefore, appear to constitute a useful index of the degree of heavy-metal stress to which the community has been subjected. Single samples, however, may produce misleading data. This is especially true for unpolluted habitats where samples may exhibit considerable spatial and temporal variability in the relative abundance of the three taxa. On the other hand, most of the temporal and spatial heterogeneity tends to be submerged in grossly polluted habitats and the taxonomic composition of samples is highly predictable. This situation is exemplified by the coefficients of variation for the percentage of chironomids in samples from the three streams (Figure 7.3.4). The CV is very low for samples from the most grossly polluted stations and increases in value for stations which are progressively less polluted. According to this index, the communities in Shayler Run at station 6 and Little Grizzly Creek at site 12 have

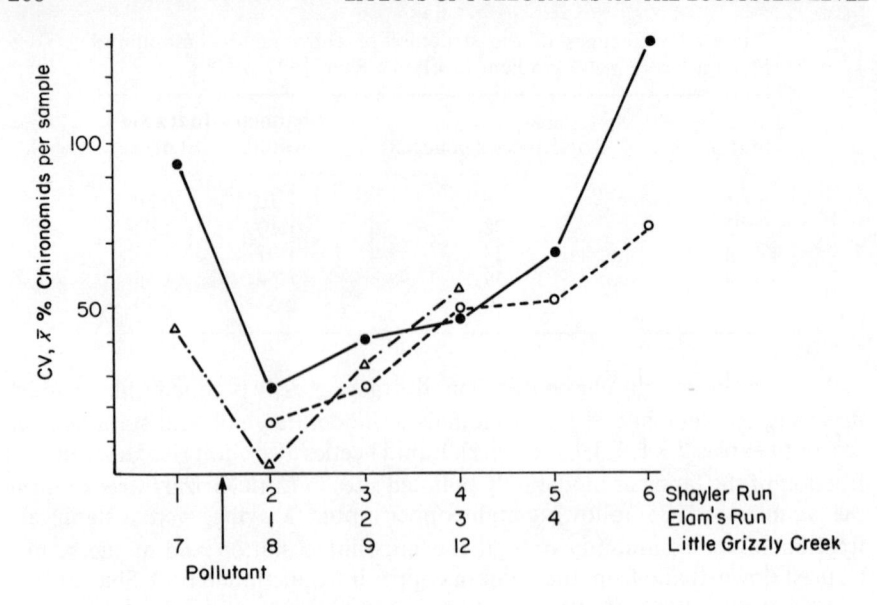

Figure 7.3.4 Changes in the coefficient of variation of mean percentage chironomids in benthic insect samples taken along heavy-metal concentration gradients in Shayler Run (●), Elam's Run (○) and Little Grizzly Creek (Δ)

recovered sufficiently to display variations in dominance similar to those of the assemblages at their respective unpolluted reference stations. However, these increases in natural variability are not dependent on a complete recovery of the communities; interpretations of the comparisons of macroinvertebrate densities and of numbers of species present in these streams do not indicate a corresponding total recovery.

7.3.5 NOTIONS OF ECOSYSTEM STABILITY RELATED TO POLLUTION STRESS

If greater stability can be associated with reduced variability in taxonomic composition, the depauperate macroinvertebrate communities under the most severe heavy-metal stress appear to be the most stable. This constancy in the chironomid-dominated composition of these communities can be interpreted as an indication of continuous, severe exposure to chemical stress swamping the spatial and temporal heterogeneity of other environmental factors. As the severity of the stress decreases, other factors such as substrate, seasonal and life-history influences begin to affect community composition. Haedrich (1975) argued that decreases in diversity would be accompanied by increases in taxonomic similarity from season to season. This relationship was exemplified by both coefficient of community and percentage similarity comparisons for Little Grizzly Creek sites 8, 9 and 12 (Sheehan, 1980).

Reduced variability in other structural parameters, under conditions of severe heavy-metal stress, has also been demonstrated. The density data for Little Grizzly Creek macroinvertebrates, from comparisons performed among samples from different summer months, showed significant differences between sampling dates within the station 7 and the station 12 measurements ($p \leq 0.005$), but no differences of significant magnitude within samples from the more polluted sites 8 and 9 ($p > 0.05$).

Community structure can be characterized, therefore, by fewer samples in polluted habitats than in those which exhibit comparable spatial and temporal heterogeneity but are unpolluted. This aspect of reduced variability is inconsistent with the idea that when comparing similar ecosystems, the one supporting the more diverse and interactive set of species should possess the greater stability.

One of the advantages of the pollutant-gradient approach is that it provides a means of assessing ecosystem vulnerability, a characteristic which is perhaps more relevant to the stability of stressed systems than are the ideas of balance or constancy. In the context of the vulnerability of polluted systems, properties which describe the degree, manner and pace of restoration to a reference-system-like structure are of primary interest and importance. Westman (1978) refers to these as the inertia and resilience of the system. Inertia is the ability of the system to resist change. It can be estimated by determining the concentration of a chemical which reduces species richness by 50 per cent. Sheehan (1980) found this concentration to be in the range of $6-18 \mu g l^{-1}$ Cu during the summer months, reflecting a runoff high of $80-100 \mu g l^{-1}$. This summer range is more than 5 times lower than the estimate from the Shayler Run data. Although this difference may infer some disparities in the buffering capacity of these ecosystems, it most certainly reflects the added severity of stress imposed on the Little Grizzly Creek fauna by high seasonal influxes of heavy-metal effluent. It is obvious that this seasonal pattern of copper exposure has overwhelmed the system's capacity to self-repair. Under current conditions, recovery to the reference structure cannot occur and, therefore, the system can be termed brittle (Westman, 1978) and is said to show a lack of resiliency (Cairns and Dickson, 1977). Although complete recovery of the Little Grizzly Creek system cannot be expected without pollution abatement, there were some short-term indications of recovery following the decreased copper fluxes in the spring of 1976. Comparisons of the composition of macroinvertebrate communities (excluding chironomid species), using Whittaker's coefficient of community (CC), showed that there was a substantial increase in similarity between moderately polluted stations and the uncontaminated reference during 1976. The CC index value between stations 7 and 12 improved from 0.34 in 1975 to 0.48 in 1976, and the CC for stations 7 and 9 increased from 0.21 to 0.33. There was no substantial improvement in similarity between sites 7 and 8 during this period. Samples from the reference station were relatively stable ($CC = 0.84$) between the high flow and drought years, while the moderately polluted sites demonstrated considerable variability in community make-up ($CC = 0.45$). Westman (1978) suggested that

85 per cent similarity would be an adequate measure of restoration. Using this criterion, it is clear that the heavy-metal exposed Little Grizzly Creek ecosystem remains substantially disturbed.

7.3.6 CONCLUSIONS

There are significant advantages in using the gradient approach to analyse pollutant-induced changes in community structure in stream ecosystems. An adequate reference site can normally be identified upstream of the outfall. Structural characteristics can be compared along the concentration gradient and correlations with the level of pollutant can be examined for significance.

There were surprising similarities in macroinvertebrate community response to heavy metals among the three streams examined even though metal effluent composition, concentration and loading were quite different, as was the duration of exposure. This observation is particularly significant in that ecosystems tests, such as those in Shayler Run, may adequately predict long-term changes in structure. However, the differences in taxonomic and density measures for the 1975 and 1976 Little Grizzly Creek samples indicate the need for a sampling period encompassing seasonal and other cyclic influences.

Not all structural indices respond equally in reflecting changes along the pollutant gradient. Species richness data provided resolution superior to density and Shannon's diversity measures, and can be used to estimate the ability of the system to absorb stress. The pattern of response provided by Shannon's index was not always consistent with other measures. Changes in the coefficient of variation of percentage chironomids per sample exhibited a consistent pattern and should provide a useful index of the severity of heavy-metal pollution.

Similarity indices provide an alternative means of evaluating taxonomic response and can be used to assess relative recovery and to estimate stability. The observation that severely stressed systems display significantly reduced taxonomic variability is important in that it shows stability, in this sense, to be inversely related to the diversity of the fauna and the well-being of the system.

ACKNOWLEDGEMENTS

Metal-sediment concentrations for 1977 Elam's Run samples were collected by Linda Yocum. Ruth E. Baker contributed to the collecting of the 1981 data while she was a National Science Foundation Undergraduate Research Fellow in the laboratory of R. W. Winner. The assistance of the California State Water Pollution Laboratory in the analysis of metal content of some Little Grizzly Creek water samples is greatly appreciated, as is the assistance and support of A. W. Knight and the University of California, Davis.

7.3.7 REFERENCES

Butcher, R. W. (1946). The biological detection of pollution. *J. Inst. Sew. Purif.*, **2**, 92–97.

Cairns, J. Jr. and Dickson, K. L. (1977). Recovery of streams and spills of hazardous materials. In Cairns, J. Jr., Dickson, K. L. and Herricks, E. E. (Eds.), *Recovery and Restoration of Damaged Ecosystems*. University of Virginia Press, Charlottesville, Virginia, pp. 24–42.
Cummins, K. W., Coffman, W. P. and Roff, P. A. (1966). Trophic relations in a small woodland stream. *Verh. Int. Verein. Limnol.*, **16**, 627–638.
Geckler, J. R., Horning, W. B., Neiheisel, T. M., Pickering, Q. H. and Robinson, E. L. (1976). *Validity of Laboratory Tests for Predicting Copper Toxicity in Streams*. Ecol. Res. Ser. Environ. Prot. Agency 600/3-76-116. Environ. Res. Lab., U.S. Environ. Prot. Agency, Duluth. MN, 192 pages.
Godfrey, P. J. (1978). Diversity as a measure of benthic macroinvertebrate community response to water pollution. *Hydrobiologia*, **57**, 111–122.
Gray, J. S. (1976). Are baseline surveys worthwhile? *New Sci.*, **70**, 219–221.
Gray, J. S. (1979). Pollution induced changes in population. In Cole, H. A. (Ed.), The assessment of sublethal effects of pollutants in the sea. *Phil. Trans. R. Soc. Lond. B.*, **286**, 545–561.
Haedrich, R. L. (1975). Diversity and overlap as measures of environmental quality. *Water Res.*, **9**, 945–952.
Heip, C. (1980). Meiobenthos as a tool in the assessment of marine environmental quality. In McIntyre, A. D. and Pearce, J. B. (Eds.), *Biological Effects of Marine Pollution and the Problems of Monitoring*. Rapp. P.-v. Réun. Cons. int. Explor. Mer., **179**, 182–187.
Heip, C. and Engels, P. (1974). Comparing species diversity and evenness indices. *J. Mar. Biol. Ass. U.K.*, **54**, 559–563.
Hynes, N. B. N. (1961). The invertebrate fauna of a Welsh mountain stream. *Arch. Hydrobiol.*, **57**, 344–388.
Jones, J. R. E. (1940). A study of the zinc-polluted river Ystwyth in North Cardiganshire, Wales. *J. Animal Ecol.*, **27**, 1–14.
Marshall, J. S. and Mellinger, D. C. (1980). Dynamics of cadmium-stressed plankton communities. *Can. J. Fish. Aquat. Sci.*, **37**, 403–414.
Morgan, N. C. and Egglishaw, H. J. (1965). A survey of the bottom fauna of streams in the Scottish Highlands. Part I. Composition of the fauna. *Hydrobiology*, **25**, 181–211.
Rosenberg, R. (1975). Stressed tropical benthic faunal communities off Miami, Florida. *Ophelia*, **14**, 93–112.
Sheehan, P. J. (1980). *The Ecotoxicology of Copper and Zinc: Studies on a Stream Macroinvertebrate Community*. Ph.D. dissertation, University of California, Davis, California.
Steel, R. G. D. and Torrie, J. H. (1980). *Principles and Procedures of Statistics—a Biometrical Approach*, 2nd ed., McGraw-Hill Book Company, New York, 633 pages.
Surber, E. W. (1959). *Cricotopus bicinctus*, a midge fly resistant to electroplating wastes. *Trans. Amer. Fish. Soc.*, **88**, 111–116.
Westman, W. E. (1978). Measuring the inertia and resilience of ecosystems. *Bioscience*, **28**, 705–710.
Wilhm, J. L. and Dorris, T. C. (1968). Biological parameters for water quality criteria. *Bioscience*, **18**, 477–481.
Winner, R. W., Scott Van Dyke, J., Caris, N. and Farrell, M. P. (1975). Responses of the macroinvertebrate fauna to a copper gradient in an experimentally-pollued stream. *Verh. Int. Verein. Limnol.*, **19**, 2121–2127.
Winner, R. W., Boesel, M. W. and Farrel, M. P. (1980). Insect community structure as an index of heavy-metal pollution in lotic ecosystems. *Can. J. Fish. Aquat. Sci.*, **37**, 647–655.
Wynes, D. L. (1979). *Predator–Prey Relationships in the Riffle Fish Community of the Little Miami River, Ohio*. Ph.D. dissertation, Miami University, Oxford, Ohio.

Effects of Pollutants at the Ecosystem Level
Edited by P. J. Sheehan, D. R. Miller, G. C. Butler and Ph. Bourdeau
© 1984 SCOPE. Published by John Wiley & Sons Ltd

CASE 7.4
Ecological Effects of Hydroelectric Development in Northern Manitoba, Canada: The Churchill–Nelson River Diversion

R. A. BODALY[1], D. M. ROSENBERG[1], M. N. GABOURY[2],
R. E. HECKY[1], R. W. NEWBURY[1] and K. PATALAS[1]

[1]*Canada Dept. Fisheries and Oceans, Freshwater Institute,
501 University Crescent, Winnipeg, Manitoba R3T 2N6*
[2]*Manitoba Dept. Natural Resources, Box 40, 1495 St. James Street,
Winnipeg, Manitoba R3H 0W9*

7.4.1	Introduction	274
7.4.2	The Ecosystems Before Development	274
7.4.3	Hydroelectric Development of the Churchill and Nelson Rivers	278
7.4.4	Ecological Effects of Hydroelectric Development—Southern Indian Lake	280
	7.4.4.1 General Physical Changes	280
	7.4.4.2 Shoreline Erosion	281
	7.4.4.3 Sediment Budgets	284
	7.4.4.4 Light and Thermal Regimes	284
	7.4.4.5 Primary Production	286
	7.4.4.6 Macrobenthic Invertebrates	287
	7.4.4.7 Zooplankton	288
7.4.5	Ecological Effects of Hydroelectric Development—Other Lakes	290
	7.4.5.1 Lakes of the Lower Churchill River	290
	7.4.5.2 Diversion Route Lakes (Rat and Burntwood River Valleys)	290
	7.4.5.3 Cross Lake	293
7.4.6	Effects of Hydroelectric Development on Fisheries	293
	7.4.6.1 Grade of Whitefish Catch in Southern Indian Lake	293
	7.4.6.2 Effects of Drawdown on Fish Populations in Cross Lake	296
	7.4.6.3 Fish Mercury Levels	297
7.4.7	Stabilization of the Ecosystems	300
	7.4.7.1 Physical Stabilization	301
	7.4.7.2 Primary Production	301
	7.4.7.3 Macrobenthic Invertebrates	302
	7.4.7.4 Zooplankton	302
	7.4.7.5 Fish Populations	302
	7.4.7.6 Conclusion	302

7.4.8	Mitigation and Planning	303
7.4.9	Summary and Conclusions	304
7.4.10	References	306

7.4.1 INTRODUCTION

Planning for hydroelectric development on the Churchill and Nelson rivers began almost two decades ago and major dams, diversions and power stations were operational by 1977. Extensive pre-development ecological studies and environmental impact predictions were carried out (Lake Winnipeg, Churchill and Nelson Rivers Study Board, 1971–1975: 1974, see especially Appendix 5). Studies on selected portions of the affected area encompassing the physical, chemical and biotic levels have continued to the present, using data from the pre-development period as ecosystem baselines. Emphasis has been on the impact of aquatic ecological changes on the harvest of fish by man. Although these studies are not strictly ecotoxicological, the study of perturbations at the ecosystem level is common to other case studies presented in this volume.

This case study has the following objectives:

1. To describe overall features of the Churchill and Nelson Rivers hydroelectric development.
2. To summarize ecological effects of the development.
3. To recommend future planning needs for hydroelectric development in subarctic areas.

7.4.2 THE ECOSYSTEMS BEFORE DEVELOPMENT

The lentic ecosystems affected by the Churchill and Nelson Rivers hydroelectric development are a set of boreal, Precambrian Canadian Shield lakes in northern Manitoba, Canada (Figure 7.4.1). Much of the bedrock in the area is overlaid with fine-grained glacio-lacustrine deposits, many of which are affected by permafrost. The area is characterized by a subarctic climate with short, cool summers and long, cold winters. Mean January daily temperatures are -22.5 to $-27.5°C$ while mean July daily temperatures are 12.5 to 17.5°C. Annual precipitation is 400–500 mm of which approximately one-third falls as snow. Annual evaporation from lakes of the area ranges from approximately one-third to three-quarters of precipitation. The ice-free season is about 5 months (June–October).

Lake surface areas varied from small ($< 10\,km^2$) to very large (approximately 2000 km^2). The lakes were generally riverine and relatively shallow for their surface area. This, together with such factors as the glacio-lacustrine shore and bottom sediments and strong winds, caused relatively high turbidity. Residence times of water ranged from a few days (lakes on the lower Churchill River) to 2–3

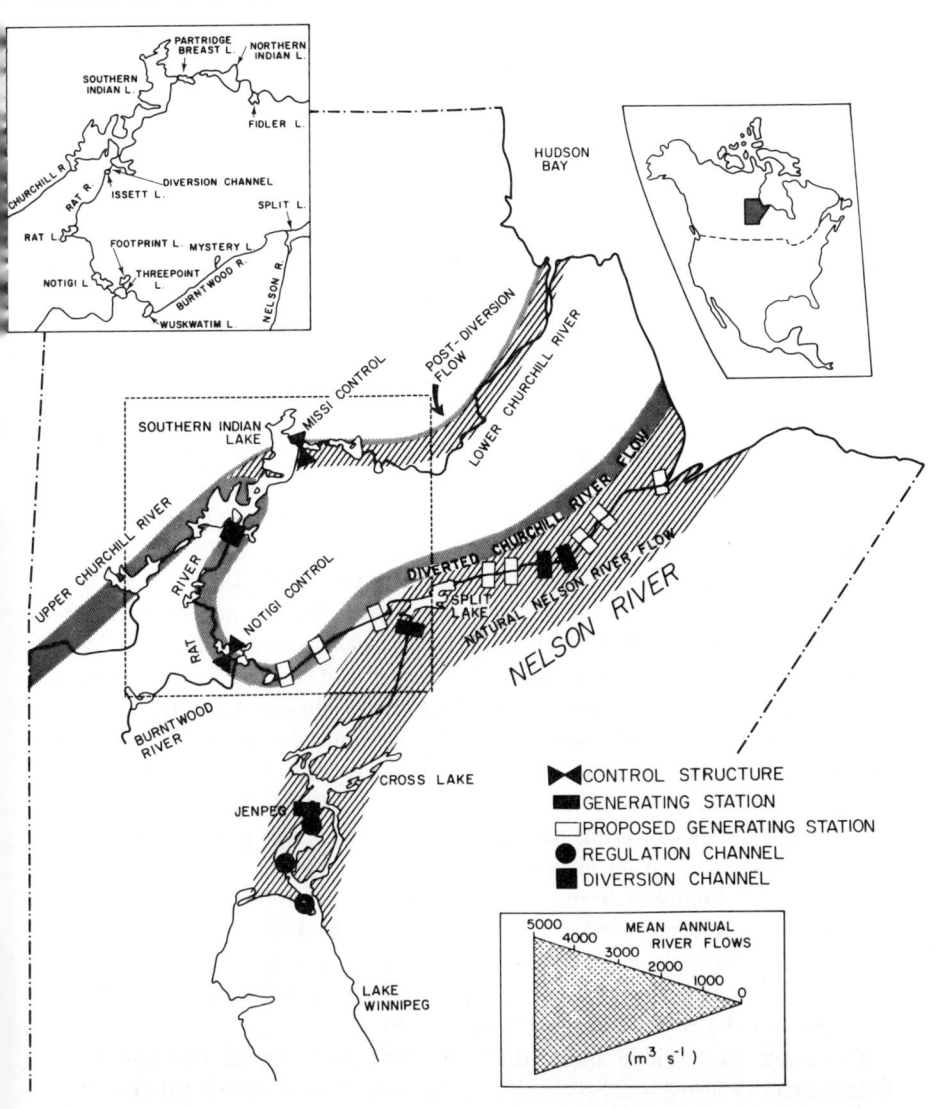

ECOLOGICAL EFFECTS OF HYDROELECTRIC DEVELOPMENT 275

Figure 7.4.1 The Churchill and Nelson rivers hydroelectric development, northern Manitoba. Pre- and post-diversion flows, the locations of generating stations, control structures, artificial channels, and study lakes are shown

years (some basins of Southern Indian Lake). The lakes were generally isothermal during the ice-free season, with water temperatures usually < 20°C. Oxygen levels generally were near saturation throughout the water column, especially during the summer.

The riverine nature of the lakes imposed a high nutrient supply per unit area but positive effects on production in most lakes were reduced because of high flushing rates. Total dissolved solids (TDS) ranged from 50 to 230 ppm. Dissolved ion compositions in the two river systems were markedly different (Table 7.4.1). The Nelson River drains a large prairie sedimentary area before it crosses the granitic Canadian Shield in northern Manitoba. Its total ion concentration is much higher than the Churchill River and concentrations of sodium, magnesium, chloride, sulphate and suspended sediments are higher because of the weathering of fine-grained, prairie soils upstream. The Churchill River primarily drains the granitic Canadian Shield, although deposits of calcareous, glacio-lacustrine clays and glacial tills are widely scattered throughout its basin. The Rat River, immediately south of Southern Indian Lake (Figure 7.4.1), drains extensive glacial deposits. Weathering of this material accounts for the relative enrichment of calcium and CO_2 in the water at Notigi Lake (Table 7.4.1). Although Notigi Lake and Southern Indian Lake lie on the Canadian Shield, their ionic concentrations and pH are substantially higher than for many others lakes on the Shield (Armstrong and Schindler, 1971) due to the proximity of abundant glacial deposits.

Variation in the rates of primary production (100–700 mg C m^{-2}) were due to differences in phosphorus loading between rapidly flushed and stagnant areas (Healey and Hendzel, 1980; Hecky and Harper, 1974; Hecky *et al.*, 1974; Hecky, 1975). Poorly flushed regions were also richer in dissolved humic substances which appear to depress primary production by binding trace metals, particularly iron (Jackson and Hecky, 1980). Algal biomass was lowest in the turbid Nelson River (approximately 300 mg m^{-3} fresh weight) and highest in lakes along the Rat and Burntwood rivers (approximately 10 000 mg m^{-3} fresh weight). High algal biomasses were generally associated with high primary production, but also resulted from importation of algae from upstream lakes (Hecky and Harper, 1974). Diatoms dominated the algal biomass in Southern Indian Lake and in the lower Churchill lakes, while cyanophytes dominated in lakes of the Burntwood and Nelson rivers. These lakes were classified as oligotrophic to mesotrophic, based on their ranges of values for primary productivity, algal biomass and chlorophyll (Vollenweider, 1968).

Chironomidae (midge flies), the amphipod *Pontoporeia brevicornis* grp., Oligochaeta (worms) and Sphaeriidae (fingernail clams) are typical of northern boreal lakes and dominated the benthic macroinvertebrate fauna. Mean standing crops of benthic macroinvertebrates were higher in lakes of the lower Churchill River (approximately 11 500 m^{-2}) than in Southern Indian Lake (about 3000 m^{-2}) or lakes along the Rat and Burntwood rivers (approximately 3000 m^{-2}). The Churchill River had a positive influence on the production of macrobenthos in the lakes through which it passed (Hamilton and McRae, 1974). Mean standing crop of macrobenthic invertebrates in Southern Indian Lake was two to ten times higher than in other large lakes in the same general region, and benthic macroinvertebrate standing crops in lakes of the lower

Table 7.4.1 Chemical composition of Churchill River water at Southern Indian Lake, the Rat River at Notigi Lake, and the Nelson River at Cross Lake during the ice-free season prior to hydroelectric development. Concentrations are mean values for the open-water season (Cleugh, 1974). TSS: = total suspended sediments

Station	Na	K	Ca	Mg	Si	Cl	SO_4	Total CO_2 (μmole L^{-1})	pH	Total N (μg L^{-1})	Total P	TSS (mg L^{-1})
	\multicolumn{7}{c}{(mg L^{-1})}					
Churchill River (Southern Indian Lake outlet, 1973)	2.6	1.1	9.4	3.5	1.0	1.1	4.3	830	7.8	400	28	2
Rat River (Notigi Lake, 1973)	2.0	1.0	16	4.3	1.6	1.1	3.8	1130	7.9	740	27	6
Nelson River (Cross Lake outlet, 1972)	15.0	2.4	29	10.7	0.7	18.2	24.7	1740	8.2	475	34	15

Churchill River were considerably higher than would be expected for lakes of that size and at that latitude (Hamilton and McRae, 1974).

The crustacean zooplankton of these lakes were typical limnetic species and were dominated by copepods. Total community abundance was strongly affected by water exchange times. A high degree of similarity existed between the fauna of lower Churchill River lakes, Southern Indian Lake and lakes of the Rat-Burntwood system. Generally, copepods comprised approximately 80 per cent of total numbers while cladocerans comprised the remaining approximately 20 per cent. In Southern Indian Lake, total abundance in the main lake basins was comparable to more southerly lakes such as Lake Ontario and Lake Winnipeg (Patalas, 1975). The lowest numbers of crustaceans occurred in the inflow regions of Southern Indian Lake (approximately $10 \, l^{-1}$) while higher numbers (approximately $100-200 \, l^{-1}$) were encountered in some well-protected bays. Short water exchange times tended to reduce severely total abundance of the community. Lakes in the Rat-Burntwood system which had water exchange times of 30–200 days tended to have lower crustacean abundance (about $15-80 \, l^{-1}$) than the main basins of Southern Indian Lake but higher than the rapidly flushed lakes along the lower Churchill River ($1-6 \, l^{-1}$) which had exchange times of 1–9 days.

The fish fauna of these lakes was typical of relatively shallow boreal Canadian lakes (Koshinsky, 1973). Diversity of fish was relatively low and the community was dominated by a relatively small number of cool-water adapted benthivores (lake whitefish *Coregonus clupeaformis*, white sucker *Catostomus commersoni* and longnose sucker *C. catostomus*), planktivores (ciscoes *Coregonus artedii* and related species), and piscivores (northern pike *Exos lucius*, walleye *Stizostedion vitreum* and burbot *Lota lota*).

Sport fishing was not important in the area but many lakes supported commercial and domestic fisheries with modest yields (up to approximately $4 \, kg \, ha^{-1}$). Walleye, pike and whitefish were the valuable commercial species. Yields were strongly influenced by economic factors such as access, cost of production and the availability of alternate employment, so over-exploitation generally was not a problem in these fisheries. Standing crops of commercially valuable species were relatively high, compared to other Canadian Shield lakes, probably reflecting the relatively shallow depths of these lakes and high nutrient loading due to riverine conditions.

7.4.3 HYDROELECTRIC DEVELOPMENT OF THE CHURCHILL AND NELSON RIVERS

Hydroelectric development of the Churchil and Nelson rivers is a major energy producing project with a total designed capacity of approximately 8400 MW. The scheme utilizes the combined flow of two of the major rivers of Canada. Most of the flow of the Churchill, the smaller and more northerly of the two rivers, has been diverted into the Nelson, the larger and more southerly river

ECOLOGICAL EFFECTS OF HYDROELECTRIC DEVELOPMENT 279

(Figure 7.4.1). The Churchill River, near the point of diversion (Missi Falls), has a drainage basin of 250 000 km^2 and a long-term mean flow of 1010 m^3 s^{-1}. It flowed through Southern Indian Lake, a large (pre-impoundment surface area: 1977 km^2), multibasin lake. A control dam was placed at the natural outlet of the lake (Missi Falls) in 1976 and the level of the lake was raised 3 m above the long-term mean level (Figure 7.4.2A, B). Water from the Churchill River now leaves the lake through a channel constructed between South Bay and the headwaters of the Rat River in the Nelson River basin. A control structure at the outlet of Notigi Lake regulates flows down the Rat River Valley. The control structure was closed in 1974, retaining local runoff water until water levels in the Rat River Valley and Southern Indian Lake were similar (Figure 7.4.2C). The diversion channel was then opened in 1976, and the diversion was operating at full design capacity by late 1977. The diverted flow joins the Burntwood River at Threepoint Lake, below the Notigi control structure, and enters the Nelson River at Split Lake. The amount of water diverted out of the Churchill River basin generally has been held constant at near 760 m^3 s^{-1} or 75 per cent of the long-term mean flow.

Hydroelectric power from the project is produced on the lower Nelson River where the combined annual mean flow of the Churchill and Nelson rivers is

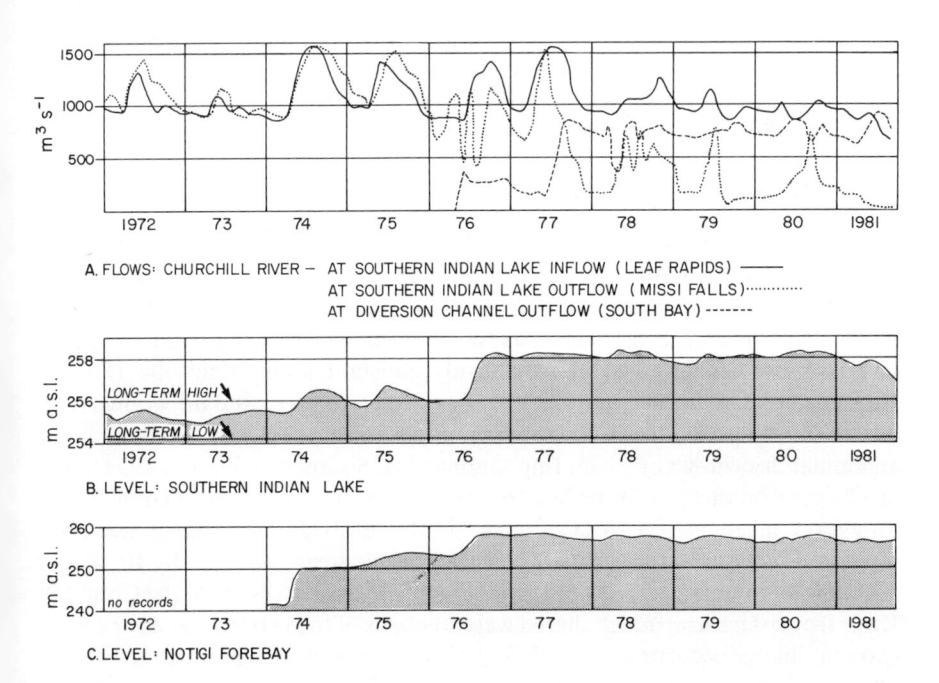

Figure 7.4.2 Churchill River flows at the Southern Indian Lake inflow, outflow and diversion channel outflow (A); water levels of Southern Indian Lake (B) and the forebay of Notigi Reservoir (C), 1972–1981

about 3500 m³ s⁻¹. Smaller generating plants are planned for the Burntwood River, below the Churchill River diversion (Figure 7.4.1). Power production also involves the regulation of Lake Winnipeg (surface area approximately 23 750 km²) for winter storage. The level of Lake Winnipeg is controlled by a generating station at Jenpeg, and midwinter flows out of the lake are regulated by various artificial channels, control dams and dikes. Downstream of Jenpeg, the three generating stations currently in place have created large reservoirs, flooding existing lakes and the Nelson River Valley. About 30 per cent (approximately 2600 MW) of the total potential of the Nelson River has been developed, with a further 1100 MW currently under construction.

7.4.4 ECOLOGICAL EFFECTS OF HYDROELECTRIC DEVELOPMENT-SOUTHERN INDIAN LAKE

Our studies of the ecological effects of the Churchill–Nelson diversion have emphasized the response of Southern Indian Lake to impoundment and diversion. Therefore, we will treat Southern Indian Lake in detail and then outline the physical changes resulting from Churchill River diversion in lakes along the lower Churchill valley and the diversion route, and briefly describe the results of some biological studies in Notigi Lake. We also include an outline of the downstream effects on Cross Lake because, although the lake is located outside of the immediate Churchill River diversion area, it demonstrates the effect that lowered water levels can have on fisheries.

7.4.4.1 General Physical Changes

The water level of Southern Indian Lake was raised by about 3 m above long-term mean levels in 1976 (Figure 7.4.2B). Water levels in the summers of 1974 and 1975 exceeded the 20-year recorded high level by approximately 0.5 m for about 5 months due to dam construction at the lake outlet. Since impoundment, lake level has been maintained at a relatively constant level. Manitoba Hydro's current operating licence restricts total drawdown to 0.9 m and drawdown over any 12-month period to 0.6 m. However, application recently has been made for an annual drawdown of 1.2 m. Impoundment of Southern Indian Lake resulted in a 20 per cent increase in surface area and a 39 per cent increase in volume over long-term mean values (McCullough, 1981). Individual basins of the lake, separated by natural topographic constrictions, increased in area by 10–45 per cent and in volume by 27–84 per cent (Figure 7.4.3). Diversion of the Churchill River from its natural outlet altered water budgets of the various basins, the most extreme change occurring in South Bay (through which the Churchill now flows) where residence time changed from 4 years to 11 days.

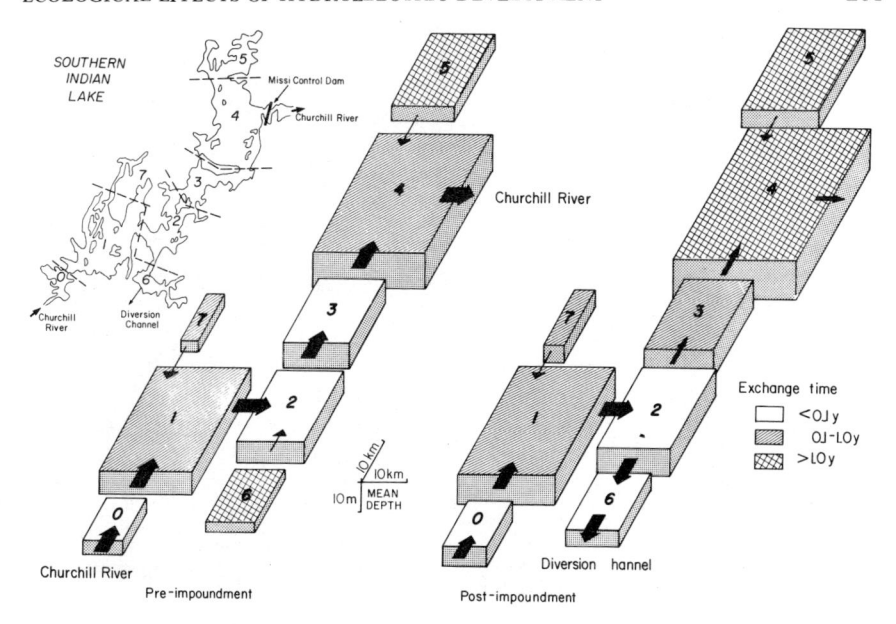

Figure 7.4.3 Schematic diagram of basins of Southern Indian Lake showing surface area, volume, mean depth and water exchange times before and after impoundment and river diversion. Each basin is represented by a three-dimensional block. Surface areas are shown by the horizontal scale, mean depths are shown by the vertical scale and water exchange times are shown as toning on the horizontal surface of each block. Inter-basin water exchanges are shown as arrows, the widths of which are proportional to mean annual flow

7.4.4.2 Shoreline Erosion

Shoreline forms and erosion rates also were altered by impoundment. Before 1976, 76 per cent of the water–land contact consisted of water-washed bedrock beaches. Shorelines along unconsolidated overburden occurred only where there was protection from long wave fetches or in deep deposits of proglacial sands and gravels and glacio-lacustrine clays. Less than 5 per cent of total shoreline length was actively eroding. Impoundment raised the lake level above the water-washed bedrock beaches and into overlying unconsolidated glacial deposits. Over 80 per cent of the shoreline immediately after impoundment occurred in fine-grained tills and lacustrine silty-clays which previously existed in permafrost. Lakewide, approximately 25 MW of power in the form of wave energy was directed against the shoreline during the open-water season (Newbury, 1981). This has caused retreats of up to $10 \, \text{m yr}^{-1}$, removing up to $25 \, \text{m}^3$ of material per metre of shoreline (Newbury and McCullough, in press). Along shores bounding large basins, erosion proceeded in a repeating sequence of the melting of

Figure 7.4.4 Eroding bank following a storm, at a shoreline affected by relatively high incident wave energies. Wave action has removed slumped material, cutting a vertical bank in frozen clay

permafrost and slumping of bank materials during periods of relative calm, followed by removal during storms (Figures 7.4.4, 7.4.5). Where the shore is exposed to smaller fetches (< 2k m), erosion of the shoreline was retarded by the protective moss and root mat of the former forest floor. In these cases, melting of permafrost in the inundated zone initiated widespread settling, indicated by slump scars and fallen trees in the backshore zone.

An average erosion index was determined by calculating wave energy using wind records from Southern Indian Lake, and by monitoring erosion rates at 20 survey sites surrounding the lake. The sites were chosen to represent a range of materials and fetches for the lake. Using shoreline survey and wind data from 1978–80, the rate of erosion was $0.00035 \, m^3$ per tonne-m of wave energy per metre of shoreline. Subtraction of peat volume and ice content from this figure yielded an index of erosion of mineral materials of $0.00012 \, m^3$ per tonne-m of wave energy per metre of shoreline.

Although grain size analyses indicated that these shoreline materials were 70–95 per cent clay, only a portion of the material eroded was carried offshore in suspension. Wave energy breaking up the clay mass did not reduce all the material to colloidal particles; rather, clay 'balls' were formed ranging from 2 cm diameter through sand and silt sizes. Bottom cores taken adjacent to eroding banks indicated that these clay balls were deposited in the nearshore zone in a pattern of decreasing size with increasing distance from shore.

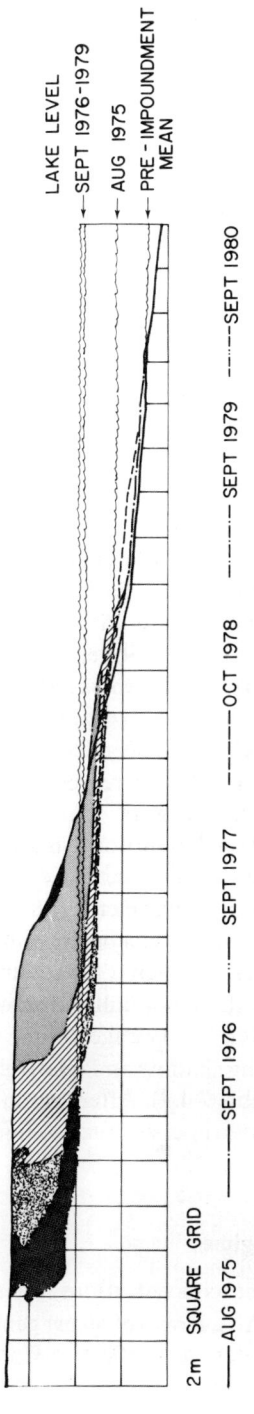

Figure 7.4.5 Surveyed profiles of the shoreline in Figure 7.4.4 indicating the quantities of material eroded annually

Table 7.4.2 Sediment budget for Southern Indian Lake before and after impoundment (from Hecky and McCullough, in press). Only sediments $> 1\,\mu\text{m}$ (nominal diameter) are considered. Net sedimentation/erosion (N) is calculated from import (I), storage in suspension (S), and export (E) as $N = I - S - E$. All values are kilotonnes

Year	Import	Storage	Export	Net sedimentation (+)/ net erosion (−)
1975	124	2	57	+ 65
1976	115	124	195	− 204
1977	148	− 83	305	− 74
1978	103	21	278	− 196

7.4.4.3 Sediment Budgets

The eastern basins of South Bay were turbid before impoundment due to resuspension of bottom sediments. Little change in suspended sediment concentration occurred in the northern-most basin of the lake because the shoreline was predominantly coarse-grained materials. After impoundment, dilution of sediment in lake waters by the Churchill River is indicated (Figure 7.4.6) by the dark plume at the inflow; export of sediment is indicated by the light tones of the outflowing waters at Missi Falls and the diversion channel. Despite the inefficient dispersion of clay from shorelines, suspended sediment concentrations rose dramatically in regions 1, 2, 3, 4 and 6. Increases were greatest in region 4 where concentrations after impoundment were 6 to 8 times higher than before impoundment (Hecky *et al.*, 1979). No substantial increases occurred in regions 5 and 7 and many small bays on the lake because either fine-grained overburden materials were lacking (region 5) or wave energies were too low to cause bank erosion or maintain sediments in suspension (region 7 and small bays).

Material from eroding shorelines altered the sediment budget of the lake. In 1975, prior to impoundment, the Churchill River was the primary source of sediment to the lake, supplying an average of 1.1×10^5 tonnes of sediment annually. The lake, despite being shallow and relatively rapidly flushed, retained > 50 per cent of this input (Table 7.4.2). After impoundment and diversion, the lake exported approximately five times as much sediment as compared to before impoundment.

7.4.4.4 Light and Thermal Regimes

Increased suspended sediment concentrations altered the underwater light regime over much of the lake. An average of 50 per cent of the suspended offshore material was $> 1\,\mu\text{m}$ in diameter, so it effectively scattered light (Kullenberg,

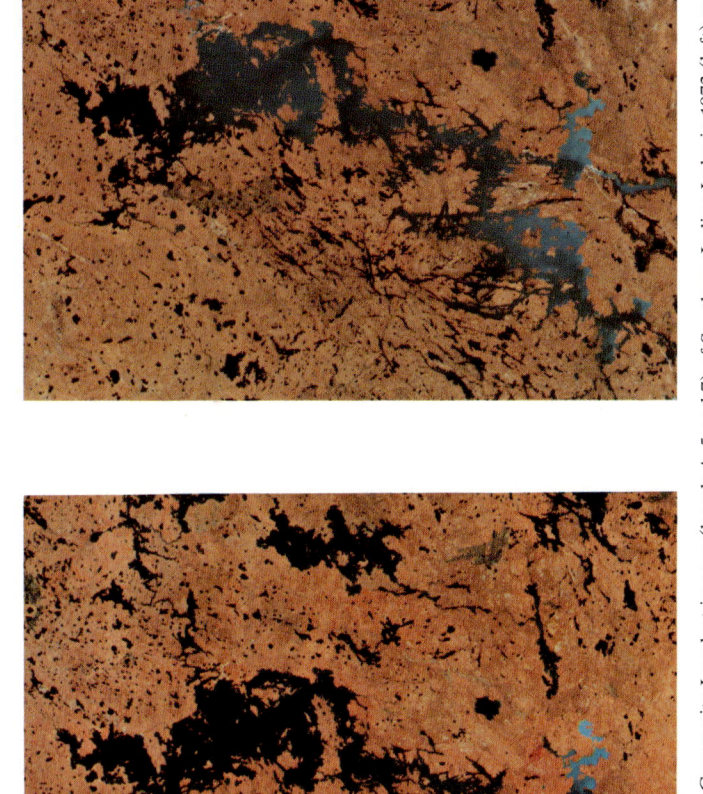

Figure 7.4.6. Composite Landsat images (bands 4, 5 and 7) of Southern Indian Lake in 1973 (left) and 1978 (right). Lighter tones indicate high reflectivity due to high concentrations of suspended sediments

1974). Increased light scattering after impoundment was apparent in Landsat photographs of the lake (Figure 7.4.6). Light penetration was reduced by mean values of 30–50 per cent in regions 2, 3, 4 and 6 (Hecky et al., 1979). Since the large regions of Southern Indian Lake are unstratified throughout the open water season, the mean water column irradiance ($\bar{I}$) defines the light regime for freely circulating phytoplankton. $\bar{I}$ is calculated using the equation:

$$\bar{I} = \frac{1}{\bar{z}} \int_0^{Z_m} I_o e^{-kz} \, dz$$

where z is depth (m), $\bar{z}$ is the mean depth of the basin, Z_m is the maximum depth of the basin, I_o is the incident photosynthetically active radiation at $z = 0$ and k is the vertical light extinction coefficient. Mean water column irradiance declined in all regions of the lake after impoundment (Figure 7.4.7). For regions in which light penetration was similar before and after impoundment (e.g. region 5), observed decreases in mean water column irradiance were due mainly to increased mean basin depth. For regions in which light penetration decreased due to increased suspended sediment levels (e.g. regions 2, 4 and 6), observed decreases in mean water column irradiance were due to the combined effect of decreased light penetration and increased mean depth (Figure 7.4.7).

Impoundment and diversion had no obvious effect on vertical thermal

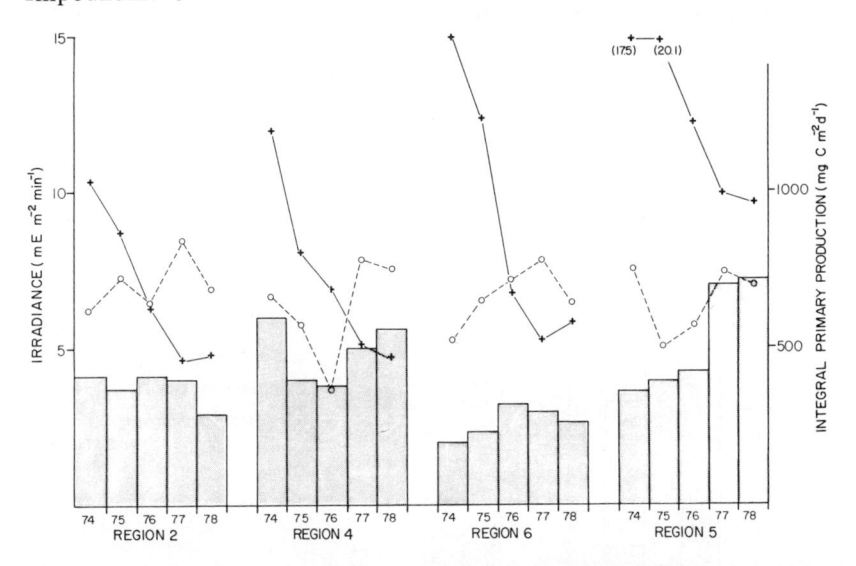

Figure 7.4.7 Mean water column irradiance during daylight ($\bar{I}$) (—+—), mean onset of light saturation (I_k) (---o---) and mean integral primary production (bars) for four regions of Southern Indian Lake, 1974–1978. Mean values are for 5 July–4 September for all five years. Mean onset of light saturation and mean integral primary production were measured using the incubator technique and digital computation programme of Fee (1973)

structure of the lake. Surface waters typically were 2–3 °C warmer than deep waters during June and July of the open-water season, but no persistent stable stratification was observed before or after impoundment. However, impoundment and river diversion did alter the horizontal distribution of heat in the system. For example, region 6, which received diversion flows, was cooled by 1 °C below its pre-impoundment temperature in the open-water season and by 2–3 °C in winter. Region 4 was cooled by 1 °C during the open-water season because of reduced heat input from the river and loss of radiant energy due to backscattered light. The ice-out pattern of the lake imposed a natural southwest to northeast thermal gradient (2 °C between regions 1 and 4 in July) and river diversion increased this gradient (Hecky *et al.*, 1979).

7.4.4.5 Primary Production

The productivity of phytoplankton in Southern Indian Lake demonstrated a variable response to impoundment. In the main regions of the lake (1, 2, 3, 4), there was no significant change in integral primary production (Figure 7.4.7). Production of phytoplankton in these regions was phosphorus limited before impoundment (Guildford, 1978; Healey and Hendzel, 1980) but was light limited after impoundment as the mean water column irradiance, $\bar{I}$, fell below the irradiance required for light saturation, I_k (Hecky and Guildford, in press). Although soluble reactive phosphorus concentrations increased in all regions in the lake (Figure 7.4.8), phytoplankton in regions 1–4 were unable to benefit because of this light limitation.

In region 5 and small bays of the lake, where light penetration did not change significantly after impoundment, integral primary production increased by

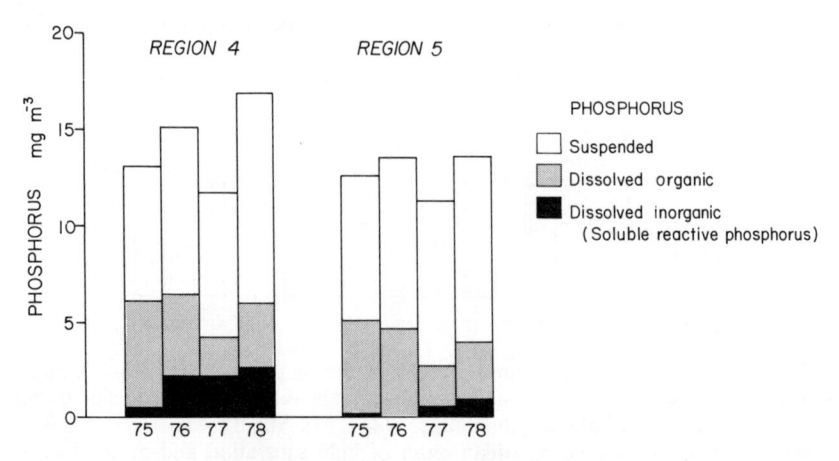

Figure 7.4.8 Concentrations of suspended, dissolved organic and soluble reactive phosphorus in regions 4 and 5, Southern Indian Lake, 1975–1978

50-100 per cent. Phytoplankton in region 5 did not become light limited because $\bar{I}$ exceeded I_k before and after impoundment (Figure 7.4.7) so the phytoplankton were able to use the increased soluble reactive phosphorus concentration (Figure 7.4.8). Primary productivity increased in region 6, despite a decline in $\bar{I}$ below I_k (Figure 7.4.7), because diversion linked this region with region 2 which was more productive than region 6 before diversion.

7.4.4.6 Macrobenthic Invertebrates

Profundal macrobenthos in Southern Indian Lake were surveyed before impact (1972), after flooding and partial diversion (1977), and after full diversion (1979) using methods described in Wiens and Rosenberg (in press). Lakewide standing crop (mean number m^{-2} ± SE) of macrobenthos increased from 3227 (369) m^{-2} before flooding (1972) to 5592 (493) m^{-2} just after flooding (1977), and decreased to 4817 (477) m^{-2} just after diversion (1979). This pattern has been observed commonly in newly created reservoirs all over the world (McLachlan, 1974; Wiens and Rosenberg, in press). Increased standing crop is ascribed to additions of nutrients and organic matter from newly flooded land (e.g. McLachlan, 1974; Baxter, 1977) while subsequent declines in standing crop may be due to depletion of the flooded organic matter (e.g. McLachlan, 1974). The response of macroinvertebrates to flooding was rapid in Southern Indian Lake as has been observed in many other reservoirs.

Mean standing crops of macrobenthos remained virtually unchanged for the three surveys in regions 1 and 6-east of Southern Indian Lake (Table 7.4.3). However, mean standing crops in the other regions increased markedly between 1972 and 1977; by 1979 densities had stabilized in regions 0 and 2, had increased in region 6-west, and had decreased in the other regions (3, 4, 5, 7), in relation to the 1977 levels. Four interrelated changes that occurred in Southern Indian

Table 7.4.3 Mean standing crops (number of individuals m^{-2} ± SE) of profundal macrobenthos in regions of Southern Indian Lake in 1972, 1977 and 1979

Region	1972	1977	1979
0	1712(401)	3587(1223)	3394(1377)
1	6239(1017)	5512(880)	5823(760)
2	2304(737)	4849(1338)	4457(1871)
3	3789(941)	6360(1335)	3468(749)
4	3832(740)	8311(955)	6903(1489)
5	2769(1350)	6123(1451)	5006(1267)
6-west	1019(588)	1503(18)	2044(770)
6-east	1770(1006)	1585(342)	1251(528)
7	3273(1231)	6717(1037)	4117(622)

Lake (additions of nutrients, additions of particulate organic matter, changes in suspended sediment concentrations and changes in integral primary production) largely explained the observed responses of macrobenthic standing crops in each region (Wiens and Rosenberg, in press).

Increased mean standing crops of macrobenthos in the first year after impoundment (1977) were proportionally highest in the 5–10 m depth zone (approximately 200 per cent) as compared to standing crops in the 0–5, 10–15 and 15–20 m depth zones. By 1979, standing crops stabilized at 10–15 m, declined at 5–10 m and 15–20 m and increased only at 0–5 m (approximately 30 per cent). Greater increases in standing crops of macrobenthos in shallower as compared to deeper zones of the lake after flooding were attributed to preferential deposition of allochthonous organic matter in these shallow areas (Wiens and Rosenberg, in press).

Four taxa of macrobenthos comprised > 95 per cent of total standing crop before and after impoundment: Diptera (mainly Chironomidae), *Pontoporeia brevicornis* grp. Oligochaeta and Pelecypoda (mainly Sphaeriidae). Responses of these major taxa to the flooding of Southern Indian Lake differed in many ways from those reported for other newly formed reservoirs (cf. Wiens and Rosenberg, in press). *P. brevicornis* grp. remained the most abundant organism; there was no evident succession of macrobenthic taxa, and a high diversity of profundal species was maintained in Southern Indian Lake. These results, together with the relatively slight changes in standing crop observed after flooding, indicate only a marginal impact on Southern Indian Lake macrobenthos. This minimal impact is probably related to the environmentally less disruptive type of reservoir formation in Southern Indian Lake (i.e. low-level flooding) compared to that normally occurring when reservoirs are formed by damming a river.

7.4.4.7 Zooplankton

Planktonic crustaceans in Southern Indian Lake were surveyed before impact (1972), during a period when water levels exceeded previously recorded high levels (1975) (Figure 7.4.2) and then yearly after diversion (1977–1980), using methods described by Patalas and Salki (in press). No dramatic changes were observed in the list of species present but lakewide average abundances (expressed as individuals per litre) of crustacean plankton following diversion decreased from 76 ind. l^{-1} in 1972 to 40–46 ind. l^{-1} in the four post-diversion years (1977–1980) (Table 7.4.4A). The abundance of plankton in 1975, an intermediate period, was 61 ind. l^{-1}. The degree to which zooplankton responded to diversion differed in various parts of the lake. No significant changes occurred adjacent to the Churchill River inflow (region 0) where post-diversion abundance was approximately 36 ind. l^{-1} compared to approximately 35 ind. l^{-1} in 1972. Abundances in the main water bodies north of the diversion route (regions 2, 3, 4) declined from 89 ind. l^{-1} in 1972 to 34 ind. l^{-1} during

ECOLOGICAL EFFECTS OF HYDROELECTRIC DEVELOPMENT 289

Table 7.4.4 Changes in abundance of planktonic crustaceans in Southern Indian Lake from 1972 to 1980. Values are lake averages for late summer in A: number of individuals l^{-1} and B: percentage composition

	1972	1975	1977	1978	1979	1980
A: Individuals l^{-1}						
Calanoida	25.2	32.7	24.8	21.0	21.2	25.2
Cyclopoida	35.2	18.9	15.6	16.4	15.0	18.5
Cladocera	15.6	9.0	3.9	4.5	3.9	2.6
Total	76.1	60.6	44.3	41.9	40.1	46.3
B: Percentage composition						
Calanoida	33.2	53.9	56.0	50.1	52.8	54.4
Cyclopoida	46.3	31.3	35.2	39.1	37.5	39.9
Cladocera	20.5	14.8	8.8	10.8	9.7	5.7
Total	100.0	100.0	100.0	100.0	100.0	100.0

1977–1980. These changes coincided with other effects of impoundment (e.g. lower midsummer chlorophyll-*a* concentrations, decreasing water transparency) and of diversion (e.g. decreasing water temperature) in these regions (Patalas and Salki, in press). In South Bay (region 6), through which the main flow was diverted, zooplankton abundance declined from 83 to 40 ind. l^{-1} because of higher flushing rates.

Not all groups of crustaceans responded in the same way to impoundment. Lakewide average numbers of cladocerans declined from 16 ind. l^{-1} before impoundment to 4 ind. l^{-1} following impoundment; cyclopoids declined from 35 to 16 ind. l^{-1}; but calanoids remained relatively stable at around 24 ind. l^{-1}. Percentage composition of cladocerans, cyclopoids and calanoids changed from 20, 46 and 33 per cent, respectively, in 1972, to 9, 38 and 53 per cent, respectively, in the four post-diversion years (1977–1980) (Table 7.4.4B). Although absolute numbers of calanoids did not change during this period, their percentage composition increased because of the decline in abundance of the other two groups.

A significant increase occurred in the abundance and distribution of some large species after diversion (e.g. *Limnocalanus macrurus, Senecella calanoides, Mysis relicta*) *Mysis relicta* was absent in pre-diversion catches, but from 1977 to 1980 it became more abundant and its distribution expanded. These large species are a preferred food item for both whitefish and cisco; they are cold-water stenotherms and they inhabit deeper layers of water. Increased abundance of these species could be due to decreased water transparency which offered better protection against predatory fish, decreased water temperatures which created more favourable conditions and increased depth of the lake which expanded the volume of deeper waters suitable for these species.

7.4.5 ECOLOGICAL EFFECTS OF HYDROELECTRIC DEVELOPMENT-OTHER LAKES

7.4.5.1 Lakes of the Lower Churchill River

The mean annual natural discharge of the Churchill River at Missi Falls has decreased, under average regulated conditions, from 1010 to 220 m^3 s^{-1} (McCullough, 1981) (Figure 7.4.2A). However, Manitoba Hydro's licence allows reduction of discharge to as low as 14 m^3 s^{-1} during the open-water period and 42 m^3 s^{-1} under ice conditions. No upper limit has been placed on the discharge released at Missi Falls. Since diversion, the annual range of releases has exceeded 600 m^3 s^{-1} (Figure 7.4.2A). Reduced mean flows have lowered lake levels dramatically on Partridge Breast, Northern Indian and Fidler lakes downstream of Missi Falls (Figure 7.4.1), and have increased annual water level fluctuations from a natural range of 1 m to post-diversion ranges of 2–3 m (Brown, 1974). Pre-diversion lake areas have been halved, exposing a total of 96 km^2 of former lake bottom on the three lakes along the lower Churchill. Natural water exchange times for each of the lower Churchill River lakes originally were only 1 to 9 days. The combination of decreased lake volumes and decreased annual flows yielded only small changes in exchange times, with a range of 1–16 days under average post-diversion flow conditions.

7.4.5.2 Diversion Route Lakes (Rat and Burntwood River Valleys)

General description

The upper Rat River Valley is now flooded by the Notigi Reservoir and water levels in lakes of the lower Rat River Valley and lower Burntwood River Valley have increased due to the Churchill River diversion. A chain of eight lakes extended along the Rat and Burntwood rivers prior to Notigi impoundment and Churchill River diversion (Figure 7.4.1). The lakes were relatively small, ranging from 3.7 km^2 (Issett, the headwater lake) to 85.3 km^2 (Wuskwatim, on the lower Burntwood River). The lakes were shallow (mean depth 1.7–5.3 m), rapidly flushed (residence times 37–136 days), and turbid (Secchi disc depths 0.5–1.5 m) because of extensive fine-grained glacial sediments in the basin, shallow mean depths, low residence times and frequent interconnecting riverine stretches.

Formation of the Notigi Reservoir united five of the lakes at a common water level and flooded 409 km^2 of terrestrial soils and vegetation. Pre-existing lake area within the reservoir boundary was 67 km^2 so 86 per cent of the reservoir surface area is underlaid by former land. Water levels of the forebay area of

Notigi Lake were raised 16 m (Figure 7.4.2C); mean depth of the reservoir after flooding was 7 m. Rat River natural mean annual discharges of 2 m^3 s^{-1} at Issett Lake and 35 m^3 s^{-1} at its confluence with the Burntwood River were augmented by 300 m^3 s^{-1} through 1976, increasing to mean monthly discharges as high as 820 m^3 s^{-1} in late 1977 (Figure 7.4.2A). Clearing was performed only for hydraulic purposes on the immediate course of the diverted water through the reservoir and also for aesthetic purposes in the vicinity of road crossings. Standing black spruce dominate this flooded area and large peat mats have floated free in headwater bog regions. Shoreline erosion within the reservoir is minimal because the dendritic shape of the reservoir results in short wind fetches (maximum 10 km) and low wave energies.

Water levels on the Burntwood River lakes below Notigi Reservoir are now 3–5 m above their pre-impoundment mean due to an increase in discharge from 100 m^3 s^{-1} to 880 m^3 s^{-1}. Winter ice-dams raise lake levels further. Erosion in the river channel below Notigi has been significant because of the dramatic increase in flow.

Notigi Lake

Limnological observations were concentrated on Notigi Lake before, during and after impoundment and diversion. Notigi Lake and two basins connected by a relatively shallow, narrow channel. The western basin was flushed rapidly by the Rat River prior to impoundment and by the Churchill River after diversion, whereas the eastern basin was relatively poorly flushed before and after impoundment (Hecky and Harper, 1974).

The western basin was isothermal throughout the open-water period (Cleugh, 1974) prior to impoundment. Greatly increased water levels in the two basins during 1974 and 1975, combined with reduced flushing and minimal change in fetch, allowed summer thermal stratification to develop (Hecky et al., 1979). With the onset of full diversion flow in 1978, the western basin became isothermal again while a thermocline persisted in the eastern basin (Hecky et al., 1979).

These changes in thermal structure, combined with the gradually increasing amounts of flooded terrestrial organic matter with its high oxygen demand, have resulted in unique oxygen stratification patterns. Prior to impoundment, dissolved oxygen in Notigi Lake was vertically homogeneous with about 75 per cent saturation during the winter and near saturation throughout the open-water period (Cleugh, 1974). After impoundment, winter dissolved oxygen concentration in the eastern basin declined with depth, indicating the interaction between warmer bottom waters and the oxygen demand of flooded substrate (Figure 7.4.9). In summer, contact of the relatively warm water of the metalimnion with flooded terrain resulted in midwater oxygen minima (Figure 7.4.9). The depths of these summer minima tended to decrease over the post-impoundment period, probably in relation to the length of time the soil had

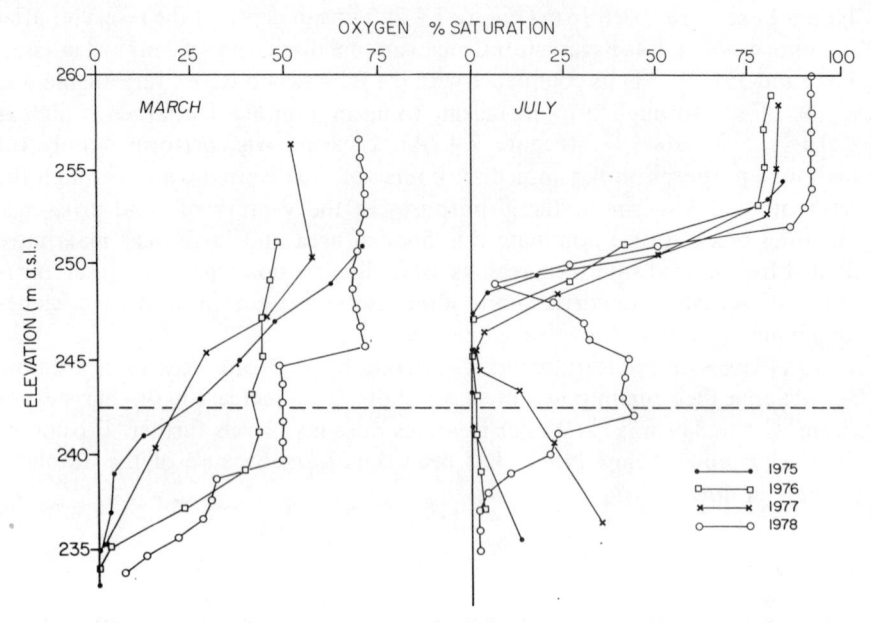

Figure 7.4.9 Winter (March) and summer (July) oxygen profiles for the eastern basin of Notigi Lake, 1975–1978. Because water levels rose steeply from 1975 through 1976, concentrations are plotted against elevations so that homologous strata may be compared from year to year. Pre-impoundment lake level is indicated by the dashed horizontal line

been flooded. The severity of oxygen depletion decreased from 1975 to 1978, during both summer and winter, perhaps indicating depletion of readily oxidizable material.

Flooding released nutrients from the soils and stimulated algal growth and accumulation. Mean chlorophyll concentrations after impoundment (16.8 mg m^{-3}) more than tripled from pre-impoundment concentrations (5.4 mg m^{-3}) and maximum chlorophyll concentrations in 1975 and 1976 (32.4–35.0 mg m^{-3}) were about four times the pre-impoundment maximum (8.6 mg m^{-3}) (Jackson and Hecky, 1980). Chlorophyll concentrations declined in subsequent years from the post-impoundment maximum. Integral primary production peaked in 1976 with rates of up to 1.76 g C m^{-2} d^{-1}. Algal biomass reached a high of 12 000 mg m^{-3} and was dominated by *Aphanizomenon flos-aquae*. Rates of primary production have been declining since 1976.

Oxygen profiles and algal production figures indicate that the reservoir experienced a rapid and intense increase in bacterial and primary production with flooding and that, subsequently, productivity declined rapidly, a response similar to more southerly reservoirs (Lowe-McConnell, 1973). The dramatic increase in flushing of the system from 1974–75 to 1978 may have hastened the decline of the productive upsurge.

7.4.5.3 Cross Lake

Cross Lake is located on the Nelson River, upstream of the confluence of diverted Churchill River water with the Nelson (Figure 7.4.1). The Jenpeg generating station was constructed at the main inflow of Cross Lake in 1974 to regulate the level of Lake Winnipeg. The flow entering Cross Lake, and the water level of the lake, are controlled by discharges through Jenpeg. Prior to regulation, Cross Lake had a mean depth of 2.4 m (open-water season) and a surface area of 460 km^2. After regulation, a minimum discharge of around 700 m^3 s^{-1} into the lake resulted in a drawdown of 1.7 m below the historic mean open-water stage of 207.1 m (Gaboury and Patalas, 1981). Drawdown decreased lake volume by 53 per cent, decreased lake area by 26 per cent and decreased mean water depth to 1.5 m.

The regulation of Cross Lake has resulted in a number of severe environmental effects. Submergent vegetation was scarce prior to regulation (Driver and Doan, 1972; Ayles *et al.*, 1974), but proliferated after regulation (Gaboury and Patalas, 1981). Oxygen depletion in winter occurred only in isolated bays prior to regulation (Driver and Doan, 1972; Koshinsky, 1973), but now the decay of extensive areas of submergent vegetation can result in low dissolved oxygen levels (< 3 mg l^{-1}) during the winter. Surface water temperatures in the lake usually were < 20 °C prior to regulation, whereas after regulation surface water temperatures of 21–26 °C are common. The frequently shallow post-regulation mean depth of the lake also apparently has resulted in high turbidity (Secchi disc transparency approximately 45 cm) through resuspension of the bottom sediments by wind (Gaboury and Patalas, 1981).

7.4.6 EFFECTS OF HYDROELECTRIC DEVELOPMENT ON FISHERIES

In this section, we present results of the following fisheries studies made along the Churchill and Nelson Rivers:

1. Catches and grade in the Southern Indian Lake whitefish fishery.
2. The effects of early spring drawdown on whitefish and cisco populations in Cross Lake.
3. Fish mercury levels in Southern Indian Lake and the diversion route lakes.

7.4.6.1 Grade of Whitefish Catch in Southern Indian Lake

Commercial catches of lake whitefish from Southern Indian Lake, the largest whitefish fishery in northern Manitoba, have changed since impoundment. The pre-diversion catch was characterized by A grade (Export) light coloured fish and relatively high mean catch per unit effort (CPE), but 5 years after impoundment, classification of the fishery was lowered to B grade (Continental) and CPE is only one-third of the pre-impoundment mean.

294   EFFECTS OF POLLUTANTS AT THE ECOSYSTEM LEVEL

Prior to impoundment, top grade was maintained by selectively fishing certain basins of the lake. Effort was concentrated in region 4 (between Sand Point and Long Point) for fish landed at the Loon Narrows packing plant which received 80–85 per cent of the fish shipped from the lake (Figure 7.4.10A). After

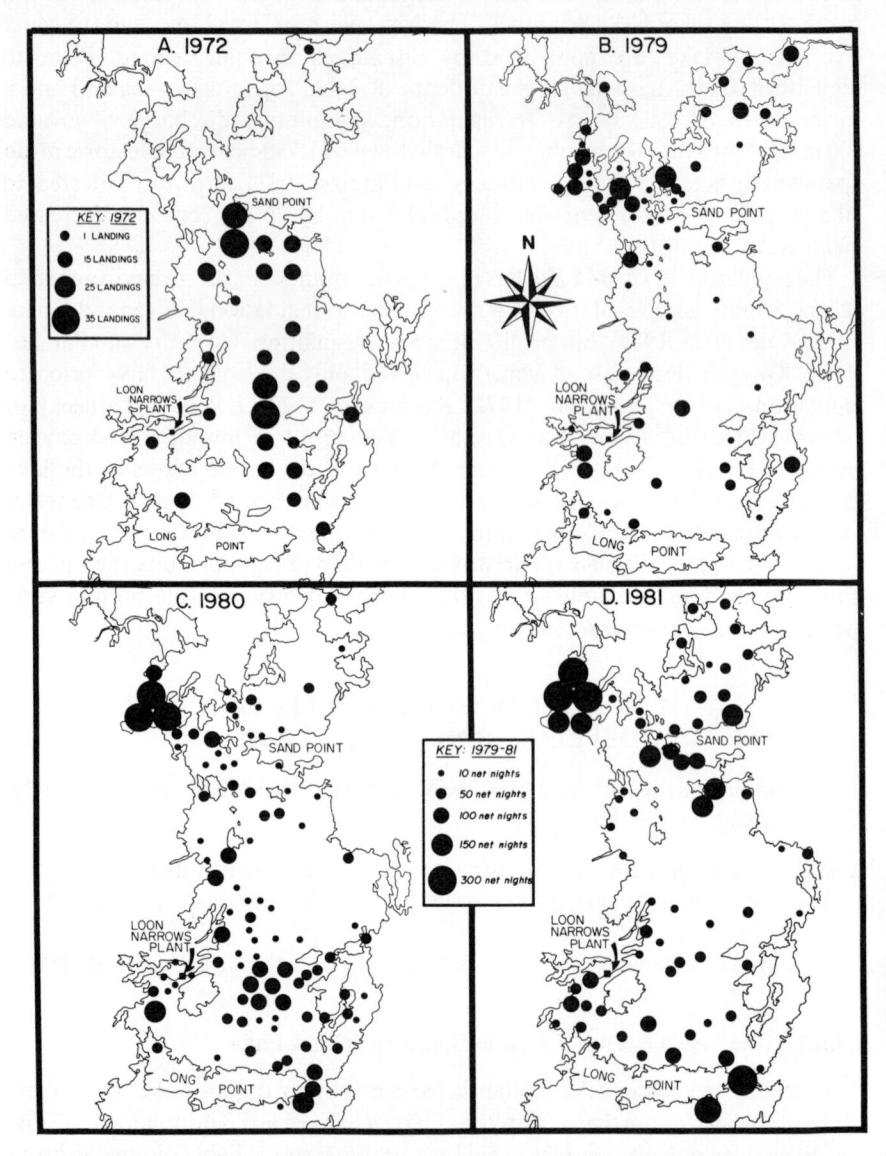

Figure 7.4.10 Geographical distribution of summer fishing effort for fish landed at the Loon Narrows packing plant, Southern Indian Lake, before impoundment (A: 1972) and after impoundment (B-D: 1979–81). From Bodaly et al. (in press, b). See Bodaly et al. (1980) for sampling methods

impoundment, this pattern of selective exploitation changed. In 1979, 1980 and 1981, the proportion of fishing effort expended in region 5 (north of Sand Point), outside of traditional areas, was 62 per cent, 30 per cent and 33 per cent, respectively (Figure 7.4.10B,C,D). Fish stocks being exploited in region 5 were composed largely of dark coloured, more heavily parasitized whitefish (Bodaly et al., 1980; Bodaly et al., in press, b). As a result of these changes, the total lake catch was composed of up to 81 per cent lower grade fish.

The change in the geographic distribution of fishing effort was a response by fishermen to sharp declines in CPE on traditional fishing grounds. Lake whitefish CPE on traditional fishing grounds for the summer fishery in 1972, prior to impoundment, was 23 kg standard net^{-1} 24 h^{-1} but declined to 14, 10.5 and 7.5 kg standard net^{-1} 24 h^{-1} in 1979, 1980 and 1981 (Figure 7.4.11). A corresponding decline has occurred in whitefish CPE for the winter fishery. Winter CPE on traditional fishing grounds decreased from 19 kg standard net^{-1} 24 h^{-1} in 1972–73 to 4 kg standard net^{-1} 24 h^{-1} in 1980–81 (Figure 7.4.11). Apparently, these catch declines are not the result of excessive fishing pressures

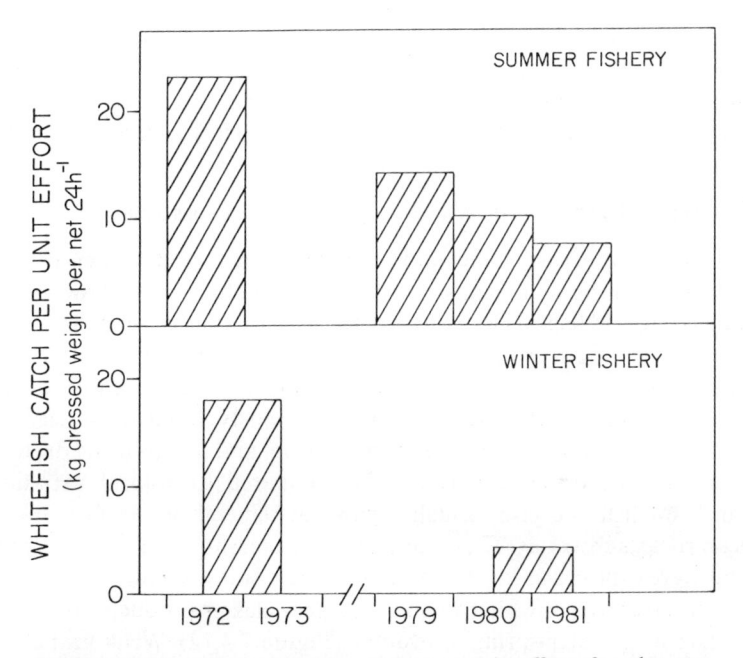

Figure 7.4.11 Lake whitefish catch per unit effort for the pre-impoundment (1972 summer and 1972–3 winter) and post-impoundment (1979, 1980, 1981 summer and 1980–1 winter) Southern Indian Lake commercial fishery. From Bodaly et al. (in press, b). Data are from traditional fishing areas adjacent to the Loon Narrows packing plant (see Figure 7.4.10). See Bodaly et al. (1980) for sampling methods

prior to impoundment (Bodaly et al., 1980). Whitefish stocks before impoundment were relatively slow growing and old, had moderate mortality rates and, therefore, showed no signs of over-exploitation (Ayles, 1976). The whitefish catch had declined in the decade prior to impoundment as a result of decreased fishing effort, making depletion of stocks due to continued high fishing effort at the time of impoundment unlikely (Bodaly et al., 1980). Furthermore, the age distribution of the post-impoundment whitefish catch has not changed significantly since flooding and the catch continues to be composed of relatively old fish with a number of year classes being represented (Bodaly et al., in press, b).

Declines in whitefish CPE may be due to major movements of fish out of Southern Indian Lake in response to changes caused by lake impoundment. Large numbers of whitefish have been found congregating immediately below the Missi Falls control dam. Similar whitefish congregations were reported at the Lobstick control dam in Labrador (Barnes, 1981). Large numbers of fish have been noted also in the Southern Indian Lake diversion channel at South Bay. Emigrations of whitefish from Southern Indian Lake may have occurred in response to reduced light penetration in major lake basins. The distribution and schooling behaviour of fishes are known to be affected by light and by suspended sediment levels (Harden Jones, 1956; Volkova, 1971; Swenson, 1978) and post-impoundment daytime light intensities are below those required by most fishes for effective feeding and schooling (Blaxter, 1970; Hecky et al., 1979).

7.4.6.2 Effects of Drawdown on Fish Populations in Cross Lake

The regulation of Cross Lake and, especially, increased winter drawdown adversely affected the year class strengths and abundance of shallow-water fall-spawning coregonid fishes. Drawdown detrimentally affected whitefish and cisco hatching success and recruitment, apparently by draining spawning areas and desiccating eggs. Under regulation, average summer water levels were lower and average winter water levels were higher than previously. Rapid decreases in water levels in late spring were common since regulation, and fall to spring drawdowns of up to 2.2 m occurred (Figure 7.4.12), compared to natural variations of < 0.5 m. Whitefish and cisco usually spawn at the end of October in shallow depths on rocky substrates. These substrates are common on the shores of Cross Lake but were exposed at low water levels. There was a significant relationship between the strengths of whitefish and cisco year classes produced in 1971–1980 and the extent of winter-spring drawdown (Figure 7.4.12). Weak year classes of whitefish and cisco resulted from years with a marked winter drawdown and strong year classes tended to result from years with little winter drawdown.

Catches of adult whitefish declined substantially after lake regulation, both in relative and absolute abundance. The relative abundance of whitefish declined 65 per cent from levels before regulation in 1965 and 1973 catches (Driver and

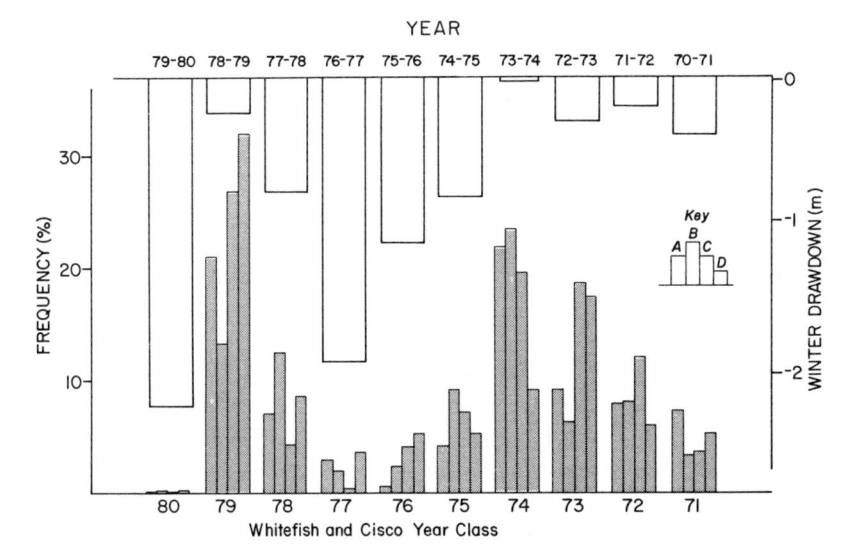

ECOLOGICAL EFFECTS OF HYDROELECTRIC DEVELOPMENT 297

Figure 7.4.12 Relationship between winter and spring drawdown (white bars) and whitefish and cisco year class success (dark bars) in Cross Lake, 1971–1980. Winter and spring drawdown calculated as minimum level between October 28 and May 31 subtracted from level on October 28. A: per cent frequency of whitefish from 1980 gillnet catches; B: per cent frequency of whitefish from 1981 gillnet catches; C: per cent frequency of cisco from 1980 gillnet catches; D: per cent frequency of cisco from 1981 gillnet catches. See Gaboury and Patalas (1981, 1982) for sampling methods and ageing procedures

Doan, 1972; Ayles et al., 1974), compared to catches in 1980 and 1981 (Gaboury and Patalas, 1981, 1982). Catch per unit effort of whitefish in 13.3 cm mesh declined from 35 fish 91 m net^{-1} night^{-1} in 1977 to 9 fish in 1980 and 4 fish in 1981 (B. Wright, personal communication; Gaboury and Patalas, 1981, 1982). Probable factors contributing to the decline in whitefish abundance include the effect of winter drawdown on year class strengths and recruitment, sudden declines in water levels which result in stranding and suffocation of fish in shallow bays and channels, and movements of fish out of Cross Lake.

7.4.6.3 Fish Mercury Levels

Reservoir formation has frequently been implicated as the cause of elevated mercury levels in fish (Potter et al., 1975; Abernathy and Cumbie, 1977; Bruce and Spencer, 1979; Meister et al., 1979). Increased fish mercury levels coincided with the creation of impoundments in the Churchill and Nelson River drainage basins and these higher levels led to restrictions on the commercial marketing of fish from ten impounded lakes.

In Southern Indian Lake, walleye muscle mercury concentrations from commercial samples increased from 0.19–0.30 ppm prior to impoundment (1971–1976) to 0.57–0.75 ppm after impoundment (1978–1981), while pike muscle mercury levels increased from 0.16–0.47 ppm prior to impoundment to 0.50–0.95 ppm after impoundment (Figure 7.4.13). Mercury levels in the muscle tissue of lake whitefish in Southern Indian Lake increased also, immediately following impoundment (Figure 7.4.13). Fish mercury levels in Wuskwatim Lake, on the diversion route below Notigi Reservoir, and in Issett Lake (Figure 7.4.1), located at the upper end of Notigi Reservoir, responded similarly (Figure 7.4.13, Table 7.4.5). No pre-impoundment fish mercury data were available for other lakes on the diversion route, but levels for predatory species (pike and walleye) exceeded the Canadian marketing limit of 0.5 ppm, and approached or exceeded the USA marketing limit of 1.0 ppm (Table 7.4.5). In general, peak mercury levels in predatory fish were highest in lakes now part of Notigi Reservoir, were moderately high in lakes on the diversion route below the Notigi control structure and were lowest in Southern Indian Lake (Table 7.4.5, Figure 7.4.13). The highest levels were found in Rat Lake, with walleye and pike having mercury concentrations over 2 ppm (Table 7.4.5).

Apparently, the fish mercury level increases observed following impoundment were due to the bioaccumulation of naturally occurring mercury. Neither agricultural activity nor known industrial mercury sources exist in the immediate vicinity of the lakes. Other lakes in the region unaffected by hydroelectric development have not experienced recent increases in the levels of mercury in fish and, therefore, atmospheric transport does not appear to be the cause of observed changes.

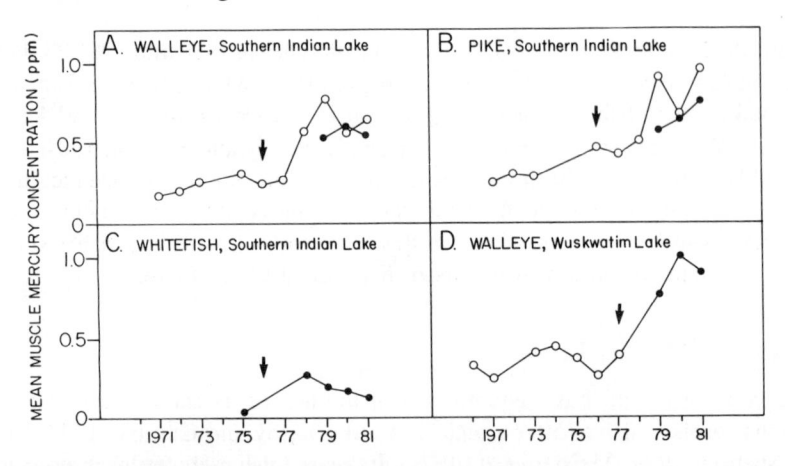

Figure 7.4.13 Fish muscle mercury levels from Southern Indian Lake and Wuskwatim Lake, 1970–1981. Arrows indicate year of impoundment. Open circles are means from one or more commercial samples; closed circles are means for survey samples (2–4 regions of the lake). See Bodaly and Hecky (1979) and Bodaly et al. (in press, a) for methods

Table 7.4.5 Fish mercury concentrations in lakes flooded by the Churchill River diversion project. S: survey sample; mercury concentration determined for individual fish. C: commercial sample; mercury concentration determined for a pooled sample of fish muscle tissue. See Bodaly and Hecky (1979) for details of sampling methods and analyses

Lake	Species	Year	Mean mercury concentration (ppm)	Number of samples and type of sample
Notigi Reservoir				
Issett	Whitefish	1975	0.15	24 S
		1978	0.32	5 S
	Pike	1978	0.61	5 S
Rat	Walleye	1978	2.56	26 S
		1979	2.32	25 S
		1980	1.15	22 S
	Pike	1978	2.05	24 S
		1980	2.32	1 C
Notigi	Walleye	1978	1.41	19 S
		1980	2.90	4 S
		1981	1.88	29 S
	Pike	1977	1.59	1 C
		1980	1.99	1 C
		1981	1.70	50 S
Below Notigi Reservoir				
Wapisu	Walleye	1977	1.17	91 S
	Pike	1977	1.08	38 S
Footprint	Walleye	1978	0.82	40 S
		1980	0.92	12 S
		1981	1.10	30 S
	Pike	1978	0.60	36 S
		1980	1.38	8 S
		1981	1.12	14 S
Threepoint	Walleye	1980	1.18	10 S
		1981	1.35	42 S
	Pike	1980	1.28	10 S
		1981	1.33	28 S
Mystery	Walleye	1979	1.13	33 S
	Pike	1979	0.79	45 S

Hypotheses concerning causes of elevated fish mercury levels in new impoundments have emphasized either increased amounts of potentially available mercury due to natural mercury present in inundated soils (Abernathy and Cumbie, 1977; Meister *et al.*, 1979) or increased retention of naturally transported mercury found on sediment (Potter *et al.*, 1975). Regional differences in peak mercury concentrations in fish from lakes affected by the Churchill River diversion suggest that the primary cause of elevated fish mercury levels is the mobilization of natural mercury from flooded soils. Highest fish mercury levels in the Churchill River diversion area occurred in lakes in Notigi Reservoir where ratios of area of flooded land to reservoir volume were greatest. The presence of flooded vegetation and soil organic matter promotes bacterial production and, therefore, may increase the rate of mercury methylation since it is well known that bacteria can methylate inorganic mercury under aerobic conditions (Jensen and Jernelov, 1969; Fagerstrom and Jernelov, 1971; Furutani and Rudd, 1980).

Recognition of enhanced mercury bioaccumulation as a result of reservoir formation is relatively recent. Mercury concentrations in predatory fish from new reservoirs can exceed 2 ppm, leading to the restriction of commercial marketing (McGregor, 1980) and to potential health problems for persons consuming large amounts of affected fish (Wheatley, 1979).

7.4.7 STABILIZATION OF THE ECOSYSTEMS

Under natural conditions, the potential energy of a river is expended by fluvial processes that carve and transport sediment from a myriad of channels that combine to form a river system. At any point in the system, the size and form of the river channel reflect the quantity of water conducted, the geological materials of the basin and the length of time that this particular combination of water and land has been in existence. The balance between the creation of land masses through crustal movements and the removal of the land surface by river erosion occurs over geological epochs that far exceed the record of man's observations. This natural rate of evolution is in strong contrast to the nearly instant change of diverting a river or flooding a lake. Such rates of rapid change are observed only during earthquakes or other catastrophic events and even these are seldom on the basin-wide scale of a river diversion. The rapid adjustment of a hydraulic regime is not followed by a rapid establishment of new stable values of other characteristics of the ecosystem. Instead, changes in the environment take place as natural forces which act on a geological time scale change the nature of their influence in the environment. Long-term instabilities occur through permafrost melting, erosion, sedimentation, turbidity changes and new flow regimes. The evolution of aquatic systems after river diversion and impoundment is not towards the original state before development, but rather towards a new state, the form of which depends on the nature of the water flow and water level manipulations as imposed by the hydroelectric utility involved.

This section summarizes our observations on the evolving, impacted ecosystems in the Churchill-Nelson diversion. The period of observation has extended for only 2–4 years after flooding and diversion and, therefore, it is difficult to predict the final form of these ecosystems or even to estimate how long it will be until they stabilize. Moreover, the ability to predict changes varies among trophic levels, being more accurate for organisms with life histories shorter than the period of observation than for more long-lived species.

7.4.7.1 Physical Stabilization

The rate of re-stabilization of the diversion systems can be estimated only at Southern Indian Lake where erosion data are available. Here, with the exception of shoreline monitoring sites in coarse granular materials (4 per cent of the total shoreline length), nearshore deposition of eroded materials has had no apparent effect in decreasing the rate of erosion. The melting, slumping and eroding sequence has not diminished for most of the shoreline and average annual erosion indices have been constant since the first full year of impoundment. Re-stabilization appears to depend upon the rate at which erosional retreat exposes bedrock underlying the backshore materials. Based on the assumption that the shoreline monitoring sites are representative of the eroding shorelines on Southern Indian Lake, a minimum restabilization period of 40 years has been estimated (Newbury and McCullough, in press).

7.4.7.2 Primary Production

The response of primary production in the reservoirs created by the Churchill-Nelson diversion has varied considerably and has depended on light, nutrients and other factors. In the two major basins of Notigi Reservoir, algal productivity followed a classical response to impoundment (Lowe-McConnell, 1973): an upsurge of productivity during flooding followed by a decline of algal biomass and oxygen demand to pre-impoundment levels within 2 years of impoundment. Backwater areas of Notigi Reservoir which have experienced poor flushing were still highly productive in 1981, relative to the main portion of the reservoir (Hecky, unpublished data), but primary production was depressed in humic back-water areas of the 3-year old Kettle Reservoir on the Nelson River (Jackson and Hecky, 1980). In Southern Indian Lake, no such surge was observed in regions with light limitation after impoundment (see section 7.4.4).

An upsurge of productivity can be expected if light conditions are adequate to allow effective utilization of nutrients from flooded soil. The intensity and duration of this productivity upsurge will depend on the relation between the area flooded and flushing time of the basin. Minimal flooding and rapid flushing will result in short periods of high production while more extensive flooding in a less rapidly flushed area will result in longer periods of high production. Humification also may occur in extensively flooded poorly flushed areas of northern reservoirs and this accumulation of dissolved humic materials may lead to lower productivity as essential trace metals become unavailable for algal growth (Jackson and Hecky, 1980). In Southern Indian Lake, where light limited conditions were caused by sediments eroding from shorelines, improvement in light conditions—and stabilization of algal productivity—will depend on shoreline stabilization.

7.4.7.3 Macrobenthic Invertebrates

It is difficult to generalize about the recovery period of macrobenthos in reservoirs (Wiens and Rosenberg, in press). Responses of individual taxa reported from other reservoirs did not occur in Southern Indian Lake but the relatively quick increase and decrease of overall standing crops after impoundment indicate that standing crops may stabilize soon. We believe this marginal impact in Southern Indian Lake to be due to an environmentally less disruptive type of reservoir formation but further monitoring should continue to determine the role of natural variation in the Southern Indian Lake Reservoir.

7.4.7.4 Zooplankton

The quantitative response of planktonic Crustacea to new conditions in Southern Indian Lake was rapid. In 1975, a period of unusually high water levels, plankton abundance decreased to 61 ind. l^{-1} from 76 ind. l^{-1} in 1972. With full impoundment in 1977, plankton abundance dropped to 44 ind. l^{-1} and remained at this level (40–46 ind. l^{-1}) from 1977 to 1980, the post-impoundment period. The zooplankton community should stabilize at its present level as long as dramatic changes are not made in present flow patterns and water levels in Southern Indian Lake.

7.4.7.5 Fish Populations

Fish species composition in temperate reservoirs depends largely on physical conditions affecting spawning success (Beckman and Elrod, 1971; Aass, 1973; Walburg, 1977). In Southern Indian Lake, spawning conditions probably will be affected by the characteristics of both flooded shores and flooded terrestrial vegetation, and sedimentation resulting from shore erosion. Since current estimates of the time until shorelines stabilize are in the range of several decades, fish populations probably will not stabilize for some time. The current period of observation on Southern Indian Lake has been too short to observe changes in recruitment due to impoundment because the fish species of interest have long generation times.

It has been hypothesized that elevated fish mercury levels resulting from lake impoundment may decline within 5–10 years as source mercurials in terrestrial material are depleted (Meister *et al.*, 1979) or become less available to the aquatic systems (Abernathy and Cumbie, 1977). However, there are still no clear trends towards decreasing fish mercury levels after 5 years of post-impoundment observations on lakes affected by the Churchill River diversion.

7.4.7.6 Conclusion

Shoreline erosion and altered light penetration have been the most dramatic physical changes resulting from creation of the Southern Indian Lake Reservoir. Biological changes have differed greatly, depending on trophic level. There have been some shifts and/or compensation to the new environment in production of algae, zoobenthos and zooplankton. In contrast, fish stocks have apparently undergone dramatic redistributions and their mercury levels have affected marketing. It has been hypothesized that these biological changes have co-occurred independently of trophic relationships, but that each is dependent directly on physical changes in the new reservoir. However, continuing research on the Southern Indian Lake Reservoir is needed to elucidate possible altered trophic relationships.

7.4.8 MITIGATION AND PLANNING

Instabilities caused by river diversion and lake impoundment can be corrected only partly by remedial works that follow project construction because of the dramatic reorganization associated with hydroelectric development (Newbury, 1981). Weirs and control structures to adjust lake levels have been suggested but not yet implemented at several lakes outlets in the Nelson and Lower Churchill rivers. Studies on Cross Lake, for example, demonstrate the importance of the maintenance of fall water levels for producing strong coregonid year classes. A weir constructed at the outlet of Cross Lake could maintain an acceptable minimum water level during the open-water season, minimize the loss in water volume and area, and improve fish production. A controlled stage could improve hatching success of whitefish and cisco in years of restricted flows and would decrease the possibility of fish kills through stranding and suffocation, by damping the rate of drawdown. A weir would not restore the natural water level regimes (increased water levels during spring), so spawning success of walleye and pike might not benefit, but weirs could be operated to manage primary production by optimizing water retention time relative to nutrient loading (Vollenweider, 1976).

Bank and shoreline protection works were successfully constructed at communities, graveyard sites and highway crossings affected by impoundment and diversion. However, the cost of extending these protective works to all lakes and channels of the diverted system was considered prohibitive. Major pre-impoundment clearing of the forested backshore was undertaken for aesthetic reasons along short sections of the Southern Indian Lake impoundment. However, this clearing was rendered useless at many sites by the erosion and elimination of the cleared backshore zone in the first 2 years after impoundment.

Independent development of hydroelectric potentials on the lower Churchill River would have displaced 2000 MW less energy than the Churchill River diversion by avoiding the impoundment of Southern Indian Lake, flooding of the

Rat and Burntwood valleys and abandonment of the lower Churchill River (Newbury, 1981). The lower Churchill River is a potentially favourable site, environmentally, for reservoir creation because it has a high, ice-scoured bank and impounded levels usually would not exceed bank height. The decision to divert the Churchill River was made solely on the basis of favourable relative cost estimates, but compensation costs for resources lost due to the project were not considered and will probably be substantial over the long term.

Many of the major environmental effects of hydroelectric development documented here were completely unpredicted or only poorly predicted by pre-development impact studies. Experience with more southerly reservoirs has not been easily transferred to this case history study. The comparison of actual to predicted impacts becomes increasingly important as hydroelectric development continues in subarctic and arctic areas. Furthermore, reservoir studies should progress beyond a case history approach and rely more on experimental manipulations to test hypotheses concerning underlying causes for observed changes.

7.4.9 SUMMARY AND CONCLUSIONS

1. Lakes impacted by the Churchill–Nelson hydroelectric development are located on the Canadian Shield, within the boreal forest zone. They were generally relatively shallow and rapidly flushed and were classified as oligotrophic or mesotrophic. The benthic macroinvertebrate faunas were dominated by chironomids, amphipods, oligochaetes and sphaeriids and standing crops were generally enhanced by rapid water renewal times. The crustacean zooplankton communities were dominated by copepods and standing crops were reduced by rapid water renewal times. Diversity of fish was low and the fauna was dominated by relatively few cool-water species. The most important commercial species were lake whitefish, walleye and northern pike.

2. Hydroelectric development of the Churchill and Nelson rivers has a potential total generating capacity of 8400 MW. Most of the flow of the Churchill River was diverted into the Nelson River for power production on the lower Nelson. The natural outlet of Southern Indian Lake, through which the Churchill River flowed, was dammed and the level of the lake was raised 3 m above long-term mean levels. The lower Churchill River valley was abandoned. A newly excavated channel diverted flow from the lake into the headwaters of the Nelson River drainage (Rat and Burntwood River valleys).

3. Abandonment of the lower Churchill River valley resulted in halving of pre-diversion mainstem lake areas. A total of 409 km^2 of land was flooded in the Rat River Valley to create the 476 km^2 Notigi Reservoir. Lakes on the Burntwood River were flooded by 3–5 m due to greatly increased diversion flows. Spring and summer drawdown on Cross Lake (upper Nelson River) resulted in a 26 per cent decrease in lake area.

4. Impoundment of Southern Indian Lake caused severe erosion of fine-grained glacial clays along shorelines. Light penetration was reduced in the main basins because of eroded clay in suspension. In contrast, shoreline erosion in the Rat and Burntwood River valleys was minimal, despite extensive glacial clay deposits throughout the area, because of relatively short wind fetches.
5. The response of primary production to impoundment and diversion depended on specific conditions within the manipulated ecosystems. In the main basins of Southern Indian Lake, increased suspended sediment levels reduced light penetration so increased concentrations of soluble reactive phosphorus from flooded shorelines could not be used and integral primary production remained unchanged. Light penetration was unaffected in Notigi Reservoir and in small bays of Southern Indian Lake and integral primary production increased significantly in response to increased nutrients originating from flooded soil and vegetation. Sequestration of essential trace metals by humic substances depressed primary production in some humic-rich backwater areas.
6. Macrobenthic invertebrate standing crops in Southern Indian Lake increased after impoundment and, within 3 years, decreased towards pre-impoundment abundances. A high degree of species diversity was maintained and there was no evidence of a succession of macrobenthic taxa due to impoundment. The crustacean zooplankton community of Southern Indian Lake responded to impoundment with reduced standing crops, especially in areas of the lake subjected to significant changes in post-diversion flows. Significant increases occurred in the abundance of some large species after impoundment.
7. Fish mercury levels increased soon after impoundment in Southern Indian Lake and lakes in the Rat and Burntwood River valleys. Post-impoundment predatory fish mercury levels ranged from approximately 0.5 ppm to >2.5 ppm, appeared to be related to the severity of flooding in the various new reservoirs, and appeared to be due to the mobilization of natural soil mercury. Mercury levels in predatory fish had not declined 5 years after impoundment.
8. The grade of the commercial whitefish catch in Southern Indian Lake declined significantly immediately following impoundment. Lower grade fish constituted from 12 to 72 per cent of catches in the 4 years following impoundment whereas they were nearly absent from pre-impoundment catches. Catch per unit effort of top grade whitefish on traditional fishing grounds decreased to one-third of pre-impoundment values.
9. Late spring drawdown of Cross Lake reduced year class strengths and abundance of shallow-water autumn-spawning coregonid fishes. There was a significant relationship between the strengths of whitefish and cisco year classes produced in 1971–1980 and the extent of winter-spring drawdown, due probably to desiccation of spawning areas and eggs.
10. A shoreline restabilization period of at least 40 years is estimated for Southern Indian Lake so elevated suspended sediment levels will continue. Stabilization of biological processes such as primary production and fish reproductive success will

depend on reduction of suspended sediment levels and sedimentation rates. Biological changes observed at different trophic levels in Southern Indian Lake appear to have co-occurred independently and in response to physical changes in the lake environment.

11. The widespread nature of changes caused by hydroelectric impoundment and river diversion renders most remedial measures impractical and uneconomic. Bank protection measures over long distances and timber clearing of land to be flooded are prohibitively expensive. As well, a cleared shore zone can be quickly eliminated by bank erosion in areas of high wave exposure. However, weirs can be effective to restore or maintain levels in lakes subject to reduced flow or drawdown.

12. Flooding and river diversion disrupt the natural rate of evolution of aquatic systems. Deleterious environmental effects can be minimized if predicted environmental impacts and alternative development schemes are seriously considered in the planning process. Alternate schemes can be evaluated by comparing the amount of energy displaced by each.

13. Many of the major environmental effects of flooding and river diversion were either poorly predicted or completely unpredicted in the Churchill–Nelson system, despite extensive pre-development studies. Comparisons of actual to predicted impacts are necessary to improve impact prediction for arctic and subarctic hydroelectric developments.

ACKNOWLEDGEMENTS

The authors gratefully acknowledge the considerable assistance in data collection, data analysis and preparation of this report provided by C. Anema, R. J. P. Fudge, S. J. Guildford, T. W. D. Johnson, G. K. McCullough, J. W. Patalas, A. G. Salki and A. P. Wiens. Many other people, too numerous to mention, also assisted with data collection and analysis. S. Ryland and L. McIver typed and processed the text and L. Taite and D. Kufflick prepared the figures.

7.4.10 REFERENCES

Aass, P. (1973). Some effects of lake impoundments on salmonids in Norwegian hydroelectric reservoirs. *Acta Universitatis Upsaliensis*, 14 pages.

Abernathy, A. R. and Cumbie, P. M. (1977). Mercury accumulation by largemouth bass (*Micropterus salmoides*) in recently impounded reservoirs. *Bull. Environ. Contam. Toxicol.*, **17**, 595–602.

Armstrong, F. A. J. and Schindler, D. W. (1971). Preliminary chemical characterization of waters in the Experimental Lakes Area, northwestern Ontario. *J. Fish. Res. Board Can.*, **28**, 171–187.

Ayles, H. A. (1976). Lake whitefish (*Coregonus clupeaformis* (Mitchill)) in Southern Indian Lake, Manitoba. *Can. Fish. Mar. Serv. Tech. Rep.*, no. 640, 28 pages.

Ayles, H., Brown, S., Machniak, K. and Sigurdson, J. (1974). The fisheries of the lower Churchill lakes, the Rat-Burntwood lakes, and the Upper Nelson lakes: present conditions and the implications of hydro-electric development. *Lake Winnipeg, Churchill and Nelson Rivers Study Board Report, 1971–75. Tech. Rep. App. 5, Fisheries and Limnology Studies*, vol. 21, 100 pages.

Barnes, M. A. (1981). *Stress Related Changes in Lake Whitefish (Coregonus clupeaformis) Associated With hydro-electric Control Structures*. M.Sc. thesis, University of Waterloo, Waterloo, Ontario, 148 pages.

Baxter, R. M. (1977). Environmental effects of dams and impoundments. *Annu. Rev. Ecol. Syst.*, **8**, 255–283.

Beckman, L. G. and Elrod, J. H. (1971). Apparent abundance and distribution of young-of-year fishes in Lake Oahe, 1965–69. In Hall, G. E. (Ed.), *Reservoir Fisheries and Limnology*. Am. Fish. Soc. Spec. Publ. no. 8, American Fisheries Society, Washington, D.C., pp. 333–347.

Blaxter, J. H. S. (1970). Fishes and light. In O. Kinne (Ed.), *Marine Ecology*, vol. 1, part 1, Wiley-Interscience, London, U.K., pp. 213–320.

Bodaly, R. A. and Hecky, R. E. (1979). Post-impoundment increases in fish mercury levels in the Southern Indian Lake reservoir, Manitoba. *Can. Fish. Mar. Serv. MS Rep.*, no. 1531, 15 pages.

Bodaly, R. A., Hecky, R. E. and Fudge, R. J. P. (1984). Increases in fish mercury levels in lakes flooded by the Churchill River diversion, northern Manitoba. *Can. J. Fish. Aquat. Sci.* (Suppl.) (in press, a).

Bodaly, R. A., Johnson, T. W. D. and Fudge, R. J. P. (1980). Post-impoundment changes in commercial fishing patterns and catch of lake whitefish (*Coregonus clupeaformis*) in Southern Indian Lake, Manitoba. *Can. MS Rep. Fish. Aquat. Sci.*, no. 1555, 14 pages.

Bodaly, R. A., Johnson, T. W. D., Fudge, R. J. P. and Clayton, J. W. (1984). Collapse of the lake whitefish fishery in Southern Indian Lake, northern Manitoba, following lake impoundment and river diversion. *Can. J. Fish. Aquat. Sci.* (Suppl.) (in press, b).

Brown, S. B. (1974). The morphometry of the Rat-Burntwood Diversion Route and Lower Churchill River lakes: present conditions and post-regulation conditions. *Lake Winnipeg, Churchill and Nelson Rivers Study Board Report, 1971–75. Tech. Rep. App. 5, Fisheries and Limnology Studies*, vol. 2D, 51 pages.

Bruce, W. J. and Spencer, K. D. (1979). Mercury levels in Labrador fish, 1977–78. *Can. Ind. Rep. Fish. Aquat. Sci.*, no. 111, 12 pages.

Cleugh, T. R. (1974). Hydrographic survey of lakes on the Lower Churchill and Rat-Bruntwood Rivers and reservoirs and lakes on the Nelson River. *Lake Winnipeg, Churchill and Nelson Rivers Study Board Report, 1971–75. Tech. Rep. App. 5, Fisheries and Limnology Studies*, vol. 2E, 230 pages.

Driver, E. A. and Doan, K. H. (1972). Fisheries survey of Cross Lake (Nelson River), 1965. *Man. Dep. Mines Resour. Environ. Manage. Res. Br. MS Rep.*, no. 73–5, 17 pages.

Fagerstrom, T. and Jernelov, A. (1971). Formation of methyl mercury from pure mercuric sulphide in aerobic organic sediment. *Water Res.*, **5**, 121–122.

Fee, E. J. (1973). Modelling primary production in water bodies: a numerical approach that allows vertical inhomogeneities. *J. Fish. Res. Board Can.*, **30**, 1469–1473.

Furutani, A. and Rudd, J. W. M. (1980). Measurement of mercury methylation in lake water and sediment samples. *Appl. Environ. Microbiol.*, 40, 770–776.

Gaboury, M. N. and Patalas, J. W. (1981). An interim report on the fisheries impact study of Cross and Pipestone lakes. *Man. Dep. Nat. Resour. MS Rep.*, no. 81–22, 190 pages.

Gaboury, M. N. and Patalas, J. W. (1982). The fisheries of Cross, Pipestone and Walker lakes, and effects of hydroelectric development. *Man. Dep. Nat. Resour. MS Rep.*, no. 82–14, 198 pages.

Guildford, S. (1978). Adenosine triphosphate concentrations and nutrient status measurements in Southern Indian Lake 1975–1977. *Can. Fish. Mar. Serv. Data Rep.*, no. 108, 24 pages.

Hamilton, A. L. and McRae, G. P. (1974). Zoobenthos survey of the lower Churchill River and diversion route lakes. *Lake Winnipeg, Churchill and Nelson Rivers Study Board, 1971–75. Tech. Rep. App. 5, Fisheries and Limnology Studies*, vol. 2H, 28 pages.

Harden Jones, F. R. (1956). The behaviour of minnows in relation to light intensity. *J. Exp. Biol.*, **33**, 271–281.

Healey, F. P. and Hendzel, L. L. (1980). Physiological indicators of nutrient deficiency in lake phytoplankton. *Can. J. Fish. Aquat. Sci.*, **37**, 442–453.

Hecky, R. E. (1975). The phytoplankton and primary productivity of Southern Indian Lake (Manitoba), a high latitude, riverine lake. *Int. Ver. Theor. Angew. Limnol. Verh.*, **19**, 599–605.

Hecky, R. E., Alder, J., Anema, C., Burridge, K. and Guildford, S. J. (1979). Physical data on Southern Indian Lake, 1974 through 1978, before and after impoundment and Churchill River Diversion (in two parts). *Can. Fish. Mar. Serv. Data Rep.*, no. 158, 523 pages.

Hecky, R. E. and Guildford, S. J. (1984). The primary productivity of Southern Indian Lake before, during and after impoundment and Churchill River diversion. *Can. J. Fish. Aquat. Sci.* (Suppl.) (in press).

Hecky, R. E. and Harper, R. J. (1974). Phytoplankton and primary productivity of the lower Churchill Lakes, the Rat-Burntwood lakes and the Nelson River lakes and reservoirs. *Lake Winnipeg, Churchill and Nelson Rivers Study Board Report, 1971–75. Tech. Rep. App. 5, Fisheries and Limnology Studies*, vol. 2F, 39 pages.

Hecky, R., Harper, R. and Kling, H. (1974). Phytoplankton and primary production in Southern Indian Lake. *Lake Winnipeg, Churchill and Nelson Rivers Study Board Report, 1971–75. Tech. Rep. App. 5, Fisheries and Limnology Studies*, vol. 1 E, 90 pages.

Hecky, R. E. and McCullough, G. K. (1984). The effect of impoundment and diversion on sedimentation in Southern Indian Lake. *Can. J. Fish. Aquat. Sci.* (Suppl.) (in press).

Jackson, T. A. and Hecky, R. E. (1980). Depression of primary productivity by humic matter in lake and reservoir waters of the boreal forest zone. *Can. J. Fish. Aquat. Sci.*, **37**, 2300–2317.

Jensen, S. and Jernelov, A. (1969). Biological methylation of mercury in aquatic organisms. *Nature (Lond.)*, **223**, 753–754.

Koshinsky, G. D. (1973). The limnology—fisheries of the outlet lakes area. Present conditions and implications of hydroelectric development. *Lake Winnipeg, Churchill and Nelson Rivers Study Board Report, 1971–75. Tech. Rep. App. 5, Fisheries and Limnology Studies*, vol. 2A, 156 pages.

Kullenberg, G. (1974). Observed and computed scattering functions, p. 25–49. In Jerlov, N. G. and Steemann-Nielsen, E. (Eds.), *Optical Aspects of Oceanography*, Academic Press, New York, N.Y.

Lake Winnipeg, Churchill and Nelson Rivers Study Board (1971–75). *Tech. Rep. App. 5, Fisheries and Limnology Studies*, vols. 1 and 2.

Lowe-McConnell, R. H. (1973). Reservoirs in relation to man-fisheries, p. 641–654. In Ackerman, W. C., White, G. F. and Worthington, E. B. (Eds.) and Ivans, J. L. (Assoc. Ed.). Man-made lakes: their problems and environmental effects. *Geophys. Monogr.* **17**, American Geophysical Union, Washington, D.C.

McCullough, G. K. (1981). Water budgets for Southern Indian Lake, before and after impoundment and Churchill River diversion, 1972–1979. *Can. MS Rep. Fish. Aquat. Sci.*, no. 1620, 22 pages.

McGregor, G. W. G. (1980). Summary of mercury levels in lakes on the Churchill-Rat-Burntwood and Nelson River systems from 1970 to 1979. *Can. Data Rep. Fish. Aquat. Sci.*, no. 195, 16 pages.

McLachlan, A. J. (1974). Development of some lake ecosystems in tropical Africa, with special reference to the invertebrates. *Biol. Rev. Camb. Philos. Soc.*, **49**, 365–397.

Meister, J. F.,DiNunzio, J. and Cox, J. A. (1979). Source and level of mercury in a new impoundment. *Am. Waterworks Assoc. J.*, **1979**, 574–576.

Newbury, R. W. (1981). Some principles of compatible hydroelectric design. *Can. Wat. Res. J.*, **6**, 284–294.

Newbury, R. W. and McCullough, G. K. (1984). Shoreline erosion and restabilization in the Southern Indian Lake reservoir. *Can. J. Fish. Aquat. Sci.* (Suppl.) (in press).

Patalas, K. (1975). The crustacean plankton communities of fourteen North American great lakes. *Int. Ver. Theor. Angew. Limnol. Verh.*, **19**, 504–511.

Patalas, K. and Salki, A. (1984). The effect of impoundment and diversion on zooplankton of Southern Indian Lake. *Can. J. Fish. Aquat. Sci.* (Suppl.) (in press).

Potter, L., Kidd, D. and Standiford, D. (1975). Mercury levels in Lake Powell. Bioamplification of mercury in man-made desert reservoir. *Environ. Sci. Technol.*, **9**, 41–46.

Swenson, W. A. (1978). Influence of turbidity on fish abundance in western Lake Superior. *U.S. Environmental Protection Agency* EPA-600/3-78-067, 83 pages.

Volkova, L. A. (1971). Daily changes in the schooling behavior of some Lake Baikal fish. *J. Ichthyol.*, **11**, 596–607.

Vollenweider, R. A. (1968). Recherches sur l'amenagement del'eau. Les bases scientifiques de l'eutrophisation des lacs et des deux courantes sous l'aspect particular du phosphore et de l'azote comme facteurs d'eutrophisation. *O.E.C.D.* DAS/CSI/68.27, 182 pages.

Vollenweider, R. A. (1976). Advances in defining critical loading levels for phosphorus in lake eutrophication. *Mem. Ist. Ital. Idrobiol.*, **33**, 53–83.

Walburg, C. H. (1977). Lake Francis Case, a Missouri River reservoir: changes in the fish population in 1954–75, and suggestions for management. *U.S. Fish Wildl. Serv. Tech. Pap.* 95, 12 pages.

Wheatley, B. (1979). Methylmercury in Canada. Exposure of Indian and Inuit residents to methylmercury in the Canadian environment. Health and Welfare Canada, Ottawa, Ontario. 200 pages.

Wiens, A. P. and Rosenberg, D. M. (1984). The effect of impoundment and river diversion on profundal macrobenthos of Southern Indian Lake, Manitoba. *Can. J. Fish. Aquat. Sci.* (Suppl.) (in press).

Effects of Pollutants at the Ecosystem Level
Edited by P. J. Sheehan, D. R. Miller, G. C. Butler and Ph. Bourdeau
© 1984 SCOPE. Published by John Wiley & Sons Ltd

CASE 7.5

Accidental Oil Spills: Biological and Ecological Consequences of Accidents in French Waters on Commercially Exploitable Living Marine Resources

C. MAURIN

French Fisheries Research Institute
(I.S.T.P.M.)
Nantes, France

7.5.1	Introduction	311
7.5.2	Main Living Marine Resources of Interest to the Fishing Industry in North and Northwest Brittany	313
	7.5.2.1 In-shore Fisheries Resources	315
	7.5.2.2 Fishing Resources in the Area Around the *Gino* Wreck	317
7.5.3	Main Factors Determining the Effects of Oil Pollution on Resources	318
	7.5.3.1 The Amount of Oil Spilled	318
	7.5.3.2 Site of the Accident	318
	7.5.3.3 Characteristics of the Oil	320
	7.5.3.4 Time of Year at Which the Accident Occurs	323
	7.5.3.5 Use and Effect of Dispersants	324
7.5.4	Short-term Consequences	326
	7.5.4.1 The *Amoco Cadiz* Accident	327
	7.5.4.2 Other Accidents—The *Gino*	336
7.5.5	Medium and Long-term Consequences	338
	7.5.5.1 The *Amoco Cadiz*	339
	7.5.5.2 The *Gino* Accident	357
7.5.6	Conclusions	359
7.5.7	References	360

7.5.1 INTRODUCTION

Torrey Canyon, The name of the 118 000-tonne Liberian tanker that went aground off Land's End in the United Kingdom on 18 March 1967, still conjures up unpleasant memories. Although far from being the first accident involving an oil tanker, it was the first time that such a large quantity of oil (over 80 000 tonnes) was spilled into the sea and caused so much damage to marine life. Since then more than twenty or so accidents of the same type have occurred all over the world. Some, like the wreck of the *Amoco Cadiz* off Brittany in 1978 or the blowout at the Ixtoc well off Campeche in the Gulf of Mexico, have each involved quantities of oil in excess of 200 000 tonnes.

Why this sudden and disturbing increase? It may be ascribed mainly to:

1. The spectacular expansion in world oil production from 280 million tonnes in 1938 to over 3000 million tonnes in 1980.
2. The increase in tanker size from 10 000 tonnes at the begnning of the century to over 500 000 tonnes now.
3. The increase in the amout of drilling and the number of production wells offshore.

One of the areas most frequently and most seriously affected by such accidents is the south coast of the English Channel, particularly the coast of Brittany, which has suffered five major spills in the last 8 years caused by the *Olympic Bravery* in February 1976, the *Bohlen* in October of the same year, the *Amoco Cadiz* on 16 March 1978, the *Gino* on 28 April 1979 and the *Tanio* on 7 March 1980 (Table 7.5.1). These accidents clearly showed that the extent of damage to marine life does not depend entirely on the amount of oil spilled. There are also several other

Table 7.5.1 Summary of some significant oil spills of the past few years. (Modified from Maurin, 1981. Reproduced by permission of Centre National pour l'Exploitation des Oceans)

Event	Date	Quantity Spilled (t)	Location
Torrey Canyon	1967	117 000	Cornwall
World Glory	1968	45 000	Durban
Ocean Eagle	1968	10 000	Puerto Rico
Arrow	1970	12 000	Canada
Polycommander	1970	13 000	Spain
Oceanic Grandeur	1970	35 000	Straits of Torres
Chryssi	1970	31 000	NE Bermuda
Texaco Oklahoma	1971	30 000	East Coast USA
Ennerdale	1971	42 000	Seychelles Islands
Trader	1972	35 000	Mediterranean Sea
Nelson	1973	20 000	Bermuda
Metula	1974	50 000	Straits of Magellan
Olympic Bravery	1976	1 000	Brittany
Bohlen	1976	10 000	Brittany
Urquiola	1976	107 000	Spain
Ekofisk	1977	20 000	North Sea
Amoco Cadiz	1978	230 000	Brittany
Gino	1979	40 500	Brittany
Ixtoc	1979	500 000	Mexico
Tanio	1980	6 500	Brittany

factors which help to aggravate or—quite the contrary—limit the damage caused by oil products. These are, in particular:

1. The site of the spill.
2. The physical and chemical properties of the oil.
3. The time of year at which the accident occurs.

To this list should also be added the effects of the agents used to treat the oil.

The aim of this case study is to assess the effects on exploitable living marine resources of those accidents which have taken place in French waters. The assessment will be based mainly on data gathered after the *Amoco Cadiz* and *Gino* accidents. These occurred in very different conditions and involved entirely distinct types of oil.

In the case of the *Amoco Cadiz*, the vessel grounded on a reef immediately off a rocky coast; its cargo tanks contained 223 000 tonnes of light crude which was less dense than sea water and not very viscous. The *Gino* on the other hand sank in 130 metres of water following collision with another ship. The bottom on which the wreck settled is made up of coarse shell sand which forms ripples known as 'ridins' by French sailors. The spot is about 24 nautical miles off Ushant and 36 miles from the Brittany coast. The vessel was carrying 40 500 tonnes of carbon black oil, a relatively heavy product slightly denser than water and fairly compact.

To provide a clearer picture of the ecological impact of these accidents we shall deal with the following points in the order given:

1. The population size of those living marine resources of interest to the fishing industry in the areas affected by the accidents, i.e. the Brittany coast around Saint-Brieuc, Paimpol, Morlaix and Brest.
2. The principal factors determining the extent of the ecological consequences of the accident.
3. The immediate effects.
4. The medium- and long-term effects.

7.5.2 MAIN LIVING MARINE RESOURCES OF INTEREST TO THE FISHING INDUSTRY IN NORTH AND NORTHWEST BRITANNY

The geographical area affected by the oil spills which took place in French waters mainly covers the coast and coastal waters between Saint-Brieuc and Brest, a distance of some 350 kilometres. It is an area of small-scale in-shore fishing businesses engaged in gathering algae and fishing for crustaceans and molluscs and one in which fish farming, particularly shellfish breeding, plays a major role. We felt that it would be useful to begin this study by describing the coastal resources affected by the *Amoco Cadiz* spill and, to a lesser extent, the *Tanio* accident, and to add some information on the sea bottom in the area where the *Gino* sank.

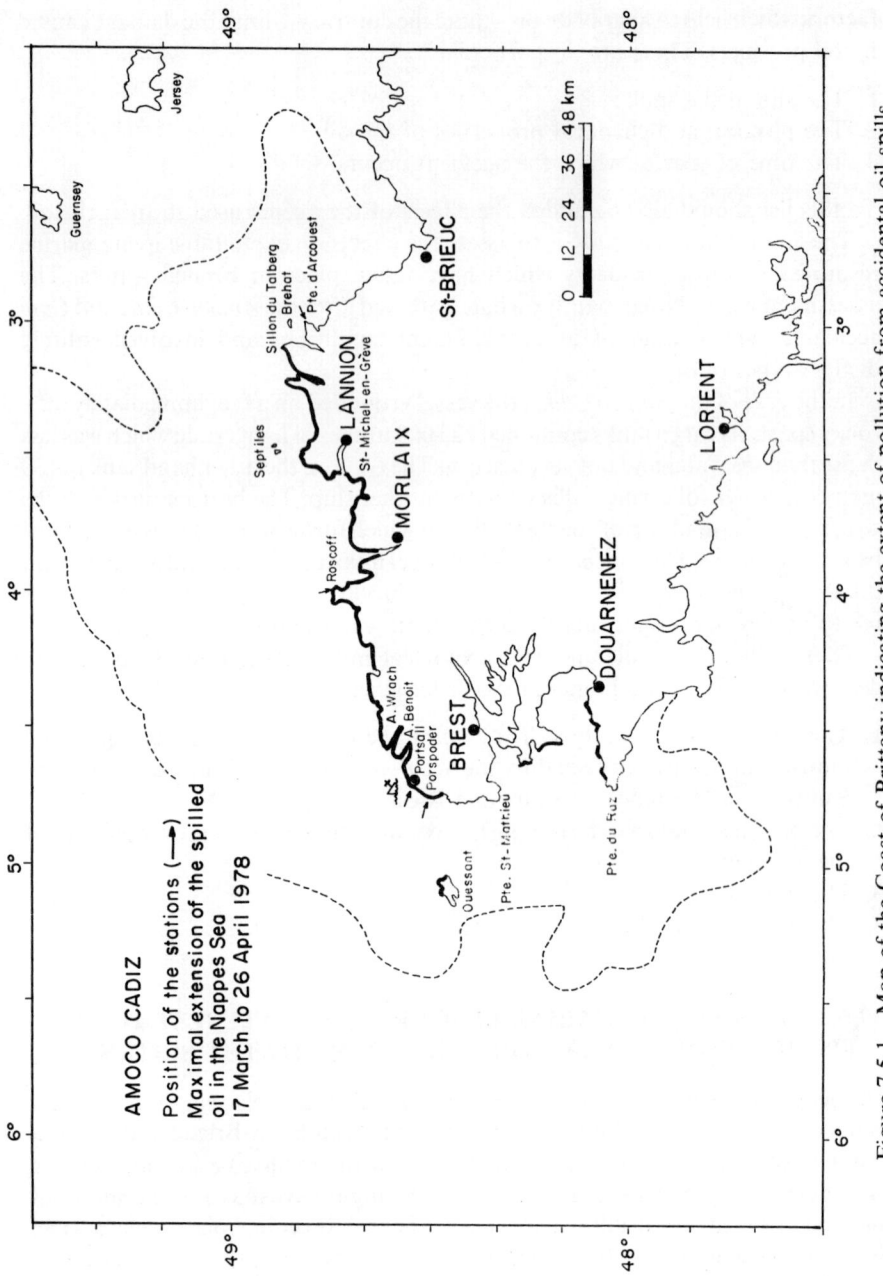

Figure 7.5.1 Map of the Coast of Brittany indicating the extent of pollution from accidental oil spills

7.5.2.1 In-shore Fisheries Resources

Algae

The Breton coast between Sillon de Talbert, which marks the northern limit, and the Saint-Mathieu headland, which marks the entrance to Brest roads, is far and away the most important French seaweed harvesting centre (Figure 7.5.1). When the *Amoco Cadiz* went aground, all coastal waters in the area were being worked. This marine vegetation is still being actively gathered, mainly in the Finistère area between Le Conquet and Goulven Bay, around Roscoff and the island group of Molène and Ushant. There are also large seaweed resources in the Department of the Côtes-du-Nord but here exploitation has only recently begun; the most important centre is between Perros-Guirec and Paimpol.

Several species are used as a basis for manufacturing products particularly sought after by the industry producing sulphated and non-sulphated polysaccharides. The species concerned can be classified in two main groups: brown seaweed and red seaweed.

Brown seaweeds or Phaeophyceae are used for the manufacture of alginic acid. The most plentiful are *Laminaria digitata* followed by *Fucus serratus*. Other brown seaweeds such as *Ascophyllum nodosum* and *Fucus vesiculosus* are also gathered mainly to prepare meal for cattle feed.

Of the red seaweeds or Rhodophyceae, *Chondrus crispus* and *Gigartina stellata* are used in the manufacture of carragheenins. In the area in question most of the algae are gathered between Paimpol and Brest roads, mainly between Portsall and Porspoder, i.e. in the area most affected by the *Amoco Cadiz* spill.

Whereas the exploitation of seaweed in France was for a long time experimental, it has developed a great deal in the last 15 years, mainly because harvesting has been

Table 7.5.2 Exploitable marine resource production (tonnes, wet weight) of maritime areas in North Brittany (from merchant marine statistics) in 1978. M: mussels, O: oysters

Maritime Area	Algae Red	Algae Brown	Crustaceans	Molluscs Harvesting	Molluscs Breeding	Fish
Saint Brieuc	–	–	277	5 320	6 400 M	483
Paimpol	625	5 236	958	1 444	2 590 O 12 M	54
Morlaix	489	6 319	3 380	333	4 613 O	837
Brest	1 198	27 311	2 010	699	1 329 O 438 M	484
Total	2 312	38 866	6 625	7 796	13 051	1 858
Percentage of national production	80.4%	95.4%	23%	21.3%	8.9%	0.6%

mechanized. The building of modern factories has allowed the exploitation of seaweed to expand rapidly into a relatively important regional, national and even international activity.

The official French fishing statistics published by the Merchant Marine indicate that in 1978 total seaweed production for the area between Saint-Brieuc and Brest amounted to just over 44 000 tonnes (wet weight); this represented 95.6 per cent of national production. The total breaks down into some 40 000 tonnes of brown seaweed, i.e. 95.4 per cent of total French production, and about 4000 tonnes of red seaweed, i.e. nearly 80 per cent of all such seaweed gathered in France (Table 7.5.2). The coastal area of Brest alone, which includes Portsall where the *Amoco Cadiz* went aground, provided over 27 000 tonnes of brown seaweed, i.e. 68 per cent of national production, and almost 2000 tonnes of red seaweed or over half the total national harvest of this species.

Crustaceans

The northern and northwestern coasts are rocky and, hence, particularly suit the large crustaceans and attract those fishing for them. On average, the coastal areas of Saint-Brieuc, Paimpol. Morlaix and Brest supply some 23 per cent of all crustaceans harvested in France in a year (7600 tonnes in 1977, 6600 tonnes in 1978, 6000 tonnes in 1979). The most productive areas are the coasts around Brest and especially Morlaix. The catch consists mainly of edible crab (*Cancer pagurus*; about 70 per cent) and spider crab (*Maja squinado*; 27.5 per cent). Lobsters account for a relatively small percentage by weight (1.6 per cent) but as they command a high price they represent 12.5 per cent of the total worth of crustaceans landed in the area.

The most active crustacean fishing ports are Primel, Moguierec and Le Conquet. Most of the year the catches come from the rocky shore areas. Spider crab catches are biggest in spring and summer, as this crustacean remains out at sea in winter.

Molluscs

Mollusc production in the area between Saint-Brieuc and up to and including Brest roads is extensive for both naturally occurring shellfish and cultured molluscs. The main natural species fishes is the scallop which is caught in Saint-Brieuc Bay, in the coastal area of Paimpol and, to a lesser degree, in Morlaix Bay and Brest roads. Stocks of this mollusc were, therefore, relatively unaffected by the oil spills which, with the exception of the *Tanio* spill, hardly touched Saint-Brieuc Bay. But the same cannot be said of other types of natural shellfish such as the variegated scallop and carpet shell and especially of cultured molluscs.

Although most of the mussels in this part of Brittany are based in the coastal area of Saint-Brieuc, oyster breeding was, and in most instances still is, a major activity in the Ria de Morlaix and the Penzé (both estuaries), in the estuaries known as the abers (Aber Benoît and Aber Wrach) and in Brest roads. In spite of the parasite infection which affected the common oyster stock (*Ostrea edulis*) from 1968 onwards, at the time of the *Amoco Cadiz* accident this oyster was still being bred in the abers, the Penzé and Morlaix estuaries, the Paimpol area and in Brest roads. By the beginning of 1978, the Pacific oyster (*Crassostrea gigas*) had been introduced in many breeding centres and was already accounting for most of the production.

Still according to official statistics, the production of uncultured molluscs in the four coastal areas concerned amounted to 7800 tonnes in 1978 (over 6500 tonnes being scallops) and this represents 21.3 per cent of total national production. Where cultured shellfish are concerned, total production for the same year came to almost 7000 tonnes of mussels (mainly in the Saint-Brieuc area) and a little over 6000 tonnes of oysters (mainly in the Morlaix and Brest areas). The figure for mussels was slightly higher than in 1977 as the mussel beds are outside the areas affected by the pollution, whereas the oyster figure was slightly lower. Taken as a whole, the volume of shellfish bred in the area represents some 7 per cent of total national production.

Fish

Apart from a few areas of Morlaix and Lannion and in Saint-Brieuc Bay, the seabed off the north and northwest Brittany coast is rocky and therefore unsuitable for trawling. This explains why few fish are landed along this coast and why, for all practical purposes, only small gill nets, drift nets or lines are used. At about 2200 tonnes in 1977 and 1860 tonnes in 1978, the total fish landed in the coastal areas of Saint-Brieuc, Paimpol, Brest and Morlaix represented only 0.6 per cent of the national catch. The species most frequently caught are either sedentary fish living in rocky areas (e.g. the Labridae, sea eel, ling, catfish and ray), open water coastal species (e.g. grey mullet and sea perch) and migratory species only present in the area at certain seasons (e.g. pollack and mackerel).

7.5.2.2 Fishing Resources in the Area Around the *Gino* Wreck

For a radius of 15 to 20 nautical miles around the Gino wreck, the water is between 100 and 200 metres deep (120 metres at the site of the wreck). The substrate is hard and is made up of coarse sands and mud which contain bryozoa and corals; the bottom is strewn with rocks. It is difficult to use a trawl in this area, which also is relatively poor in expolitable species of fish, so the area is only occasionally visited by fishing vessels—trawlers from Concarneau or Guilvinec, 'caseyeurs' (lobster boats) from Le Conquet and Ile de Sein. Catches are small. This being said, the most representative species are the selachians (catfish, ray), the gades (haddock, blue whiting, pollack, pout, ling, young hake) and gurnards. Mackerel and horse mackerel may be abundant in season. Crustaceans, particularly edible crabs, are present but in relatively small numbers. There are many scallops where there are no rocks but they are too scattered for regular commercial fishing.

7.5.3 MAIN FACTORS DETERMINING THE EFFECTS OF OIL POLLUTION ON RESOURCES

7.5.3.1 The Amount of Oil Spilled

The magnitude of a spill is certainly a major factor but it is far from being the only one which determines the extent of the damage the oil will inflict on exploitable living resources. A few details will illustrate this fact (Table 7.5.1).

In 1967 the *Torrey Canyon* spilled 117 000 tonnes of oil; the oil, a light crude, spread out over a large section of the Eastern Channel and affected both the English and French coasts. The 1000 tonnes of oil spilled by the *Olympic Bravery* in 1976 caused only local pollution. In the same year the *Bohlen*'s 10 000 tonnes did not spread very far either, because the product was highly viscous; nevertheless, the effect on marine resources, particularly the larger crustaceans, was perceptible.

The 20 000 tonnes spilled into the North Sea in 1977 from the Ekofisk field had no major visible consequences because of the existence of converging currents and the distance from the coast. By contrast, the *Amoco Cadiz*'s 230 000 tonnes, a third of which luckily evaporated because the product was light, affected nearly 350 kilometres of coastline. The *Gino*'s 40 000 tonnes of heavy product which settled on the bottom around the wreck and covered an area with a radius of only about four to five sea miles, caused detectable contamination of scallops within a radius of some 20 sea miles.

As for the spill from the Ixtoc well to the North of Campeche on the Yucatan Peninsula of Mexico, the volume of oil spilled from the beginning of June 1979 onwards probably totals some 500 000 tonnes, but this was spread over several months. The pollution, which was carried northwest and then north by various currents, reached the Mexican coast north of Tampico and the south coast of the United States in Texas, the latter being nearly 1000 kilometres from the well.

Finally, in March 1980 there was the *Tanio* accident which occurred 35 nautical miles off Batz Island near Roscoff. As it proved possible to tow the after section of the vessel—containing 10 000 tonnes of No. 2 fuel—to Le Havre, it is estimated that only 6500 tonnes of oil was spilled in the sea while 10 000 tonnes remained in the wreck at a depth of 87 metres. Although the product concerned was relatively viscous, the pollution affected 195 kilometres of the coastline of the Finistère and Côtes-du-Nord Departments, some sections suffering more than others (Berne, 1980).

7.5.3.2 Site of the Accident

The site of an accident plays a major part in determining the amount of damage done to living marine resources. To begin with, it should be pointed out that the productivity of the oceans is not everywhere the same. It is generally accepted that of approximately 100 000 known marine species almost 95 000 live mainly in the coastal areas, on or near the bottom, where the depth is less than 200 metres.

In spite of the very important role of plankton, particularly phytoplankton, in maintaining the biological and physico-chemical balance of the oceans, it may be said that an accident in the open sea has less direct consequences on commercially expoitable resources than one occurring close to the coast. Furthermore, pelagic fish species which are found in the open sea can and do escape from pollution. This was demonstrated when, following the *Amoco Cadiz* accident, open water species such as mackerel and pollack temporarily disappeared from the area.

There are also certain other factors which prove more helpful than not when an accident occurs in the open sea rather than close to land. To begin with, it is easier to lighten and tow a tanker in open sea than in shallow water, particularly if the coastline is dotted with reefs. Compare, for example, what happened to the *Amoco Cadiz*, which grounded on the Portsall rocks, with what happened to the *Tanio*, which was sliced in two at a spot 35 miles from the coast where the water was 87 metres deep. In the first instance towing was impossible, as any ships coming to the rescue of the stricken tanker would in their turn have gone aground; whereas with the *Tanio*, it proved possible to tow the after section of the vessel to Le Havre, thereby preventing 10 000 tonnes of oil from being spilled into the sea.

It should be noted that a wreck still containing oil in its cargo tanks and resting in water sufficiently deep to prevent movement by the swell has every chance of remaining intact, whereas a grounded vessel is quickly broken up by the waves. This fact, too, was illustrated by the *Amoco Cadiz* and the *Tanio*; in the first instance the entire cargo was spilled into the sea, but this was not so in the second.

The greater the depth and volume of water at the site of a wreck, the easier it is for biodegradation to take place. The non-volatile residues have time to oxidize and form tar balls which have little effect on marine life. Also, when the oil slick spreads out far from the coast, as was the case in the Ixtoc and particularly the Ekofisk blowouts, the lighter and more toxic fractions can evaporate completely before the rest is dispersed in the body of water. If a spill occurs in the open sea there is plenty of time for the oil to spread out in a very thin layer which facilitates evaporation and makes treatment with dispersants more effective.

Another helpful factor is the presence of concentric vortex currents which frequently occur in the open sea and whose action prevents pollution reaching the coast. This was the case in the Ekofisk blowout. Dominant currents flowing in a given direction, which frequently occur at some distance from the coast, may also help prevent contamination of the coastal area closest to the site of the accident. As in the Ixtoc case, this does not prevent all coastal pollution, but part of the oil which has been spilled can evaporate and be biodegraded, so becoming less toxic.

Excluding blowouts of oil wells at sea, it may be said that most accidents occur close to the shore mainly because of the presence of reefs and of the heavier traffic. In Europe this is true of the sea around Ushant (at the gateway to the Channel), in Dover Strait and at Cape Finistère.

Quite apart from the absence of the above-mentioned advantages, oil spills close to land cause major damage. As wave action quickly causes the oil and water to mix, the oil may be rapidly emulsified in the water before a proportion of the toxic volatile fractions can evaporate. In the case of the *Amoco Cadiz* this was probably a major factor in the high mortality rate among coastal bottom-living creatures immediately after the accident (Amphipoda, periwinkles, cockles, sea urchins and rock-dwelling fish).

Shoreline species such as seaweed are sedentary or less mobile. They therefore are exposed for the duration of the contamination.

Finally, evaporation of the volatile fraction is followed by the biological or photochemical oxidation of the oil and its adsorption to mineral particles; this causes sedimentation of the oil, particularly in certain sheltered spots along the coastline. Degradation of the product is considerably slowed down by the lack of oxygen so that the toxicity may persist for several years. The aftermath of the *Amoco Cadiz* spill has demonstrated that oil has a maximum toxic effect when it remains trapped in sediment (Gouygou and Michel, 1981a).

The experience has also confirmed earlier observations as to the varying degrees of sensitivity of different types of coast in the sublittoral zone. From this it would seem that a coastline with plenty of wave action is the type which is most quickly decontaminated, as indicated by Owens in 1978. Rocky areas pounded by the sea, followed in decreasing order by firm bottoms covered with pebbles or gravel, and coarse sandy bottoms, are the types of coast which will most quickly and easily return to a state of biological equilibrium because they are more quickly decontaminated. For instance, by November 1978 almost all the oil had disappeared from the subtidal zone of Morlaix Bay and, a year after the catastrophe, there were no longer any traces between Portsall and Batz Island (Beslier *et al.*, 1981).

It is a different matter in sheltered areas in which fine sediments accumulate, for instance in the estuaries of the North Breton coast known as the abers or in marshland. These areas are subject to the most serious and most lasting damage. Some authors expect the pollution to persist there for up to ten years (d'Ozouville *et al.*, 1981).

7.5.3.3 Characteristics of the Oil

The physical and chemical characteristics of the oil are cardinal factors with regard to both the direct consequences of the pollution on the marine resources and the effectiveness of treatment by dispersants.

Volatile fraction

It is thought that oil fractions with a boiling point below 300 °C quickly evaporate from the surface of the water owing to the action of the wind. From a quantitative

Table 7.5.3 Some physiochemical characteristics of oil from accidental spills. (from Maurin, 1981. Reproduced by permission of Centre National Pour l'Exploitation des Océans)

	% Distillation at 210°C	% Saturated HC	% Aromatic HC	Density at 20°C	Flow point
Torrey Canyon	21.3	31.1	33.7	0.866	
Boehlen	0.9	2.5	35.0	1.000	−12°C
Amoco Cadiz	26.6	31.8	28.0	0.853	
Gino	0	11.5	60.9	1.083	−30°C
Ekofisk	26.5	42.2	21.7	0.844	
Ixtoc	15.2	35.7	37.6	0.833	

point of view this is important because it helps considerably in the on-site elimination of a pollutant. Column 1 of Table 7.5.3 gives figures for several accidents showing the percentage of hydrocarbons contained in the oil which boil at 210 °C or less. Extrapolating from this, it may be assumed that between 30 and 50 per cent of most light crudes is likely to volatilize rapidly (P. Michel, personal communication). This applies particularly to the benzenes and alkyl benzenes, and has practical consequences, as the light hydrocarbons are the most toxic.

Some crudes, such as the Venezuelan crude carried by the *Bohlen* and *Boscan*, contain no light fraction. This is also true of some residual oils such as those carried by the *Olympic Bravery* (Bunker C) or the *Gino* (carbon black oil). It should also be noted that the evaporation of light fractions is moderated if the oil is emulsified in water and, particularly, if the formation of water-in-oil emulsions is due to water penetrating the oil. This phenomenon is often described as 'chocolate mousse' and is promoted by the presence of impurities such as particles of asphalt. As opposed to a true emulsion which can be biodegraded fairly easily, the mousse is stable and may persist for several months (Nelson-Smith, 1973).

It should be pointed out that while heavy crudes are, generally speaking, less toxic in the short term than light fractions, they may have long-term mutagenic or carcinogenic effects. This is true of polycyclic aromatic hydrocarbons such as benzopyrene or the benzanthracenes. However, as Nelson-Smith (1973) has pointed out, there are many polycyclic aromatics which have no harmful effects; besides, small but significant amounts of benzopyrene occur naturally in the marine environment.

Saturated hydrocarbon content

See Table 7.5.3, column 2.
The presence of saturated hydrocarbons, like that of paraffins, causes no major problems in an oil spill at sea. Only in exceptional cases do paraffins have a toxic effect. Furthermore, their biodegradation is rapid, at least where straight chain

compounds are concerned. Oil with a high saturated hydrocarbon content such as that from the Ekofisk field (42.2 per cent) would therefore be less harmful in the short and long terms than oil from the Ixtoc well (35.7 per cent), the *Amoco Cadiz* (31.8 per cent) or, particularly, the products carried by the *Gino* (11.5%).

Aromatic hydrocarbon content

See Table 7.5.3, column 3.

Aromatic hydrocarbons are much more persistent than saturated hydrocarbons. The lightest (the benzenes and alkyl benzenes, the naphthalenes and alkyl naphthalenes) have a high acute toxicity. The heavier ones (aromatics with four or more benzene rings) may have mutagenic or carcinogenic effects. The best known is 3, 4-benzopyrene, for which carcinogenic effects were demonstrated 40 years ago. Fortunately only traces of this compound are present in crude oils.

This is not always true of oil products. For instance, the cargo of the *Gino* contained 60.9 per cent aromatic hydrocarbons (compared with 28 per cent for the *Amoco Cadiz* and 21.7 per cent for Ekofisk oil) and 400 parts per million of benzopyrene.

As we shall see later, in the case of the *Amoco Cadiz*, Gouygou and Michel (1981a) demonstrated that compounds with three and four rings, frequently highly alkylated, are the most persistent. This is true of dibenzothiophene and naphthobenzothiophene which, after a period of 3 years, account for 56.4 and 13.3 per cent, respectively, or a total of almost 70 per cent of the residue.

Specific gravity

See Table 7.5.3, column 4.

Generally, oil is lighter than water and most of it therefore stays on the sea surface. This was the case, for instance, with the oils spilled from the *Torrey Canyon*, the *Amoco Cadiz* and the Ekofisk and Ixtoc wells where the specific gravities at 20 °C were effectively the same: 0.866, 0.853, 0.844 and 0.883, respectively.

Yet, the density of the products carried by the Bohlen (1.0 at 20 °C), which sank in 100 metres of water, was such that some of the oil could float between two masses of water. Some by-products have a density greater than 1. In the case of the *Gino*—whose wreck lies in 120 metres of water—all that has come to the surface is the oil and fuel for the vessel's own use. The *Gino*'s carbon black oil has a density of 1.083 at 20 °C.

Nevertheless, it should be remembered that sooner or later, aging, oxidation and the formation of water-in-oil emulsions make even the lightest oils settle.

Viscosity

The more fluid an oil, the greater its immediate impact on marine resources because it spreads further and mixes more throughly with the water and the sediments, particularly fine sediments.

In the case of some oils or derivatives with a flow-point above 0 °C, this property tends to prevent the oil slick spreading and, therefore, in theory, also tends to limit the size of the contaminated area. This was true of the *Bohlen* and *Olympic Bravery*. As we have seen, although the slick from the *Gino* was limited in extent, there was a certain amount of contamination at quite a distance from the wreck. It should be remembered that the *Amoco Cadiz* oil had a viscosity of some 10 centistokes at a temperature of 20 °C; at the same temperature the equivalent viscosity for the *Bohlen* was 10 000 centistokes while for the *Tanio* it was some 15 000. In connection with the latter, Berne (1980) pointed out that this high viscosity prevented the oil from penetrating massively into the fine sediment, but that it did not prevent infiltration to a depth of 40 centimetres into the coarse granitic sands, a process which was helped along by the higher summer temperature.

Finally, it must be noted that the initial viscosity of the oil increases, partly in proportion to the amount that evaporates and partly in proportion to the formation of water-in-oil emulsions. This is a further factor which suggests that, given comparable quantities, oil spills at sea have less serious consequences than those close to the shore.

7.5.3.4 Time of Year at Which the Accident Occurs

It should be remembered that one factor which affects the consequences of an accidental oil spill on living marine resources is the time of year at which the accident occurs. Its importance lies in the meteorological and physical, chemical and biological aspects.

As far as the meteorological circumstances are concerned, one cannot help noticing that the major accidents involving oil tankers in European waters in the last few years have all happened in March—the *Torrey Canyon* on 18 March 1967, the *Amoco Cadiz* on 16 March 1978 and the *Tanio* on 7 March 1980. This is no coincidence, as this time of the year is a period of particularly violent storms in the Channel and its neighbouring waters.

Apart from the fact that meteorological conditions of this type cause accidents, they also make lightening and towing operations and the treating of oil slicks more difficult and sometimes impossible.

Although the wind helps to spread the oil and encourage evaporation, which is beneficial, it can also lead to rapid mixing of the oil with the water before evaporation can take place; this facilitates the formation of stable water-in-oil emulsions and slows down the process of biodegradation. Unfavourable meteorological conditions of this type also ensure that a bigger geographical area is affected by the pollution. In this connection it has been noted that when there is a high wind the rate of displacement of the slick is higher than its rate of spreading. Its speed of movement is some 2.5–3.5 per cent of the wind speed (Nelson-Smith, 1973).

Where physical and chemical factors are concerned, sunshine and, more generally, heat facilitate evaporation. Similarly, heat helps to lower the density of a product and make it more fluid. We have seen the consequences this had for the oil from the *Tanio*.

Greater biological harm is caused if, for example, the accident occurs in a period of algal bloom (i.e. at the beginning of spring or in autumn), at a time when marine organisms, particularly algae and molluscs, are going through a period of active growth (spring and summer), or during a reproductive period, especially if this period is very short. It would seem that algae are less sensitive to oil spills during their macroscopic phase than during their microscopic one, when the reproductive systems are at their most vulnerable (R. Pérez, personal communication). Observations on biota at the time of the *Amoco Cadiz* accident illustrate these points.

7.5.3.5 Use and Effect of Dispersants

As Pierre Michel pointed out in 1979, if we had methods for the physical recovery of oil spilt in the sea as a result of an accident, the problem of dispersants would never arise. In fact, the mechanical methods tried so far (pumping, collecting in nets, etc.) can rarely be used, due to unsuitable meteorological conditions or because of the nature of the pollution.

On the other hand, although the damage caused by the dispersants in use several years ago was extensive, if difficult to assess, a great deal of progress has been made since then. It may be said that in the last 10 years the toxicity of these products has been reduced by a factor of nearly 1000 (Michel, 1979).

Before summarizing what we know about the advantages and disadvantages of oil dispersants, we should explain what they are. This category of products comprises binders, precipitants and emulsifiers or dispersants. Binders such as rubber crumb are completely innocuous. They can be fairly effective but are difficult to use as they are very light and wind makes it difficult to spread the product. Precipitants such as sawdust and chalk are not toxic either, but are also difficult to use. When not very effective, they cause the salting-out of oil from the water and do not prevent shore-line pollution. When effective, they settle the oil on the bottom and trap it in the fine sediments. At the time of the *Amoco Cadiz* accident, it was thought that precipitants slowed down the biodegradation of the oil. In fact, they were only rarely used and then only to prevent the oil reaching particularly sensitive areas such as Brest roads.

The toxicity of emulsifiers or dispersants varies a great deal with the product. They contain surface-active agents including a lipophilic fraction which fixes the oils and a water-absorbent fraction which fixes the water. Thus, they help break down an oil slick into fine droplets made up of water and oil. In the currently used dispersants these agents are usually mixed with a solvent.

Surface-active agents may be divided into three main categories.
1. Cationic surfactants which, in addition to the lipophilic fatty acids common to

other surfactants, contain a water-absorbent ammonium radical. They are highly toxic, practically nonbiodegradable and are mainly used as bactericides.
2. Anionic surfactants which contain a sulphonic water-absorbent radical. They are very active, not very biodegradable and highly toxic, particularly when mixed with a solvent rich in aromatics. This category includes what we normally call detergents.
3. Nonionic surfactants whose water-absorbent element is a primary alcohol. These agents are more easily biodegradable and their action is less harsh. They are properly known as dispersants.

Generally speaking, the biological effect of surface-active agents is that they make the cell membrane more permeable which increases the penetration of the agent into the cell. When the product contains an anionic surfactant, there occurs a chemical reaction with the proteins of the cell membrane which can cause it to burst. The polycyclic hydrocarbons or light aromatics then penetrate the membrane and concentrate there as shown in Figure 7.5.2 (Nelson-Smith, 1973). On the other hand, if non-ionic surfactants are used there is no chemical reaction with the proteins of the cell membrane so that there is no bursting. But the non-ionic products are still surface-active and as such help the oil penetrate the membrane. This explains why, on the one hand, non-ionic surfactants may be regarded as a thousand times less toxic than their anionic counterparts and why, on the other hand, they can have quite a harmful effect on living organisms.

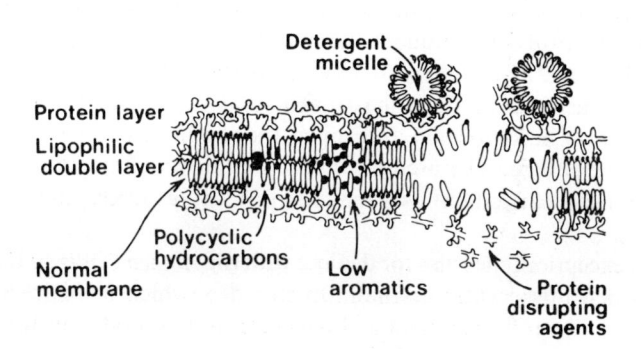

Figure 7.5.2 The effects of various agents on the plasma (cell) membrane. To the left is the normal structure, with a double layer of fatty molecules stabilized on each side by an outer protein layer. The large molecules of polycyclic hydrocarbons penetrate slowly and push the fatty molecules apart, while low-boiling aromatics penetrate rapidly and become solubilized in the fatty membrane, disrupting its spacing. Detergents (surface-active agents) and some other agents not discussed here strip off or break up the protein layer. (Reproduced by permission of Dr A. Nelson-Smith, 1973)

It would seem useful to set out the possible advantages and disadvantages of using dispersants (i.e. non-ionic dispersants).

As far as the advantages are concerned, these products encourage the spreading out of the oil slick and the formation of very fine droplets; they do not apparently prevent the lighter and more toxic fraction from evaporating; they speed up the biodegradation of some oils, can limit damage caused to birds, often prevent deposits forming on solid surfaces and help to slow down the formation of water-in-oil emulsions (Canevari, 1978).

Nevertheless, the disadvantages of using these products justify avoiding their use in coastal areas with a rich population of seaweeds and marine organisms. The products are in fact quite toxic, particularly in their role as surface-active agents. In spite of observations to the contrary made in a laboratory (Canevari, 1978), they do not seem to prevent oils settling on mobile substrates or their penetration of sediments. They do speed up biodegradation in those families of hydrocarbons which are less toxic and degrade easily, e.g. paraffins, but it is far from certain whether they do the same for the aromatics (P. Michel, personal communication). The *Amoco Cadiz* accident showed that water-in-oil emulsions can form almost immediately, before any treatment can be given.

Finally, all treatments are totally ineffective when the oil slicks are thick (*Amoco Cadiz*, Ixtoc) or when the oils have a high viscosity, as was the case with the *Tanio* and the Bohlen.

It should be remembered that after the *Amoco Cadiz* accident, as opposed to the *Torrey Canyon*, only non-ionic surfactants were used at sea. Research carried out in France by l'Institut Scientifique et Technique des Pêcheries Maritimes (ISTPM) between 1970 and 1975, and thereafter by various bodies acting under the auspices of an inter-ministry committee, has meant that the use of the least harmful products can now be advocated. In France, the conditions under which the least harmful dispersants may be used have been laid down after numerous tests on marine organisms carried out as a function of local resources and geographical position.

With one exception, the rules for dealing with the *Amoco Cadiz* spill prohibited the use of dispersants in water less than 50 metres deep which, in this area, is more or less the limit of the euphotic zone, and this was certainly a good thing for the marine resources. The rules did not obstruct the carrying-out of a very effective treatment which helped prevent the pollution from reaching the west coast of the Cotentin, the Channel Islands and the western coasts of Brittany.

7.5.4 SHORT-TERM CONSEQUENCES

This section sets out to deal with the short-term consequences of oil spills in French waters while deliberately limiting the survey to exploitable living resources—algae, crustaceans, molluscs and fish. The problem will be approached from various angles.

1. Direct damage and mortality.
2. The level of contamination.
3. Irregularities in reproduction and growth.

The observations which follow apply mainly to the consequences of the *Amoco Cadiz* accident. However, in an attempt to bring out the differences which can result from differences in the types of oil spilled or the conditions in which an accident of this kind occurs, we shall add some information on the consequences of the wreck of the *Gino*.

7.5.4.1 The *Amoco Cadiz* Accident

Direct damage, mortality

Certain species of algae such as *Fucus vesiculosus* and *Ascophyllum nodosum* live fairly high up in the intertidal zone, just where the oil slicks were at their thickest after the grounding of the *Amoco Cadiz*. Hence, as René Pérez (1978b) commented, these algae were heavily coated with oil. They remained coated with the brown water-in-oil emulsion known as 'chocolate mousse' for several weeks and were only very gradually cleaned by the waves. In September, their surfaces still showed the characteristic sheen which indicates the presence of oil. Nevertheless, observations made on these two species at various sites along the shore provided no evidence of necrosis or abnormal mortality (Pérez, 1978b; Cross *et al.*, 1978).

In order to assess what damage the oil had done to the most widely exploited species, *Laminaria digitata*, Pérez (1978b) established a method for measuring the density and assessing size distribution. To estimate density this author recorded the number of seaweed plants visible to the naked eye in one square metre of shoreline. For size distribution he analysed the structure and population density by using histograms based on the frond length of 250 specimens of *Laminaria*. He repeated these observations every month at seven sites along the coast including the Arcouest headland which had not been polluted and was used as a control. To prevent errors of interpretation, the selected sampling points were in sectors not exploited in 1977. The extent of new mortalities also had to be calculated; to do so the author compared his results with those he had obtained for the same shoreline between 1969 and 1971. From this he calculated that the normal population density of *Laminaria* was 72–95 plants per square metre in February and an average of 65 in March, 47 in April and 43 in May. He attributed this natural mortality mainly to predatory herbivorous organisms, particularly gastropods of the limpet family such as *Helcium pellucidum* (Table 7.5.4).

The author's observations in March, April and May 1978 showed that, in the control area, the figures corresponded closely to these averages. In the more polluted areas the averages were very similar in March and May except in the Portuval area, where the explanation for the slightly lower figures is the extreme exposure of the site to waves. In April the figures were in general above average.

Table 7.5.4 Study of density in algae m^{-2}. Mean values one shown, obtained for *Laminaria digitata* and *Chondrus crispus* at the different sites studied (March to May, 1978). Arcouest is the control station. (From Pêrez, 1978b.)

Location	Number of *Laminaria* m^{-2} March	April	May	*Chondrus* weight, g m^{-2} March	April	May
Normal average	65	47	43	4300	5500	3000
Arcouest	63	48	45	4050	4300	4900
Talbert	67	47	42	4100	4600	5000
Perros-Guirec	63	47	43	3900	4400	4800
Pte de Primel	64	48	41	3700	4150	4600
Roscoff	69	51	43	4100	4600	5100
Portuval	60	47	38	3900	4700	5050
Trémazan	67	52	45	4000	4700	5250
Porspoder	65	50	43	3800	4400	4950

Pérez thinks that the explanation lies in the high mortality among predators, as during these months he counted only five *Helcium* per kilogram of algae at Trémazan and three at Roscoff, compared with 18 at the unpolluted control site on the Arcouest headland.

An analysis of the histograms highlights the fact that no necrosis of the thalli was observed. Furthermore, the absence of unusually large quantities of seaweed detritus leads the author to the conclusion that there was no abnormal mortality of *Laminaria* in the polluted area in spite of the high concentration of hydrocarbons.

The red seaweed, *Chondrus crispus*, occupies the lower section of the intertidal zone; therefore, it was highly contaminated by the oil spill. Figures for the North Brittany population of these Rhodophyceae provided by Kopp (1975) indicate that in normal times the *Chondrus* biomass increases rapidly in March and April and the drops in May because the older fronds disappear. Pérez showed that in 1978 the volume of this biomass was considerably below average but that it continued to increase in May. A comparison between the control site and the Trémazan area, which is close to the site of the wreck, indicates that this was mainly a natural phenomenon due to high insulation and lack of nitrogen. The phenomenon was also observed on the French Atlantic coast that year (Figure 7.5.3).

The data provided by the National Association for Marine Algae clearly illustrate the fact that *Laminaria* production in 1978 was good in spite of the net deficit in May, the latter being due to the disturbances caused by the accident. Taken as a whole, production was some 30 per cent up from the 1977 level. By contrast, the red seaweed harvest was very much lower than that for the previous year. It is difficult to be certain about the reason for this drop. We have to distinguish between how much was due to climatic and harvesting conditions and how much can be attributed to the pollution. This is very difficult if not impossible. Nevertheless, the fading and withering of a proportion of the *Chondrus* population

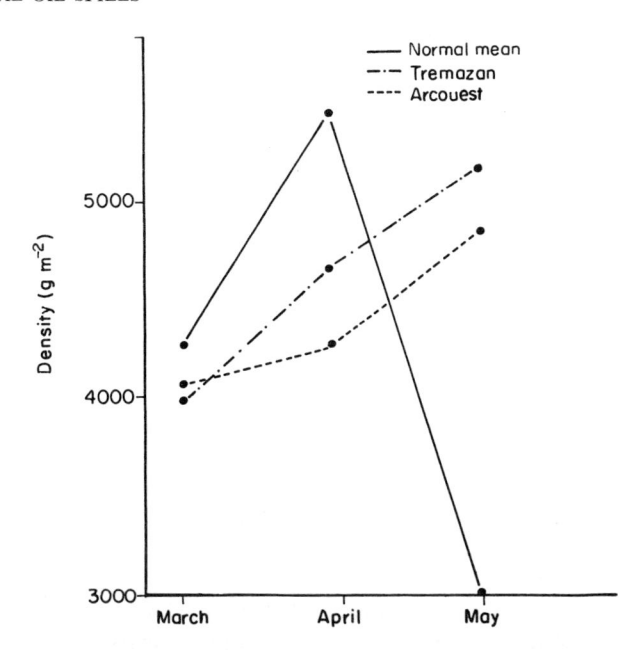

Figure 7.5.3 Variation in biomass of *Chondrus crispus* populations; production was relatively poor in March and April, better in May; the maximum that year coincided with the beginning of harvesting. (From Pérez, 1978b)

during the weeks following the accident may in part be due to the latter. Finally, the *Fucus* and *Ascophyllum* harvests dropped by almost 60 per cent.

No immediate mortality was noted in commercially important crustaceans such as lobster, edible crab (*Cancer pagurus*), spider crab (*Maja squinado*) and crayfish. It is not impossible that crustaceans are among the species which were destroyed at the beginning of the spill when it was not possible to make any observations because the layer of oil was so thick. As we shall see in more detail later, a number of crustaceans caught during fishing experiments in April 1978 between Batz Island and the wreck had a distinctly oily taste.

In May of the same year, tests in the areas around Portsall set aside for crustaceans gave a very good lobster yield. The catches included mature individuals (54.3 per cent of males and 45.7 per cent of females) with an average weight of 1.084 kg (Léglise and Raguenès, 1981).

Elsewhere, Desaunay and his colleagues (1979) pointed out that the common shrimp (*Crangon vulgaris*) disappeared from the coastline and from Morlaix and Lannion bays for several months in 1978. They attribute this either to the mortality which may have affected these crustaceans or to a natural reflex to escape from the pollution.

It should be said that the tanks where the larger live crustaceans are usually kept before being marketed were swamped with oil, which caused enormous damage.

The small species of molluscs living on the rocky bottom along the shore (limpets and winkles) suffered a very high mortality rate in the area between the wreck and Lannion Bay. The same was true of species which burrow down into the sea bottom (cockles, clams, sand clams, etc.) which suffered enormous losses (Cabioch et al., 1978, 1981).

Chassé and Guenole-Bouder (1981) estimated that the destruction among molluscs in the sea off Saint-Efflam and in Lannion Bay alone amounted to 36 million cockles (20 million *Echinocardium cordatum* and 16 million *Cardium edule*), 14 million Mactridae (*Mactra corallina*) and 6 million sand clams (*Pharus legumen* and *Ensis siliqua*).

But scallops, which live further out to sea, were not similarly affected. The first test in April 1978 in Lannion and Morlaix Bays did not indicate any abnormal mortality although there was some contamination. A second series of catches in Lannion Bay in July showed that mortality did not exceed 14 per cent, a figure regarded as nearly normal (Léglise, 1978).

While the mussel beds to the east of Sillon de Talbert were not affected, this did not hold for the oysters bred in the Penzé and Morlaix estuaries, in the abers (Aber Benoît and Aber Wrach), and in the other breeding areas to the west of Sillon de Talbert.

In order to assess to the amout of damage done, l'Institut des Pêches monitored all oyster breeding areas in northern Brittany and the western Cotentin from March 1978 onwards, paying regular visits at every major tide between March and May. Surveillance continued in the abers (two stations per aber), the Penzé estuary (five stations), and in the Ria de Morlaix (11 stations). The pollution had spared all the centres east of Sillon de Talbert (Figure 7.5.1). To the west of this point, oil was present in the oyster beds in the form of sheen, tar balls or more or less thick puddles. All three forms were present in the abers where the pollution was enormous. Contamination appeared to be less marked in Morlaix Bay and at Penzé but in fact it turned out to be high. Pollution also seriously affected the Saint-Efflam, Trébeurdan and Ile Grande areas around Lannion and, to a lesser degree, the beds in the Ria de Tréguier (Grizel et al., 1978).

Nevertheless, it turned out that mortality was patchy, particularly in the abers. Mainly affected were those oysters covered by a thick layer of oil. It is difficult to put a figure to the mortality rate. For Aber Benoît it has been estimated at 50 per cent of all oysters which had not been transferred and for Aber Wrach at 20 per cent (Maurin, 1978). Elsewhere the mortality rate was low and nearly normal in spite of the fact that large quantities of hydrocarbons were found during analysis (Grizel et al., 1978).

During the days immediately following the grounding, the species of fish living among the rocks immediately around the site of the accident (wrass, ling, sea-eel) suffered high mortality. The same was true of coastal species such as the grey mullet (*Mugil*) and seaperch (*Dicentrarchus labrax*). The populations of trawlable species in Morlaix Bay do not appear to have been affected by the pollution. By contrast, in Lannion Bay the piscine fauna appeared to be depleted in spring 1978 during the acute phase of the pollution. This depletion or disappearance affected mainly catfish, whiting, pout (*Trisopterus luscus*), silver-sides and horse mackerel (*Trachurus*).

Other strictly benthic fish like *Gobius*, gurnard, plaice and dab disappeared in December. This disappearance may be regarded as abnormal and connected with the pollution, either directly or because of the lack of food. Sole and yellow sole (*Buglossidium*) were equally abundant in April and December which would seem to indicate that the pollution had a selective effect, as stated by Desaunay *et al.* (1979). The same authors marked 76 plaice taken from Lannion Bay at the end of April 1978 and released them in Morlaix Bay. They reported that ten were recaptured, seven in the first 3 months. Two were taken close to Roscoff, one in Morlaix Bay, three off Beg and Fry and four in Lannion Bay. According to Desaunay, this shows that the plaice tended to return to shallow coastal waters in spite of their being very polluted.

In summary, it may be said that in the short term, catches of fish decreased immediately after the accident. Throughout the spring and autumn, catches made by professional fishermen were not abundant, the fish being few in number and small in size.

As we shall see in the next section, many observations and analyses by l'Institut des Pêches showed that in April 1978 several types of fish (pollack, gurnard, ray) were contaminated by the oil. Yet the percentage of contaminated fish rarely seems to have exceeded 15 per cent of total catches (Maurin, 1978). From this point of view, the situation seemed to be almost normal in May.

Short-term contamination

With a view to detemining whether the hydrocarbon content of substances extracted from *Laminaria* was too high to permit their use, Pérez (1978b) extracted alginates from samples which were assumed to be polluted. He analysed the alginates and compared them with a control alginate extracted from *Laminaria* collected in December 1977. Gas chromatography analysis provided additional results. To summarize his findings, the author noted that the hydrocarbon content of the alginate extracted from algae collected in April and May on the Primel Headland and at Trémazan, close to the site of the accident, amounted to 60 ppm and was only slightly higher than the level found in the controls (50–55 ppm). Extracts from *Fucus serratus* gathered on the shore at Trémazan contained hardly more hydrocarbons than the controls, and the same was true of samples of carragheenins obtained from *Chondrus crispus* (45–55 ppm).

While these analysis figures show that, after treatment, the hydrocarbon content of the seaweed extract was almost normal, it is nonetheless true that the analyses were not representative of the contamination of the tissue of the seaweed. An analysis of aliphatic hydrocarbons carried out in June and August 1979 at nine sites along the Breton Coast by Topinka and Tucker (1981) demonstrated a high and persistent level of contamination in *Fucus vesiculosus, Fucus serratus, Fucus spiralis* and *Ascophyllum nodosum* along the coast near the site of the wreck and, particularly, in the abers, which suggests that the level of pollution was high at the start.

Anylyses by l'Institut des Pêches carried out in March and April 1978 demonstrated that the flesh of crustaceans (edible crab, spider crab), which smelled slightly of oil, contained only slightly more hydrocarbons than normal (40–67 ppm). By contrast, high levels of hydrocarbons were found in the liver—up to 296 ppm whereas the average is between 100 and 150 ppm. It appears that not more than 5 per cent of the crustaceans caught were contaminated (Maurin, 1978).

The scallops sampled in Morlaix and Lannion Bays had a total hydrocarbon content of between 44 and 76 ppm—three to four times higher than normal (Maurin, 1978).

Although the recorded mortality among oysters was relatively localized and confined to the abers, the level of contamination was fairly high in this area and sufficiently high in the Morlaix and Penzé estuaries to justify destruction of the stock.

In the days immediately following the accident, the total hydrocarbon content of the oysters in the abers was around 300 ppm. Two months after, in mid-May, this level increased to a maximum 444 ppm in samples coming from Aber Benoît and to 643 ppm in those from Aber Wrach, as reported by Grizel *et al.* (1978). In 1981, Laseter and colleagues provided figures that were fairly comparable in view of the much later date at which the samples were taken. The total hydrocarbon content found by these authors in Aber Wrach three months after the accident was 360 ppm (± 170) for the Pacific oyster and 290 ppm (± 35) for the flat or (European) oyster.

Grizel *et al.* consider that the increase they observed could be attributed to the penetration into the abers of treated oil slicks mixed with the mass of water, or to the salting-out of the hydrocarbons trapped in the sediment. Later evidence suggested that this hypothesis was highly probable.

Still according to the same authors, the initial hydrocarbon content was around 220 ppm in the Penzé estuary. Later it tended to decrease by a process which we shall describe when we deal with the medium and long-term effects. The same was true of Morlaix Bay where an analysis of a sample of oysters immediately after the accident yielded a level of 248 ppm.

The first series of analyses on fish was made by the ISTPM on 90 samples taken in April and May 1978 in the polluted area; the analyses were supplemented by organoleptic tests (Michel, 1978) and demonstrated that, generally speaking, the total hydrocarbon content was low, the highest levels being contained in the flesh of rays (up to 85 ppm of total hydrocarbons). The additional organoleptic tests made on this occasion showed that rays, gurnards and pouts tasted strongly of oil. Later more detailed gas chromatography tests using a capillary column were made on 36 further samples of fish from the polluted area. By this method it is possible to distinguish between biogenic hydrocarbons such as squalene and pristane and exogenic hydrocarbons originating in the oil spill. Only two of the 36 samples contained exogenic hydrocarbons, namely heavy alkanes (C_{23} to C_{37}); the first came from a young pollack caught close to Portsall and contained 2.6 μg of alkane per gram of freeze-dried flesh. The flesh of the second, a sole taken in the vicinity of Primel-Trégastel, contained 1.1 μg of alkanes per gram of flesh (ISTPM, 1979).

Short-term disruptions in reproduction and growth

Biological laboratory tests carried out by Steel (1977, 1978) would seem to indicate that the process of fertilization had been upset in the *Fucus* exposed to pollution. Later observation did not confirm these fears. Early on in May, Pérez (1978a) reported:

1. The presence of dark brown spots on the fronds of *Laminaria hyperborea*, which meant that spores were being formed.
2. The fertility of stalks of *Gigartina, Fucus* and *Ascophyllum*.
3. The presence of cystocarps on 35 percent of the stalks of female *Chondrus crispus* plants, which proved that they had been fertilized.

Shortly thereafter, *Laminaria digitata*, in its turn developed spores which the author germinated in the laboratory from June onwards. In the polluted area, 80 per cent of the algae were fertile (Pérez, 1978b).

The appearance of a new generation of plantlets confirmed that reproduction was progressing satisfactorily in this species. At the end of 1978 the new generation accounted for 44 per cent of the *Laminaria digitata* population at Bréhat Island in an unpolluted area, a similar 44 per cent at Roscoff and Porspoder, and 53 per cent at Portsall. An estimate made at the same time of year in 1972 at Porspoder set the figure at 42 per cent (Pérez, 1979).

The same author found that in both the polluted area and the control area there were 31 to 35 per cent of new plants of *Chondrus crispus*. He estimated that in the case of *Fucus* the equivalent figure was 31 per cent.

In order to determine the comparative trends in the growth of *Laminaria digitata* in the polluted area and an uncontaminated one, Pérez (1978a, b) ringed several samples of this species which were about 2 years old. He recorded that there was relatively less growth in April than the average recorded for previous years. Nevertheless, as the difference was about the same for both the control group and the plants in the contaminated area (Trémazan) there is reason to think that this phenomenon was not due to pollution but rather to climatic conditions, which were not very favourable that month.

In any event, in May the metabolism rate increased. Plants grew as much as 31 cm at Trémazan and 32.6 cm in the control area, the usual figures varying between 28 and 34 cm. Thereafter, growth was first normal and then above average. For instance, in October, fronds in the area close to the wreck grew by another 10 cm whereas generally growth during this month is between 4 and 7 cm (Figure 7.5.4).

In the case of *Chondrus crispus* the growth rate turned out to be good. After the harvest, Pérez noted that shoots began to grow from the basal discs. These shoots were more numerous on algae at Portsall and Roscoff than in the control area.

It was not possible to monitor the short-term trends in the reproduction and growth of the large crustaceans. However, Léglise and Raguenès (1981) reported that the percentage of berried female lobsters caught at the end of May 1979 in the Portsall beds was abnormally low for the time of year (25 per cent). Observations made in the Méloines beds near Roscoff prior to the accident showed that the

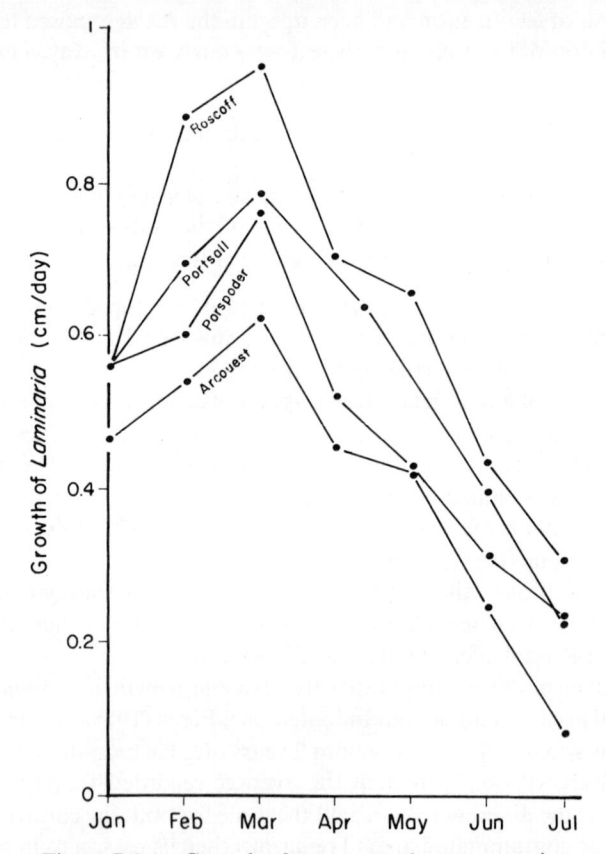

Figure 7.5.4 Growth of *Laminaria digitata* at various stations during the year 1978. Growth was distinctly higher in polluted areas. (From Maurin, 1979; Data collected by R. Pérez)

percentage of females with eggs was 87 per cent in April, 73 per cent in May, 56 per cent in June and 28 per cent in July (Audouin et al., 1971). This led Léglise and Raguenès to the conclusion that some females lost their eggs prematurately or dumped them, a course which they had earlier been observed to take when the eggs were no longer viable.

Balouet and Poder (1981) observed an atrophying of the gonadal cells in oysters from the abers, examined shortly after the 'Black tide'. The figures for the trends in the gonadic indices of the common oyster (*Ostrea edulis*) from Aber Benoît were very much lower for the period May to September 1978 than for the same months in 1977 and were respectively 4.1 to 9.2 and 11.6 to 24.6. The authors did not think there was any reproduction. For the Pacific oyster (*Crassostrea gigas*), the figures seemed to follow a normal trend; in Aber Benoît and at Carantec, still for the period May to September, they were fairly high—11.6 to 31.4 for Aber Benoît and 7.6 to 36.4 for Carantec. Generally this oyster does not reproduce in this region.

The same authors examined the figures indicating the condition of the oysters (dry weight × 100/intervalvular volume) by sampling ten oysters out of each of 142 lots. The results show that in Aber Wrach and Aber Benoît the condition index for 1978 gives an annual curve comparable to that for 1977 for both *Ostrea edulis* and *Crassostrea gigas*. By contrast, in the Morlaix Bay area a certain drop in the condition index was recorded, from the beginning of May, for Pacific

Table 7.5.5 Condition index of oysters after the *Amoco Cadiz* oil spill (from Balouet and Poder, 1981, by permission of Centre National pour l'Exploitation des Oceans)

	Aber Benoît		Aber Wrach		Carante	
	ACP Edulis	ACP Gigas	ACP Edulis	ACP Gigas	ACP Gigas	ACA Gigas
May 1978	102	130	85	161	177	120
June	164	140	160	138	179	76
July	132	223	220	172	130	101
August	151	198	159	165	125	101
September	108	95	118	111	90	110
October	112	154	118	153	99	
November	118	166	244	114	89	82
December				100	66	
January 1979		178	143	160	70	
February	85	150	110	130		62
March	74	134	130			73
April	155	133	176	154		90
May	128	162	176	160	67	71
June	117	187	179	195	75	78
July	166	236	224	160		93
August						
September	128			142		131
October	197			192		120

oysters present in the area at the time the pollution occurred or transferred into this sector immediately thereafter (Table 7.5.5).

Linear growth is considered by Grizel and colleagues (1981) to have been interrupted in 1978 in the abers. They monitored developments in the comparative lengths (in millimetres) of Pacific oysters (*C. gigas*) raised in the Ria d'Auray which is in south Brittany and, therefore, some distance from the affected area, and in the polluted area of the Ria de Morlaix. The results obtained show that growth began late in both the Bay of Morlaix and the Ria d'Auray but that this was made up later. In October, the average length of 100 average sized individuals, which had initially been all the sample size (55 mm), was 83 mm in Morlaix and 88 mm in Auray. Also, 18-month-old oysters weighing 20 to 25 g, which were placed in the beds in Morlaix Bay in February and March 1978, weighted 40 to 60 g in November to December of the same year, so that these results are similar to those obtained for previous years.

While studying the population structure of plaice in Lannion and Morlaix Bays, Desaunay *et al*. (1979), and Desaunay (1981) observed that the structure in April 1978 was similar to that recorded in other Channel areas. In the next few months there were abnormal developments. In particular there were no young at the end of the year, which meant that there had been no reproduction or that the eggs and larvae had been destroyed. In addition, the one-year-olds grew very little (2 cm in eight months) by comparison with the normal rate. In the case of sole, the most obvious factor was the absence of juveniles in 1978. Growth of adults did not appear to have been significantly affected. No recruitment in the dab population was observed.

Miossec (1981a, b) reported irregularities in the reproduction of plaice in the abers in 1978. The gonado-somatic ratio was very low compared with the reference areas. These anomalies were confirmed by a histological study: in the vitellogenesis period the ovocytes were heterogeneous in size, relatively small and widely dispersed in the gonads. The author reported atresia, large numbers of blood vessels and the development of conjunctive fibres. Finally, Miossec reported very little recruitment in the population at the end of the year.

Conan and Friha (1979, 1981) estimated trends in growth for sole and plaice in Aber Benoît. To do so they compared figures for the standard length of the fish, and those for otoliths observed over a period of 4 to 5 years prior to the wrecking of the *Amoco Cadiz*, with the figures for 1978. This showed that there was a clear difference between 1978 and previous years. In 1978, growth was abnormally slow in all age groups. Growth in sole dropped to 25 per cent of normal in young and 88 per cent of normal in adult fish. But the opposite was noted for plaice: growth in young fish was 88 per cent of normal while in adults it was 30 per cent.

7.5.4.2 Other Accidents: The *Gino*

We have fewer biological and ecological records of which fishing resources were affected by other accidents occurring in Brittany waters than we have for the

Amoco Cadiz. There are various reasons for this, particularly for the *Bohlen* and the *Gino*, as these wrecks lie in relatively deep water—around 100 m—and the latter is quite far from the coast. Where the *Tanio* is concerned, her pollution affected almost the same sections of the coast as the *Amoco Cadiz* spill, which made it difficult to identify what damage was done. We will therefore note only a few facts which will permit some comparisons.

Mortality

The *Bohlen*'s cargo was a highly viscous crude which included approximately 60 per cent asphalt, some polymerized products, 35 per cent aromatic hydrocarbons and 2.5 per cent saturated hydrocarbons. Although relatively small, the quantity involved (about 10 000 tonnes) caused significant mortality in various species, particularly edible crab (*Cancer pagurus*). The reason for this mortality may be that, with a density almost equal to that of water, the highly toxic product spread throughout the water column and blocked up the branchiae of the animals.

In the case of the *Gino*, the carbon black oil in its cargo tanks contained 60.9 per cent toxic aromatic hydrocarbons, 11.5 per cent saturated hydrocarbons, 15.8 per cent resins and 11.8 per cent asphaltenes.

The earliest observations made on board the *Roselys* and the *Thalassa* at the beginning of May 1979 showed that in the area around the wreck the scallops were alive, but that some of them were stained with a brown, flaky, viscous material which came from the tanker's cargo. The piscine fauna, comprising mainly gurnard (*Trigla lucerna*), horse mackerel (*Trachurus*), pout and capelin (*Trisopterus luscus* and *minutus*), Blue whiting and catfish (*Scylliorhynus*) appeared to be unaffected by the pollution (ISTPM, 1979, 1980).

A year later, within a radius of some 7 miles of the wreck, Michel observed evidence of recent mortality in scallops, frequent atrophy of the gonads, and absence of reaction in some living individuals.

Contamination

After the *Gino* spill, in the fish analysed, even in the viscera, there were no signs of significant amounts of accumulated hydrocarbons. Traces of low-level contamination were found in crustaceans (*Cancer pagurus*). Microscopic examination of eggs in the females of this species showed no detectable contamination (Michel and Abarnou, 1981).

Scallops, which make excellent indicators of contamination in the open sea because they are active filterers, were highly contaminated (up to 100 ppm of aromatic hydrocarbons in the muscles) within a radius of up to 5 to 10 miles of the wreck, depending on the compass bearing. Up to a distance of 15 miles (and occasionally more than that), the aromatic hydrocarbon content in the muscles was some 20 ppm. At a distance of 15 to 30 miles the content was generally 5 ppm or less.

7.5.5 MEDIUM AND LONG-TERM CONSEQUENCES

The public is most deeply moved by the immediate consequences of an oil spill. But these consequences do not always reflect the actual situation. In fact, in order to have any real idea of the impact on resources, the ecological situation has to be monitored over a period of years. Although the period separating us from the *Amoco Cadiz* accident is too short and that for the *Gino* even shorter, and it is therefore still too early to make any final pronouncements, it has been possible to make some observations. These are interesting enough to merit an attempt to summarize them.

We focused attention on three points which we found important, namely:

1. The spread of the contamination.
2. The ecological, teratological and pathological effects.
3. Developments in stocks.

We will begin by giving examples for the *Amoco Cadiz*. For the *Gino* we shall keep mainly to how contamination progressed, as this is the only factor about which we have any definite knowledge so far.

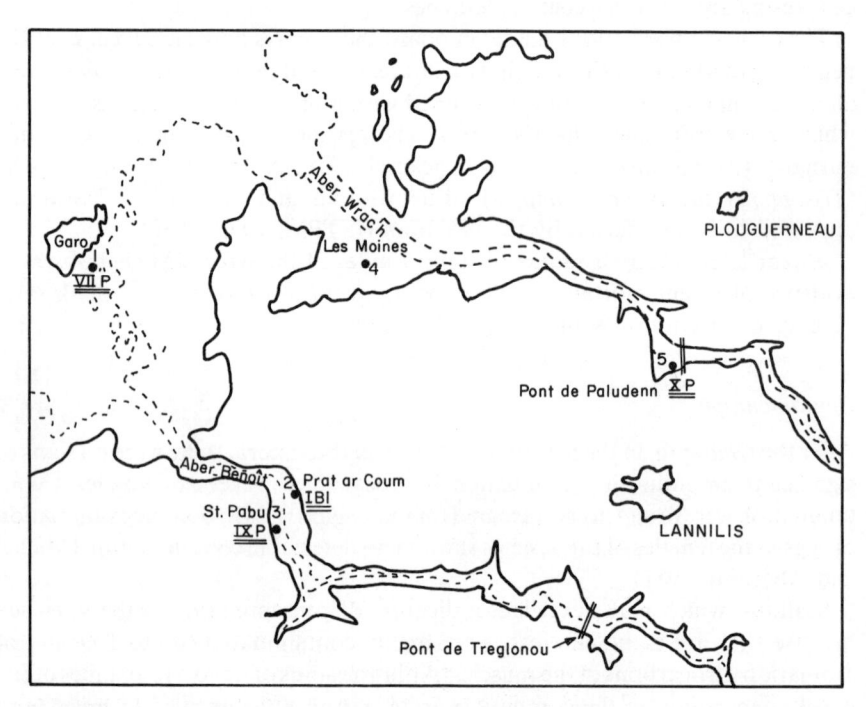

Figure 7.5.5 Map of stations observed in Abers Benoît and Wrach. (Reproduced from Grizel *et al.*, 1981, by permission of Centre National pour l'Exploitation des Oceans)

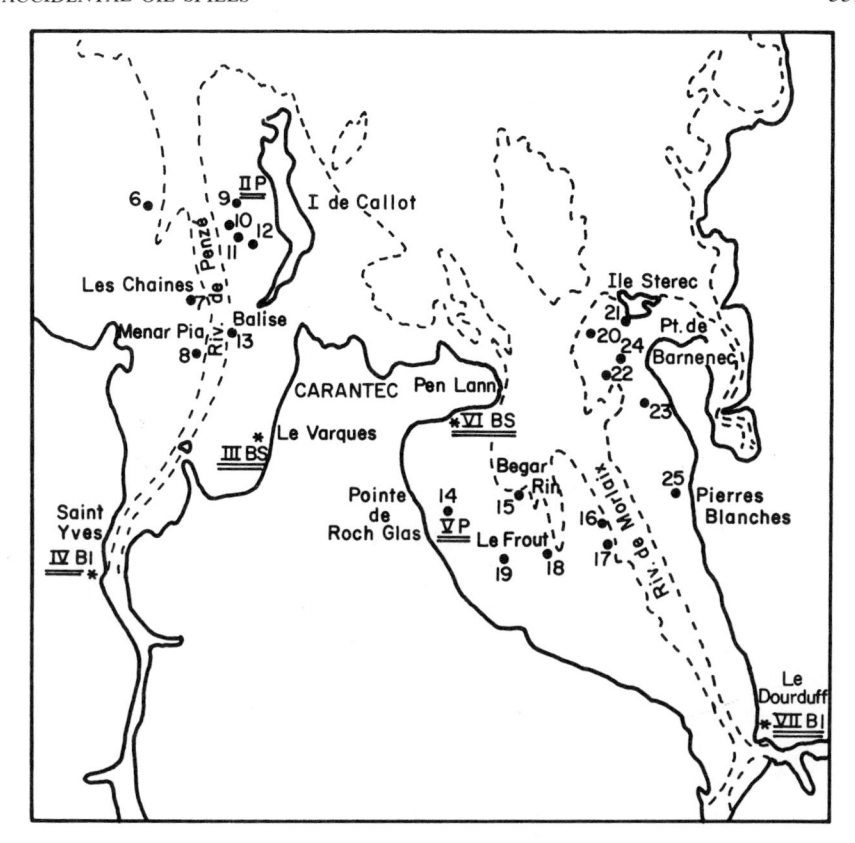

Figure 7.5.6 Map of stations observed in Penzé and in Morlaix Bay. (Reproduced from Grizel *et al.*, 1981, by permission of Centre National pour l'Exploitation des Oceans)

7.5.5.1 The *Amoco Cadiz*

The spread of the contamination

We have taken oysters as an example of how contamination caused by the *Amoco Cadiz* oil spread, since these molluscs, which are easy to sample or transfer, struck us as highly representative.

First Grizel, Michel and Abarnou (1978) and then Grizel *et al.* (1981) monitored this development in respect of oysters from the North Brittany shellfish breeding areas. The authors made observations on 5325 Pacific oysters (*C. gigas*) from 213 lots of 25 individuals each, and on 400 common oysters (*O. edulis*) from 16 lots of 25 each. The samples came from three stations in Aber Benoît, two in Aber Wrach, eight in the Penzé estuary (right and left banks) and 12 in Morlaix Bay (also the right and left banks) (Figures 7.5.5 and 7.5.6).

Table 7.5.6 Level of contamination of abers oysters shown by ppm total petroleum hydrocarbons per kg wet weight. (From Maurin, 1979; compiled from data collected by P. Michel and H. Grizel)

Stations	1978									1979				
	5 Apr	25 Apr	25 May	26 Jun	20 Jul	17 Aug	17 Sep	19 Oct	15 Nov	15 Dec	1 Feb	27 Feb	28 Mar	20 Apr	25 May	20 Jun
Aber Wrach	293	131	643		304	208	277	106		140	260	155	188	154	159	150
Aber Benoit	310	298	444	154	155	188	222	193	200	203	247	114	162	167	154	187

The authors state that the average reference value for eight different regions not affected by pollution from the *Amoco Cadiz* was around 60 ppm ($\pm$ 20) of total hydrocarbons.

The authors' preliminary data are for the kinetics of the *in situ* decontamination of oysters. In Aber Benoît, the highest and lowest averages established per station went from 298–310 ppm in April 1978 to 275–444 ppm in May and 154–188 between June and August. In October to November, when the sediments were again stirred up by the swell, the levels were 193–203 ppm. There was then a slight drop for the period between the end of February and April 1979 (114–167 ppm). In June the level was again some 187 ppm (Table 7.5.6).

In Aber Wrach the average levels for each station went from 142 and 293 ppm, in April, to 131 ppm for the station at the mouth of the aber and 643 ppm for the station upstream, in May. Levels of between 304 and 208 ppm remained unchanged from July to September and then started decreasing, beginning in November, particularly at the entrance to the aber (106 ppm). The contamination level rose again (to 260 ppm, upstream) and then settled at around 160 ppm from April to June.

Thirteen months after the accident, Laseter *et al.* (1981) found 225 $\pm$ 23 ppm of total hydrocarbons (of which 131 $\pm$ 12 ppm were aromatics) in the Pacific oyster and only 59 $\pm$ 9 ppm (of which 30 $\pm$ 6 ppm were aromatics) in the common oyster. This means that decontamination was much less rapid in the Pacific oyster

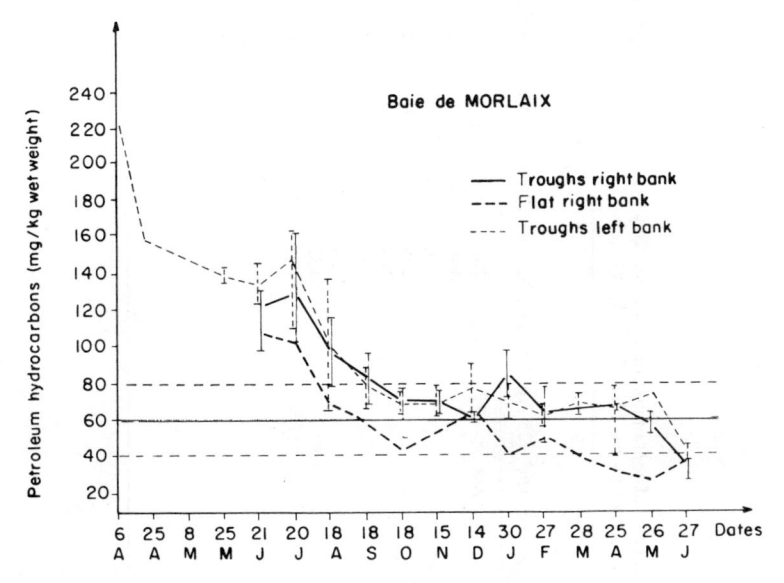

Figure 7.5.7 Purification kinetics of Morlaix Bay oysters. (Reproduced from Grizel *et al.*, 1981, by permission of Centre National pour l'Exploitation des Oceans)

Table 7.5.7 Level of contamination of Morlaix Bay oysters shown by ppm total petroleum hydrocarbons per kg wet weight (From Maurin, 1979; compiled from data collected by M. Michel and H. Grizel)

	1978										1979					
Stations	5 Apr	25 Apr	25 May	26 Jun	20 Jul	17 Aug	17 Sep	18 Oct	15 Nov	14 Dec	1 Feb	27 Feb	30 Mar	20 Apr	25 May	20 Jun
9	248	161	146	146	111	118	86	78		79	28	68	77	79	75	38
2				140	145	120	71	57	45	64	77	62		53		50
5				100	136	99	80	39	50	70		60		45		39
6				175	113	91	71	55	50	65		56	57	52	73	42
10				141		105	80	76	75	92	57	69	62	74		
11				126	164	111	87	77	61	83	72	63	77	79		48
12				134	191	138	90	71	80	68	63	71	73	40		45
17				135	129	116	97	74	78	59	74					
18					102	78	69	76	67							
1					106	95	64	56	56	56	74	48	67	54		32
19						69	58	50	62	64	71			66	53	28
12*						66	88									
13						120	68	65								
14						88		67								
15						116	79									

*(repeat)

(*C. gigas*) than in the common oyster (*O. edulis*), since the values the authors found in May 1978 were some 360 ± 170 ppm of total hydrocarbons (of which 210 ± 100 ppm were aromatics) for the former species and 290 ± 35 ppm (of which 150 ± 12 ppm were aromatics) for the latter.

For Morlaix Bay, as for the Penzé estuary, the only figures available are those provided by Grizel and colleagues (1978, 1981). In Morlaix Bay the Pacific oysters, whose flesh contained between 204 and 220 ppm in April 1978, decontaminated themselves fairly completely between May and October of that year. The hydrocarbon content was 145 ppm on the right bank and 142 ppm on the left bank, in May, and remained stable in June and July (120 to 128 ppm on the right bank and 136 to 151 ppm on the left bank). Contamination levels then dropped again, to 71 ± 12 ppm on the right bank and 72 ± 6 ppm on the left bank. The authors reported a slight increase in December 1978 and January 1979 (62 ± 3 to 80 ± 10 and 85 ± 20 to 73 ± 8 ppm, respectively) followed by a stable period. Average values in May 1979 were between 58 ppm and 75 ppm and then fell in June to 33–44 ppm (Table 7.5.7, Figure 7.5.7).

Grizel *et al.*, reported findings similar to those of Laseter and his colleagues in observing that the figures for the Pacific oyster are always higher than those for the common oyster. In the latter species, average contamination levels decreased from 110 ppm in June 1978 to 30 ppm in May 1979 (right bank).

In the Ria de Renzé the kinetics of decontamination were similar to those to Morlaix Bay, for both the right and left banks. The average contamination level of 200 ppm in April 1978 dropped to about 40 ppm in June 1979 (Figure 7.5.8).

These authors also reported on the operations to transfer oysters from contaminated to clean areas (south Brittany) and from clean to contaminated areas. In the first instance, the transfer took place 20 days after the grounding. In the case of oysters from Aber Benoît whose initial hydrocarbon content was 320 ppm, the level dropped to 66 ppm in 18 days. One month later other oysters from Morlaix Bay, which originally contained 248 ppm, contained no more than 54 ppm of total hydrocarbons. In the second instance, oysters from the Auray estuary in South Brittany, with an initial hydrocarbon content of some 47 ppm, had levels of 89 ppm 2 months later. Finally, samples from South Brittany initially containing 27 ppm and transferred to Aber Wrach in January 1979, contained 120 ppm by June that year.

Similar experiments in the abers have been published by Friocourt *et al.* (1981). Here the oysters kept in a polluted area had partially decontaminated themselves after 16 months. Residual contamination levels were still high, particularly as regards aromatic hydrocarbons (32 ± 14 to 66 ± 13 ppm total hydrocarbons, 8.8 ppm aliphatics, 24 ± 13 ppm aromatics). In the polluted oysters transferred to uncontaminated sites, the level of total hydrocarbons decreased rapidly, falling to 16 ± 4 ppm in 6 months, but the aromatic compounds persisted.

Healthy oysters were placed in polluted parts of the abers in May 1978 and again the rapidity with which hydrocarbons accumulate was demonstrated. The

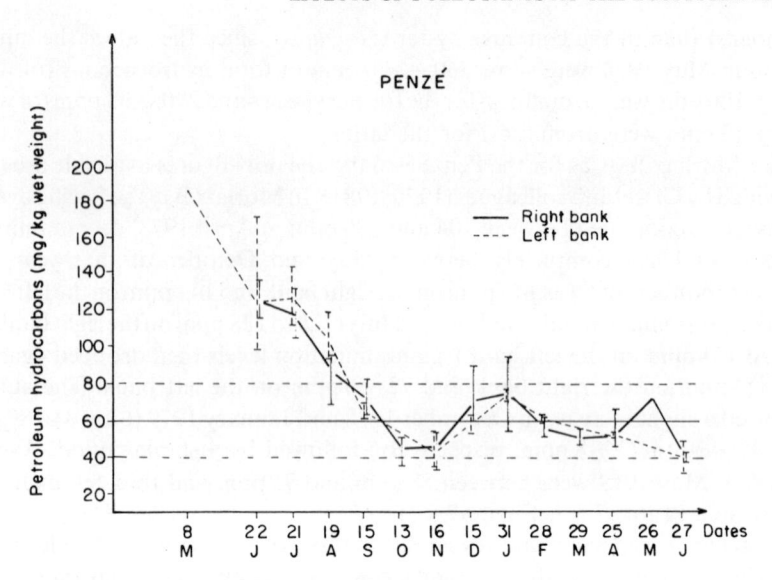

Figure 7.5.8 Purification kinetics of Penzé oysters. (Reproduced from Grizel et al., 1981 by permission of Centre National pour l' Exploitation des Oceans)

aliphatic and aromatic content increased from 4 to 53 ppm in 15 days and then steadily but slowly declined during the following 16 months, as the level of pollution in the environment decreased.

Three years after the pollution occurred, Gouygou and Michel (1981a, b) took up the matter again by studying a batch of health Pacific oysters from Quiberon Bay which were placed in an oyster bed at Saint-Pabu in Aber Benoît in March 1979. The site selected has a mud and sand bottom contaminated by hydrocarbons. The batch of oysters remained in this contaminated area for almost two years.

In February 1981, the total aromatic hydrocarbon content of the oysters was some 49.5 mg kg^{-1} of dry flesh. Analysis showed that the hydrocarbons consisted of nearly 70 per cent persistent thiophene compounds (54.4 per cent alkyl dibensothiophene or DBT, and 13.3 per cent naphthobenzothiophene or NBT), about 21 per cent phenanthrene and 4.9 per cent fluoranthenes, pyrenes and their alkylated derivatives. An examination of the results showed that the saturated hydrocarbons initially present in the oil were absent and that the aromatic hydrocarbons such as alkyl benzene and alkyl naphthalene had disappeared. The absence of both clearly demonstrated that no significant chronic or incidental pollution was present to interact with the oil spilled by the *Amoco Cadiz*.

Having clarified this point, it was important for the authors to ascertain the kinetics involved in the process of eliminating the hydrocarbons. Therefore, in

February 1981 they undertook a decontamination experiment in an unpolluted area. The figures they obtained were:

48 mg kg^{-1} after 4 days
41.2 mg kg^{-1} after 18 days
24.4 mg kg^{-1} after 32 days
17.7 mg kg^{-1} after 53 days
14.2 mg kg^{-1} after 69 days.

The composition of the hydrocarbons did not vary significantly from what it had been at the outset.

The authors believe that the compounds they found are present in the oil initially, but in relatively small amounts, and that these may be masked during analysis. This therefore points to the need for both general and specific measurement of aromatics as part of the process of monitoring contamination (Michel, 1981a, b). The experiment also showed that healthy oysters transferred into a contaminated area one year after the accident may, two years later, be flushed of the persistent hydrocarbons they then contained by reintroducing them into an uncontaminated area. The authors consider that decontamination in this case takes longer than after contamination by fresh oil but that, in the case in point, the oysters can again achieve a satisfactory quality after a prolonged stay in clean water.

However that may be, the observations of both Gouygou and Michel, and the other authors referred to here, call for comment on our part. If there were any need for justification, the presence and persistence of thiophenic hydrocarbons in the oysters in the abers, three years after the accident, fully justifies the decision taken shortly after the grounding of the vessel to destroy the remaining stocks in the area. Although, as yet, there is very little data available on the toxicology of these derivatives, we would like to remind readers that Hermann and his collaborators (1979) found a correlation between mutagenesis and the polynucleated aromatic hydrocarbons.

Ecological, teratological and pathological effects

Continuing where Pérez had left off, Kaas (1980, 1981) studied the development between January 1979 and May 1980 of the algal fields on the Brittany Coast between Porspoder and the Arcouest headland south of Bréhat Island. He selected four sites at which he monitored growth. From west to east these were:

1. The St. Laurent Peninsula at Porspoder.
2. The shingle between Trémazan and Portsall.
3. La Rochezu at Roscoff.
4. The unpolluted Arcouest headland, which was used as a control.

During the 16-month period beginning 10 months after the spill occurred, Kaas monitored the apparent average growth of the fronds of *Laminaria digitata*, i.e. actual new growth minus that eroded by wave action. He noted that, on the whole, apparent growth was greater among the algae in the polluted areas than those in the control zone. The difference was particularly marked at the Trémazan site which is closest to the wreck. In the algae at this site, average monthly growth of the fronds achieved a maximum of 210 cm in April 1979, while at Arcouest the highest average (registered in June) was some 50 cm.

For actual growth, calculated by means of the progress of a reference marker, the differences were even greater. For instance, average daily rates of growth at Roscoff reached a maximum of nearly 1 cm (per day) in April 1979 and a minimum of around 0.30 cm in July to August. At Portsall and Porspoder the respective averages were 0.75 cm and 0.25 cm for the same months, whilst at Arcouest growth hardly exceeded 0.60 cm in April and dropped to 0.08 cm in July to August. It should be stated that these figures are borne out by those given by various authors, particularly Pérez (1968), prior to the accident, at least for the summer months in Normandy (Figure 7.5.9).

For the stipe of the seaweed the results are similar. Although figures vary significantly for the various contaminated sites, it may be said that the stipes of

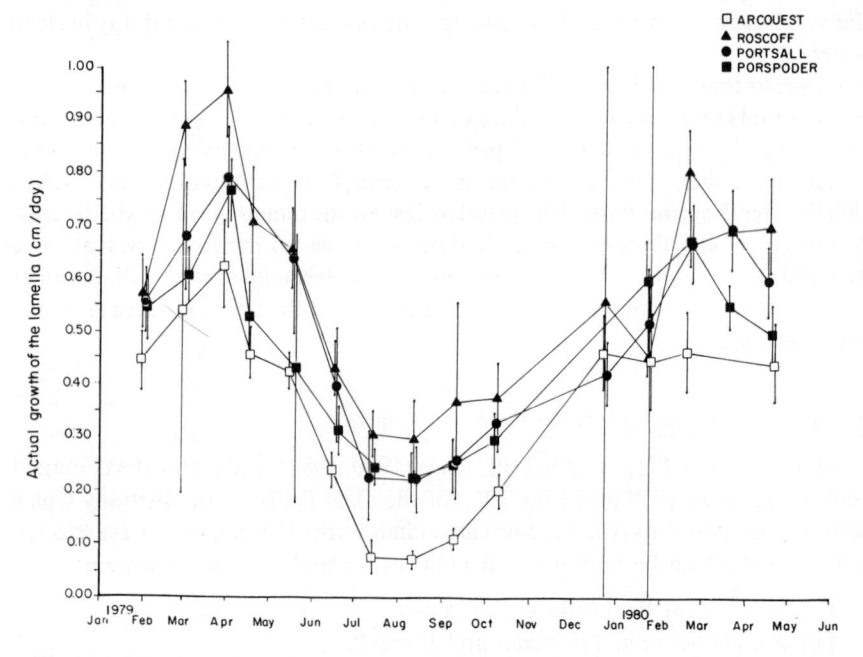

Figure 7.5.9 Actual growth of the lamina of *Laminaria*, from January 1978 to June 1980, at various sites in the contaminated zone and at the control station. (From Kaas, 1980. Reproduced by permission of ISTPM, B.P. 1049, 44037, Nantes Cédex)

seaweed polluted by hydrocarbons grew faster during the period in question than those of seaweed in the control zone.

In all, and in spite of local differences which may be ascribed to the hydrodynamic conditions, it may be said that the seaweed in the polluted areas grew much faster than the *Laminaria* in the control zone. The explanation for this apparently favourable phenomenon lies in the eutrophication caused by an enrichment of the environment by organic matter. We must say 'apparently', as Kaas states that the metabolic activity of *Laminaria* was promoted only up to a certain point. Once this point was reached, physiological deterioration followed, taking the form of increased fragility in the plants. The author stresses the fact that the phenomenon occurred one year after the accident and that two years after the accident the *Laminaria* population of the polluted area was more heterogeneous than that of the control area.

With respect to reproduction, Kaas confirmed the preliminary observations of Pérez reported in Section 7.5.4, that young plants appeared even in the most contaminated areas, and the presence of these young individuals in the algal fields in 1979 is a clear sign that reproduction in 1978 was successful. On some sites young plantlets account for more than 50 per cent of the total population.

This reproductive success was confirmed by further experiments to get the spores of polluted *Laminaria* to germinate in the laboratory, in 1979. The experiments showed that the reproductive potential of the algae did not appear to be significantly impaired.

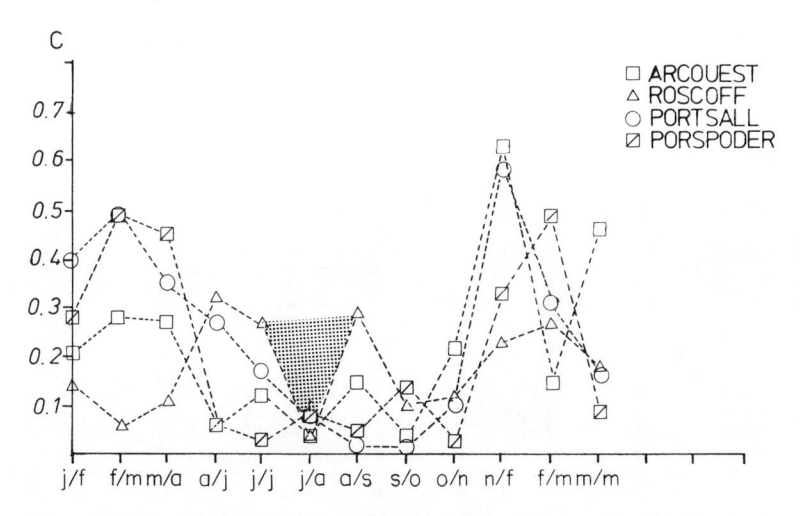

Figure 7.5.10 Variation of the coefficient of contingency for *Laminaria* at three contaminated sites and at the control station (Arcouest). The diagram shows the period of limited fertility (shaded area) of algae at the Portsall site as compared to that at the other stations. (From Kaas, 1980. Reproduced by permission of ISTPM, B.P. 1049, 44037, Nantes Cédex)

However, the author indicated that the number of plantlets was lower at polluted sites than at others. He also noted that the period of fertility was shorter at Roscoff than in other sectors (Figure 7.5.10).

Where the seaweed *Chondrus crispus* is concerned, Kaas noted a net drop in biomass as compared with the years preceding the accident, but he does not attribute this to the pollution or, at least, not entirely. A similar decrease was reported in other geographical areas, particularly Normandy. In conclusion, Kaas considers that, although the oil spilled by the *Amoco Cadiz* may have had a harmful effect on the population of *Chondrus*, this merely accelerated the process of deterioration (Figure 7.5.11).

Where crustaceans are concerned, it is worth drawing attention to one fact reported by Ramirez-Pérez *et al.* (1980). In the course of their study on the pathology of marine crustaceans of the Atlantic coast of Europe, the authors noted a pathological symptom, never before recorded, which manifested itself mainly in skin ulcers in the populations of *Carcinus maenas*, common shore crab, of the Roscoff area. The skin ulcers were frequently associated with internal lesions which caused generalized fatal syndromes. These complications were of two types: septicaemia or penetration of the visceral cavity by the dense substance covering

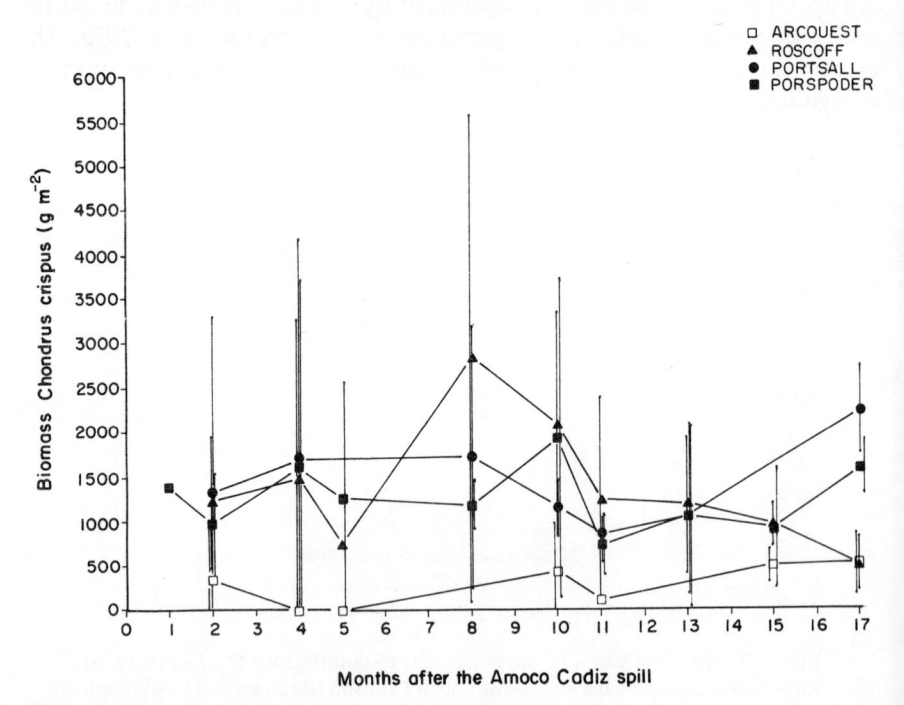

Figure 7.5.11 Development of biomass of red alga *Chondrus crispus* in the contaminated zone and at the control station (Arcouest). (From Kaas, 1980. Reproduced by permission of ISTPM, B.P. 1049, 44037 Nantes Cédex)

the ulcerated area. In the latter instance, irregularly shaped particles of this substance are surrounded by lesions and multicellular haemocyte reactions. The authors noted that, without exception, all the affected crustaceans were from a sector throughly contaminated by the *Amoco Cadiz* spill and that the syndrome had not been noted prior to this accident. They consider that the syndrome is significant as an indicator of pollution and that it could be taken as a demonstration of the delayed pathological effect of such pollution.

In the course of their study on the biological effect of the pollution caused by the *Amoco Cadiz* oil on shellfish breeding in North Brittany, Balouet and Poder (1981), working on oysters from the abers, noted a degenerative cellular impairment which was particularly marked in the epithelial cells and the digestive diverticulum. They also noted the frequency of inflammation, indicated by the presence of macrophagic cells, and found many large pigmented cells in oysters observed 6 months after the black tide. These authors clearly state that the lesions are not very specific but that they are interesting because of their spread, their frequency and their development. It should be added that, even outside the polluted areas, this type of lesion is not rare in oysters which have suffered stress due to major fluctuations in temperature or salinity or due to having been left out of water for too long, etc.

Balouet and Poder therefore tried to quantify their observations. To do so they used a necrosis index for each type of tissue (branchiae, interstitial tissue, gonads, alimentary tract) and another index for the total incidence of lesions. A score reflecting the seriousness of the lesion was given to each tissue when it was examined under the microscope. In the case of oysters, the total index obtained in this way was 18.8 for batches taken directly from polluted areas, 27.9 for batches transferred to other areas (the high figure could be explained by the phenomenon of extra stress) and 23.1 for healthy oysters bedded in a polluted area. Samples examined in 1977, before the oil spill, were used as controls. Their necrosis index, at 5.3, was very much lower. There was no significant difference between the common oyster and the Pacific oyster.

When changes over a period of time were considered, the most typical fluctuations in the necrosis index were obtained for *Ostrea edulis* in Aber Benoît. There was a considerable increase between May and August 1978, a decrease in October–November of the same year, and a certain degree of stabilization between February and July 1979, although the figures were always higher than those for the control batches. Nevertheless, Balouet and Poder consider that the level of damage in the oysters was not as high as had initially been feared.

Frequent anomalies in fish taken from areas polluted by the *Amoco Cadiz* oil have been reported by several authors. These are not specific either, as similar anomalies have been observed in other regions. Nevertheless, the large numbers of anomalies and their frequency would seem to indicate that in this instance they could have some connection with the pollution.

During the sampling carried out after the *Amoco Cadiz* was wrecked, Desaunay

(1981) noted that in Lannion and Morlaix bays anomalies were observed in several species of fish, from the end of summer 1978 onwards, i.e. some 6 months after the acute phase of pollution. Most of the anomalies were found in flatfish such as brill (*Scophthalmus rhombus*), turbot (*S. maximus*), flounder (*Platichthys flesus*), Dover sole (*Solea vulgaris*), and particularly plaice (*Pleuronectes platessa*). Anomalies also occurred in black sea-bream (*Spondyliosoma centharus*), pollack (*Pollachius pollachius*) and cod (*Gadus morhua*).

Figure 7.5.12 A young turbot (*Scophthalmus maximus*) and two plaice (*Pleuronectes platessa*) caught in Lannion Bay at the end of 1978, showing 'fin rot'. Insert: 'bent fin ray' phenomenon in a plaice caught in Lannion Bay in May 1979. (From Desaunay, 1981. Reproduced by permission of Centre National pour l'Exploitation des Océans)

The symptoms were excessive blood flow to the fins which congested with blood (frequently noted in sole), total or partial fin erosion or fin rot mainly affecting the caudal fin in Gadidae or the posterior part of the fish, and the caudal fin in plaice, or else bent fin rays. In the latter case the fin looked as if it was broken (Figure 7.5.12). For plaice, the fish in which he observed these deformations most frequently, Desaunay calculated the percentage of individuals affected in the coastal area in which the trials were being carried out. This was 0 per cent in April 1978, 9 per cent in December of the same year, 73 per cent in May 1979, and 2.5 per cent in October 1979.

Conan and Friha (1979, 1981) reported frequent occurrences of the phenomenon of fin and tail rot in plaice and sole from Aber Benoît.

Starting in November 1978, Miossec (1981a, b) made a detailed study of this fin necrosis on a population of plaice in Aber Wrach and Aber Benoît. The study highlighted the fact that there were big fluctuations over a period of time and between different places.

In Aber Benoît, the percentage of fish affected increased from 43.6 per cent in December 1978 to 81 per cent in February 1979, fell to around 63 per cent in April–May of the same year, rose to 86.4 per cent in July, decreased again to 63 per cent in August, increased slightly in September and fell to 15.3 per cent in November 1979. Thereafter there were further fluctuations as follows:

53.8% in December 1979
28% in January 1980
37.5% in March 1980
47.6% in April 1980

Out of a total of 891 plaice captured from Aber Benoît between December 1978 and April 1980, 533 (59.8 per cent) had damaged fins. However, signs of scar formation appeared beginning in September 1979 in 33.3 per cent of the fish observed. Scars were present in 77.3 per cent of fish in February 1980 and in 42 per cent of fish in April of the same year.

In Aber Wrach, the development of fin damage was fairly similar. Starting at 38.7 per cent in November 1978, the figures reached their first peak in January and February 1979 (93.3 and 95.5 per cent) and thereafter a second peak in March 1980 (84.2 per cent). In April 1980 some 57.1 per cent of the fish were still affected. Of a total of 694 plaice fished out of this aber, 369 (53.2 per cent) exhibited anomalies. In this area, the author also observed evidence of scarring from September 1979. Here maximum scar formation occurred in January 1980 (88.9 per cent).

Miossec noted that in both areas, the dorsal, caudal and anal fins were more frequently damaged than the pelvic and pectoral fins.

The author also pointed out the considerable differences in the geographical spread of this necrosis. In a preliminary balance sheet drawn up in October 1979 she noted that, in the case of Aber Wrach, the percentages obtained for all fish observed at this period were 56.3 per cent at the mouth, 63.3 per cent in the middle

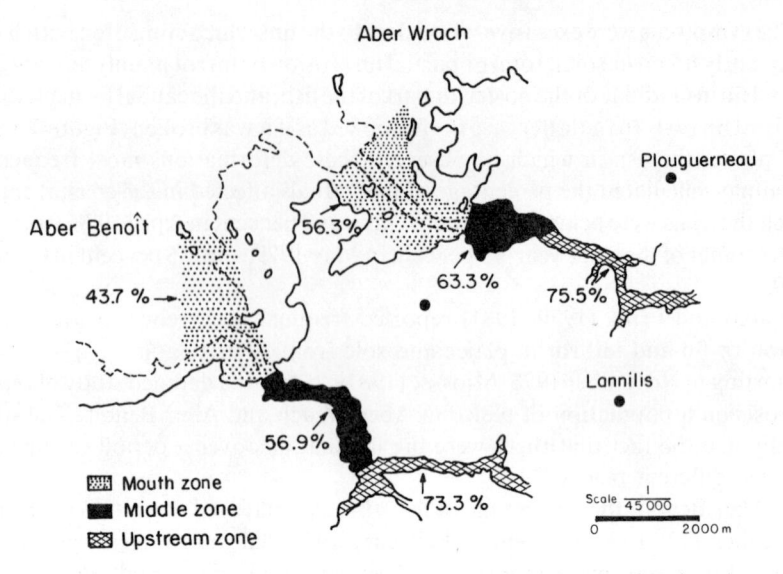

Figure 7.5.13 Percentage of fish (*P. platessa*) affected by fin deterioration in Aber Benoît (left) and in the Aber Wrach (right). (From Miossec, 1981. Reproduced by permission of Centre National pour l'Exploitation des Océans)

of the aber and 75.5 per cent upstream. For Aber Benoît (Figure 7.5.13) the respective figures were 43.7 per cent, 56.9 per cent and 73.3 per cent.

At the end of the study in April 1980 the overall percentages covering 18 months of observations were lower but showed a similar distribution. The highest figure (66.9 per cent) was recorded in the upstream section of Aber Benoît.

It is not impossible that the fluctuations during the period covered by the author reflected the level of agitation in the environment and the availability of food. Where spatial variations are concerned, the explanation could be the nature of the substrate, the finest substrates being the better absorbers of oil.

Finally, Miossec also noted damage to fins in flounders (25 per cent in Aber Wrach and 6.7 per cent in Aber Benoît) as well as in sole (37.2 per cent in the first aber and 41.2 per cent in the second). No anomalies were noted in the fins of dab.

Following a slightly different line, but one which is also connected with the teratological and pathological effects of pollution, Haensly *et al.* (1979) studied the long-term effects of the *Amoco Cadiz* oil spill on plaice taken in Aber Benoît. The conclusion of this major study is that plaice gathered in this estuary had a large number of histopathological anomalies. A comparison with fish of the same species taken in Douarnenez, which was hardly affected by the accident, allowed the authors to establish a relationship between these phenomena and the pollution, at least for a large proportion of the anomalies.

In particular, they consider that hyperplasia and hypertrophy of the lamellar epithelium of the mucous cells of the branchiae are the result of the reaction of tissue to the irritant effects of the oil and an attempt by the animal to avoid absorbing the toxic substances. This reaction had a deleterious effect on respiration, and the dilation of the capillary lamellae was an attempt to balance this effect.

Also according to these authors, the oil constituents caused degeneration of the gastric glands and the pancreas. Lesions occurring in these organs enable the toxic substances to penetrate into the portal system and possibly into the liver.

The oil also affects the skin causing significant hyperplasia and hypertrophy of the mucous cells of the epidermis.

Haensly and colleagues (1979) consider it as a proven fact that the liver of these fish metabolizes the toxic substances. The process requires so much energy that the cellular or hypodermal lipid reserves are used up and the subjects lose weight. The authors consider that the absence of lesions in the kidneys seems to prove that the detoxification process is effective.

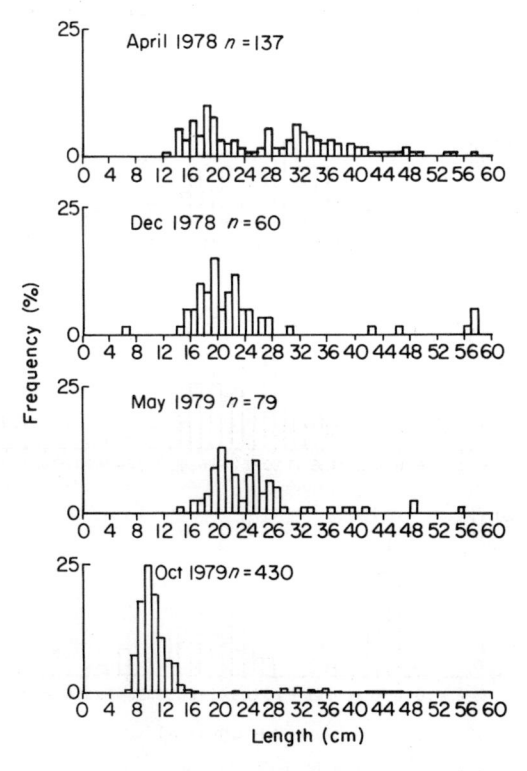

Figure 7.5.14 Demographic structure of plaice populations in Lannion and Morlaix bays. (From Desaunay, 1981. Reproduced by permission of Centre National pour l'Exploitation des Océans)

Developments in the level of stocks

As of mid-1982, the most specific and significant data available on developments in the level of stocks of interest to the fishing industry are on flatfish. Because of the difficulties inherent in conducting a study of this type in the open sea the data are also limited spatially and temporally.

In 1978 and 1979 Desaunay and colleagues (1979) and Desaunay (1981) monitored the composition of the stocks of this type of fish in a coastal area in

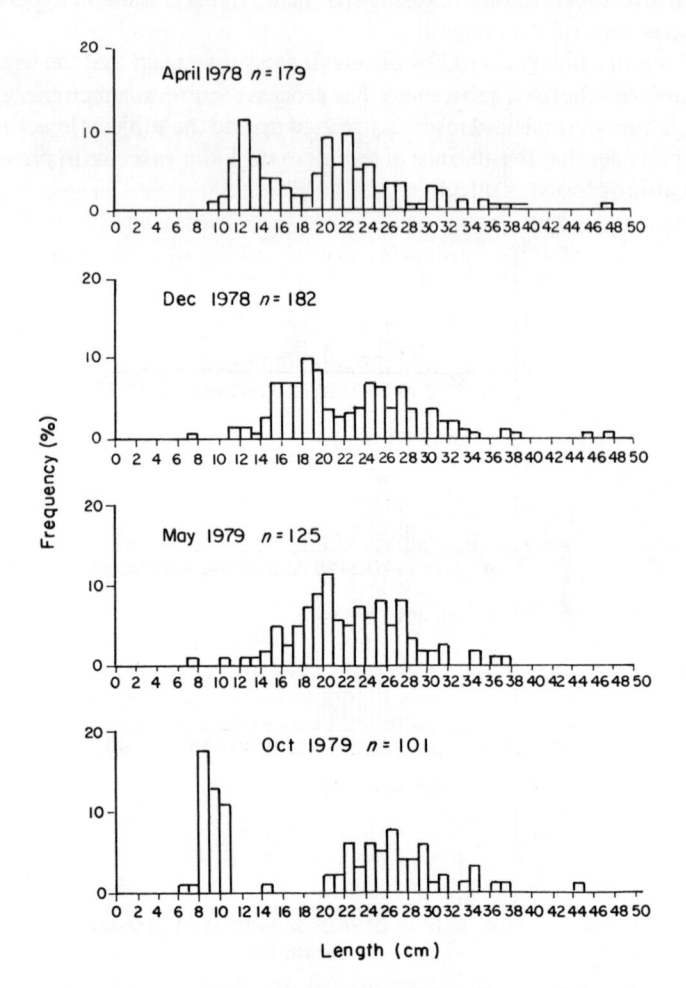

Figure 7.5.15 Demographic structure of sole populations in Lannion and Morlaix bays. (From Desaunay, 1981. Reproduced by permission of Centre National pour l'Exploitation des Océans)

Lannion and Morlaix Bays. Their observations were limited to the southeastern section of Lannion Bay, from the coast to a depth of approximately 20 metres, and to two areas in Morlaix Bay with approximate depths of 10 to 20 metres. This limitation is due to the fact that the bottoms are strewn with rocks and therefore frequently cannot be trawled. The study covered plaice (*Pleuronectes platessa*), Dover sole (*Solea vulgaris*) and dab (*Limanda*).

For plaice, the histograms show that in April 1978, the population of the study area was composed of young fish measuring 12 to 22 cm and of individuals 2 to 3 years of age and over, measuring 28 to 56 cm (Figure 7.5.14).

No young appeared in December 1978, which is abnormal for this time of year. Most of the adults left the study area. We have already seen in the previous section that there was little growth in one-year-old individuals; they grew only 2 cm in 8 months.

One year after the black tide, its effects were confirmed: there were no fish representing the 1978 reproductive season and there had been practically no growth in fish born in 1977. In October 1979 the situation was beginning to return to normal as there was good recruitment of juveniles of 6 to 16 cm and the two-year-olds were growing well. However, there are still very few of the latter.

The absence of juveniles was also noted for sole at the end of 1978 and the beginning of 1979. This could be due either to lack of reproduction or to the destruction of larvae. Growth in adults, on the other hand, appeared to be normal (Figure 7.5.15). There was recruitment of young in October 1979 but it appeared to be less abundant than for plaice. The stock of adults was relatively abundant.

In April 1978 the only representatives of dab in this area were young individuals of 9 to 15 cm. They then practically disappeared from the study area for several months. Only two individuals of the species were caught in December 1978 (Figure 7.5.16). In May 1979 the population, which was larger than in April 1978, was made up of two size groups: 7 to 13 cm and 17 to 22 cm. In October, population growth was good but there were few adults present.

In short, Desaunay therefore considers that the unfavourable consequences of the pollution for these three species is very clearly demonstrated. In his opinion stocks will probably not be back in balance before the end of 1981.

Miossec (1981a, b) provided detailed data on the composition of the stock of plaice in the abers.

A frequency analysis of the size of fish of this species, taken in Aber Benoît, demonstrated the absence during the winter 1978–1979 of the juveniles which should have been spawned in 1978.

Only a few young of this age group appeared in spring 1979. According to the author this absence of young was probably due to the destruction in 1978 of most of the larvae, as spawning had taken place before the black tide. In 1979, two-year-old plaice were well represented but in 1980 a drop in the number of plaice of reproductive age was noted.

Observations for Aber Wrach are similar, but here there were fewer fish born in

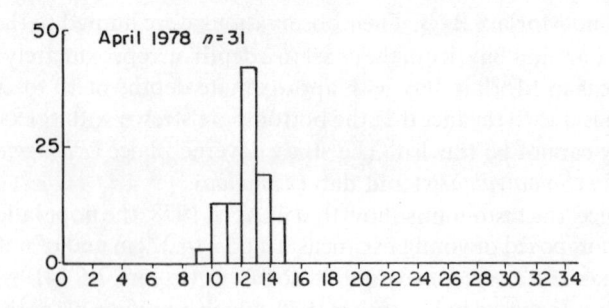

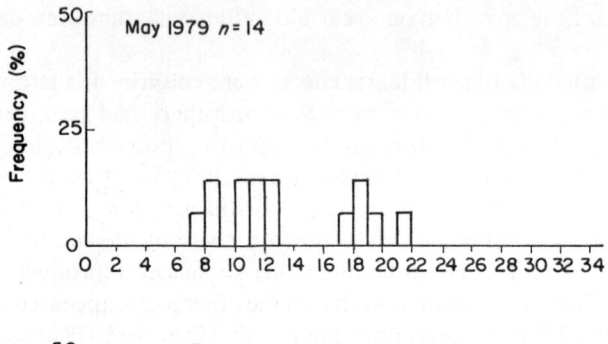

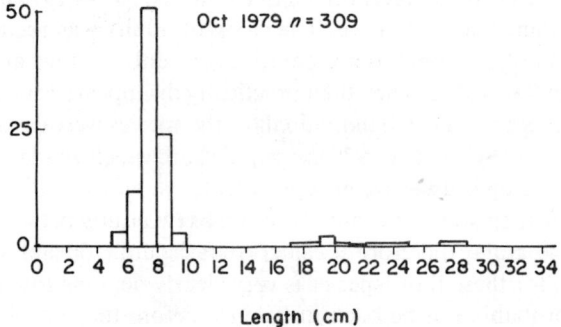

Figure 7.5.16 Demographic structure of flatfish populations in Lannion and Morlaix bays. (From Desaunay, 1981. Reproduced by permission of Centre National pour l'Exploitation des Océans)

1977 than in Aber Benoît. This age group seems to have suffered damage comparable to that which affected Plaice born in 1978.

As other species of flatfish are usually absent from the abers for part of the year, no specific conclusions could be drawn as to the level of their stocks.

7.5.5.2 The *Gino* Accident

Using the oceanographic vessel *Pélagia*, l'Institut des Pêches Maritimes made a third series of observations in April 1981 (Michel and Abarnou, 1981). This research project and the laboratory analysis provided backup for some of the observations made in March 1980, during the previous project.

First of all, contamination of the sediments was low everywhere but clearly present within a radius of 5 nautical miles around the wreck. Slight traces of pollution were still present up to about 8 miles. In the area under examination it was noted that, apart from the actual slick itself, there were well-oxygenated biogenic sediments. This led the authors to the conclusion that biodegradation was progressing actively and was affecting all the hydrocarbons including the aromatics.

An analysis was made of the liver—the organ most likely to accumulate hydrocarbons—of 45 fish taken at a distance of less than 20 miles from the wreck. In spite of this proximity, the average figures obtained were some 2.17 mg kg^{-1} of dry weight with a standard deviation of about 1.3 mg kg^{-1}. These figures may be regarded as indicating an absence of contamination as the liver contains natural compounds which cause interference during analysis.

The first explanation given by Michel and Abarnou (1981) for the fact that the piscine fauna had not suffered the consequences of the pollution is that in this open sea area the fauna comprised adult individuals capable of covering large distances. This would not have been the case if the accident had happened close to the shore where the young are concentrated. The authors think that it may also be due to the fact that, unlike molluscs, fish have a detoxification mechanism (see above, Haensly *et al.*, 1979). They also note that no fish caught in the study area showed anomalies comparable to those described in connection with the *Amoco Cadiz* accident, i.e. skin necrosis, fin erosion, etc.

The results of an analysis of five edible crabs (*Cancer pagurus*) showed that the eggs of these crabs did not contain any significant quantities of aromatic hydrocarbons. Examination under the microscope confirmed that they showed no anomalies and appeared to be viable. The muscle tissues were no longer contaminated. The hepatopancreas of the females contained very small quantities of hydrocarbons (0.9 to 2.9 mg kg^{-1}) while that of the males contained larger quantities (5.6 and 21.3 mg kg^{-1} in one individual).

Where scallops are concerned, 76 stations up to some 25 nautical miles from the wreck were examined in 1981. It proved possible to catch this type of mollusc at 50 of the stations. It was noted first of all, that hydrocarbon levels in the muscles were much lower than in the previous year. The content was low between 8 to 10 and 12 miles from the wreck (13.7 to 3.40 mg kg^{-1}), while beyond that distance the content was normal. This demonstrates that the contaminated area is definitely shrinking.

Figure 7.5.17 Detail of a scallop shell collected in the vicinity of the wreckage of the *Gino* in April 1981. Note the presence of oil residues under the mother-of-pearl

The total absence of living scallops close to the *Gino* indicates that mortality was general. It was in this area, within a radius of some 5 miles, that individuals in a poor physiological state had been observed the previous year (Figure 7.5.17).

In addition, the use of mass spectrometers to identify the hydrocarbons enabled Michel and Abarnou (1981) to establish a comparison between the hydrocarbons of the *Gino* and those of the *Amoco Cadiz*. The authors reported two findings:

1. A lower proportion of dibenzothiophene and its alkylated derivates.

2. A higher proportion of aromatic compounds with 4 and 5 rings, as well as their corresponding alkylated derivates.

Compounds with 5 rings were present only in samples taken at a distance of less than 8 miles from the wreck.

To summarize, whereas the pollution caused by the *Gino* seems to have had no harmful consequences for fish, and its impact on crustaceans was small, this cannot be said for scallops which, it should be remembered, are fortunately too scarce in the region affected by the oil spill to be exploited. Nevertheless, their progress should continue to be monitored.

7.5.6 CONCLUSIONS

This study (which was voluntarily limited to exploitable living resources) does not pretend to be an exhaustive analysis of the consequences, for those resources, of oil spills occurring along the coast of Brittany during the last few years or even of the consequences of the *Amoco Cadiz* spill. The study is an attempt to bring together some of the observations made by different kinds of specialists or bodies, particularly l'Institut Scientifique et Technique des Pêches Maritimes. It also aims to point out certain lessons that may be learned from these accidents which were among the biggest to have occurred anywhere in the world.

The first point to be noted is that the pollution caused by the two accidents chiefly investigated did not have an equal effect on all the organisms involved. The physical disturbance to the algae was not as far reaching as might have been feared after the grounding of the *Amoco Cadiz*, even though they were throughly doused in oil. Reproduction was almost normal. Growth and sometimes even population density increased up to a certain level of contamination. Beyond that threshold, certain irregularities in growth and population homogeneity were observed.

It is more difficult to measure the effect on the big crustaceans. Following the *Amoco Cadiz* accident, eggs and larvae appeared to have suffered more than adults. After the *Bohlen* incident, an appreciable mortality rate was noted, mainly caused by the density and viscosity of the oil.

In the event of pollution, fish react either by fleeing if they are not trapped in their habitat or by making use of their mechanisms for eliminating toxic matter. Here too the question of a threshold plays a part. Beyond a certain level the pollution causes mortality, as was seen after the *Amoco Cadiz* accident in the case of species living very close to the shore. Whatever the circumstances, the young, the eggs and the larvae are more sensitive than the adults.

Without question it was the molluscs, particularly the filtering molluscs, which were most seriously affected, due to the accumulation of toxic substances and their persistence. Among the economically valuable species, striking examples of this were provided by oysters after the *Amoco Cadiz* and scallops after the *Gino* spills.

The second important point concerns medium- and long-term damage. This has turned out to be more serious than the immediate effects. Examples are the absence of young in the case of flatfish in the abers and in Lannion and Morlaix bays and the medium-term interruption of reproduction and growth in plaice. Neither should we forget the appearance of frequent anomalies of the skin, the fins and the branchiae, several months after the *Amoco Cadiz* spill.

The same applies to the disturbing persistence of aromatic hydrocarbons in oysters and scallops, and the skin blemishes noted in *Carcinus maenas* at Roscoff along with the later complication of lethal bacterial septicaemia which accompanied these blemishes.

The site of the accident may be a determining factor, as shown by a comparison between the apparently limited effects of the Ixtoc well blowout and the damage caused by the *Amoco Cadiz* spill. Similarly, it may be said that the consequences of the wreck of the *Gino* would certainly have been more serious if it had happened near the coast.

The great importance of the type of oil involved should also be re-emphasized. In the case of the *Amoco Cadiz*, the evaporation of the most volatile fraction definitely helped to limit the damage, at least in the short term. But in the *Gino* accident the stability of the product and its high aromatics content were the cause of the heavy contamination of scallops in an area considerably bigger than that covered by the slick itself, although the *Gino* spilled six times less oil than the *Amoco Cadiz*.

The use of anti-oil agents was a significant factor in limiting the pollution caused by the latter vessel. For instance, it helped avoid contamination of the shores of the Cotentin, the Channel Islands, the Brest roads and the major part of the southwest coast of Brittany. Nevertheless, the ban on the use of the anionic and the more toxic non-ionic surfactants, and the limitation of their use to areas sufficiently far off the coast, also helped restrict the consequences of the surface-active effects on living resources.

Certain accidents which occurred in Brittany have confirmed that dispersants are totally ineffective against heavy hydrocarbons and residual oils.

All these observations taken as a whole mean that we are in a position to advocate certain measures, in particular those outlined below.

1. For each of the areas most likely to be affected by an accident, determination of which species and areas are biologically most sensitive and development of ways and means of ensuring maximum protection for them.
2. The monitoring of the persistence of aromatic hydrocarbons with the help of specific methods of analysis and by paying special attention to the filtering molluscs.
3. Research into and perfection of mechanical methods for removing oil without the use of chemicals.

7.5.7 REFERENCES

Audouin, J., Campillo, A. and Léglise, M. (1971). Crustacean colonies along the French Atlantic and Channel coast. *Science et Peche, Bull. Inst. Pêches Marit.*, **205**, 1–19 (in French).

Balouet, G. and Poder, M. (1981). Biological effects of hydrocarbon pollution from the 'Amoco-Cadiz' on oyster-culture in Northern Brittany. *Amoco-Cadiz*, Actes Coll. Int. Centre Océanol. Bret., Brest, 19–22 November 1979. CNEXO Paris: 703–714 (in French).

Berne, S. (1980). Mapping of coastal pollution by hydrocarbons from the 'Tanio' and its impact on the sediments. Rapport Centre Océanol. Bret. CNEXO, 96 pages (in French).

Beslier, A., Birrien, J. L., Cabioch, L., Douville, J. L., Larsonneur, C. and Leborgne, L. (1981). Pollution of sublittoral sediments in Northern Brittany by hydrocarbons from the 'Amoco-Cadiz': distribution and development. *Amoco-Cadiz*, Actes Coll. Int. Centre Océanol. bret., Brest, 19–22 November 1979. CNEXO Paris: 95–106 (in French).

Cabioch, L., Dauvin, J. C. and Gentil, F. (1978). Preliminary observations on pollution of the sea bed and disturbance of sub-littoral communities in Northern Brittany by oil from the 'Amoco-Cadiz'. *Mar. Poll. Bull.*, **9**, 303–307.

Cabioch, L., Dauvin, J. C., Gentil, F., Retière, C. and Rivain, V. (1981). Disturbances in composition and activity of sublittoral benthic populations induced by hydrocarbons from the 'Amoco-Cadiz'. *Amoco-Cadiz*, Actes Coll. Int. Centre Océanol. Bret., Brest, 19–22 November 1979. CNEXO Paris: 513–525 (in French).

Canevari, G. P. (1978). Some observations on the mechanism and chemistry aspects of chemical dispersion. In McCarthy, L. T. Jr., Lindblom, J. P. and Walter, H. F. (Eds.), *Chemical Dispersants for the Control of Oil Spills*, American Soc. for Testing and Materials, Philadelphia, Special techn. publ., **659**, 5–17.

Chassé, C. and Guenole-Bouder, A. (1981). Quantitative comparison of benthic populations of the St. Efflam and St-Michel-en-Grève beaches before and after the shipwreck of the 'Amoco-Cadiz'. *Amoco-Cadiz*, Actes Coll. Int. Centre Océanol. Bret., Brest, 19–22 November 1979. CNEXO Paris: 347–357 (in French).

Conan, G. and Friha, M. (1979). Impact of hydrocarbon pollution from the 'Amoco-Cadiz' on the growth of sole and plaice in the inlets of Northern Brittany. *Cons. Int. Explor. Mer (ICES)*, E: 54, 22 pages (in French).

Conan, G. and Friha, M. (1981). Effects of hydrocarbon pollution from the oiler 'Amoco-Cadiz' on the growth of sole and plaice in the Aber Benoît estuary. *Amoco-Cadiz*, Actes Coll. Int. Centre Océanol. Bret., Brest, 19–22 November 1979. CNEXO Paris: 749–773 (in French).

Cross, F. A., Davis, W. P., Hoss, D. E. and Wolfe, D. A. (1978). Biological observations. In *The 'Amoco-Cadiz' Oil Spill*, a preliminary scientific report, N.O.A.A./E.P.A. **5**, ronéo., 17 pages.

Desaunay, Y. (1981). Development of flatfish stocks in the zone contaminated by the 'Amoco-Cadiz'. *Amoco-Cadiz*, Actes Coll. Int. Centre Océanol. Bret., Brest, 19–22 November 1979. CNEXO Paris: 727–735 (in French).

Desaunay, Y., Bellois, P., Dorel, D. and Lemoine, M. (1979). Effect of pollution ensuing from the foundering of the 'Amoco-Cadiz'. State of salvageable resources; Morlaix and Lannion bays. Rapport Inst. Pêches Marit., ronéo., 32 pages (in French).

Friocourt, M. P., Gourmelun, Y., Berthou, F., Cosson, R. and Marchand, M. (1981). Effects of pollution from the 'Amoco-Cadiz' on oyster-culture in Northern Brittany: chemical study of pollution, purification and adaptation. *Amoco-Cadiz*, Actes Coll. Int. Centre Océanol. Bret., Brest, 19–22 November 1979,. CNEXO Paris: 617–631 (in French).

Gouygou, J. P. and Michel, P. (1981a). Three years after the 'Amoco-Cadiz': identification of persistent hydrocarbons. *Cons. Int. Explor. Mer (ICES)*, E: 53, 5 pages (in French).

Gouygou, J. P. and Michel, P. (1981b). Three years after the 'Amoco-Cadiz': persistent hydrocarbons in oysters and kinetics of purification. Rapport Inst. Pêches Marit., ronéo. pp. 1–20 (in French).

Grizel, H., Michel, P. and Abarnou, A. (1978). Primary data on *in situ* purification kinetics of oysters contaminated by 'Amoco-Cadiz' effluents. Useful experimental results for oyster-growers. In The "Amoco-Cadiz' accident, initial studies of the observations made by the I.S.T.P.M. *Science et Pêche, Bull. Inst. Pêches Marit.*, **283-284**, 7-15 (in French).

Grizel, H., Michel, P. Abarnou, A. and Guégan, B. (1981). Repercussions of the 'Amoco-Cadiz' oil spill on the oyster-growing industry. *Amoco-Cadiz*, Actes Coll. Int. Centre Océanol. Bret., Brest, 19-22 November 1979. CNEXO Paris: 715-726 (in French).

Haensly, W. E., Neff, J. M. and Sharp, J. R. (1979). Long term effects of the 'Amoco-Cadiz' crude oil spill on *Pleuronectes platessa* of 'l'Aber Benoît', Brittany, France: preliminary histopathologic observations. Document Mimeo., College of Veterinary Medicine, College of Science, Texas, 27 pages.

Hermann, M., Durand, J. P., Charpentier, J. M., Chaude, O., Hofnung, M., Petroff, N., Vandecasteele, J. P. and Weill, N. (1979). Correlations of mutagenic activity with polynuclear aromatic hydrocarbons content of various mineral oils. *Fourth International Symposium on Polynuclear Aromatic Hydrocarbons*, Columbus, Ohio.

ISTPM (1979). Impact of the 'Amoco-Cadiz' accident on exploitable marine resources, work carried out by the ISTPM in 1979; crustaceans, fishes, algae, molluscs. Rapport Inst. Pêches Marit., ronéo., 15 pages (in French).

ISTPM (1980). Report on the 'Gino'. by Desaunay, Y. Michel, P. and Fontaine, B., Rapport Inst. Pêches Marit., ronéo (in French).

Kaas, R. (1980). Effects of the stranding of the 'Amoco-Cadiz' on exploitable algae populations. *Rev. Trav. Inst. Pêches Marit.*, **44**, 157-194 (in French).

Kaas, R. (1981). Development of exploitable algae populations since the wreck of the 'Amoco-Cadiz'. *Amoco-Cadiz*, Actes Coll. Int. Centre Océanol. Bret., Brest, 19-22 November 1979. CNEXO Paris: 687-702 (in French).

Kopp, J. (1975). *Contribution to the study of the alga Chondrus crispus Stack. Biochemistry of the carrageenins extracted from it.* Ph.D. thesis, Université de Bretangne Occidentale, 93 pages (in French).

Laseter, J. L., Lawler, G. C., Overton, E. B., Patel, J. R., Holmes, J. P., Shields, M. I. and Maberry, M. (1981). Characterization of aliphatic and aromatic hydrocarbons in flat and Japanese type oysters and adjacent sediments collected from l'Aber Wrach following the 'Amoco-Cadiz' oil spill. *Amoco-Cadiz*, Actes Coll. Int. Centre Océanol. Bret., Brest, 19-22 November 1979. CNEXO Paris: 633-644.

Léglise, M. (1978). 'Amoco-Cadiz' black tide data. Rapport Inst. Pêches Marit., Ronéo., 1 page (in French).

Léglise, M. and Raguenès, G. (1981). Study of the effect of the 'Amoco-Cadiz' shipwreck on commercially expoitable crustaceans in the polluted zone. *Amoco-Cadiz*, Actes Coll. Int. Centre Océanol. Bret., Brest, 19-22 November 1979. CNEXO Paris: 775-787 (in French).

Maurin, C. (1978). Observations on the effect of the grounding of the 'Amoco-Cadiz' on living and expoitable marine resources. *Cons. Int. Explor. Mer (ICES)*, E: 50, 10 pages (in French).

Maurin, C. (1979). Accidental oil spills and their effect on living exploitable marine resources with reference to the 'Amoco-Cadiz' accident. *Cons. Int. Explor. Mer (ICES)*, conférence d'ouverture, 67 ème réunion statutaire, Varsovie, 12 pages (in French).

Maurin, C. (1981). Effects of the 'Amoco-Cadiz' accident on living exploitable marine resources; synthesis. *Amoco-Cadiz*, Actes Coll. Int. Centre Océanol. Bret., Brest, 19-22 November 1979. CNEXO Paris: 667-686.

Michel, P. (1978). 'Amoco-Cadiz' observations on fish contamination after 3 months. In The 'Amoco-Cadiz' accident, initial studies of the observations made by the I.S.T.P.M. *Science et Pêche, Bull. Inst. Pêches Marit.*, **283-284**, 3-6 (in French).

Michel, P. (1979). Possibilities and limitations in the use of dispersants in the struggle against oil pollution. Union des Villes du Littoral Ouest-Européen, Colloque de Brest, 28-30 March 1979, 8 pages (in French).

Michel, P. (1981a). Global measurment of aromatic hydrocarbons in marine products by H.P.L.C. *Cons. Int. Explor. Mer (ICES)*, E: 54 (in French).
Michel, P. (1981b). Liquid chromatographic estimation of aromatic hydrocarbons in living matter and in sediments. Rapport Inst. Pêches Marit., Ronéo., annexe 1, 5 pages (in French).
Michel, P. and Abarnou, A. (1981). Impact study of the 'Gino'; 'Gino II' campaign results. Rapport Inst. Pêches Marit., ronéo., pp. 1–25 (in French).
Miossec, L. (1981a). Effects of the 'Amoco-Cadiz' pollution on the morphology and reproduction of plaice (*Pleuronectes platessa*) in the Wrach and Benoît estuaries; primary results. *Amoco-Cadiz*, Actes Coll. Int. Centre Océanol. Bret., Brest, 19–22 November 1979. CNEXO Paris: 737–747 (in French).
Miossec, L. (1981b). *Impact of Oil Pollution Due To the 'Amoco-Cadiz' on the Biology of Flat fishes of the Aber Benoît and Aber Wrach*. Ph.D thesis, Université de Bretagne Occidentale, vol. 129, 143 pages (in French).
Nelson-Smith, M. (1973). *Oil Pollution and Marine Ecology*, Plenum Press, New York, 260 pages.
Owens, E. H. (1978). Mechanical dispersal of oil stranded in the littoral zone. *J. Fish. Res. Board Can.*, **35**, 563–572.
Ozouville, L. d', Berné, S., Gundlach, E. R., and Hayes, M. O. (1981). Progression of pollution of the coast of Brittany from hydrocarbons from the 'Amoco-Cadiz' between March 1978 and November 1979. *Amoco-Cadiz*, Actes Coll. Int. Centre Océanol. Bret., Brest, 19–22 November 1979. CNEXO Paris: 55–78 in French).
Pérez, R. (1968). *Laminaria digitata* Lamouroux growth studied during 3 consecutive years. 6 ème Symp. Int. Santiago de Compostela (in French).
Pérez, R. (1978a). Impact of the wreck of the 'Amoco-Cadiz' on exploitable algal populations; initial appraisal. Rapport Inst. Pêches Marit., ronéo., 8 pages (in French).
Pérez, R. (1978b). Initial results on the impact of the stranding of the 'Amoco-Cadiz' on exploitable algal populations. In the 'Amoco-Cadiz' accident, initial studies of the observations made by the I.S.T.P.M. *Science et Pêche, Bull. INST. Pêches Marit.*, **283–284**, 17–27 (in French).
Pérez, R. (1979). State of exploitable algal populations in January 1979. Rapport Inst. Pêches Marit., ronéo., 3 pages (in French).
Ramirez-Pérez, E., Amargier, A. and Vago, C. (1980). Cuticular ulceration in marine decapod crustaceans. *C.R. Acad. Sci. Paris*, **291**, 437–439 (in French).
Steel, R. L. (1977). Effects of certain petroleum products on reproduction and growth of zygotes and juvenile stages of the alga *Fuscus edentatus* de la Pilay. In Wolfe, D. A. (Ed.), *Fate and Effects of Petroleum Hydrocarbons in Marine Organisms and Ecosystems*, Pergamon Press, New York, pp. 138–142.
Steel, R. L. (1978). In Progress report on 'Amoco-Cadiz' oil spill. Environmental Protection Agency.
Topinka, J. A. and Tucker, L. R. (1981). Long term oil contamination of macroalgae following the 'Amoco-Cadiz' oil spill. *Amoco-Cadiz*, Actes Coll. Int. Centre Océanol. Bret., Brest, 19–22 November 1979. CNEXO Paris: 393–403.

Effects of Pollutants at the Ecosystem Level
Edited by P. J. Sheehan, D. R. Miller, G. C. Butler and Ph. Bourdeau
© 1984 SCOPE. Published by John Wiley & Sons Ltd

CASE 7.6
Impact of Airborne Metal Contamination on a Deciduous Woodland System

MALCOLM HUTTON

Monitoring and Assessment
Research Centre Chelsea College
459A Fulham Road London SW10 0QX

7.6.1 Introduction... 365
7.6.2 Ecosystem Description... 366
7.6.3 Pollutant Input.. 366
7.6.4 Pollutant Behaviour... 367
7.6.5 Effects of Metals on the Contaminated Woodland System............ 369
    7.6.5.1 Effects on Individuals and Communities................... 369
    7.6.5.2 Effects on Ecosystem Function in Contaminated Woodlands........ 371
7.6.6 Conclusions... 373
7.6.7 References... 374

7.6.1 INTRODUCTION

Non-ferrous metal smelting has been carried out in the Avonmouth area, near Bristol in southwest England, since 1929. In 1967, an Imperial Smelting Furnace (ISF) lead–zinc smelter was commissioned at this site. Currently, this plant has an annual production capacity of about 100 000 tonnes zinc and 400 t cadmium. In 1979, the smelter produced 77 000 t zinc and 414 t cadmium. The lead bullion produced at Avonmouth is refined overseas.

    The smelter is equipped with modern particulate control devices but significant stack emissions of metals still occur. Estimates of annual emissions are 3–4 t cadmium, 40–60 t zinc and 20–30 t lead. These values will vary with the composition and quantity of ore processed in any one year.

    It is clear from the above that the smelter is a major source of heavy metals in the area. The effect of these emissions on adjacent terrestrial systems has been the subject of a continuing research effort at the University of Bristol. This case study is concerned with one aspect of the work carried out at the University, namely, the effects of these metal emissions on nearby woodland systems. In some instances, information obtained from other studies of metal-contaminated woodlands is also referred to.

7.6.2 ECOSYSTEM DESCRIPTION

Several deciduous woodlands located at various distances from the smelter served as the study sites. Most of the sites were mixed oak *Quercus robur* and hazel *Corylus avellana* woods. Stands of beech *Fagus sylvatica*, ash *Fraxinus excelsior* and sycamore *Acer pseudoplatanus* were also commonly found. The attendant shrubs and ground flora included *Rubus, Ligustrum, Euonymus, Crataegus, Mercurialis, Milium, Holcus* and *Festuca*.

Most work was carried out in Hallen Wood, located about 3 km northeast of the smelter.

7.6.3 POLLUTANT INPUT

Metal deposition in the Avonmouth area has been examined in detail by Little (1974) and Little and Martin (1974). The pattern of deposition was found to be strongly influenced by the prevailing wind, which is generally from the southwest. Estimates of current deposition rates for lead, zinc and cadmium at Hallen Wood are given in Table 7.6.1. The corresponding values for uncontaminated areas in the UK are also shown. Hallen Wood is located downwind of the smelter. Not surprisingly, deposition rates of the metals were markedly elevated. Cadmium in particular shows the most pronounced increase above background, being thirty-fivefold higher than the control site.

Present-day inputs of metals are considered to be lower than those experienced in the past. This is because of the more stringent control activities adopted at present, together with the replacement of the outmoded vertical retort plant with an ISF smelter in 1971.

Table 7.6.1 Deposition rates ($g\ ha^{-1}\ yr^{-1}$) of selected heavy metals at Hallen Wood, Avonmouth and at background areas in the UK

Soil	Lead	Cadmium	Zinc
Hallen Wood[1]	2900	92	6000
Background areas	230[2]	2.6[3]	510[2]

[1] Source: Martin and Coughtrey (1981).
[2] Plynlimon, rural Wales. Source: Cawse (1980).
[3] Rural Norfolk. Source: Horler and Barber (1979).

7.6.4 POLLUTANT BEHAVIOUR

Concentrations of lead, cadmium and zinc have been measured in the major vegetation types, leaf litter and soils from the contaminated woodlands around Avonmouth. Representative values from Hallen Wood are given in Table 7.6.2. The concentrations of all three metals are found to increase in the order vegetation < soil < litter. As found for the deposition rates, cadmium levels were consistently lower than lead and zinc in all components.

The elevated concentrations of lead, cadmium and zinc present in trees and shrubs appear to be derived mainly from aerial deposition rather than soil uptake. In most cases, differences in the metal content of foliage can be related to position in the canopy and leaf surface morphology. Metal deposition rates have been shown to increase with height in woodland systems while leaf surface characteristics have a strong influence on metal retention (Little, 1974).

The elevated metal status of litter results partly from contamination of foliage by aerial deposition and partly from subsequent inputs to litter from throughfall and washoff. The accumulation of metals in surface soil is probably a consequence of rainfall leaching metals from the litter layer.

Martin and Coughtrey (1981) observed that most of the lead, cadmium and zinc accumulated in the upper few centimetres of soil in the Avonmouth woodlands. Similar findings have been reported in other systems where soils are subject to airborne contamination. The efficient retention of metals by soils is a result of several complex and interacting processes, involving cation exchange with organic matter and clays and adsorption from soil solution by oxides of manganese and iron.

The standing pools and fluxes of metals in the contaminated woodland system

Table 7.6.2 Metal concentrations ($\mu g\,g^{-1}$ dry weight) in vegetation, litter and soil from Hallen Wood, Avonmouth, UK. (Source: Martin and Coughtery, 1981)

	Lead	Element Cadmium	Zinc
Quercus robur			
Leaves	61	2.7	149
Bark	606	13	715
Corylus avellana			
Leaves	132	3.4	257
Bark	864	1.8	128
Holcus lanatus	1646	5.0	457
Milium	63	3.8	323
Dryopteris	184	8.2	971
Litter	2559	45	2267
Surface soil	1432	23	1010

EFFECTS OF POLLUTANTS AT THE ECOSYSTEM LEVEL

Metal Inputs

Cd	92
Pb	2900
Zn	6000

Trees: Leaves

Cd	4.6
Pb	90
Zn	354

Branch

Cd	319
Pb	14×10^3
Zn	16×10^3

Bark

Cd	212
Pb	10×10^3
Zn	11.7×10^3

Wood

Cd	80
Pb	24
Zn	1100

Shrubs: Leaves

Cd	1.9
Pb	73
Zn	142

Branch

Cd	6
Pb	2000
Zn	558

Bark

Cd	3.2
Pb	1500
Zn	230

Wood

Cd	9.2
Pb	4400
Zn	655

Ground flora

Cd	16
Pb	181
Zn	913

Total litterfall

Cd	23
Pb	1100
Zn	1500

Ground flora uptake

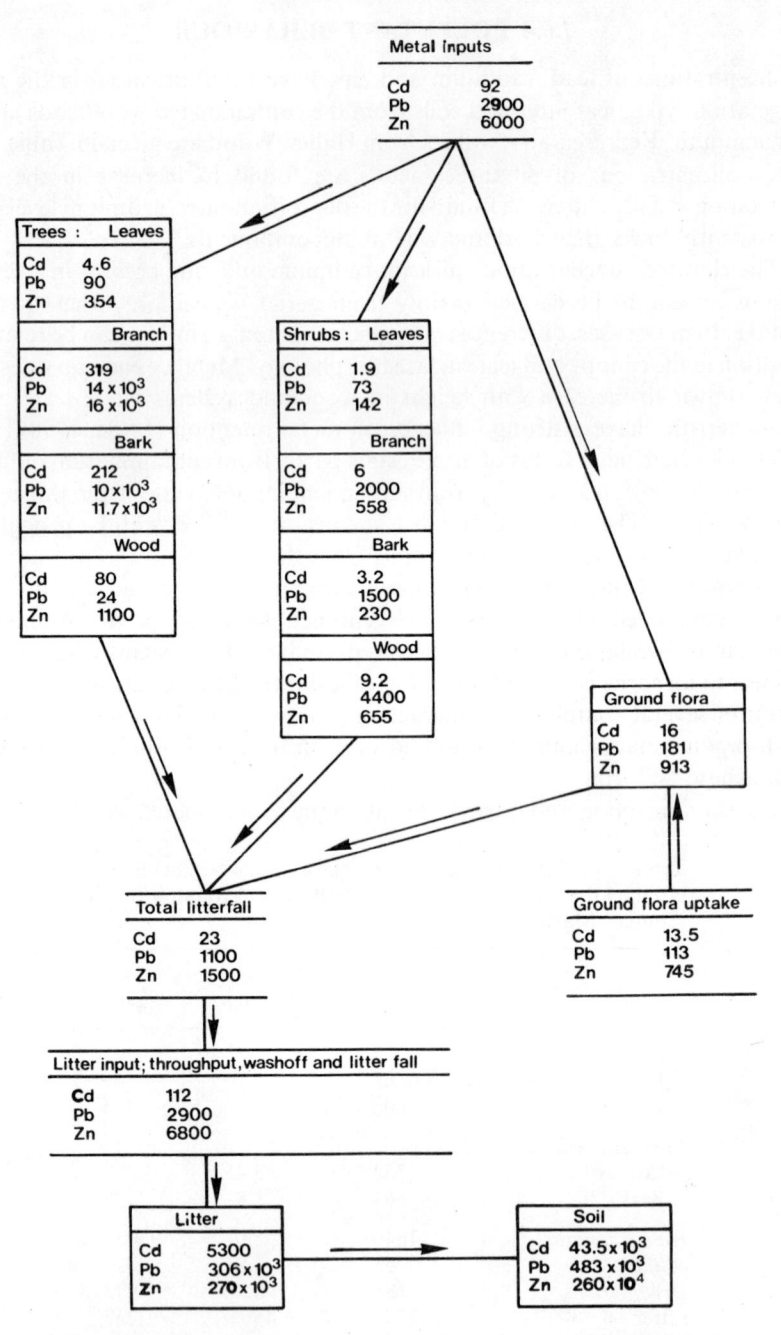

Cd	13.5
Pb	113
Zn	745

Litter input; throughput, washoff and litter fall

Cd	112
Pb	2900
Zn	6800

Litter

Cd	5300
Pb	306×10^3
Zn	270×10^3

Soil

Cd	43.5×10^3
Pb	483×10^3
Zn	260×10^4

Figure 7.6.1 Summary of metal inputs, fluxes and standing pools in a deciduous woodland close to the Avonmouth zinc smelter. Standing pools (□) are in g ha^{-1}. Inputs and fluxes are in g ha^{-1} y^{-1}. (Source: Martin and Coughtrey, 1981)

have also been estimated. Figure 7.6.1 shows that the majority of the total metal load is associated with soil, followed by litter. Metals in vegetation account for only a small fraction of the total burden in the system. Figure 7.6.1 also indicates that annual inputs are much smaller than the quantities of metals associated with standing pools, particularly in the case of soil and leaf litter.

Overall, these findings reveal the importance of the soil as the major long-term sink for metal inputs. One consequence of the retention and accumulation of metals in surface soils is the possibility that recycling of metals will occur via root uptake. Some evidence for such an effect was obtained at Avonmouth, as shallow rooting ground flora contained significantly higher concentrations of cadmium and zinc that the foliage of deep-rooted tree species (Table 7.6.2). Indeed, this flux appears to be relatively important at Hallen Wood, for, as Figure 7.6.1 indicates, root uptake by ground flora corresponds to about 10 per cent of the total annual inputs of zinc and cadmium.

7.6.5 EFFECTS OF METALS ON THE CONTAMINATED WOODLAND SYSTEM

7.6.5.1 Effects on Individuals and Communities

Metal tolerance

It is well known that plant species growing on soils heavily contaminated with zinc, lead or copper are tolerant to the effects of these metals. Much less attention has been paid to the possible development of cadmium tolerance in plants. The relatively high soil cadmium levels at woods in the Avonmouth area prompted such an investigation by Martin and his co-workers (Coughtrey and Martin, 1977; Martin *et al.*, 1980). The two species examined, the grass *Holcus lanatus* and dog's mercury *Mercurialis perennis*, both displayed a distinct tolerance to cadmium. It is likely that other species from these woods, particularly the ground flora components, will also show tolerance to cadmium. It is also possible that those species without the ability to adapt to contaminated conditions will not be able to survive in these woods. Hitherto, however, there have been no published studies of the species composition of these woods to support this suggestion.

Studies of bacteria and fungi from soils in the Avonmouth area also indicate that a greater proportion of the soil microflora are tolerant to heavy metals compared with control sites (Martin *et al.*, 1980). Similarly, leaf surface microflora from oak *Quercus* sp. 0.5 km from the smelter contained a greater proportion of metal tolerant fungi and also reduced numbers of pigmented yeasts and bacteria (Bewley, 1980).

Metal accumulation in soil and litter invertebrates

Invertebrates from contaminated woods close to Avonmouth contain elevated concentrations of lead, cadmium and zinc; representative data are shown in

Table 7.6.3 Metal concentrations ($\mu g g^{-1}$ dry weight) in soil and litter-dwelling invertebrates from a contaminated woodland close to the Avonmouth smelter. (Source: Martin and Coughtery, 1981)

Species		Lead	Element Cadmium	Zinc
Earthworms	*Lumbricus spp.*	32.8	41.0	1502
Slugs	*Arion hortensis*	63.6	57.0	1430
	Arion fasciatus	118	30.0	1250
Snails	*Helix aspersa*	39.0	52.5	404
	Clausilia bidentata	245	53.0	613
Woodlice	*Oniscus asellus*	568	232	462

Table 7.6.3. When Martin *et al.* (1980) expressed values as a ratio of the metal content of the estimated diet (Table 7.6.4), it became apparent that cadmium is accumulated more efficiently than either lead or zinc. The greater accumulation of cadmium compared with lead or zinc has been observed in another contaminated terrestrial food chain (Roberts and Johnson, 1978). This characteristic may be associated with the long retention times of cadmium in animals, related to its binding to the low molecular weight protein, metallothionein.

The toxic significance of the metal burdens in woodland invertebrates from Avonmouth is unclear at present. In this respect, it is of interest to note that earthworms from a pasture site, also close to the Avonmouth smelter, contained metal levels similar to those given in Table 7.6.3 for *Lumbricus*, but showed no reduction in population density (Wright and Stringer, 1980).

There are at present no published studies of species diversity or population densities of woodland invertebrates from the Avonmouth area. Relevant data

Table 7.6.4 Ratios of metal concentrations in invertebrates to concentrations in presumed food items. (Source: Martin *et al.*, 1980)

Species		Lead	Element Cadmium	Zinc
Earthworms	Animal: soil	0.3	7.6	0.5
	Animal: litter	0.4	5.0	0.5
Slugs (*A. reticulatus*)	Animal: vegetation	0.4	5.3	1.9
Snails (*H. aspersa*)	Animal: vegetation	0.4	41.6	2.9
	Animal: litter	0.1	4.5	0.1
Woodlice (*O. asellus*)	Animal: litter	0.5	5.3	0.4
	Animal: litter	0.8	6.7	1.3

are, however, available for litter arthropods from three woodland sites at varying distances from a zinc smelter in the USA (Strojan, 1978a). Densities of all taxonomic groups decreased with proximity to the smelter; mean densities being 35 000 m^{-2} at the 1 km site, 72 000 m^{-2} at the 6 km site and 160 000 m^{-2} at the 40 km control location. The most pronounced decrease occurred in mites (Acarina). The density of this group at the 1 km site was less than a quarter of the control value. Similar reductions in numbers of Acarina have also been observed in the contaminated litter of a lead-zinc mine (Williams *et al.*, 1977). Additionally, when Strojan (1978a) divided Acarina into Oribatid and non-Oribatid mites, it was found that the decrease in numbers was most striking in the former group. These findings suggest that mites, and Oribatid mites in particular, may be rather sensitive to metal contamination.

Whether such changes in population density have taken place in the invertebrate communities from woods close to Avonmouth remains to be established; at present there is no published information on the subject. In this context, it should be noted that the extent of metal contamination in litter around the smelter studied by Strojan (1978a) was much greater than that found at Avonmouth.

7.6.5.2 Effects on Ecosystem Function in Contaminated Woodlands

Litter decompostion

Woods close to the Avonmouth smelter have been found to contain elevated standing crops of litter (Coughtrey *et al.*, 1979). Table 7.6.5 shows the litter standing crops and metal levels from several woods around Avonmouth. It is apparent that a marked accumulation of litter occurs at these locations; the value at Hallen Wood, for example, is nearly ten times higher than that from the uncontaminated Wetmoor Wood. It is likely that the stand composition of

Table 7.6.5 Litter mass and metal concentrations (μg g^{-1} dry weight) at woodlands various distances from the Avonmouth smelter. (Source: Coughtrey *et al.*, 1979)

Wood	Litter mass (g m^{-2})	Distance from smelter (km)	Cadmium	Lead	Zinc
Moorgrove	8 345 ± 2 131	2.5	23	1052	764
Blaise	13 160 ± 1 411	2.9	32	721	1844
Hallen	8 343 ± 598	2.9	62	2179	2469
Haw	7 910 ± 640	3.1	98	1545	2814
Leigh	1 784 ± 157	6.8	7.2	191	169
Wetmoor	913 ± 100	23.0	1.5	44	80
Midger	3 104 ± 268	28.5	5.7	103	202

woodlands has some influence on the standing crop of litter, but this is not considered to be important at the Avonmouth sites as most were of similar composition, being mixed oak–hazel woodlands. Excessive litter accumulation has been recorded previously in other woodland sites close to non-ferrous metal smelters (Tyler, 1972; Strojan, 1978b). Compared with these earlier studies, however, metal concentrations at Avonmouth, particularly cadmium and zinc, are relatively low.

Litter accumulation at such sites has generally been assumed to result from a reduction in the rate of litter decomposition. Actual evidence for this suggestion was obtained by Martin and Coughtrey (1981) at Avonmouth by leaf litter bag experiments. Percentage decomposition of litter bags placed in contaminated sites was much lower than at sites distant from the smelter. Additionally, the possibility of a larger leaf fall was ruled out, as annual litter fall at the contaminated Hallen Wood was similar to values from other deciduous woods in uncontaminated areas (Coughtrey et al., 1979).

In addition to an accumulation of litter in woods around Avonmouth, there are also changes in the particle size distribution of litter from these sites (Coughtrey et al., 1979). Thus, when compared with control sites, a much larger proportion of the litter mass consisted of particles in the small size range (0.5–2 mm). This finding led Coughtrey et al. to suggest that the influence of metals on litter decomposition does not take place at the initial stages of litter fragmentation. Rather it was proposed that the effect occurs at a later stage of breakdown, involving particles in the size range of 0.5–2 mm.

It has not been fully established which metals affect the accumulation of litter at contaminated woodland sites. However, cadmium is strongly implicated at Avonmouth, as Coughtrey et al. (1979) observed that the weight of accumulated litter in the lower size range was most closely correlated to the cadmium content of this fraction.

At present, it is unclear whether reduced litter decomposition is caused by adverse effects on either litter-dwelling invertebrates or on microorganisms or both. Strojan's studies (Strojan, 1978a, b) indicate that depletion of arthropod numbers, particularly mites, may be responsible for litter accumulation at an American smelter site. Mites are known to play an important role in the mechanical breakdown of litter. Indeed, it has been estimated that the mass of material passing through the adult Oribatid population in one year represents about half the total annual leaf fall (Butcher et al., 1971). Additionally, mites and other decomposer invertebrates are thought to make litter more accessible to microorganisms as a consequence of microbial inoculation by faeces (Butcher et al., 1971). Nevertheless, Jordan and Lechevalier (1975) are of the opinion that reduced microbial activity was the cause of litter accumulation at the same smelter site studied by Strojan. This was based on the finding that the numbers of bacteria, fungi and actinomycetes were also reduced close to the smelter.

In contrast to the American smelter site, there is no evidence of reduced soil microbial populations at contaminated woodland sites around Avonmouth (Martin and Coughtrey, 1981). Nevertheless, the possibility exists that heavy metals have an adverse effect on the gut microflora of decomposer invertebrates. These microorganisms have an important function in the breakdown of ingested litter and Coughtrey *et al.* (1980) have suggested that this may enhance metal release in both gut and faeces and thus inhibit subsequent microbial colonization of faeces.

Ecological significance of excessive litter accumulation

One possible consequence of litter accumulation in Avonmouth woodlands is that there may be a reduction in the cycling of organic matter and, therefore, essential minerals. Indeed, the tenfold increase in the litter standing crop at one such site (Table 7.6.5) strongly suggests that a significant fraction of the nutrient pool has been removed from circulation. One implication of this suggestion is that a reduction in nutrient availability could conceivably lead to a decline in primary production. At present, no published data on either nutrient cycling or primary production are available for these, or other contaminated woodlands. Until such information is available, it is only possible to speculate on the long-term consequences of litter accumulation at such sites.

7.6.6 CONCLUSIONS

This case study has summarized one of the more comprehensive investigations into the impact of metals on a terrestrial system. The observed increase in litter mass is considered to be the most significant finding of this study, because it is indicative of a fundamental effect on ecosystem function. Additionally, this finding also raises the following points of general significance to terrestrial systems subject to metal contamination.

1. Decomposer organisms in the litter compartment appear to be relatively sensitive to metals. This is probably a consequence of their intimate association with litter particles, in which elevated metal levels are encountered. It is thus proposed that those attempting to detect early effects of metals in terrestrial systems should pay close attention to both the status of decomposer organisms and to the rate of litter decomposition.

2. It is commonly found in polluted systems that contamination is caused by several metals. This situation causes difficulties when attempts are made to link an effect observed in the system to a single metal contaminant. This case study has, however, shown that with careful analysis it is possible to establish such an association. Thus, statistical analysis of metals in size-fractionated litter particles strongly implicated cadmium as the cause of reduced litter decomposition in contaminated woodlands around Avonmouth.

3. An attempt was made in this study to evaluate the ecological significance of

reduced litter decomposition in contaminated woodlands. However, in the absence of relevant information, it was not possible to comment on the influence of this effect on the overall well-being and stability of these systems. It is envisaged that this situation will be a common feature of other studies, where changes at the structural or even functional level are found, but their overall ecological significance is not apparent. This problem will only be resolved when it becomes possible to recognize critical changes in key processes essential to the long-term stability of ecosystems.

In the case of metal-contaminated woodlands, it is considered that the biogeochemical cycling of nutrients is the key process subject to potential disruption by metal inputs. Thus, in order to evaluate the long-term implications of metal contamination, it is essential to quantify the size of nutrient pools in these woodlands and the fluxes between them. This basic data base may be used to construct a dynamic transport model, consisting of compartments representing the important nutrient reservoirs, and linked by the fluxes through the system. The model could then be used to simulate the long-term implications of reduced litter decomposition on nutrient cycling in these woodlands. In this way, it should be possible to predict whether the nutrient status of soils from these woodlands will become sufficiently depleted to reduce plant productivity in the future.

7.6.7 REFERENCES

Bewley, R. J. F. (1980). Effects of heavy metal pollution on oak leaf microorganisms. *Appl. Environ. Microbiol.*, **40**, 1053–9.

Butcher, J. W., Snider, R. and Snider, R. J. (1971). Bioecology of edaphic Collembola and Acarina. *Ann. Rev. Ent.*, **16**, 249–88.

Cawse, P. A. (1980). Deposition of trace elements from the atmosphere in the U.K. In *Inorganic Pollution and Agriculture*, Ministry of Agriculture, Fisheries and Food, Reference Book 326, HMSO, London, pp. 22–46.

Coughtrey, P. J. and Martin, M. H. (1977). Cadmium tolerance of *Holcus lanatus* L. from a site contaminated by aerial fallout. *New Phytol.*, **79**, 273–80.

Coughtrey, P. J., Jones, C. H., Martin, M. H. and Shales, S. W. (1979). Litter accumulation in woodlands contaminated by Pb, Zn, Cd and Cu. *Oecologia (Berl.)*, **39**, 51–60.

Coughtrey, P. J., Martin, M. H., Chard, J. and Shales, S. W. (1980). Micro-organisms and metal retention in the woodlouse *Oniscus asellus*. *Soil Biol. Biochem.*, **12**, 23–27.

Horler, D. N. H. and Barber, J. (1979). Relationships between vegetation and heavy metals in the atmosphere, In *Management and Control of Heavy Metals in the Environment*, CEP consultants Ltd, Edinburgh, pp. 275–278.

Jordan, M. J. and Lechevalier, M. P. (1975). Effect of zinc smelter emissions on forest soil microflora. *Can. J. Microbiol.*, **21**, 1855–65.

Little, P. (1974). *Airborne Zinc, Lead and Cadmium Pollution and Its effects On Soils and Vegetation*. PhD thesis, University of Bristol.

Little, P. and Martin, M. J. (1974). Biological monitoring of heavy metal pollutants. *Environ. Pollut.*, **6**, 1–19.

Martin, M. H., Coughtrey, P. J., Shales, S. W. and Little, P. (1980). Aspects of airborne cadmium contamination of soils and natural vegetation. In *Inorganic Pollution and Agriculture*, Ministry of Agriculture, Fisheries and Food, Reference Book 386, HMSO, London, pp. 55–69.

Martin, M. H. and Coughtrey, P. J. (1981). Impact of metals on ecosystem function and productivity. In Lepp, N. W. (Ed.), *Effects of Heavy Metals on Plants*, vol. II, Applied Science Publishers, Barking, pp. 119–158.

Roberts, R. D. and Johnson, M. S. (1978). Dispersal of heavy metals from abandoned mine workings and their transference through terrestrial food chains. *Environ. Pollut.*, **16**, 293–310.

Strojan, C. L. (1978a). The impact of zinc smelter emissions on forest litter arthropods. *Oikos*, **31**, 41–6.

Strojan, C. L. (1978b). Forest leaf litter decomposition in the vicinity of a zinc smelter. *Oecologia (Berl.)*, **32**, 203–12.

Tyler, G. (1972). Heavy metals pollute nature, may reduce productivity. *Ambio*, **1**, 52–9.

Williams, S. T., McNeilly, T. and Wellington, E. M. H. (1977). The decomposition of vegetation growing on metal mine waste. *Soil Biol. Biochem.*, **9**, 271–5.

Wright, M. A. and Stringer, A. (1980). Lead, zinc and cadmium content of earthworms from pasture in the vicinity of an industrial smelting complex. *Environ. Pollut. Ser. A*, **23**, 313–21.

Effects of Pollutants at the Ecosystem Level
Edited by P. J. Sheehan, D. R. Miller, G. C. Butler and Ph. Bourdeau
© 1984 SCOPE. Published by John Wiley & Sons Ltd

CASE 7.7
A Case Study of the Use of Fenitrothion in New Brunswick: The Evolution of an Ordered Approach to Ecological Monitoring

M. F. MITCHELL and J. R. ROBERTS
Environmental Secretariat
Division of Biological Sciences
National Research Council of Canada
Ottawa, Ontario, Canada K1A 0R6

7.7.1	Introduction	377
7.7.2	Development of the Problem	379
	7.7.2.1 Lifecycle of the Budworm	379
	7.7.2.2 Crop Protection	380
	7.7.2.3 Efficacy of the Control Program to Date	380
	7.7.2.4 Perspectives on the Spray Program	384
7.7.3	Principles of an Ordered Approach to Ecological Monitoring	386
7.7.4	Conclusions	398
7.7.5	References	400

7.7.1 INTRODUCTION

Fenitrothion is an organophosphorus insecticide used to control a number of closely related moths that are collectively called spruce budworm. The spruce budworm inhabits and feeds on coniferous forest in areas throughout North America, particularly on pure or mixed stands of balsam fir, alpine fir, and white, red and black spruce. Its preferred habitat is a mixed stand of balsam fir and spruce; such stands comprise approximately 15×10^6 acres in New Brunswick. They are the mainstay of the pulp and paper industry and spruce is also utilized for sawlogs. In addition, forest areas support or interact with a very large number of organisms, all of which fulfil important, interrelated ecological roles as well as being of aesthetic value.

The budworm is a natural part of such a complex ecosystem. It, and the forest it inhabits, are thought to have coexisted for thousands of years. There appear to have been at least seven major budworm outbreaks in Eastern Canada in the past 200 years, each extending over millions of acres (Baskerville, 1976). Such outbreaks have allowed the survival of both the insect and the forest since an outbreak tends to destroy the mature forest and a forest fire can follow, burning the dead trees and leaving ideal conditions for the germination of surviving seeds (Baskerville, 1976). A cycle induced by the fires exists whereby firspruce forests naturally tend to become even-aged and therefore present ideal conditions for future destructive budworm outbreaks. At the same time, the renewal processes lead to a very healthy young forest.

The spruce-budworm problem arose when man wished to compete with the budworm for the coniferous forest harvest. After an outbreak in 1913–1919, it seemed that the budworm might severely limit the industrial use of the forest that was gradually being established in New Brunswick. Spruce trees were replacing the large, old, white pines that had previously been heavily cut for sawlogs (Baskerville, 1976). In the 1950s, the intensity of the spruce harvest increased further as the industry became involved in supplying pulp for paper mills. The technology of the paper mills was designed specifically to utilize a continuous annual harvest of spruce and fir trees, which was incompatible with the sporadic but, to all intents and purposes, complete harvest of the budworm.

A crop-protection programme was introduced in 1952. Since then, vast areas of New Brunswick have been sprayed with assorted biocides each year. This case study will evaluate the usefulness of the ecological monitoring programme that has been conducted for fenitrothion. The principles of experimental strategies which can be applied to maximize the usefulness of monitoring studies will be identified.

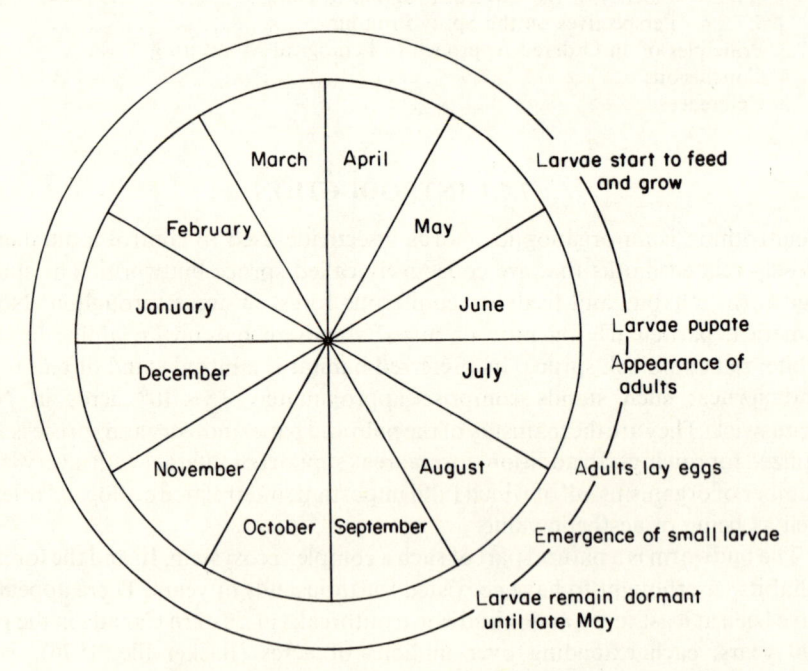

Figure 7.7.1 Lifecycle of spruce budworm in New Brunswick

7.7.2 DEVELOPMENT OF THE PROBLEM

7.7.2.1 Lifecycle of the Budworm

The annual life cycle of the budworm in New Brunswick is represented in Figure 7.7.1. Soon after emerging, in late June or early July, female moths lay their eggs. If the population is small, eggs will be laid locally. However, when the population reaches a critical size and local host trees have become heavily defoliated, females disperse over an area as large as 40–160 km and their subsequent egg-laying spreads the outbreak (Baskerville, 1976). Small larvae emerge in late July and August but do not begin feeding until the following spring. The actual dates of the budworm life history vary from year to year and from place to place, depending on climatic conditions. Under normal conditions, parasites, predators and disease cause a 99 per cent total mortality of immature life stages. Consequently, although each female lays about 200 eggs, most of the time the budworm population naturally remains low.

However, it is obvious that very small reductions in mortality can cause explosive increases in population levels. This is illustrated in Figure 7.7.2. For example, if the mortality rate falls to 90 per cent, a 10 000-fold increase could result

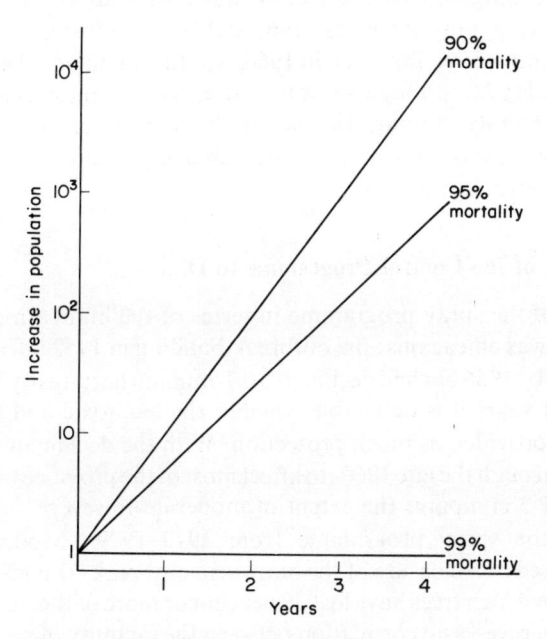

Figure 7.7.2 Effect of decreased generation mortality on budworm population levels over four generations (assuming each female lays 200 eggs and the male:female ratio is 1 : 1)

in 4 years. If this happens to take place in a highly susceptible older forest, an outbreak occurs which results in the severe defoliation of trees. Budworm larvae normally prefer current foliage and, since fir and spruce trees retain a portion of each year's needles for at least eight years, it can take several years to kill a tree. When budworm numbers are exceptionally high and new foliage is limited, the budworm will consume old foliage and this hastens the death of the tree. Mature forests are thus particularly susceptible to destruction. Outbreaks have a natural time limit; factors such as starvation, lack of egg-laying sites or prolonged cool and wet climatic conditions have been postulated to eventually cause budworm populations to collapse back to endemic levels.

7.7.2.2 Crop Protection

New Brunswick has engaged in a spruce–fir crop protection programme involving the aerial spraying of larvae with assorted biocides since 1952. The history of this programme (summarized in Table 7.7.1) is complex because different spray strategies have been used. On average, 17 per cent of the total forest area has been sprayed annually since 1952. Until 1962, mainly DDT was used, but as the environmental problems caused by DDT and the extreme persistence of its metabolites were recognized (Brooks, 1974), new chemicals were tested. DDT and phosphamidon were sprayed between 1963 and 1968, the last year in which DDT was applied. Fenitrothion, first used in 1966, was the dominant chemical sprayed between 1968 and 1978, although phosphamidon, trichlorfon and aminocarb were also utilized. Recently, aminocarb and the biological control agent, *Bacillus thuringiensis* (Bt), have been more widely used but fenitrothion remains the predominant control agent.

7.7.2.3 Efficacy of the Control Programme to Date

The net result of the spray programme in terms of the budworm, is difficult to evaluate. DDT was efficacious; the outbreak building in 1952 affected about 2.5 million hectares by 1956 but had declined to 0.16 million hectares by 1962 (Prebble, 1974). In recent years it is debatable whether the less toxic and less persistent pesticides have provided as much protection. With the decline in DDT use, the outbreak rose through the late 1960s to affect most of the province by 1973 (Figure 7.7.3). Table 7.7.2 compares the extent of moderate-to-severe defoliation with the extent of the spray programme from 1970–1979. 'Moderate-to-severe defoliation' is used as a measure of the budworm outbreak (Davidson, 1981) and indicates areas in which trees have lost 30 per cent or more of their current foliage. This comparison reveals no correlation between the quantity of pesticide applied and the extent of defoliation in subsequent years (see also Desaulniers, 1977).

Because the growth in budworm population is suppressed by the spray programme, the natural pulsing cycles and population collapses may no longer

Table 7.7.1 History of the spray programme in New Brunswick[1]

Year	Area sprayed[2] per year (ha)	Primary[3] pesticides	Application[4] rate (g ha^{-1})	Solvents	Emulsifier(s)
1952–1962	$8 \times 10^4 - 2 \times 10^6$	DDT	290–1165		
1963–1966	$3 \times 10^5 - 8 \times 10^6$	DDT Phosphamidon	290		
1967	4.0×10^3	DDT Phosphamidon Fenitrothion	290		
1968	2.0×10^5	Phosphamidon Fenitrothion DDT	290		
1969	1.3×10^6	Fenitrothion Phosphamidon	145	Atlox 3409F	Arotex 3470
1970	1.7×10^6	Fenitrothion Phosphamidon	145–218	Atlox 3409F Toximol MP 8	Arotex 3470
1971	2.4×10^6	Fenitrothion	145–218	Atlox 3409F Toximol MP 8	Arotex 3470
1972	1.9×10^6	Fenitrothion Phosphamidon	145–290	Atlox 3409F Toximol MP 8	Arotex 3470
1973	1.8×10^6	Fenitrothion Phosphamidon	145–218	Atlox 3409F Toximol MP 8	Arotex 3470

Table 7.7.1 (*Cont.*)

Year	Area sprayed[2] per year (ha)	Primary[3] pesticides	Application[4] rate (g ha^{-1})	Solvents	Emulsifier(s)
1974	2.3×10^6	Fenitrothion Phosphamidon Trichlorfon	145–218 466	Atlox 3409F Toximol MP 8	Arotex 3470
1975	2.7×10^6	Fenitrothion Phosphamidon	180	Atlox 3409F Toximol MP 8	Arotex 3470
1976	4.2×10^6	Fenitrothion Trichlorfon	217–280 580	Atlox 3409F	Arotex 3470
1977	1.6×10^6	Fenitrothion Aminocarb Trichlorfon	210 70 580	Atlox R No. 2 fuel oil Water	Arotex R Nonyl Phenol
1978	1.5×10^6	Fenitrothion Aminocarb	217–290 70	Atlox R Diluent 585	Arotex R Nonyl Phenol
1979	1.6×10^6	Aminocarb Fenitrothion	70–90 218	Diluent 585 Atlox R	Nonyl Phenol Arotex R

[1] Data from: NRCC, 1975; Schneider, 1976; Baskerville, 1976; Varty, 1977, 1980.
[2] Only includes major spray programmes.
[3] Does not include pesticides applied to small areas for experimental reasons.
[4] Number of applications varied (1–5 times).

.... 1950
- - - - 1960-1967
— — 1970
—·—· 1973

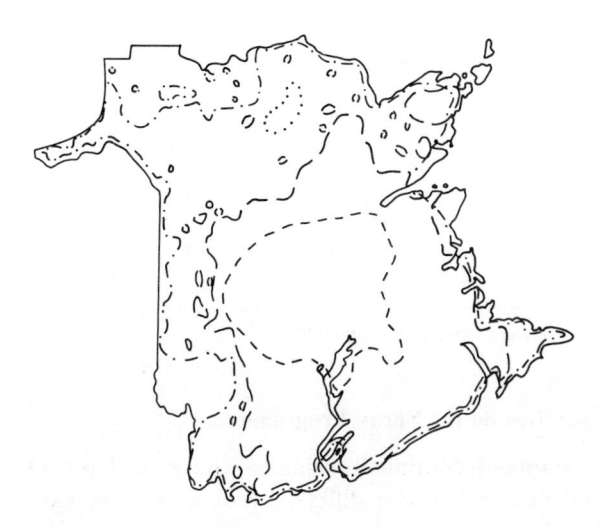

Figure 7.7.3 Margin of areas of generally persistent outbreaks in New Brunswick (adapted from Prebble, 1974)

take place. The spray programme helps to maintain an abundant food source and ample egg-laying sites. In fact, populations now tend to migrate from defoliated areas to areas where defoliation has been limited by previous protective measures. 'Return migration' has become possible, making the opportunity for persistence of the outbreak essentially limitless (Baskerville, 1976). As well, insecticides possibly reduce predator numbers (NRCC, 1975; Varty, 1980) and recent improvements in forest-fire control have favoured a higher component of balsam fir in forest stands (Newfoundland Royal Commission, 1981). The budworm outbreak has spread to include virtually the entire host forest in New Brunswick (Figure 7.7.3) and appears able to persist indefinitely.

Such a persistent, semi-outbreak condition has been tolerable to the forest industry. Millions of acres of wood have been lost, about as much as would have been lost in a single uncontrolled epidemic, but the loss has been spread over many years and has allowed an annual harvest to be perpetuated.

The problem is really a question of forest management. Is it acceptable, economically or ecologically, to continue annual spraying of biocides indefinitely? If not, what alternatives exist?

Table 7.7.2 Comparison of total area defoliated with total area sprayed; 1970–1980

Year	Area sprayed (ha)	Moderate-severe[a] defoliation (ha)
1970	1.7×10^6	5.7×10^5
1971	2.4×10^6	1.6×10^6
1972	1.9×10^6	1.7×10^6
1973	1.8×10^6	3.2×10^6
1974	2.3×10^6	3.4×10^6
1975	2.7×10^6	3.5×10^6
1976	4.2×10^6	6.9×10^5
1977	1.6×10^6	4.7×10^5
1978	1.5×10^6	6.7×10^5
1979	1.6×10^6	1.3×10^5

[a] Data from Davidson (1981).

7.7.2.4 Perspectives on the Spray Programme

The spray programme is continually being assessed at two levels. One level involves the emotional debate between different groups with vested interests in the situation; the other concerns scientific environmental monitoring and risk-benefit analysis.

Vested-interest groups include the wood industry and secondary forest product users on one side, and farmers, fishermen and public-interest groups on the other. Issues that have been raised include the following:

1. Does the spray programme have any adverse effect on the fishing industry? There is considerable evidence that DDT (at 290–1165 g ha^{-1}) caused severe economic loss to the Atlantic salmon industry (Logie, 1975). Fenitrothion at the application rates used (210–290 g ha^{-1}), is not directly toxic to fish but is toxic to aquatic invertebrates (NRCC, 1975). The potential long-term secondary effect of this reduction in invertebrate populations on fish productivity is one of the critical issues.
2. To what extent do pesticides damage other beneficial invertebrates, including pollinators, and in this case, what is the resulting damage to fruit crops? It is now clear that fenitrothion applied at 210–290 g ha^{-1} can cause up to a 100 per cent reduction in fruit crop (e.g. Bridges Brothers Ltd versus Forest Protection Ltd, The Supreme Court of New Brunswick, 1976; Thaler and Plowright, 1980) and the overall long-term implications have raised concern.
3. Do pesticides adversely affect other species, particularly birds and man? It is now clear that small crown-canopy birds can be affected by fenitrothion at present application rates (210–290 g ha^{-1}) (NRCC, 1975; Pearce et al., 1976).
4. What is the net efficacy of the spray program?

These issues are obviously important but the concept of discontinuing the spray program also raises questions, including the following:

1. What would be the general economic, social and environmental damage of a full-scale, widespread budworm epidemic?
2. The budworm can severely damage the seed-bearing capacity of trees. Could this, in extreme situations, affect the natural replacement cycle?
3. After an outbreak, could a viable forest industry ever be established in a way that could avoid or deal with the possibility of another outbreak?
4. What will be the effects of increased fire-hazard in defoliated stands?
5. Would the tourist industry survive the aesthetic damage to recreation areas and would this damage be acceptable to the public at large?

These issues, and others that have been raised, need not contradict one another. Is it to anyone's advantage to allow the entire forest ecosystem to self-destruct? Equally, is it rational to disperse toxic chemicals, which may have deleterious effects on the ecosystem and may not even be effective as control agents, into the environment? Budworm control alternatives have been discussed by Baskerville (1976). They conclude that the idea, perpetuated by many professionals as well as by the public at large, that a simple solution to the problem exists, has been one of the greatest impediments to understanding and effectively dealing with the situation.

Interest groups can be expected only to voice questions; it is the scientists' responsibility to try to answer the particular issues raised by the interest groups, as well as to monitor and define general ecological effects.

By 1977, the Canadian scientific community was aware that these complex issues were not going to be resolved if a *laissez-faire* approach to monitoring continued (Varty, 1976; NRCC, 1977). It was recognized that indicative ecological monitoring requires recognition of the complex interrelations between forest management and ecological interactions. Since 1976, forest managers in Canada have been encouraged to cooperate with researchers in an effort to delineate the nature of spray programmes more precisely and to include the needs of the scientist in developing operational programmes, since monitoring and subsequent risk prediction apply only to relatively specific situations (NRCC, 1977). It has been accepted that, for economic and pragmatic reasons, the scope of ecological and toxicological monitoring of a pesticide used in large-scale operations must be limited and that risk-benefit analyses can take place only after the scenarios from the multiplicity of possible spray regimes are simplified to a point where indicative monitoring is possible (Varty, 1976).

This case study will concentrate on establishing principles which can be applied to ecological monitoring to maximize the indicative nature of the data obtained. Since 1977, scientists have adopted a much more organized approach to ecological monitoring in New Brunswick. Fenitrothion will be used as an example to clarify these principles.

Other aspects of the spray program, such as forest management scenarios and cost-benefit analysis, will not be discussed here. For discussion of such issues, the reader is referred to Baskerville (1976).

7.7.3 PRINCIPLES OF AN ORDERED APPROACH TO ECOLOGICAL MONITORING

Monitoring studies yield maximum information when focused on the critical path for a pollutant's fate, i.e. on the environmental compartments in which significant quantities of a pollutant are likely to accumulate and on the degradation processes most likely to produce toxicants. Establishing a critical path for pollutant fate involves the systematic characterization of the interactions between a chemical and the environment. The relative sensitivity of the different organisms that inhabit critical compartments can then be determined so that key indicator species may be identified and specifically monitored. A strategy for such an analysis (summarized in Figure 7.7.4) may be simplified to the sequential steps now described.

Step one: Characterize (1) the nature of the formulation initially loaded into the environment and (2) its basic physical and chemical properties

1. Characterization of a pesticide includes a complete description of the formulated solution as well as the designated active ingredients. Toxicity is a function not only of the amount of active ingredients, but also of the amount of contaminants, the nature of the adjuvants used in the preparation of the formulated product applied in the field and the age and the storage history of both the pure chemical and the formulation. The nature of the formulation also affects its volatility, deposition and degradation patterns. Frequent changes in the formulation make it difficult to establish patterns of effects from monitoring programs. Data obtained during the first year in which a chemical is monitored only suggest possible patterns of effects and behaviour, for observed effects may arise coincidentally to the spray programme and not as a result of the programme. If any aspect of the spray programme is changed in the second year, patterns can be masked by the change and the level of uncertainty is high. Patterns can be meaningfully established only if the formulation is kept constant over reasonable time periods.

The various fenitrothion formulations used in New Brunswick since 1968 are listed in Table 7.7.1. Fortunately, in New Brunswick the formulation was consistently water-based until 1977. In other regions (e.g. Quebec), the use of an oil-based formulation, which may have contained polyaromatic hydrocarbons, further complicates the assessment. Most early monitoring studies in New Brunswick neglected to indicate which formulation was used and formulations often changed; as well, laboratory studies focused almost exclusively on the active ingredient. These problems have been suggested as factors contributing to inconsistencies in the early literature (NRCC, 1975; Varty, 1976; NRCC, 1977).

A CASE STUDY OF THE USE OF FENITROTHION IN NEW BRUNSWICK 387

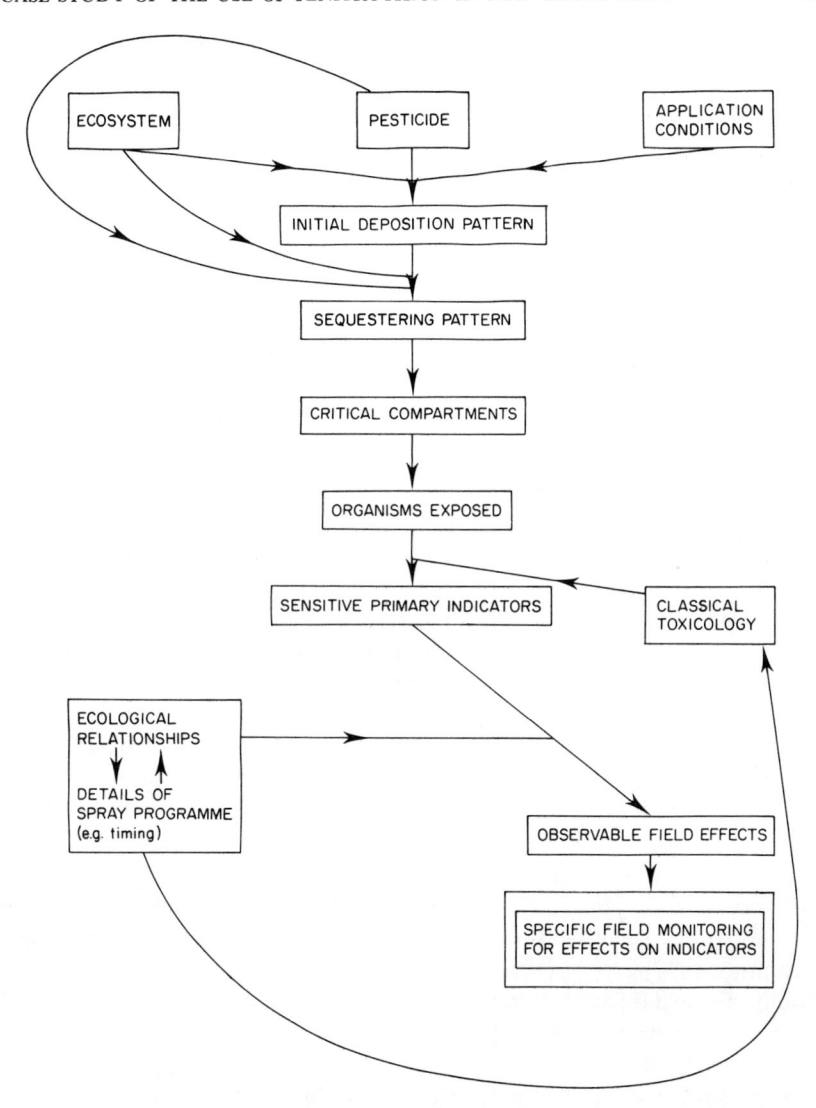

Figure 7.7.4 Strategy for an organized approach to ecological monitoring. Arrows indicate a rational sequence of investigation

One reason for the need to characterize the formulation stems from the variation in the toxicity of batches of technical fenitrothion (NRCC, 1975). In part, the variability has been shown to arise because of the formation of the S-methyl structural isomer of fenitrothion which may be more toxic than the parent compound (Greenhalgh and Shoolery, 1977). Also, the relative amount of this isomer, and possibly those of other degradation products and contaminants, tends to increase with the age of the sample. The toxicity of S-methyl fenitrothion has received much attention (Hladka *et al.*, 1977) as a result of these variations.

Some questions are so complex that they cannot be decisively answered unless overall population trends are monitored over a number years. Such monitoring can be meaningful only if the formulation and application technology remain relatively constant during the study.

2. A compound's basic physical and chemical properties include its molecular weight, melting point, solubility, vapour pressure, octanol–water partition coefficient, etc, as well as information on its susceptibility to hydrolysis, photolysis, oxidation and biodegradation (e.g. microbial). Such properties, in part, determine the behaviour and persistance of a chemical in a specific environment. With an adequate data base, it is possible to make limited predictions of a chemical's behaviour in specific environments and to suggest worst-case situations. Several computer models have been constructed on this principle (e.g. Mackay, 1979; NRCC, 1981b).

Fenitrothion is relatively nonvolatile (NRCC, 1975). It is stable to acid hydrolysis (Zitco and Cunningham, 1974) and susceptible to alkaline hydrolysis (Truchlik *et al.*, 1972). At pH ≤ 9, fenitrothion does not hydrolyse at an appreciable rate (NRCC, 1975). Hydrolysis could usually account for only a very small fraction of the known disappearance rate as New Brunswick soil is generally slightly acidic (Roberts and Marshall, 1980). Photodegradation, e.g. to carboxy fenitrothion and amino fenitrothion, is probably the most important degradation route (Lockhart *et al.*, 1973; NRCC, 1975; Greenhalgh *et al.*, 1980). Based on these properties, it has been suggested that sorption to sediment and dispersive processes could compete with these degradative processes and at least partially account for the initial disappearance of fenitrothion from pond waters in New Brunswick.

There are only limited data on microbial degradation although, in a qualitative sense, degradation to amino fenitrothion (and subsequently to nitroso-compounds) or 3-methyl-4-nitrophenol is known to occur in certain soils (Miyamoto, 1977).

All of the above data apply to relatively pure fenitrothion. There is still very little information on properties of the solvent (Dowanol) or the emulsifier (Atlox 3409F) now in use.

Step two: Establish the nature of the loading of pollutant into the environment

This is the first step toward establishing the primary distribution pattern of a chemical in the environment, and hence towards elucidating which species are likely to be exposed to toxicologically significant levels of the compound. Forest pesticides are applied in the form of an aerial spray and such techniques are not amenable to fine control. After the pesticide is released from the aircraft, it forms a cloud which disperses as it descends. Atmospheric patterns (e.g. wind speed and direction), temperature, spray droplet size and the topography of the area all affect the amount of spray deposited and the deposition pattern (Armstrong, 1977). The amount deposited can be extremely uneven (at least ± 100 per cent of the theoretical deposition) and spray drift can cause a significant fraction of the total application to be deposited in nontarget areas (Crabbe *et al.*, 1980 a, b).

The early history of fenitrothion use has been fraught with application problems. It has been applied over millions of hectares of forest (Table 7.7.1) and the deposition pattern has been uneven; some areas have received doses much higher than the supposed maximum (290 g ha^{-1}) (Carrow, 1974). In New Brunswick, spray drift has caused increases of up to $3 \mu g\,m^{-3}\,d^{-1}$ in the total phosphorus content of the air in nearby towns downwind from spray plots (Yule *et al.*, 1971). Areas have been accidently sprayed twice with fenitrothion (NRCC, 1975) and, in at least one case, a plot was sprayed with the wrong compound—phosphamidon instead of fenitrothion (NRCC, 1975). Under ideal conditions, only an average 50 per cent of the emitted spray is deposited in the target area (Armstrong, 1977). The need for a standard test protocol involving measurement of spray deposit, weather conditions, equipment and other aspects of aerial application has been recognized (Armstrong, 1977) in recent years attempts have been made to even out the deposition patterns. Computer models are being developed to facilitate prediction of deposition patterns and drift potential (e.g. Crabbe, 1980a, b), but human error and other unpredictable variables will still cause a high degree of uncertainty.

Step three: Characterize the ecosystem factors which affect initial deposition, sequestering and persistence patterns

The distribution of a chemical in the environment depends on the properties of the environment as well as on the properties of the chemical. The initial deposition pattern can be affected by topographical factors, the weather or the geological characteristics of the area. Similarly, the primary sequestering pattern will depend on ecosystem factors such as floral structure, soil type or geological characteristics. The rates of most degradative processes, and hence the persistance of a pesticide, are also directly affected by the environment and the sequestering pattern of the compound (Step four). Some examples will clarify these principles. Pesticide aerially applied to a densely foliated area will, at least initially, be intercepted by the foliage. If application is over a more exposed area, most of the chemical will reach the ground. If deposition is onto absorbent soil, it is likely that the compound will be rapidly absorbed but, if the ground is rocky, there will be little absorption and the compound might volatilize or photolyse from the surface at a much faster rate.

Fenitrothion is mostly applied over foliated areas and up to 80–90 per cent of the initial deposit is intercepted by the crown canopy of trees. Species inhabiting areas not protected by the crown canopy (e.g. small crown-canopy birds, meadow voles) are likely candidates to suffer immediate exposure and may be designated as indicators (see Step seven).

Step four: Establish the nature of secondary mobility, sequestering patterns and biological accumulation patterns

The initial deposit into a given environmental compartment will be transferred into other compartments at rates determined by both the nature of the chemical and the environment. It is, therefore, necessary to establish mobility patterns and, from this, secondary sequestering patterns. For example, a compound initially deposited onto crown canopy foliage might eventually reach the forest floor and sequester in the soil litter or accumulate in soil invertebrates.

The amount of compound that will be sequestered depends on the relative rates of degradative and removal processes operative within the compartment receiving the initial load (see Step five). It is necessary, at this point, to determine whether the compound will simply be degraded upon deposit or be transferred in a manner that can lead to its accumulation in various other compartments including flora and fauna. As well, it is necessary to consider whether toxic degradation products are formed and sequently accumulated in specific compartments.

We have stated that a major amount of the fenitrothion initially deposited on crown canopy foliage and open areas is lost quite rapidly (NRCC, 1975), probably

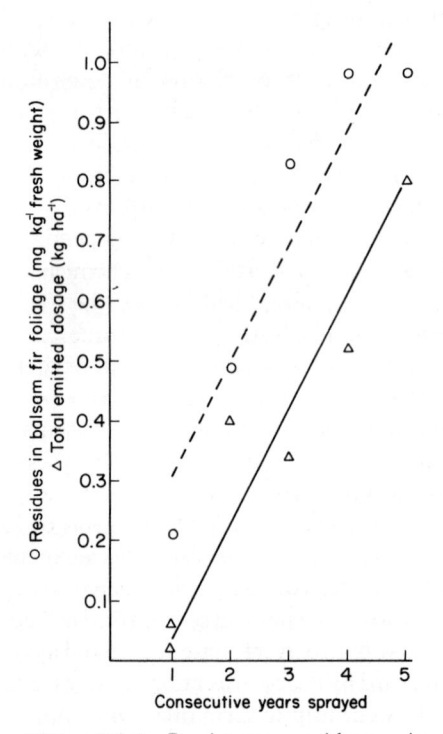

Figure 7.7.5 Persistent residues in mixed-age balsam fir foliage, and total dosage, versus the number of consecutive years a location in New Brunswick was sprayed with fenitrothion. (Adapted from Yule, 1974)

via photodegradation, but about 10 per cent is absorbed by the coniferous foliage (Yule and Duffy, 1972). Material sequestered in this matrix is relatively stable because waxy leaf pigments attenuate light and limit photolytic degradation. If there are annual applications, these residues accumulate from year to year (Yule, 1974) and are proportional to the number of annual applications and the total dosage of fenitrothion (Figure 7.7.5). This is one of the few environmental compartments in which fenitrothion is known to be persistent from one year to the next. It is likely, but has not been established, that forest needle litter will also contain such residues. Organisms inhabiting these compartments, e.g. birds, defoliating insects and possibly soil organisms (including decomposers), are likely to receive higher than average exposure to fenitrothion. Such organisms are thus possible indicators and are being monitored for acute and chronic toxic effects and for population shifts.

Fenitrothion reaching the ground, or the surface of water bodies, will be transferred to and sequestered in other compartments. The rates and extent of such transfers will depend on the properties of the local environment. For example, the extent of absorption into soil or aquatic sediments will depend on their sorptive capacity, i.e. their organic content. Since fenitrothion is only moderately hydrophobic, this effect will be less significant than for a very hydrophobic compound such as DDT. Still, conceptually, benthic organisms and filter feeders could be exposed to higher than average concentrations of fenitrothion and they are being studied as indicators. So far, there is no evidence that fenitrothion is bioconcentrated in the food chain to higher mammals (Miyamoto, 1977). However, organophosphorus insecticides are accumulated in tadpoles up to 60 times the levels in water (Hall and Kolbe, 1980) and may possibly be concentrated in the lower levels of the aquatic food chain. Animals consuming dead budworm or other defoliators may also be exposed to higher than average levels of toxicants.

Step five: Characterize the situations which will maximize and minimize persistence

The relative persistence of a pollutant is dependent upon its rates of degradation by various processes. The rates of these degradative processes can vary in different environments. Degradative processes can be directly influenced by the environment or by the partitioning pattern of the pollutant which can affect the fraction of pollutant accessible for degradation through a given process. It is theoretically possible to identify the environmental factors which will maximize or minimize the persistence of a given pollutant, although interacting effects make this extremely complex. For example, photolysis will be maximized in summer and in situations where the pollutant is exposed to direct sunlight over a large surface area (e.g. leaves or a rocky surface) and minimized in winter and in situations where light attenuation is high. The extent of microbial degradation is likely to be maximal where the bulk of the pesticide is sequestered in organic soil, organic sediment, eutrophic water bodies, etc. However, the situations which maximize microbial degradation tend to minimize photolysis. Since the extent of partitioning of pollutant into, e.g. organic soil, depends on the lipophilic nature of the chemical as well as on the soil, the effects of the environment on persistence will not be the same for different chemicals. However, within a chemical class (e.g. the class of extremely hydrophobic compounds) the patterns should be comparable.

In the case of fenitrothion, we know that microbial degradation and photolysis are the important identified degradation processes, even though microbial breakdown is not well characterized. If we assume that photolysis is often the dominant process, then fenitrothion should be least persistent when deposited on a rocky surface with little sorptive capacity. Persistence would be maximum when the compound was deposited on a sorbant material. However, if microbial degradation is dominant, the situation could be just the reverse.

Step six: Identify general sensitivity patterns and potential indicator organisms

We have discussed the importance of establishing which compartments are most likely to receive a transient or persistent exposure to a pollutant or a toxicologically significant degradation product. The next step is to analyse the organisms that live in these compartments, in terms of their relative susceptibility to damage by the pesticide or associated chemicals. Initially, this analysis involves the use of data obtained through classical laboratory toxicological studies to identify primary indicators (organisms most likely to be directly affected by contact with the toxicant). The lethal-effect dose and concentration are useful in establishing the relative sensitivity of different organisms to lethal effects. For economic reasons, it may not be practical to establish these parameters for all the organisms that are potentially exposed. Known trends in toxicological response may be used to guide the choice of laboratory tests. For example, relationships between size and sensitivity may exist for certain types of response. It must be remembered, however, that such trends are tenuous and variations in response may occur even within groups of very similar organisms. The response can also vary depending on the route of exposure and it is necessary to examine likely routes before defining relevant laboratory studies.

At the ecosystem level, effects on an individual organism are less important than effects on a species (see Chapter 4) and individual effects may be difficult to detect in the field because of natural fluctuations. Thus, individual effects are often not the best indicators of ecosystem perturbation. Indicators should be selected to reflect the actual doses likely to be encountered in the field. *It is of no use to study a highly sensitive species that is not exposed*! Subacute and chronic effects will be important at the ecosystem level only if they have the potential to cause population shifts, alter the behavioural pattern of a species or start a chain of effects. For example, teratogenic or reproductive effects could be significant if spraying coincided with the reproductive season. These ideas will be discussed in Step seven. Ideally, dose-response curves should be established for likely indicators since, if the curve is steep, small changes in deposition rates could drastically alter effect patterns. In practice, it is usually too expensive and time consuming to do this for many organisms. At best, laboratory programmes identify only a few likely sensitive primary indicators which can serve as pointers to the most productive approach to indicative field monitoring. The population responses and life cycles of the primary indicators, as well as their interactions with other components of the ecosystem, must be considered to determine the most sensitive field indicators (Step seven). It is necessary to develop a loop of laboratory–field–laboratory monitoring which allows the continued improvement of the sampling matrix and hence the quality and precision of indicative monitoring.

In the case of the fenitrothion spray programme in New Brunswick, several primary indicators have been identified. The crown canopy is a critical compartment in terms of primary exposure since much of the initial deposit is intercepted by the canopy and fenitrothion can accumulate in coniferous foliage (Step four). Organisms which inhabit this compartment, e.g. birds, will potentially suffer various levels of adverse effects as a function of their sensitivity to fenitrothion and the dose they encounter. Sensitivity varies dramatically between and within species (Figure 7.6.6). Initial field monitoring programmes found fenitrothion to have little impact on birds because these programmes did not monitor the species which, in terms of sensitivity, would be good indicators, i.e. small crown-canopy birds (NRCC, 1975). For example, no effect was

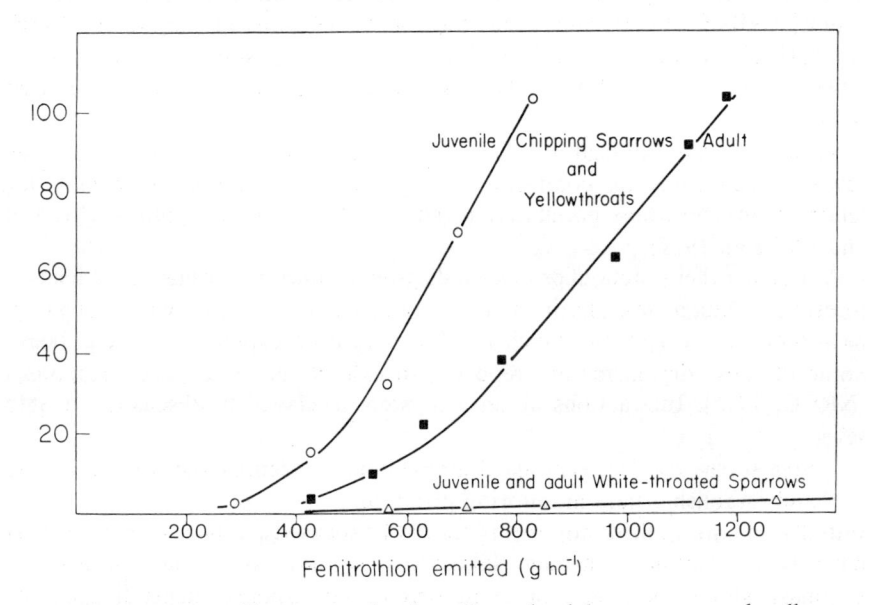

Figure 7.7.6 Mortaility curves for juvenile and adult sparrows and yellow-throats. (From Buckner, 1975. Reproduced by permission of National Research Council of Canada)

Table 7.7.3 Effect of fenitrothion on number of singing males (mean, standard deviation) recorded on transects[1] (data from Pearce et al., 1976)

Species	Prespray count	Post-first spray count (175–280 g ha^{-1})	Post-second spray count (175 g ha^{-1})
Cape May warbler	14.3 ± 3.3	6.2 ± 3.7[3]	–
	11.6 ± 2.7	4.3 ± 1.2[3]	6.2 ± 2.0
Tennessee warbler	11.3 ± 6.1[2]	32.5 ± 9.5	–
	43.5 ± 5.3	38 ± 4.0[3]	23.2 ± 3.4[3]
	4.6 ± 2.4	0.3 ± 0.5[3]	0.0 ± 0.0

[1] Counts taken in several spray areas.
[2] Tennessee warbler numbers were very low pre-spray, indicating that much of the population had not yet arrived.
[3] Reduction is significant at 95 per cent level.

observed on populations of sapsucker (Rushmore, 1971), a relatively large bird which does not generally nest in or inhabit the forest crown. Another influence which may explain the original failure of monitoring programmes to detect effects is that census techniques have an unavoidable, built-in negative bias (Pearce, 1968; NRCC, 1975). Despite this, patterns of effects have been reported in small songbird populations, particularly those which inhabit the crown canopy (NRCC, 1975). Some examples of the 1975 monitoring data (Pearce et al., 1976) are presented in Table 7.7.3. These studies show an unambiguous adverse effect on Cape May and Tennessee warblers at the application rates used today.

Another sensitive indicator in the crown canopy compartment is the jack-pine sawfly which feeds on coniferous foliage (as noted, an ultimate sink for fenitrothion) and whose population is reduced after use of fenitrothion (McNeil and McLeod, 1977).

As noted, other potential primary indicators include organisms which feed on forest litter, fauna which frequent areas not protected by the crown canopy (e.g. meadows), and organisms which inhabit or feed on exposed flora (e.g. bees). Some of these organisms are reported to exhibit individual toxic responses (NRCC, 1975). Interactions at an ecosystem level will be discussed in Step seven.

Step seven: Analyse ecological interactions to identify the best indicators and establish a field monitoring program

Initially, of course, field monitoring should be focused on those organisms that have been identified, through laboratory studies, as primary indicators. Emphasis should be placed on acute or chronic responses depending on the organism studied and the nature of the exposure (see Steps three to six). However, as noted, primary indicators may not be the most sensitive or useful indicators of ecosystem perturbation. Population and ecosystem level responses can be quite different from, and sometimes more dramatic than, those detected at an individual level. A field monitoring programme requires consideration of the many possibilities involving species affected via a chain of effects including the following:

1. Species that are particularly sensitive due to an interaction between their life cycle, e.g. migratory or reproductive period, and the timing of the spray programme.
2. Species that are actively foraging at the time of application.
3. Species with a critically low population (e.g. endangered species) where even a small percentage increase in mortality could be critical in terms of stability of the species.
4. Situations where an entire population might be temporarily concentrated in a small area (e.g. migrating bird flocks) making them particularly vulnerable. In such cases, unusually high loading due to variable deposition patterns could have a devastating effect on the entire flock if the dose-response curve is steep and the average application rate is close to an effect level.
5. Synergistic interactions with other pollutants or with natural factors.
6. Species with hierarchical social organization where certain individuals may be fundamental to the structure and survival of a population.

It is clear that ecological effect monitoring is extremely complex. The lack of baseline data on flora–fauna sturcture is perhaps the biggest problem. However, it is not necessary to adopt a random approach to field monitoring. Some of the interactions which help identify indicators can be rationalized. Often, field monitoring will indicate a need for further laboratory assessment.

Eventually, a sampling matrix will evolve which maximizes the opportunity for detecting small changes in complex environments. Ultimately, the most difficult point of the analysis is the prediction of long-term ecosystem consequences of apparently small perturbations, in the light of natural catastrophies, etc. (e.g. NRCC, 1981a).

The fenitrothion spray programme presents good examples of the points outlined above. As already noted, bees are often exposed to lethal concentrations of fenitrothion when it is applied to nontarget pollen and nectar sources. The overall impacts on specific populations of pollinators can range from negligible to drastic, depending on interactions between the life cycle and behaviour patterns of the species, the specific ecology of the area and the timing of the spray programme. Furthermore, it is difficult to distinguish between pesticide and environmental effects on the populations because of the lack of baseline data on temporal trends in populations.

Inital studies on the effects of fenitrothion on colonies of honeybees (Buckner, 1975) indicated that the application of fenitrothion at $275\,g\,ha^{-1}$ had few long-term effects on hive stability. This result was interpreted to indicate that the pollinating force remained unaffected. Substantial numbers of dead bees were found adjacent to the hives but, since they represented only 1 per cent of the total hive population, the net impact was assumed to be very low. The key questions, the percentage of active foraging bees killed and hence the short- and long-term impact on pollinating activity, were not addressed in the discussion. Although the observed mortality was most likely restricted to the foraging bees directly exposed to the spray, experience with other pesticides has now shown that the level of kill observed in these studies is indicative of severe mortality in the foraging component (NRCC, 1981a).

While there were initial inconsistencies in the results from different field monitoring programmes (NRCC, 1981a), it is now apparent that when fenitrothion is sprayed at operational rates, significant population reductions may occur in nearly 100 species of native bees in New Brunswick (Thorpe, 1979; NRCC, 1981a). Plowright (cited in NRCC, 1981a) reports that pollinating bees are almost non-existent in the centre of spray blocks soon after spraying. Our poor understanding of the deposition patterns associated with particular spray programmes, and the limitations of the census techniques now available for pollinators, make direct quantification of the level of impact impossible. Analysis of the problem has primarily relied upon the recognition of perturbations in ecological patterns involving bees, their individual sensitivity, preferences for pollen sources and the specific pollination requirements of some plants.

Bumblebees have received particular attention in this spray programme because the overwintering queen emerges alone early in the spring, i.e. usually just before the spray programme. Bumblebee colonies are reestablished annually and, if the queen is killed before a colony is established, there would be a drastic effect on population levels over the season. Plowright *et al.* (1978) observed up to 100 per cent mortality when caged bumblebees were sprayed with fenitrothion at 206 g ha^{-1} (operational levels). The degree of mortality depended on the amount of coverage afforded by the forest canopy. Populations of bumblebees in spray areas remained depressed for several months after the spray. Solitary bees will potentially be at high risk because they forage in open meadows for long periods during the spray season.

The identification of appropriate secondary indicators of pollinator effects requires careful analysis of the preference of pollinators for different plants, the period of bloom relative to the spray period and the dependence of the plants on native bees for pollination. Preference of pollinators for specific plants has been demonstrated by the difference in the variety of pollen types collected from captured bumblebees in sprayed and unsprayed areas (Figure 7.7.7). Fewer pollen varieties were collected in sprayed areas, suggesting a bias towards a few preferred species of plants as pollinator density, and hence competition, decreased (Plowright *et al.*, 1978). All other factors being equal, the less-preferred species of plants should be the best indicators of when the pollinator force is reduced. As already suggested, the variation in blooming periods may (at

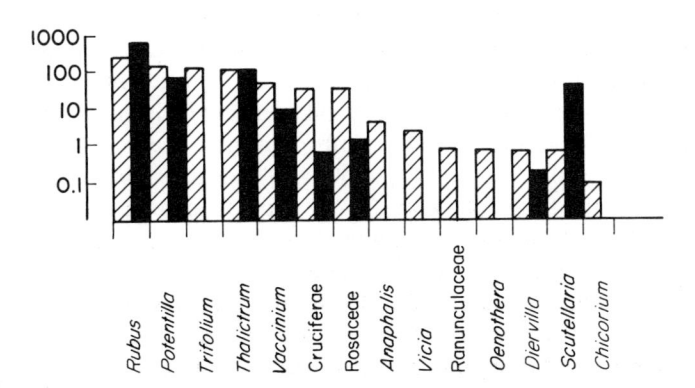

Figure 7.7.7 Ranked abundance of pollen types collected by bumblebees in fenitrothion sprayed and unsprayed areas of southwest New Brunswick. (From Plowright et al., 1978)

least partly) account for contradictions in the reports about fruit-set. Thaler and Plowright (1980) have found unambiguous reductions in fruit-set of plants that bloomed shortly after application of fenitrothion. Buckner (cited in NRCC, 1975, 1981a) did not see any reduction in fruit-set in plants that bloom at periods which do not coincide with the spray period. One would anticipate that plants with short blooming periods which coincide with the spray period would be particularly vulnerable and likely indicators. The low-bush blueberry, dependent on bees for pollination, is now known to be adversely affected by the fenitrothion spray programme to such an extent that compensation has been paid to growers (e.g. $58,000 to Bridges Brothers Ltd; The Supreme Court of New Brunswick, 1976).

The implications of these observations in terms of floral diversity and stability have not been assesed, mainly because of the lack of demonstrated protocols for assessment of the problem.

The NRCC Panel (NRCC, 1981a) concluded that:

> It is in this area that we should be most concerned about our current ignorance: we are not yet able to forecast the persistence of the secondary effects of reductions in genetic variance in plant populations, although we may list some possible effects.

Pesticide-pollinator interactions emphasize the importance of an organized, systematic approach to ecological monitoring.

The fenitrothion spray programme presents other examples of possible secondary indicators. Pine foliage in which fenitrothion has accumulated is eventually shed to form a significant part of forest floor debris. Although it has not been documented, this debris may contain relatively high levels of fenitrothion. Organophosphorus insecticides in general have been shown to be toxic to certain soil organisms, including invertebrates (Edwards and Thompson, 1973). Such organisms are closely associated with the decomposition of dead plant material into its organic and inorganic constituents and in the incorporation of these materials into the soil structure (Edwards and Thompson, 1973). If fenitrothion decreases the numbers of these decomposer organisms, a decrease in soil fertility might eventually result. Edwards and Thompson (1973) report that fenitrothion (4.5 kg ha^{-1}) is toxic to certain soil invertebrates. Spillner *et al.* (1979) report that fenitrothion does not qualitatively affect the microorganisms which degrade leaf litter and cellulose in forest soild, but these questions have not been carefully addressed in the field. Conceivable indicators for such effects include decreases in the nitrogen content of soil, the rate of decomposition of total organic matter in surface soil, etc.

The possibility that fenitrothion may accumulate in tadpoles has already been raised (Step four). If this happens, potentially toxic levels of fenitrothion could accumulate in amphibians and consequently carnivorous species could be exposed to toxic levels (Hall and Kolbe, 1980). Spray programmes aimed at the adult budworm, and hence conducted later in the summer, might be particularly hazardous to young ducks, etc. This group of organisms, which are not exposed to any direct source nor to any obvious secondary exposure vector, have not been studied.

7.7.4 CONCLUSIONS

Experience in New Brunswick demonstrates the need for a close liaison between the forest manager who mounts the control programme and the scientists who assess its impact on the forest ecosystem. The manager has an array of choices to make, e.g. which application technique to use, the nature of the product(s) to be used, the formulation, the timing and extent of the programme, etc. As well, the harvest patterns chosen by the manager strongly influence population dynamics and the effectiveness of natural control mechanisms and hence, eventually, the need to spray (Baskerville, 1976).

The integral relationship between management's decisions and the impact of the control strategy on both the pest and other organisms is depicted in Figure 7.7.8. Ultimately, each decision feeds back to the forest manager who is thus the only person who can create a stable situation conductive to the development of a productive assessment program. In the past, the absence of a stable programme with an effective liaison between managers and scientists made early efforts at assessment generally equivocal. Since the liaison has improved, a point has been reached where there is sufficient evidence to conclude that a

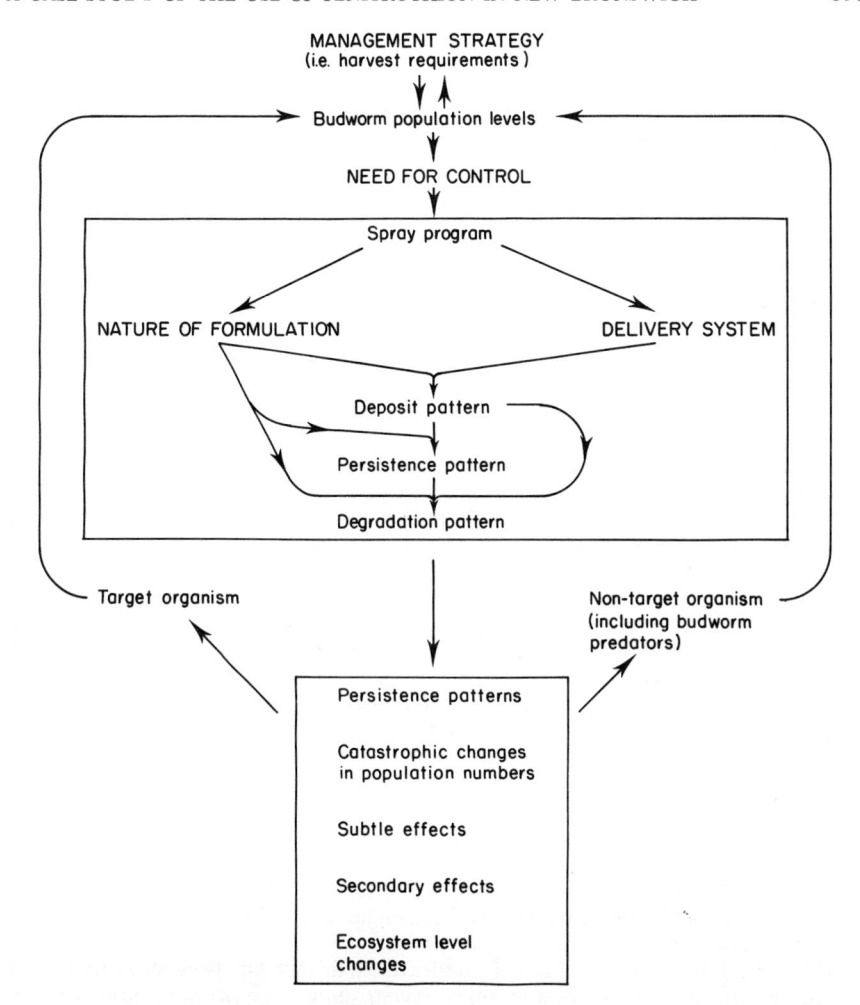

A CASE STUDY OF THE USE OF FENITROTHION IN NEW BRUNSWICK

Figure 7.7.8 Management decision scheme. Decision points are capitalized. Arrows indicate possible consequences of decisions

number of indicator organisms are being affected. These include pollinators and certain other beneficial insects, small crown-canopy birds and aquatic invertebrates. To date, the overt impact of the spray programme has been judged tolerable by the forest managers (Varty, 1980). The critical question is whether undesirable long-term side-effects will be associated with a harvest policy that ecologists believe perpetuates the outbreak and hence the need to spray (Baskerville, 1976).

Unfortunately, it is also the conclusion of the ecologists that our understanding of the specific fundamental ecological relations operative in a given forest system are too rudimentary to permit assessments of the long-term implications of such programmes at the ecosystem level (NRCC, 1981a; Varty, 1980). In the face of the sheer complexity of the problem, it has been suggested that the first productive step is the development of management strategies that minimize the need for large-scale spray programmes and hence the need to consider the long-term consequences of such programmes at the ecosystem level. In New Brunswick, alternate strategies for the harvest have been suggested (Baskerville, 1976) that could fundamentally alter the dynamics of the spruce budworm, bringing natural biocontrol mechanisms more nearly into balance, and significantly reduce the need for artificial pest control strategies.

7.7.5 REFERENCES

Armstrong, J. A. (1977). Relationship between the rates of pesticide application and the quantity deposited on the forest. In *Proceedings of Symposium on Fenitrothion: the Long-term Effects of Its Use in Forest Ecosystems*. Associate Committee on Scientific Criteria for Environmental Quality, National Research Council Canada. NRCC No. 16073, pp. 183–202.

Baskerville, G. (Task-Force Leader) (1976). Report of the task-force for evaluation of budworm control alternatives. Prepared for The Cabinet Committee on Economic Development, Province of New Brunswick.

Brooks, G. T. (1974). *Chlorinated Insecticides*, vol. II, *Biological and environmental aspects*, CRC Press, Cleveland, Ohio.

Buckner, C. H. (1975). Cited in Fenitrothion: the effects of its use on environmental quality and its chemistry. Associate Committee on Scientific Criteria for Environmental Quality, National Research Council of Canada. NRCC No. 14104, 56 pages.

Carrow, J. R. (1974). Aerial spraying operations against block-headed budworm on Vancouver Island—1973. Pacific Forest Research Centre. Inf. Rep. BC-X-101, 56 pages.

Crabbe, R., Elias, L., Krzymien, M. and Davie, S. (1980a). New Brunswick forestry spray operations: filed study of the effect of atmospheric stability on long-range pesticide drift. NRCC Rep. No. LTR-UA-52, 66 pages.

Crabbe, R., Krzymien, M., Elias, L. and Davie, S. (1980b). New Brunswick spray operations: measurement of atmospheric fenitrothion concentrations near the spray area. NRCC Rep. No. LTR-UA-56, 43 pages.

Davidson, G. (1981). Personal communication to *Can. For. Ser., Environ.*, Canada, Ottawa.

Desaulniers, R. (1977). Comparison of forest treatment programs carried out in Québec and in New Brunswick. In *Proceedings of symposium on Fenitrothion: the Long-Term Effects of Its Use in Forest Ecosystems*. Associate Committee on Scientific Criteria for Environmental Quality, National Research Council Canada. NRCC No. 16073, pp. 135–159 (in French).

Edwards, C. A. and Thompson, A. R. (1973). Pesticides and the soil fauna. *Res. Rev.*, **45**, 1–81.

Greenhalgh, R. and Shoolery, J. N. (1977). The use of ^{31}P-NMR for the rapid analysis of fenitrothion formulations and technical products. In *Proceedings of symposium on Fenitrothion: the Long-term Effects of Its Use in Forest Ecosystems*. Associate Committee on Scientific Criteria for Environmental Quality, National Research Council Canada. NRCC No. 16073, pp. 77–95.

Greenhalgh, R., Dhawan, K. L. and Weinberger, P. (1980). Hydrolysis of fenitrothion in model and national aquatic systems. *J. Agric. Food Chem.*, **28**, 102–105.

Hall, R. J. and Kolbe, E. (1980). Bioconcentration of organo-phosphorus pesticides to hazardous levels by amphibians. *J. Toxiocol. Environ. Health*, **6**, 853–860.

Hladka, A., Batora, V., Kovacicova, J. and Rosival, L. (1977). Occupational health hazards and significance of technical fenitrothion and its contaminant S-methyl fenitrothion in toxicology of formulations. In *Proceedings of Symposium on Fenitrothion: the Long-term Effects of Its Use in Forest Ecosystems*. Associate Committee on Scientific Criteria for Environmental Quality, National Research Council Canada. NRCC No. 16073, pp. 415–451.

Lockhart, W. L., Metner, D. A. and Grift, N. (1973). Biochemical and residue studies on rainbow trout following field and laboratory exposures to fenitrothion. *Manit. Entomol.*, **7**: 26–36.

Logie, R. R. (1975). Effects of aerial spraying of DDT on salmon populations of the Miramichi River. In M. L. Prebble (Ed.), *Aerial Control of Forest Insects in Canada*, Environment Canada, Ottawa.

Mackay, D. (1979). Finding fugacity feasible. *Environ. Sci. Techn. P.*, **13**, 1218–1223.

McNeil, J. N. and McLeod, J. M. (1977). Apparent impact of fenitrothion on the waine jack-pine sawfly, *Neodiprion swainei*, Midd. In *Proceedings of Symposium on Fenitrothion: the Long-term Effects of Its Use in Forest Ecosystems*. Associate Committee on Scientific Criteria for Environmental Quality, National Research Council Canada. NRCC No. 16073, pp. 203–216.

Miyamoto, J. (1977). Long-term toxicological effects of fenitrothion in mammals including carcinogenicity and mutagenicity. In *Proceedings of Symposium on Fenitrothion: the Long-term Effects of Its Use in Forest Ecosystems*. Associate Committee on Scientific Criteria for Environmental Quality, National Research Council Canada. NRCC No. 16073, pp. 459–496.

Newfoundland Royal Commission (1981). Report of the Royal Commission on Forest Protection and Management. Province of Newfoundland.

NRCC (1975). *Fenitrothion: the Effects of Its Use on Environmental Quality and Its Chemistry*. Associate Committee on Scientific Criteria for Environmental Quality, National Research Council Canada. NRCC No. 14104, 162 pages.

NRCC (1977). *Fenitrothion: the Long-term Effects of Its Use in Forest Ecosystems*. Current Status Report. Associate Committee on Scientific Criteria for Environmental Quality, National Research Council Canada. NRCC No. 15389.

NRCC (1981a). *Pesticide-Pollinator Interactions*. Associate Committee on Scientific Criteria for Environmental Quality, National Research Council Canada. NRCC No. 18471, 190 pages.

NRCC (1981b). *A Screen for the Relative Persistence of Lipophilic Organic Chemicals in Aquatic Ecosystems—an Analysis of the Role of a Simple Computer Model in Screening*, part. 1: *A Simple Computer Model as a Screen for Persistence*. Associate Committee for Scientific Criteria for Environmental Quality, National Research Council Canada. NRCC No. 18570, 106 pages.

Pearce, P. A. (1968). Effects on bird populations of phosphamidon and sumithion used for spruce budworm control in New Brunswick and hemlock looper control in Newfoundland in 1968: a summary statement. *Can. Wildl. Serv. MS. Rep.*, no. 14.

Pearce, P. A., Peakall, D. B. and Erskine, A. J. (1976). Impact on forest birds of the 1975 spruce budworm spray operation in New Brunswick. *Can. Wildl. Serv. Progr. Notes*, 62, 7 pages.

Plowright, R. C., Pendrel, B. A. and McLaren, I. A. (1978). The impact of aerial fenitrothion spraying upon the population biology of bumble bees (*Bombus* Latr. Hym.) in southwestern New Brunswick. *Can. Entomol.*, **110**, 1145–1156.

Prebble, M. L. (1974). Aerial control of forest insects in Canada. Environment Canada, Ottawa.

Roberts, J. R. and Marshall, W. K. (1980). Retentive capacity: an index of chemical persistence expressed in terms of chemical-specific and ecosystem-specific parameters. *Ecotoxicol. Environ. Safety*, **4**, 158–177.

Rushmore, F. M. (1971). Effects of accothion on birds. In Nash, R. W., Peterson, J. W. and Chansler, J. F. (Eds.), *Environmental Studies of Accothion for Spruce Budworm Control*. A cooperative study by the State of Maine, U.S. Dept. Agric. and U.S. Dept. Interior, pp. 88–100.

Schneider, W. G. (Chairman) (1976). Forest spray program and Reye's syndrome, report of the Panel Convened by the Government of New Brunswick, p. 4a.

Spillner, C. J., Thomas, V. M. and DeBaun, J. R. (1979). Effect of fenitrothion on microorganisms which degrade leaf-litter and cellulose in forest soils. *Bull. Environ. Contam. Toxicol.* **23**, 601–606.

Thaler, G. R. and Plowright, R. C. (1980). The effect of aerial insecticide spraying for spruce budworm control on the fecundity of entomophilous plants in New Brunswick. *Can. J. Bot.*, **58**, 2022–2027.

The Supreme Court of New Brunswick (1976). Queen's Bench Division, between: Bridges Brothers Limited (Plaintiff) and Forest Protection Limited (Defendant), 1973–1975. 14 June 1976.

Thorpe, E. A. (1979). *The Effects of Forest Spraying on Solitary Bee (Hymenoptera: Apordia) Populations in New Brunswick*. M.Sc. thesis, University of New Brunswick.

Truchlik, S., Drabek, I., Kovac, I. and Gager, S. (1972). Metathion. A new low-toxicity organophosphorus insecticide. In Chemistry and application of organophosphorus compounds. Proc. Third Conference. NTIS JPRS 57825, U.S. Dept. Commerce. P. 129.

Varty, I. W. (1976). In Environmental effects of the spruce budworm spray program in New Brunswick, 1976. Maritimes Forest Research Centre, Fredericton, N.B. Inf. Rep. M-X-67, 21 pages.

Varty, I. W. (Chairman) (1977). Environmental surveillance of insecticide spray operations in New Brunswick's budworm-infested forests. An interim report by The Committee for Environmental Monitoring of Forest Insect Control Operations (EMOFICO).

Varty, I. W. (Chairman) (1980). Environmental surveillance in New Brunswick, 1978–1979. Effects of spray operations for forest protection against spruce budworm. The Committee for Environmental Monitoring of Forest Insect Control Operations (EMOFICO).

Yule, W. N. (1974). The persistence and fate of fenitrothion insecticide in a forest environment. II. Accumulation of residues in Balsam fir foliage. *Bull. Environ. Contam. Toxicol.*, **12**, 249–252.

Yule, W. N. and Duffy, J. R. (1972). The persistence and fate of fenitrothion insecticide in a forest environment. *Bull. Environ. Contam. Toxicol.*, **8**, 10–18.

Yule, W. N., Cole, A. F. W. and Hoffman, I. (1971). A survey for atmospheric contamination following forest spraying with fenitrothion. *Bull. Environ. Contam. Toxicol.*, **6**, 289–296.

Zitco, V. and Cunningham, T. D. (1974). Fenitrothion derivatives and isomers, hydrolysis, adsorption and biodegradation. *Fish. Mar. Serv. Tech. Rep.*, no. 458, 27 pages.

Effects of Pollutants a the Ecosystem Level
Edited by P. J. Sheehan, D. R. Miller, G. C. Butler and Ph. Bourdeau
© 1984 SCOPE. Published by John Wiley & Sons Ltd

CASE 7.8
Rehabilitation of Mine Tailings: A Case of Complete Ecosystem Reconstruction and Revegetation of Industrially Stressed Lands in the Sudbury Area, Ontario, Canada

T. H. PETERS

Agricultural Department INCO Limited
Copper Cliff, Ontario, Canada

7.8.1 Introduction 403
7.8.2 Historical Ecology 404
7.8.3 Mine Tailings Problem. 405
7.8.4 Revegetation. 408
7.8.5 Present and Future Considerations 419
7.8.6 Conclusions 420
7.8.7 References. 420

7.8.1 INTRODUCTION

Mining, agriculture, forestry and fisheries, the four basic resource industries, on which all wealth is built, are extractive processes from the planet earth. They all have had and will continue to have an impact on the earth's surface. Modification of the future impact and amelioration of the past impact of extractive techniques on the earth's environment are a major concern of citizens, industries and governments in many areas of the world today.

Mining, and the subsequent ore beneficiating processes associated with extractive metallurgical techniques, impinge on the earth's surface in various ways which are specific to the ore being mined, the site and the waste products from the metallurgical processes. This case study will deal with the field solution of some of the problems which arise in hard rock mining and with sulphide ores in particular.

In the Sudbury area of Ontario, Canada, Inco Limited, is involved in a progressively expanding programme initiated some 70 years ago, to modify the physical effects of mining, concentrating, smelting and refining, on the local environment.

This region is one of the most ecologically disturbed in Canada. It is of interest to note that Hooker, in the report of a geological survey dated November 1886, described the appearance of the area as follows:

> The streams are small and sluggish, impeded by beaver dams, and the water rendered impure by decayed vegetation. The timber has almost been entirely destroyed by forest fires, and the charred trunks still standing, tottering, or thrown to the ground, impart a dismal aspect to the scenery, and together with the thick underbrush which prevail render passage beyond the beaten trails very difficult. There appears to be little land suitable for cultivation. The thin layer of earth covering the hills has likewise in places, suffered from the fires, permitting it to be readily washed away and exposing the bare rock underneath.

The region's importance, as an example of man's impact upon the environment, is reflected by the many more recent investigations which have been carried out in the area (Gorham and Gordon, 1960a, b; Whitby *et al.*, 1976; Hutchinson and Whitby, 1977; Freedman and Hutchinson, 1980).

7.8.2 HISTORICAL ECOLOGY

The original topography of the area now occupied by tailings was undulating, with small valleys in between rounded rock hill tops. The Wisconsin glaciation in the late Pleistocene era was responsible for many of the geomorphic formations and soil types in the area. The land was heavily scoured and much of the soil was removed, with the result that numerous rock outcrops were and remain exposed.

Soils of the region are humo-ferric podzols with glacial surface deposits of water-modified tills, lacustrine silts, and sands located in the valleys. The most common soils are those developed on stony, sand tills. These soils are usually no more than 1 m over bedrock and in many cases, bedrock appears at the surface. Organic soils occupy many of the depressions in the area. Very little of the soil in the area is suitable for agriculture, however, certain parts of the area are moderately high producers of agricultural crops (such as potatoes, spring grain and hardier market garden crops). Soil pH values are 6.0 or less (Heale, 1980).

In the valleys are lakes, swamps and, in some cases, areas of land sufficiently well drained to permit the establishment of native species of trees and shrubs. The vertical difference in elevation between hill tops and valley bottoms is generally less than 60 m. Drainage in lower areas is often very poor and as a consequence soils in specific areas warm very slowly in the spring. The lack of drainage in these soils adversely affects their chemical and physical structure, nutrient capacity, aeration and porosity. These factors significantly influenced the climax vegetation found in the different localized areas.

Braun (1950) included the climax vegetation of the immediate Sudbury area in the Laurentian Upland section of the Great Lakes—St. Lawrence Division of the Hemlock – White Pine–Northern Hardwoods Forest. Due to the close proximity of the Boreal Forest Region, intrusions of this forest type do occur in the Sudbury area.

The natural ecosystem in the area was a pine forest. Three species, eastern white pine (*Pinus strobus* L.), red pine (*Pinus resinosa* Ait.) and jack pine (*Pinus banksiana* Lamb.), grew on the soil types which were suitable to the particular species. Eastern hemlock (*Tsuga canadensis* Carr.) and white spruce (*Picea glauca* Voss) were the other main coniferous species in the climax forest. Extensive logging of the virgin forest commenced around 1872 (de Lestard, 1967) with the white pine (*Pinus strobus* L.) being the most valuable economic species. Although some of the pine are returning naturally, the principal forest cover is that of a succession forest composed of American white birch (*Betula papyrifera* Marsh.), trembling aspen (*Populus tremuloides* Michz.) and large tooth aspen (*Populus grandidentata* Michx.) (Rowe, 1972).

Mean annual temperature for the Sudbury Climatic Region of Northern Ontario is 4.4 °C (Chapman and Thomas, 1968). The area has an average frost free period of 112 days and an average annual growing period of 183 days. Mean annual precipitation is 0.84 m.

7.8.3 MINE TAILINGS PROBLEM

The discovery of deposits of nickel and copper sulphide ores in the Sudbury District occurred in 1883 during the building of Canada's first transcontinental railway. The International Nickel Company was formed by the merger of several smaller companies in 1902. Over the years, the Company has grown. At the present time, Inco Limited is the largest mining, milling, smelting and refining complex in the non communist world. In the Sudbury area, the Company operates eight mines, two mills, a smelter, a sulphuric acid plant, a copper refinery and a nickel refinery. At Shebandowan, in Northwestern Ontario, a mine and mill are operational, and a nickel refinery is located at Port Colborne in Southern Ontario. A total work force of approximately 12200 is employed in these operations. Altogether 15 elements are extracted from the ores mined. The list includes nickel, copper, iron, sulphur, gold, silver, cobalt, platinum, osmium, iridium, selenium, tellurium, palladium, rhodium and ruthenium.

In the Sudbury area, the rated capacity of the two mills, Clarabelle and Frood-Stobie, is 53 524 tonnes per day. This figure will give an indication of the size of the daily milling operation. At full production, the milling and concentrating is a continuous process operating 24 hours per day, 7 days per week. Production operations vary with market requirements.

Replacement of the roast yard system of smelting by the Herreschoff Roaster and the reverberatory furnace in the new smelter in 1930, necessitated the development of a finely ground concentrate. Finding disposal sites for the tailings, produced as a waste by-product of the new system, became a necessity. Nearby valleys located between the rock hills provided these areas and the rock outcrops were used to act as buttresses for the tailings dams and as dams themselves.

Currently the dams are built by first constructing a starting dam, using waste

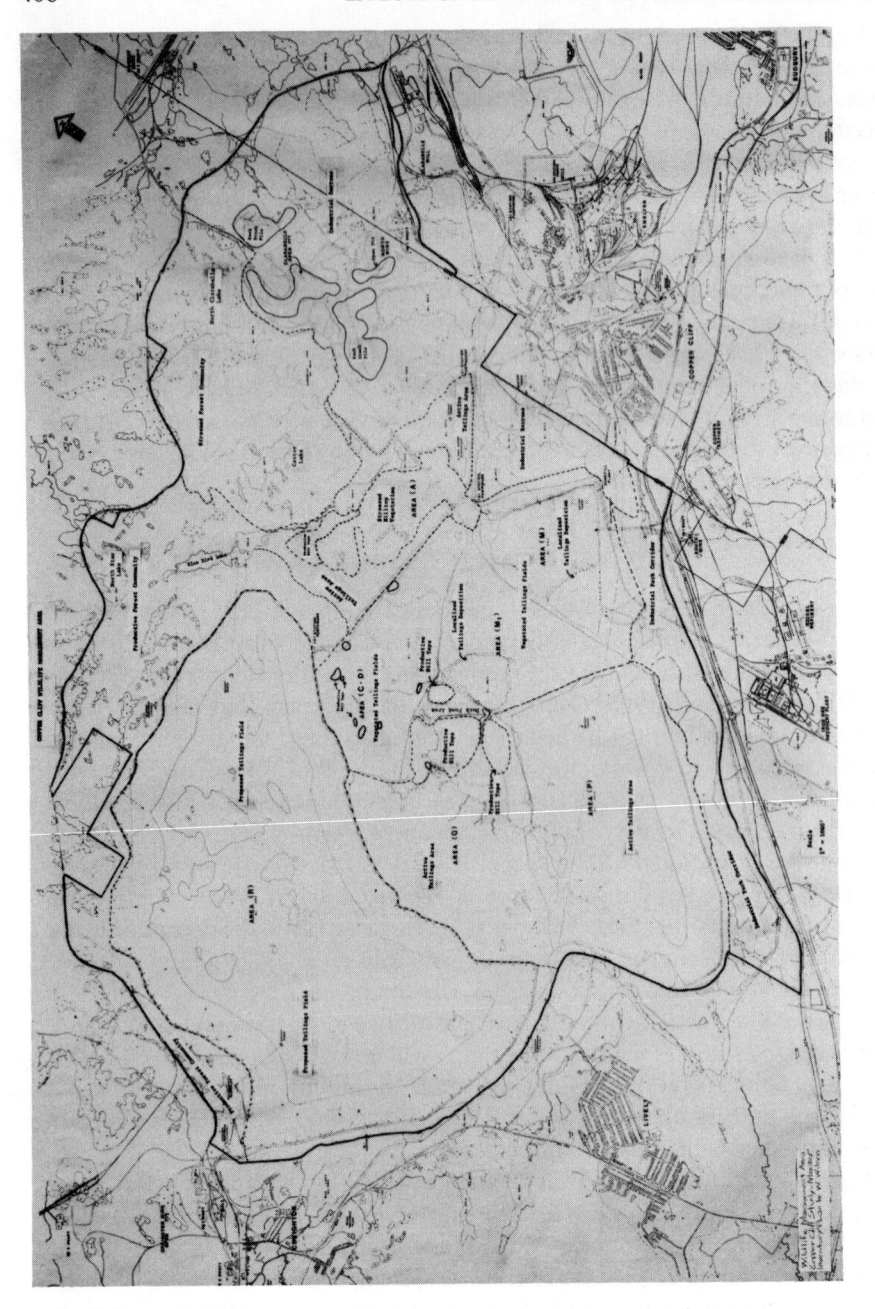

Figure 7.8.1 Map of the Copper Cliff mine tailings area showing the site of the revegetation study

rock of a suitable size. The height of the dams is then increased by raises of 3 m to 5 m using the tailings on site and by stepping in towards the centre with each raise. Dam heights vary with local requirements but are currently built to a height of approximately 50 m in the Copper Cliff Tailings Disposal Area at Sudbury. At present, the 'A', 'CD' and 'M' areas (Fig. 7.8.1) are at their final elevation and are at various stages in their reclamation and revegetation programmes. Their total area is approximately 485 ha. The 'P' and 'O' areas are nearing completion, and in areas where they have reached their final elevation, grass has been established. These two areas cover approximately 600 ha in total. The 'R' area, of approximately 1120 ha, is in the development stage. It will bring the total tailings area at Copper Cliff to approximately 2226 ha. Two smaller tailings areas totalling approximately 80 ha, are located adjacent to the mill at Levack Mine and at the Frood–Stobie complex.

Although it is difficult to come up with an average composition of tailings, since the composition will vary with the source and nature of the ores being mined at any given time, the following general figures may be utilized (Montreal Engineering Company, 1975; Wilson, 1976).

Mineralogical composition of copper cliff tailings
Sample size not reported

Feldspar	+50%
Amphiboles (chlorite)	20%
Quartz	10%
Pyroxenes	7%
Biotite	7%
Pyrrhotite	5.6%
Magnetite	0.6%
Pentlandite	0.5%
Chalcopyrite	0.3%

Trace element analysis of tailings soils (available concentrations). Range found in 12 sampling sites. Single sample taken at each site.

pH	3.7–6.2
Cu	1–81
Ni	1–87
Fe	59–441

Expressed in ppm on a dry weight basis.
Technique: 2.5 per cent acetic acid leach for 30 minutes.

In the early stages, when the elevation of the tailings is low, the surrounding hills minimize the effect of winds and thus limit the amount of dust generated from the tailings beaches. As the elevation of the disposal area increases, the wind protection afforded by the surrounding hills decreases, until a point is reached where this protection becomes non-existent.

The necessity of controlling dust from these sources became apparent to the Company in the late 1930s. In addition to being a nuisance to the residents of Copper Cliff, the dust contaminated certain electro-metallurgical refining processes and reduced the effectiveness of lubricants used to minimize normal wear in operating machinery.

Therefore, the immediate concern was to stabilize the surface of the tailings. At first, various stabilizers such as chemical sprays, limestone chips, bituminous sprays and timed water sprays pumped through an irrigation system were tried. These methods either proved to be ineffective or uneconomical. In the mid 1940s some experimental seedings of grasses were made but were unsuccessful, largely due to the unavailability of agricultural and ecological expertise at that time and the consequent failure to consider such factors as species selection and adjustment of the substrate pH.

7.8.4 REVEGETATION

In 1957 an experimental programme of revegetation plots was established on the tailings area. The successful stabilization of mine tailings with vegetation was the first step in developing new ecosystems. These systems were designed and developed to complement, as far as possible, the stage of development of the ecosystems in the surrounding area. With the knowledge gained from this and subsequent experiments over the years the present programme for establishing vegetation on tailings has evolved as follows:

1. The seeding should be established on that portion of the area which is closest to the source of the prevailing wind during the growing season to minimize covering or damaging of young plants by the drifting tailings.
2. Agricultural limestone, as required, should be applied at least 6 weeks prior to seeding; this permits sufficient time for reaction to raise the tailings pH to approximately 4.5–5.5.
3. In the Sudbury area late summer is the best time to seed grasses; after July 21, rates of success of seed germination and seedling establishment are enhanced due to more suitable temperatures and the increased availability of moisture.
4. Although this is the optimum time for seeding grasses, the short period which is left of the growing season is insufficient for legume seedlings to establish a sufficiently deep and strong root system to withstand the heaving effects of the repeated surface freezing and thawing the following spring.
5. The use of a companion crop, to reduce surface winds and provide some shade, is beneficial.
6. Nitrogenous fertilizer should be applied several times, as required during the establishment period, to ensure maximum uptake.

7. Slopes with southerly and southwesterly exposures should be mulched to provide shade for seedlings and to reduce evaporation of moisture from the soil during the critical period of seedling establishment.

As the preceding facts became obvious from experimental data and field programmes, it appeared essential to review the entire project from an ecological point of view in order to establish new project procedures.

Our review indicated that in order to establish a self-sustaining system, several ecological considerations should be taken into account:

1. Establishing initial plant communities using available species that are tolerant of drought, low soil pH, poor soil texture, the lack of organic materials and nutrients, and other factors characteristic of metal-extracted tailings.
2. Modifying the localized microclimates to benefit plant establishment.
3. Re-establishing soil invertebrate and microbial communities to ensure natural organic decomposition, essential to the rebuilding of soils.
4. Re-establishing essential nutrient cycles and conservation.
5. Establishing a vegetative habitat suitable for wildlife colonization.
6. Establishing the climax plant communities for the area by manipulating species competition.

Similar considerations have been outlined in other studies dealing with the problems of land restoration (e.g. Bradshaw et al., 1978).

Since the start of the tailings rehabilitation programme, the grass mixture used has been changed from time to time to increase the percentage content of the species that show hardiness and ability to persist under local conditions. The mixture currently used is:

    25% Canada blue grass (*Poa compressa* L.)
    25% Red top (*Agrostis gigantea* Roth.)
    15% Timothy (*Phleum pratense* L.)
    15% Park Kentucky blue grass (*Poa pratensis* L.)
    10% Tall fescue (*Festuca arundinacea* Schreb.)
    10% Creeping red fescue (*Festuca rubra* L.).

Other grasses which have been tried and are no longer used include crested wheat grass (*Agropyron cristatum* Gaertn.) and some varieties of fescue including sheep fescue (*Festuca ovina* L.). They have not shown comparable persistence over the years under the local conditions.

The practice of late summer seedings has made it difficult to establish good stands of legumes for reasons previously explained. Various methods to overcome this problem have been and continue to be researched. One alternative method which has yielded positive results is to seed the tailings with legumes in the spring. Alfalfa (*Medicago sativa* L.) has persisted for over twenty years from an original spring planting. Other legumes with which successful stands have been established experimentally include white blossom sweet clover (*Melilotus alba* Desr.), yellow blossom sweet clover (*Melilotus officinalis* Desr.), alsike (*Trifolium hybridum* L.) and bird's-foot trefoil (*Lotus corniculatus* L.).

The current revegetation programme follows an orderly sequence starting early in the summer. Agricultural limestone is spread at the rate of 4.4 tonnes per hectare in late May or the month of June whenever possible. The limestone is then disced into the tailings' surface. The major portion of our seedings is carried out during the last week of July and early August. Recent experience has taught us the necessity of waiting to seed until after the first heavy rain following July 20. This rainfall provides moisture for a sufficient period for germination and encourages the companion crop, fall rye (*Secale cereale* L.), to grow rapidly to a height providing early shade and reducing the velocity of surface winds. These two factors make a significant contribution to the successful establishment of the slower growing grasses.

At seeding time, an additional 4.4 tonnes per hectare of limestone are spread and the area disced, and then, 450 kg ha^{-1} of 5–20–20 fertilizer (5 per cent nitrogen, 20 per cent phosphate, 20 per cent potash) are broadcast over the area and harrowed into the top 5–8 cm of the surface. A conventional farm seed drill follows, seeding 94 kg ha^{-1} of fall rye (*Secale cereale* L.), 22 kg ha^{-1} of grass seed mixture and an additional 392 kg ha^{-1} of 5–20–20 fertilizer. Then, 9–11 kg ha^{-1} of brome grass (*Bromus inermis* Leyss.) are broadcast and a Brillon seeder is used to plant an additional 22 kg ha^{-1} of grass seed mixture. This machine compacts the soil and presses both the brome grass seed and the final seeding into the surface of the tailings where it germinates best.

After the grass germinates, light top dressings of nitrogen are applied, as required, during the balance of the season. Limestone and fertilizer are applied in subsequent seasons, on the basis of soil tests, to maintain and encourage growth.

In an effort to introduce legumes into the resultant sward, the inoculated, desired legume is seeded as early as possible in the following spring using a powertill seeder. Bird's-foot trefoil (*Lotus corniculatus* L.) is the legume currently being used.

These seeding practices generally provide the best germination and establishment, and are used whenever contours permit the use of standard agricultural machinery. Outside tailings berm slopes, and other slopes (sand pit, clay pit, or roadside) encountered in our reclamation work, do not usually permit the use of this equipment. Unique tillage implements, such as Klod-Busters, are utilized and a hydroseeder is necessary for fertilizing and seeding. It has been our experience that a mulch is very beneficial, if not essential, in promoting early germination and survival of seedlings on southerly and southwesterly facing slopes. Mulches and chemical stabilizers can be applied with a hydroseeder or straw mulches may be applied by using a mulcher.

It has become apparent that a new ecosystem, with an altered substrate was being established. The rock materials were finely ground but had not had sufficient time to weather. The physical sizing of the particles was in a smaller range, the drainage was different and organic matter was completely absent. The site had been raised up to 46 m and had become a flat saucer-shaped basin in contrast to the original topography of undulating rocky hills interspersed with small valleys.

The 'CD' area, the first area to be revegetated, has been the subject of many studies related to soil and plant development. Watson (1970) commented on the high surface and reflected temperatures in artificial waste areas. This reflected heat from the tailings' surface necessitates the use of a fast growing companion crop and shows the necessity of getting some vegetation on the surface to permit natural forest regeneration. Labine (1971), in his study of monoliths of the 'CD' area soils, found the development of an organic horizon (A°) of 2–3 cm and the beginning of a podzolic soil profile. It was found that drainage at different slope elevations quantitatively affected soil profile development and the formation of an iron pan at varying depths (Swanson, 1977). Root penetration of the pan was through cracks. However, single hair roots penetrated into the iron pan from thick tangles at root stop.

The physical ability of soil to retain moisture and nutrients is essentially a function of particle size. Tailings, which are generally deficient in clay-sized particles, behave much like a sandy loam and may be prone to moisture deficiency (Dimma, 1981). Lacking colloidal moisture adsorption, it would appear that the water which is retained in tailings has resulted from capillary action, which is most pronounced in the silt-sized particles (Pitty, 1979).

Under the above conditions, the rate of leaching of plant nutrients is high. Thus, in the early years' continuous nitrogenous fertilizer applications have been necessary to ensure that sufficient nitrogen was available for the successful growth of grasses.

Also, due to the fixation of phosphates by the high level of iron oxides found in the tailings, the amout of phosphorus which is actually available for plant growth may be significantly lower than analytical results indicate (Dimma, 1981).

Light annual applications of general fertilizers were made to supplement the availability of phosphorus and other nutrients.

We have found that the establishment of grass is only the first step in the return of a reclaimed area to the climax vegetation which is to exist for each site. Our early thoughts had been agriculturally oriented, the first step being the cutting of grass for hay. However, before long, seedling trees were observed voluntarily establishing themselves in the grassed areas. By the mid-1960s, a programme to gradually reduce the area being mowed, in order not to cut the seedlings, was initiated.

The voluntary invasion of birch trees (*Betula papyrifera* Marsh.) which started in the mid-1960s continues in all areas. Trembling aspen (*Populus tremuloides* Michx.) and willow (*Salix spp.*) constituted less than 10 per cent of the total voluntary tree seedlings in the early stages of forest invasion. The Trembling aspen, due to their vegetative method of propagation, are spreading in the established 'CD' site where they are becoming the numerically dominant species in localized areas. A few oak (*Quercus rubra* L.) have begun to appear, probably due to rodents or birds transporting acorns from the adjacent hills. Experimental plots of hybrid poplar and black locust (*Robinia pseudo-acacia* L.) have been established.

Replicated plots, to study the potential for the introduction of coniferous species, were established in 1972 with the following species: jack pine (*Pinus banksiana* Lamb.), red pine (*Pinus resinosa* Ait.), scots pine (*Pinus sylvestris* L.), white spruce (*Picea glauca* Voss) and black spruce (*Picea mariana* Mill.) The jack pine (*Pinus banksiana* Lamb.) has proven to be the most adaptable to date. Seven years after planting, as forestry seedlings, they have reached a height of 3–3.6 m and produced seed cones.

These plots indicated that in specific locations red pine (*Pinus resinosa* Ait). and white spruce (*Picea glauca* Voss) are the next most adaptable. Since 1976, 5000–15 000 forestry seedlings of these particular species have been planted annually. Although these seedlings must compete with the grasses for water and nutrients, a 60 per cent success rate for planting has been achieved.

A recent (1980) study of the plant species found growing in the 'CD' area showed the slow rate of volunteer introductions and the gradual development of dominant species (McLaughlin *et al.*, 1982) (see Table 7.8.1).

Along with the invasion of flora, various insects, grasshoppers, spittle bugs, ants and small mammals (deer-mice and voles) have colonized the new ecosystem. Meadow birds attracted by the small insects became resident, and sparrow hawks, marsh hawks and owls, along with foxes, began to provide predatory control of the small mammals. Waterfowl and shorebirds began making stops during their spring and fall migration and some even began to raise their broods in the area.

In 1974, after consultation with local Rod and Gun Clubs, a decision was made to develop the tailings and adjacent land as a Wildlife Management Area. A wildlife biologist was then hired to design a plan of development for the Wildlife Management Area and negotiations were started with the Ontario Ministry of Natural Resources to have the area so designated. The plan was completed in May 1976, and since that time investigations of various components of the plan have been carried out to determine the final route which will be taken. At present, we are investigating whether or not heavy metals accumulate in plants and small mammals which inhabit the area. Studies are also being conducted on birds in the area, including tests for heavy-metal levels in the blood, feathers and muscles of waterfowl. A general bird species census and a kestrel nesting and banding programme are also in progress (see Table 7.8.2).

The period from 1976 to the present (1981) is relatively short, and this makes the measurement of recognizably significant differences almost impossible. However, if these measurements are examined as being indicative of developing trends, they are of interest to researchers observing the broad spectrum of ecosystem development.

Table 7.8.1 Checklist of plants found on the 'CD' area in 1979–1980

			1	2	3	4
Trees and Shrubs	Red Maple	*Acer rubrum*	+	+	+	+
	Paper Birch	*Betula papyrifera*		+	+	+
	Honeysuckle	*Diervilla lonicera*		+	+	+
	Jack Pine	*Pinus banksiana*	+		+	+
	Trembling Aspen	*Populus tremuloides*		+	+	+
	Red Oak	*Quercus rubra*		+	+	+
	Black Locust	*Robinia pseudo-acacia*	+		+	+
	Long-beaked Willow	*Salix bebbiana*		+	+	+
	Common Elder	*Sambucus canadensis*			+	
	Blueberry	*Vaccinium angustifolium*		+	3.4	1.2
Grasses	Redtop	*Agrostis alba* (= *A. stolonifera* L., *A. gigantea* Roth)	13	+	50.	30.
	Hairgrass, Ticklegrass	*Agrostis scabra* (= *A. hyemalis* Walt)			2.7	0.7
	Crested Wheat Grass	*Agropyron cristatum*	6	+	+	+
	Witch Grass, Couch Grass	*Agropyron repens*		+	+	+
	Awnless Brome	*Bromus inermis*	24	+	17.	6.9
	Red Fescue	*Festuca rubra*	6	+	+	+
	Timothy	*Phleum pratense*	6	+	2.7	0.1
	Reed Canary Grass	*Phalaris arundinacea*		+	+	+
	Wiregrass	*Poa compressa*	38	+	2.1	0.3
	Kentucky Bluegrass	*Poa pratensis*	6	+	50.	34.
	Alkali-grass	*Puccinellia distans*			3.4	1.8
	Rye	*Secale cereale*	+		+	+

Table 7.8.1 (*cont.*)

			1	2	3	4
Sedges, Rushes and Cat-Tails	Sedge	*Carex siccata*			0.7	0.2
	Rush	*Juncus filiformis*		+	+	+
	Bulrush	*Scirpus lacustris*	+	+	+	+
	Cat-tail	*Typha latifolia*	+	+		
Horsetails	Meadow Horsetail	*Equisetum pratense*		+	+	+
Legumes	Bird's-foot Trefoil	*Lotus corniculatus*	+	+	2.1	1.7
	Alfalfa, Lucerne	*Medicago sativa*	+	+	2.7	0.6
	White Melilot	*Melilotus alba*	+		+	+
	Yellow Sweet Clover	*Melilotus officinalis*	+		+	+
	Alsike Clover	*Trifolium hybridum*	+			
Field Weeds	Pearly Everlasting	*Anaphalis margaritacea*		+	+	+
	Milfoil	*Achillea millefolium*		+	+	+
	Lamb's-quarters	*Chenopodium album*		+	+	+
	Sheep Sorrel	*Rumex acetosella*		+	+	+
	Goldenrod	*Solidago graminifolia*		+	+	+
	Spiny-leaved Sowthistle	*Sonchus asper*				+
Mosses		*Pohlia nutans*			2.1	0.9
		Leptodictyum riparium			+	+

1: Plants seeded; numbers are per cent composition of the original seed mixture (rate of seeding = 38 kg ha^{-1}).
2: Species found by Wilson in 1975.
3 and 4: Species found in this study; per cent frequency and per cent cover, along the transects, respectively (McLaughlin *et al.*, 1982).

Table 7.8.2 Record of bird species sited, 1973–1981 (no records available for 1978–79)

			1973	74	75	76	77	80	81[1]
Loons	Common Loon	*Gavia immer*						+	+
Herons	Great Blue Heron	*Ardea herodias*						+	
Waterfowl	Whistling Swan	*Cygnus columbianus*					+		
	Canada Goose	*Branta canadensis*	+	+	+	+	+	+ + +	+ +
	Snow Goose	*Chen caerulescens*	+	+				+	
	Mallard	*Anas platyrhynchos*		+ +	+	+		+ + +	+
	Black Duck	*Anas rubripes*	+	+ +					
	American Widgeon	*Mareca americana*		+ +					
	Pintail	*Anas acuta tzitzihoa*		+ +	+	+	+ + + +	+ +	
	Green-winged Teal	*Anas carolinensis*	+	+ + +					
	Redhead	*Aythya americana*	+	+ +				+ +	
	Ringed-neck Duck	*Aythya collaris*	+	+ +	+ + +				
	Lesser Scaup Duck	*Aythya affinis*	+	+ +	+				
	Golden-eye	*Bucephala clangula*							
	Bufflehead	*Glaucionetta albeola*				+ +			
	Old-squaw	*Clangula hyemalis*							
	White-winged Scoter	*Melanitta fusca*							
	Surf Scoter	*Melanitta perspicillata*	+ +						
	Hooded Merganser	*Lophodytes cucullatus*	+ +						
Hawks	Red-tailed Hawk	*Buteo jamaicensis*							+ + +
	Sparrow Hawk	*Falco sparverius*		+		+		+ +	
	Marsh Hawk	*Circus cyaneus*							+ + +
	American Kestrel	*Falco sparverius*	+ +				+		
	Merlin	*Falco columbarius*	+ +	+					+
Cranes	Sandhill Crane	*Grus canadensis*						+	
Rails, Gallinules, Coots	Coot	*Fulica americana*			+				

Table 7.8.2 (cont.)

			1973	74	75	76	77	80	81[1]
Shorebirds	Semi-palmated Plover	*Charadrius semipalmatus*	+	+	+	+	+	+ +	
	Killdeer	*Charadrius vociferus*	+	+		+			+
	Golden Plover	*Pluvialis dominica*	+	+	+ +	+ + +	+ +	+ + + + +	+ +
	Black-bellied Plover	*Squatarola squatarola*	+ +	+ +	+ +				+ +
	Ruddy Turnstone	*Arenaria interpres*		+ +		+			
	Spotted Sandpiper	*Actitis macularia*		+ +					
	Solitary Sandpiper	*Tringa solitaria*		+					
	Greater Yellowlegs	*Totanus melanoleucus*	+	+ +	+ +	+ + +	+ + +	+	+ + +
	Lesser Yellowlegs	*Totanus flavipes*	+ +	+ +	+ + + +		+ +	+ +	+ +
	Redknot	*Calidris canutus*		+ +					
	Pectoral Sandpiper	*Erolia melanotos*	+ + + +	+ + + +	+ + + +	+ + +	+ +	+	+
	White-rumped Sandpiper	*Erolia fuscicollis*		+ + + +			+		
	Bairds's Sandpiper	*Erolia bairdii*		+ + + +					
	Least Sandpiper	*Erolia minutilla*		+ + + +					
	Dunlin	*Erolia melanotos*		+ + + +					
	Dowitcher	*Limnodromus griseus*	+	+		+	+	+ +	+
	Stilt Sandpiper	*Micropalama himantopus*	+		+ +	+	+	+	+ +
	Semi-palmated Sandpiper	*Ereunetes pusillus*		+ + +				+	
	Western Sandpiper	*Ereunetes mauri*							
	Buff-breasted Sandpiper	*Tryngites subruficollis*							
	Hudsonian Godwit	*Limosa hoemastica*							
	Sanderling	*Crocethia alba*							
	Wilson's Phalarope	*Steganopus tricoler*							
	Northern Phalarope	*Lobipes lobatus*							
Gulls, Terns	Herring Gull	*Larus argentatus*				+ +		+	+ +
	Ringed-billed Gull	*Larus delawarensis*						+	
	Caspian Tern	*Hydroprogne caspia*						+	
Woodpeckers	Common Flicker	*Colaptes auratus*							+
	Hairy Woodpecker	*Dendrocopus villosus*							

		1973	74	75	76	77	80	81[1]
Tyrant Flycatchers								
Eastern Kingbird	*Tyrannus tyrannus*						++	+
Eastern Phoebe	*Sayomis phoebe*	+					+	
Horned Lark	*Eremophila alpestris*			+	+	+	++	
Swallows								
Tree Swallow	*Iricloprocne bicolor*			+	++	+	++	+
Bank Swalow	*Riparia riparia*						+++	++
Jays, Crows								
Blue Jay	*Cyanocitta cristata*						++++	
Common Raven	*Corous coxax*							
Common Crow	*Corous brachyrhynchos*							
Titmice, Wrens Thrushes								
Black-capped Chickadee	*Parus atricapilla*							
House Wren	*Troglodytes aeclon*							
Brown Thrasher	*Toxostonia rufrum*							
American Robin	*Turclus migratorius*							
Hermit Thrush	*Hylocichla guttata*							
Pipits, Waxlings, Starlings								
Water Pipit	*Anthus spinoletta*	+		+		+	++	+
Starling	*Sturnus vulgaris*							
Vireos, Woodwarblers								
Philadelphia Vireo	*Vireo philadelphius*						+++++	+++
Solitary Vireo	*Vireo solitarius*							
Yellow-rumped Warbler	*Dendroica coronceta*							
Bay-breasted Warbler	*Dendroica castanea*							
Palm Warbler	*Dendroica planarum*							

Table 7.8.2 (cont.)

		1973	74	75	76	77	80	81[1]
Meadowlarks, Blackbirds, Orioles								
Bobolink	*Dolichonyx oryzivorus*							+
Eastern Meadowlark	*Sturnella magna*						+	+
Western Meadowlark	*Sturnella neglecta*		+					+
Red-winged Blackbird	*Agelaius phoenicius*						+ +	
Brown-headed Cowbird	*Molothrus atu*						+ +	+ + +
Baltimore Oriole	*Icterus galbula*							
Finches, Sparrows, Buntings								
American Goldfinch	*Spinus tristis*							
Savannah Sparrow	*Pasberculus sandwichensis*	+				+ + +	+ + + + +	
Vesper Sparrow	*Pooecetes gramineus*							
Chipping Sparrow	*Spizella passering*					+	+	
White-throated Sparrow	*Zonotrichia albicollis*				+			
Slate-coloured Junco	*Junco hyemalis*							
Lapland Longspur	*Calcarius lapponicus*	+ +	+	+ +				
Snow Bunting	*Plectrophenax nivalis*	+ +						
Whimbrel	*Numenius phoepus*							

[1] July and August only (1981).

The revegetation of tailings areas, by establishing a grass cover crop as the initial step, in a region which is at the succession forest stage in the natural ecological process, effectively creates an islands of vegetation cover, contrasting with and differing from the surrounding plant communities. This is also reflected in the variety of different animal species which invade the newly revegetated stage. This difference is apparent in the species of insects, birds and mammals which invade the new area under a natural succession. Bird censuses (Nicholson, 1974–1976; Hendrick, 1980; Laing, 1981) described a collection of grassland species unusual to the surrounding climax vegetation areas. Waterfowl, shore birds, and other bird species not usually seen in the Sudbury area were observed, identified and recorded (Table 7.8.2). A study of the nesting of kestrels (*Falco sparverius* L.), over a period of several years, noted a change in the source of food for nestlings from predominantly insects to frogs in locales adjacent to the 'CD' area pond where bulrushes (*Scirpus spp.* L.) and cat-tail (*Typha spp.* L.) have successfully been established. These plant species have permitted the leopard frogs (*Rana pipiens*) to flourish (J. Lemmon, personal communications, 1980).

The investigation of heavy-metal levels has not as yet indicated the existence of a problem with these particular elements. Rutherford and Van Loon (1980) found that the mineral content of two species of grass, red top (*Agrostis alba* L.) and Canada blue grass (*Poa compressa* L.), growing on the tailings fell within the normal range for grasses, although the concentration of iron and copper appeared to be somewhat higher than the mean. Studies of duck species nesting and raising young in the area have yielded preliminary results which indicate that blood levels of nickel and copper in ducks are not significantly affected. Our findings correspond to those of Rose (1981) who in his research on ruffed grouse (*Bonasa umbellus* L.) in the Sudbury area found no significant change in body tissue metal content.

7.8.5 PRESENT AND FUTURE CONSIDERATIONS

Although the overall programme is in its early stages, with respect to the successional development of an ecosystem, research continues to further the progress of constructing new ecosystems. This is to ensure, as far as possible, that the directions taken are beneficial to the evolving wildlife communities. Current research activities include a detailed look at the soil development processes which are occurring, lime application studies, investigating methods of promoting the development of the microflora and microfauna responsible for the breakdown of organic matter and the subsequent release of nutrients, tree growth studies, monitoring nickel and copper levels in a variety of waterfowl species, and the introduction of new species of flora and fauna.

Industrial process changes are of concern, as we look to the future. The possible change in concentrating processes which may be implemented to lessen pyrrhotite content, in order to reduce sulphur dioxide emissions from the smelter to the atmosphere, will provide a challenge to be met. This change would increase the amout of pyrrhotite in the tailings, with a subsequent increase in acid generation potential. Therefore, a method to allow revegetation under this condition may have to be developed.

7.8.6 CONCLUSIONS

The development of a Wildlife Management Area on the Tailings Areas of Inco Limited at Sudbury, as the end use for this land, has been and will continue to be an attempt by man to influence the rate and direction of the development of a new ecosystem.

To date, there has been progress in the establishment of ecological processes on the tailings area as evidenced by the following:

1. The visual observations of soil horizons are clear evidence that soil development processes are occurring.
2. The stabilization of the tailings surface with the establishment of a grass cover is providing a satisfactory environment for colonization by native flora and fauna.
3. The voluntary invasion of trees into the grassed areas is a positive step in the natural vegetation succession to the climax forest state which existed before mining development.
4. The successful establishment of trees, both naturally and in plots, indicates that the environment is suitable for growth and development, and will likely support seedlings of other native tree and shrub species.
5. The number of birds nesting and being sighted on the tailings is evidence of a habitable environment and an adequate food supply.
6. Legumes are being introduced successfully and are accelerating the development of the natural nitrogen cycle.
7. To the best of our knowledge, the environment is not toxic to species living and reproducing on the Copper Cliff tailings.

7.8.7 REFERENCES

Bradshaw, A. D., Humphries, R. N., Johnson, M. S. and Roberts, R. D. (1978). The restoration of vegetation on derelict land produced by industrial activity. In Holdgate, M. W. and Woodman, M. J. (Eds.), *The Breakdown and Restoration of Ecosystems*, Plenum Press, New York and London, pp. 249–278.

Braun, E. L. (1950). *Deciduous Forests of Eastern North America*. Hafner Publishing Co. Inc., New York.

Chapman, L. J. and Thomas, M. K. (1968). The Climate of Northern Ontario. In *Climatological Studies No. 6*, Department of Transport, Meteorological Branch, Toronto, Canada.

de Lestard, J. P. G. (1967). A History of the Sudbury Forest District. *History Series*, Published Report no. 21 of the Sudbury District, 90 pages.

Dimma, D. E. (1981). *The Pedological Nature of Mine Tailings Near Sudbury, Ontario.* M.Sc. thesis, Queens University, Kingston, Ontario, 184 pages.

Freedman, B. and Hutchinson, T. C. (1980). Pollutant inputs from the atmosphere and accumulations in soils and vegetation near a nickel-copper smelter at Sudbury, Ontario, Canada. *Can. J. Bot.*, **58**, 108–131.

Gorham, E. and Gordon, A. G. (1960a). Some effects of smelter pollution northeast of Falconbridge, Ontario. *Can. J. Bot.*, **38**, 307–312.

Gorham, E. and Gordon, A. G. (1960b). The influence of smelter fumes upon the chemical composition of lake waters near Sudbury, Ontario and upon the surrounding vegetation. *Can. J. Bot.*, **38**, 477–487.

Heale, E. L. (1980). *Effects of Nickel and Copper on Several Woody Species.* M.Sc. thesis, University of Guelph, Guelph, Ontario, 220 pages.

Hendrick, A. (1980). *Bird Census, Copper Cliff Tailings Area.* Internal Report, Agricultural Department, Inco Limited. Unpublished.

Hooker, W. A. (1886). *Geological Survey of Canadian Copper Company's Properties in Sudbury.* Internal Report, Inco Limited. Unpublished.

Hutchinson, T. C. and Whitby, L. M. (1977). The effects of acid rainfall and heavy metal particulates on a boreal forest ecosystem near the Sudbury smelting region of Canada. *Water Air Soil Pollut.*, **7**, 421–438.

Labine, C. L. (1971). *The Influence of Certain Seeded Grasses on the Evolution of Mine Tailings.* B.Sc. thesis, Biology Department, Laurentian University, Sudbury, Ontario.

Laing, C. M. (1981). *Bird Census, Copper Cliff Tailings Area.* Internal Report, Agricultural Department, Inco Limited. Unpublished.

Montreal Engineering Company (1975). *Report on Sulphide Mine Tailings in Noranda, Timmins and Sudbury Areas.*

McLaughlin, B. E., Crowder, A. A., Rutherford, G. K. and Van Loon, G. W. (1982). Site Factors Affecting Semi-natural Herbaceous Vegetation on Tailings at Copper Cliff. *Reclamation and Revegetation Research* (in press).

Nicholson, J. (1974–1976). Manitoulin and Sudbury Districts, Monthly Bird Reports. Private distribution; available on request.

Pitty, A. F. (1979). *Geography and Soil Properties*, Methuen and Co. Ltd, London, 287 pages.

Rose, G. (1981). *Effects of Smelter Pollution on Temporal Changes in Feather and Body Tissue Mental Content of Ruffed Grouse Near Sudbury, Ontario.* M.Sc. thesis, Biology Department, Laurentian University, Sudbury, Ontario.

Rowe, J. S. (1972). *Forest Regions of Canada.* Canadian Forestry Service, Report No. 1300, Department of Fisheries and the Environment, Ottawa, Canada, 172 pages.

Rutherford, G. K. and Van Loon, G. W. (1980). The pedological properties of some tailings in the Sudbury rim area and the composition of vegetation associated with them. *Proceedings, Fifth Annual Meeting of the Canadian Land Reclamation Association*, Timmins, Ontario, pp. 43–55.

Swanson, S. (1977). *Tailings Field Monoliths and Tree Root Distribution.* Internal Report, Agricultural Department, Inco Limited. Unpublished.

Watson, W. Y. (1970). *Reclaiming Pollution-Scarred Lands.* Laurentian University. Unpublished.

Whitby, L. M., Stokes, P. M., Hutchinson, T. C. and Myslik, G. (1976). Ecological consequence of acidic and heavy-metal discharges from the Sudbury smelters. *Can. Mineral.*, **14**, 47–57.

Wilson, W. R. (1976). *Master Plan for the Development of the Copper Cliff Wildlife Management Area.* Internal Report, Agricultural Department, Inco Limited. Unpublished.

Part III
Conclusions

Part III

Conclusions

Effects of Pollutants at the Ecosystem Level
Edited by P. J. Sheehan, D. R. Miller, G. C. Butler and Ph. Bourdeau
© 1984 SCOPE. Published by John Wiley & Sons Ltd

CHAPTER 8
Conclusions and Recommendations

Patrick J. Sheehan[1], Donald R. Miller[1]
Gordon C. Butler[1], Philippe Bourdeau[2]

[1] Division of Biological Sciences
National Research Council of Canada
Ottawa, Ontario, Canada K1A 0R6
[2] Directorate General for Science, Research and Development
Commission of the European Communities, Brussels, Belgium

The most obvious conclusion from the survey and case studies is that ecotoxicologists have been slow to apply an ecosystem-level approach to the assessment of the effects of toxic pollutants even though, in some cases, theory and methodologies are sufficiently developed to support such activity. It is hoped that this volume will encourage further long-term research on stressed systems and on the evaluation of techniques for distinguishing and quantifying pollutant-related effects at levels of biological organization above that of individuals and populations.

The first question to be answered in assessing the pollution of an ecosystem is: 'How polluted is it, or might it be in the future?' This has led to concern because of the lack of available data on production and release of chemicals which might achieve levels sufficient to cause a regional or even a global hazard. Only for a few very specific chemicals, in restricted areas, is anything like complete information available. Even an elementary hazard ranking scheme based on production and environmental persistence of a large number of potential environmental contaminants would be useful.

Another basic question that ecotoxicologists must ask is whether or not a change is real and whether it can be attributed to the pollutant. If a quantitative assessment of impact is to be made, far greater care must be taken to define natural variability in population and ecosystem parameters, and to determine the amplitude and patterns of natural fluctuations in structure and function.

Individual and population characteristics (behaviour, biochemistry, physiology, etc.) are quite sensitive to toxic chemicals but the meaning of such adverse responses in relation to the long-term success of the inhabitants or functioning of the ecosystem is not fully understood. In cases such as that of low reproductive success associated with pollutant-induced behavioural abnormalities, the significance of individual response to the well-being of the population is obvious. On the other hand, the implications of the loss of a particular population in a particular area, as a result of pollution, remain largely unresolved.

Special consideration in the interpretation of the initial community response to pollution must be directed towards such factors as life history stages of the resident species, potential for emigration, selection for tolerance and exposure potential associated with particular habitats, since these can influence the long-term adaptation of populations to pollution stress.

Studies of multispecies systems have been used to verify theories on ecology regarding the importance of competition, predation and other relationships as factors in shaping community response to a perturbation. The application of data on competition, to pollutant impact assessments, has been particularly productive in both aquatic and terrestrial plant studies, and information on predation has been valuable in the interpretation of certain aquatic trophic linkages. However, the influence of population interactions on community response to toxic stress needs further clarification, since this may provide the mechanistic information for predicting changes.

There is an important need to determine key species, in relation to ecosystem function. These may be organisms with unique functional roles, such as pollinators, or those which through competitive or predatory interactions exert a major influence on community composition and on the movement of energy and materials. The dynamics of such species may be the factor that determines the rate of whole system change.

There are a variety of quantitative structural indices describing community organization which provide conflicting appraisals; therefore, a good deal of disagreement exists over their utility in monitoring stress. This is particularly evident for diversity indices with regard to their sensitivity and consistency in measuring structural change. In many instances, simple indices requiring little or no calculation (such as species richness) appear to be more sensitive to the presence of a pollutant than do more complex indices.

The concept of ecosystem stability is basic to any assessment of changes brought about by environmental pollutants. Various measures of the dynamic performance of an ecosystem are available to describe the response to perturbations. There is evidence suggesting that the buffering capacity (homeostatic ability) of a system can be measured and used to predict the amount of pollutant which can be absorbed before unacceptable changes result. Contrary to the prevalent idea that stress increases fluctuations in ecosystem indices, it has been found that gross levels of pollution often suppress both temporal and spatial variability in community composition. Furthermore, recovery of an ecosystem following pollution abatement should not be expected to follow the reverse sequence of degeneration, nor should the new state be expected to match the original configuration.

The implications of long-term structural changes for ecosystem function are not clearly defined. In addition to examining such changes in detail, it is important to examine processes which reflect the functioning of the whole system (e.g. primary productivity) and to identify key processes which are pivotal to the functioning of particular types of ecosystems. Furthermore, it is necessary to consider the yield of the entire system in economic terms.

The importance of assessing interference with organic decomposition and nutrient recycling processes must be emphasized. In this regard, quantitative data on nutrient pools and fluxes and on the relationship of element 'leakage' to long-term productivity are essential to the refining of interpretations of biogeochemical disruptions.

Enclosure techniques are providing an important experimental link between laboratory and field studies. Such methods should enable researchers to consider environmental and population interactions in their experimental designs. A further refinement of enclosure methodologies is required in order that results adequately forecast ecosystem responses.

Similarities between the results from short-term controlled exposure studies in natural systems and those from chronically contaminated systems suggest that ecosystem experiments (at least in streams) may have predictive value.

The usefulness of examining changes along a pollution gradient has been clearly demonstrated. The problem of locating an appropriate reference is minimized, and experimental difficulties related to temporal changes in system parameters can be partly overcome, through the built-in properties of the gradient approach to monitoring relative changes. These factors enhance the utility of this approach in providing both spatially and temporally comparable data.

In addition to the recommendations implicit in the above conclusions, special attention is directed to the need for developing procedures for making integrated quantitative assessment of the effects of pollutants on ecosystems, including the relative importance of these effects to problems of impact assessment, abatement and control. To permit realistic integration requires that all the most important responses be identified and expressed quantitatively in the most meaningful form. Moreover, the response should be quantitatively related to the causative exposure.

Since many polluted systems are contaminated by several toxic substances, it is essential that techniques be developed to link responses to the combined effects of specific groups of substances as well as to individual substances. Combined effects can also result from the addition of pollutants, at levels not normally injurious, to systems already endangered by a broad spectrum of existing chemical or physical stresses. It is imperative that these types of interactions be recognized and taken into account in ecosystem studies.

It is natural that biologists feel more comfortable studying the effects of single pollutants on individuals and populations of single species. Nevertheless, those with an interest in ecology must move beyond this foundation and address the more complex problems involved in assessing the well-being of whole ecosystems.

Subject Index

Abundance
   changes in, 48, 56, 86, 209–210, 261–264, 289
   estimation of, 18–20, 55
   factors in decline of whitefish, 297
   fluctuations in, 20–22, 55
   index, 55
   of species, 56, 57, 67, 70, 206, 224, 264
   relative, 65, 257
   zooplankton, 289
Accumulation of toxic materials, 144
Acid
   deposition of, 231
   precipitation (*also* acid rain), 9, 13, 60, 109, 113, 115, 117, 128, 131–136
   waters, 47
Acidic leaching, 111
Acidification
   changes due to, 193
   effect on aquatic systems, 56, 135, 136, 241–243
   effect on biomass, 237
   effect on forests, 109–110, 134–135
   effect on microbial activity, 114–117
   of lakes, 57, 59, 111, 136, 233–236, 243
   reversibility of effects, 230
Acidity and heavy metals, 118, 233, 366
Adsorption, 256
Agricultural insecticides, 58
Airborne pollutants, 47, 230
AlA-D activity, lead effects on, 31
Algae
   and metal tolerance, 237
   density of, 328
   effect of copper on, 49
   effect of oil on, 327, 345, 347, 359
   populations, 45, 237, 315
   resource, 315
Algal
   biomass, 56, 302
   growth, and flooding, 292
   production, 241, 292, 301

Alginates, hydrocarbon content, 331
Allele, 49
Aluminium
   and acidity, 118, 233
   buffering system, 112, 248
Ammonia production, 117
Ammonification, and pesticides, 127
Amplitude, 84, 89
Amylase activity, and metals, 122
Analysis
   of macroinvertebrate community, 255
   of physico-chemical data, 219
   of variance techniques, 257
   techniques, community changes, 193
Analytical techniques, 71, 81
Antagonistic interaction, 42
Antipredator behaviour, 45
Aquatic
   ecological changes, 274
   invertebrates, effect of fenitrothion on, 384
   systems, 11, 104, 107
Arthropods, metals, 58, 260, 263
Assessment, of pollutants at ecosystem-level, 48, 425
Autotrophic maintenance, 130
Avoidance behaviour, 29, 58
Avoidance response and copper, 30, 262

Bacteria
   and acidity, 114, 116, 240
   and base additions, 243
   symbiotic relationship with coral, 45
   tolerance to metals, 369
Bacterial colonization, acidity, 115–116
Bactericides, 325
Base additions, effect of, 243
Baseline data, 52, 54, 395
Baseline levels, 18–20, 89
Bees, effect of fenitrothion on, 395–396
Behaviour response, 29, 30, 44
Benthic macroinvertebrate fauna, 240, 276
Benthos, and pH, 113–114, 250

SUBJECT INDEX

Bicarbonate, and acidity, 233
Bioaccumulation
 of mercury, 46, 298, 300
 of pollutants, 26, 28, 53
Bioassay
 natural, 64, 256
 techniques, 25, 88
Biochemical
 interactions, 28
 response to pollutants, 31–32
Biocides, use of, 380–382
Biodegradation, effect on hydrocarbons, 319, 321, 323, 357
Bioenergetic processes, 103
Bioindicators, 66
Biological
 changes and trophic level, 303
 equilibrium, return to, 320
 indices, 69
 processes and sediment levels, 305–306
Biomagnification, 10
Biomass, 49, 51
 accumulation in, 131
 accumulation ratio, 104
 diversity, 73
 indices, 76
 of phytoplankton, 138, 139
 production, 103
 reduction in, 54, 56
 storage, 107
Biosphere Reserves, 20
Biota, distribution of, 197
Biotio
 interactions, 51, 94
 score, 67
 stress, 59
Biotransformation, of xenobiotics, 31
Bioturbation, 142
Birds, and fenitrothion, 393
Bray–Curtis index, 75, 81
Brood size, index use, 39
Buffering capacity, of ecosystems, 87, 269

Cadmium
 accumulation of, 370
 exposure, 45
 tolerance in plants, 369
Calcium, leaching, 110, 111, 118
Carbon, cycling of, 105
Carbon dioxide
 evolution, 117
 use by phytoplankton, 238

Carbon–nitrogen ratio, 121
Carcinogenic effects, aromatic hydrocarbons, 322
Carcinogens, 34
Cell division rate, 36
Cellulose, decomposition of, 115
Changes
 assessment of, 52
 detection of, 20
 in environmental conditions, 223
 in environmental quality, 79
 pollution-induced, 52, 80, 217, 256
 prediction of, 300
 temporal and spatial, 79
Chemical
 composition of rainfall, 232
 composition of water and sediments, 233, 277
 degradation, 13
 energy, 105
 form, 10
 injury, 47
 substance, attributes of, 7
 transformation, 11, 12
Chemicals
 absorption and retention by biota, 11
 distribution of, 10
 effect on growth rate, 37
 entries into environment, 8–9
 in aquatic systems, 11
 physical and chemical properties of, 388
 priority, 7
 sensitivity to, 43
 types of, 9
Chironomid
 communities, changes in structure, 267
 larvae and heavy-metals, 29, 266
Chlorophyll
 concentrations, and impoundment, 292
 vertical profiles, 236
Chlorosis, 137
Chronic low level exposure, 55
Climate, effect on recovery, 93
Cluster analysis, 71, 79, 217–219, 224
Co-evolve, 52
Cobalt, contamination, 46
Coefficient of variation, 249, 267
Community
 abundance and water exchange, 278
 changes in, 81, 213, 256, 266, 270, 426
 composition, 64, 67, 70, 266, 268, 426
 control, 54

SUBJECT INDEX

definition of, 51
dynamics, 67, 242
effect of metals, 71, 80, 81, 261, 369
elasticity, 88
equilibrium, 207
interactions, importance of, 193
interactions, 241–242
interpretation of initial response to pollution, 426
level, 49
plant morphometry, 82
recovery of, 88, 92, 95–98, 409
statistical methods for comparisons, 71
structure, characteristics, 52, 58, 79, 209, 269, 270
structure, indices of, 55
Competitive
  inhibition, of metalloenzymes, 137
  interactions, pollutant effects on, 47
  success, 70
Concentration gradient, 255, 256, 270, 427
Constancy, 84, 86, 268
Consumers
  abundance of, 70
  interaction, 143
  regulation of energy, 143
Contamination, oil, 337, 339, 340
Copper
  and photosynthesis, 241
  atmospheric input, 230
  contamination, 57, 120, 257–260
  effect on stream ecosystems, 257–262, 264–266
  effect on flies, 48
  effect on growth, 36
  effect on marine copepod, 34
  effect on phytoplankton, 34, 49
  effect on zooplankton, 238
  mining of, 405
  toxicity and fish, 245
  vegetation parameters, 63
Corticosteroid response, 32
Courting and mating, 37–38
Crustaceans
  and impoundment, 289
  effect of oil on, 44, 329, 332, 337, 357, 359
  hydrocarbon content, 332
  production, 316
  resource, 316
  skin ulcers, 348

Cycling
  index, definition and use of, 145
  of metals, 243

D-glucose uptake, effect of oil on, 34
Dab population, effect of oil on, 355
DDT, 9, 10, 39, 46, 140, 380–381
Death, premature, 47
Decomposer organisms
  effect of fenitrothion on, 398
  effect of metals on, 120–124, 373
  sensitivity to pH, 108, 113, 114
Decomposition
  and acid rain, 115
  and metals, 122–123, 371–374
  and pesticides and oil, 124
  of litter, 123, 125, 371, 372, 374
  organic, 50, 427
  pollutant effects on, 128
  system, 61, 105
Degradation
  of chemicals, 13, 386
  of fenitrothion, 388
Dehydrogenase activity, 122
Density
  estimates, 264
  of populations, and metals, 261–264
  population, 256
Density, reductions in, 58, 59
Density-dependent regulation, 94
Detergents, synthetic, 203
Detoxification, of oil in fish, 353
Detrital materials, 105, 108
Diatoms
  analysis of, 235
  community structure, 61
  dominance, 70
Diet alterations in, 46
Dietary strategy, 143
Diflubenzuron, 46
Dilution, 256
Dispersants
  effect of, 324
  of oil, 360
Dispersants, surface-active agents, 324–325
Dissolved oxygen, 224, 291
Distribution, changes in, 48
Diversity
  factors affecting, 73
  indices of, 72–75, 97, 209, 256

SUBJECT INDEX

Diversity (*continued*)
   measures, sensitivity of, 79
   of profundal species, 288
   pollutant effects on, 76
   Shannon's value, 73
   taxonomic, 52, 87
   trophic, 54
DNA, damage by toxicants, 32
Dominance
   changes in, 49, 73, 87, 97, 208
   component, changes in, 266
   indices of, 71
   patterns, 64, 67
   pollution effect on, 52, 70
   relative, 213–215
   switches, 76
   trophic, 71
Drawdown
   and fish populations, 296–297
   rate of, 303
Drought, 42
Dynamic simulation models, 103
Dynamics
   community, 67, 242
   measures of, 85

Ecological
   considerations for self-sustaining system, 409
   interactions, 49, 52, 61, 394
   monitoring, 385–387, 395
   succession, definition of, 92
   tolerance, 65
Ecology
   impact of aquatic changes, 274
   impact of hydroelectric development, 280
   impact of oil spills, 313
   master-factors, 197
   recovery of, 200
Ecosystem
   amount of stress, 89
   and energy, 104
   assessment of changes, 1, 2, 103, 145, 193, 269, 425, 427
   buffering capacity, 269
   change, models of, 248
   changes, 1, 47, 256
      detection of, 20
   characteristics, 1, 82, 86, 101, 389
   definition of, 3, 49, 51
   development
      of new, 410, 419, 420
         strategies, 103
   dynamics of polluted, 24, 426
   effect of metals on, 371
   fluctuations in, 84
   freshwater, 64
   function, 61, 241–243
   growth perturbation, 131
   homeostasis of, 105
   indices of welfare, 2
   inertia of, 87
   measurement of dynamics, 85
   model studies, 18
   persistence of, 104, 105
   perturbation, 109, 274, 392, 394–395
   pollution studies, assessment of, 103
   processes, changes in, 50, 109
   productivity, 131, 138
   quantitative examination of, 15
   recovery, 93, 95, 243
   regulation of, 141
   response to pollution, 2
   restoration, 91, 403
   stability, 72, 268, 426
   stabilization of, 301
   state of, measurement of, 246–247
   stream, 270
   structural changes, implications of, 426–427
   structure
      and function, 9, 44, 53, 103
      fluctuation of, 86
   variation of properties, 246
Ecotoxicological assessment, 49
Effluents, industrial, levels of, 201, 202
Egg shells, effect of DDT on, 47
Eggs, number per female, 37
Elasticity, 84, 89
Electrophoretic analysis, 49
Element outputs, 131
Elements, uptake of, 137
Embryo development, 37, 38, 43
Energy
   and ecosystems, 104
   and material movement, 104, 106
   availability, 130
   balance, 139
   chemical, 105
   flow, 71, 141, 144
   metabolism, 105
   of a river, 301
   pool, 130
   regulation, 141
   studies, 103
   transfer, 242
   wave, 282

SUBJECT INDEX

Environment
 changes in, 16, 223, 300
 effects of flooding on, 306
 effects of hydroelectric development, 304
Environmental
 factors, stress, 89
 fluctuations, 59, 87, 101
 homogeneity, 255
 impact assessment, 49, 65
 quality, 66, 79
Epilimnion, 206
Equilibrium
 community, 207
 structures of ecosystems, 83
Equitability, 72, 74, 77
Erosion
 and physical stabilization, 301
 index, 282, 301
 shoreline, 281, 291, 305
Essential elements, and organic pollutants, 126
Eutrophication, 347
Evenness
 indices, 73, 75, 209, 212, 224, 256
 related to stress, 73, 87
Evolution, rate of, 306
Exposure, chronic low level, 55
Extinction, coefficients, 236
Extinction
 gradual, 60
 population, 58, 59, 62

Fecundity, 48
Feeding
 activity, 34
 effect of pollutants on, 43, 46
 interactions, 47
Fenitrothion
 effect of, 61, 377
 effect on bees, 395, 396
 effect on birds, 393
 effect on decomposer organisms, 398
 effect on food chain, 391
 effect on pollinators, 396
 formulations of, 386
 isomers of, 387
 microbial degradation, 392
 persistent residues, 390
 photolysis, 392
Fertilization, 37, 38
Fertilizers, 11
Filtering, rates of, 242
Filtration-rate, effect of oil on, 34

Fire
 and heavy metals, 137
 forest, 378
Fish
 and acidification, 240–241
 and oil spills, 330, 349–353, 355, 359
 communities, 217, 302
 detoxification mechanism, 357
 extinction of, 59, 248
 fin and tail rot, 47, 351
 hydrocarbon content, 332, 337
 mercury concentrations, 293, 297, 299, 300, 302, 305
 populations, and drawdown, 296
 production, 303
 recovery of populations, 209
 resource, 317
 spawning, conditions of, 302
 species
  reductions in, 60
  seasonal cycle, 211
 stocking, effect of, 244
 stocks, 206, 211, 216, 295, 354
 structure of flatfish populations, 356
Fisheries
 and hydroelectric development, 293
 resources, 315, 317
Fishing
 distribution of, 294
 effect of insecticide spraying on, 384
 effort of, 295
Flooded soils and mercury, 300
Flooding
 and evolution of aquatic systems, 306
 and macrobenthic invertebrates, 287
 and nutrients, 292, 301
 environmental effects of, 290, 306
Fluctuations
 amplitude of, 213
 ecosystem indices, 86, 426
 natural, 15, 16
 of numbers of individuals, 209
 seasonal, 55, 212
Foliar
 injury and heavy metals, 118
 leaching, 109, 133
Food
 conversion efficiency, 35
 resources, 71
 supply, 45
 web (*also* Food chain), 49, 105, 141, 143, 207, 208
  and fenitrothion, 391
  and methylmercury, 46

Food, web (*continued*)
    length of, 141
    reduction of supply, 143
Function, disruption of, 52
Fungi, tolerance to metals, 369
Fungicides, effects of, 126

Gametes, development of, 37, 38
Gamma radiation, 88
Gas chromatography analysis, 331
Gene pool, change in, 48
Generalists, 71
Genetic effects of oil, 48, 49
Gill respiration rate, 35
Global approach, 84
Gradient
    concentration, 270
    studies (*also* gradient approach), 255, 256, 270
Grazing population, 46
Groundwater, 11
Growth
    and sulphur dioxide, 133, 134
    effects of stress on, 49
    of algae, actual and apparent, 346
    rate, 36, 37, 46, 48, 334

Habitat type, 55
Hatching success, 37, 38
Hazard assessment, 55
Heat
    generation of, 13
    bil evaporation, 323
Heavy metals
    accumulation of, 42, 412
    and acidity, 14, 115, 117, 245
    and decomposer invertebrates, 120, 121, 373
    concentration gradients, 257
    contaminated streams, 266–267
    decomposition, 118, 128
    effect on productivity and respiration, 136–139
    effect on skeletal form, 33
    environmental levels of, 419
    index of stress, 70, 256, 267
    pathological effects, 31
    pollution, 193, 255, 257
        graded response, 256
Hepatic cells, 34
Hepatomas, 34
Herbicides, 82, 140
Heterogeneity, spatial and temporal, 82, 267, 268, 269

Heterotrophic development, 130
Heterozygosities, 49
Homeostasis
    nutrient effect on, 92
    of ecosystem, 1, 54, 103–105
Homeostatic mechanisms, importance of, 245
Homogeneity, environmental, 255
Horizontal patchiness, 88
Host-parasite interactions, 105
Humification and flooding, 301
Hydraulic regime, 300
Hydrocarbons
    aromatic, 322, 345
    biodegradation of, 357
    elimination of, 344–345
    in alginates, 331, 332
    in crabs, 357
    in oil, 321
    in scallops, 357
    level of, 332, 341, 343
    saturated, 321
Hydroelectric
    development, environmental effect of, 274, 278, 304, 380
    impoundment, nature of changes caused by, 304, 306
Hydrogen ion, deposition of, 193, 230, 231
Hypertrophy, of lamellar epithelium, 352–353
Hysteresis, definition of, 84, 91

Indicator
    organism concept, 65, 66
    species, 67, 251
    physiological, 66
Indices
    assessment of pollutant effects, 74, 75, 101, 144
    biological, 67–69
    cycling index, 145
    diversity and evenness, 72–76, 212, 224, 256
    for incidence of lesions, 349
    for necrosis, 349
    of community structure, 52, 54, 55, 426
    of dominance, 69, 71
    of erosion, 302
    of heavy-metal stress, 70, 256, 267
    of nutrient cycling, 145
    of welfare, 1
    qualitative, 65
    similarity, 75, 270
    taxonomic, 55, 265

SUBJECT INDEX

Individual characteristics, and responses, 23, 27, 425
Individuals, effect of metals on, 31, 32, 34, 35, 36, 369
Industrial effluents, 201, 202
Inertia
  assessment of, 88
  definition of, 87, 269
  of an ecosystem, 84, 87
Infectious hematopoietic necrosis virus, 47
Ingestion, copper effects on, 34
Inhibition, of metalloenzymes, 137
Insect communities, density, 261
Insecticides
  effect on predators, 383
  fenitrothion, 377
  organophosphorous, 30, 31, 143, 398
  physical and chemical properties of, 388
Instabilities, environment, 301, 304
Interactions
  biochemical, 28
  community, 241–242
  competitive, 47, 58
  ecological, 49, 52
  feeding, 45, 47
  importance of, 44, 193
  indices of, 29
  of key organisms, 61
  of pollutants, 40, 42
  pollutant-disease, 47
  population, 44, 58, 104
  predator–prey, 45
  trophic, 44, 58, 105, 143
Intermoulting period, 43
Interspecies competition, 47
Intertidal
  communities, resilience, 246
  pollutant gradients, 78
  systems, 94
Intraspecific competition, 47
Invertebrates
  and metal accumulation, 369
  decomposer effect of heavy metals, 373
  densities, 371
  diversity, 370
  effect of fenitrothion on, 384
  predators, role of, 242
  re-establishment of communities, 409
  surface dwelling, 125
Ion, concentration, 276
Irradiance, water column, 286, 289
Irradiation, chronic, 82
Isothermal, 292

Key populations, 49, 60, 61, 426

Lake
  acidity and metal contamination, 236, 250
  impoundment, 296
  sediment chemistry, 234
Lakes
  acidification of, 233, 241, 243
  clarity of, 236
Larval development, time of, 44
Leaching
  and pesticides and oils, 127
  effects of, 112
  of essential element, 108, 118, 134
  of nutrients, 111, 411
Lead
  and growth, 118
  and nutrient concentrations, 119
  effects on skeletal form, 33
Lethal response, 23
Liapunov Direct Method, 91
Life history, 43
Light
  conditions, and productivity, 301
  penetration, 296, 303, 305
  regimes, 284
Lipid solubility, 12
Litter
  accumulation
    and metals, 122–123, 371–373
    ecological significance, 372–373
  decomposition
    effect of metals on, 123, 137, 371, 372
    effect of pesticides on, 125–127
    nutrient cycling and, 107, 374
    particle size, 372
Liver
  hydrocarbon contamination, 357
  pathology, 34
Local releases, chronic, 8
Locomotor impairments, 30
Log-normal distribution, 78, 98–99, 213
Long-term trends, 17
Lysosomal membrane, effect of toxicants on, 32

Macrobenthic invertebrate
  and diversion, 287
  and flooding, 287
  recovery period, 302
  standing crop, 287–288, 305

Macrofauna, recovery of, 207
Macroinvertebrates
    communities and heavy metals, 57, 255, 260, 264, 276
    density, 56, 57, 261, 262
    diversity, 57, 73, 76, 264–265
    structural characteristics, 263
Macrophytes
    and acidity, 239–240
    biomass, 239
Malleability, 84, 91
Margalef index, 57, 74, 213
Marine, resources, 312, 315
Mathematical model
    of populations, 17
    of stability, 92
Mathematical procedures, 16, 17
Mercury
    bioaccumulation of, 46, 298, 300
    levels in fish, 297, 298, 300, 302
    methylation of, 11, 300
    organic, 12
    reduction in, 202, 203
Metabolic
    expenditures, and sulphur, 132
    index, 144
    processes, response to pollutants, 34
Metallothionein-like proteins, 26, 31
Metallurgical processes, 403
Metals (see also Heavy metals)
    accumulation in soil and litter, 367, 369
    and organic content, 261
    atmospheric input, 193
    behaviour, 30
    concentrations
        in invertebrates, 370
        in sediments, 234, 235, 258, 259
        in vegetation, 367
        profiles of, 234
    corticosteroid response, 32
    deposition of, 231, 232, 234, 366, 367
    effect of base additions on, 243
    effect on communities, 57, 121, 261, 264, 266–267, 369
    effect on osmoregulatory processes, 35
    effect on woodlands, 365, 368, 369, 373
    effect on zooplankton, 80, 238, 250
    effects of accumulation, 120
    factors affecting concentrations, 237
    form of and toxicity, 233
    non-ferrous smelting, 365
    polluted lakes, 29, 236, 250
    regulation of, 26
    reversibility of effects, 230
    symptoms of, 137
    tolerance to, 48, 237, 369
    transformation of, 11
Methylparathion, 45
Microbial
    activity and oil, 126
    community and metals, 120–121
    mineralization, 108
    re-establishment of communities, 409
Microorganisms, and pH, 116, 250
Migration, 142
Mine tailings
    rehabilitation program, 405, 409
    stabilization of surface, 408
Mineral cycling, 50, 103
Mining, 403, 405
Mites and decomposition of litter, 113, 372
Mixed function oxygenase, 31
Model
    studies, ecosystems, 18
    concentration–response, 24
Molluscs
    effect of oil on, 330, 359
    production, 316
    resource, 316
Monitoring systems, 22
Morphological effects, 32, 53
Moulting cycle, 43
Multifactor interactions, 41
Mussels
    effect of oil on, 48, 330
    production of, 317
Mutagenic effect, aromatic hydrocarbons, 245, 322
Mutations, 32

Natural
    bioassay, 64, 256
    changes, 55
    factors, effects of, 79, 224
    predation stress, 46
    variability, importance of assessment, 20–22, 48, 55, 425
    water exchange times, 290
Necrosis index, 349
Nitrate, imputs of, 233
Nitrification rate and acidity, 117
Nitrified effluent, production of, 199
Nitrogen
    atmospheric input, 230
    budget, 233
    cycle
        and acidification, 115, 116, 117
        and metals, 123

and pesticides, 127
fixation, 107
  and acidity, 116, 117, 132
  and oil, 140
  and polluted air, 66
loss, 91
regeneration, 107
Nitrogen–phosphorus ratio, 127
Number of
  individuals, 56–57, 209, 224
  phytoplankton genera, 237
  species, 58, 61, 209, 211, 224, 264, 265
Numbers, reduction in, 54
Numerical
  classification, 71
  indices of dominance, 69, 71
Nutrient cycling
  and ecosystem persistance, 91
  and population studies, 61, 103, 243
  disruptions of, 109, 128–129
  indices of, 145
  re-establishing, 409
  regulation of, 141–144
Nutrient
  availability, 106, 142, 373
  biogeochemical cycling of, 374
  concentration in plants, and metals, 137
  conservation
    and acid precipitation, 117
    and heavy metals, 118–119
    and pesticides and oil, 124
  dynamics, 131, 141
  factors affecting availability, 243
  leaching, 110, 112, 118–119, 127, 411
  pool, 373, 374
  recycling, 91, 92, 107, 427
  uptake, and heavy metals, 136
Nutrition, low pH effect on, 238
Nutritional value of foods, 43

Oil
  and specific gravity, 322
  aromatic hydrogen content, 322
  biodegradation of, 319–321, 323, 324, 326
  characteristics of, 320, 321
  contamination
    indicators of, 337
    of sediments, 29
  effect of, 124–130
  effect of constituents, 353
  effect of dispersants, 324, 326
  effect on algae, 327, 345
  effect on benthic invertebrates, 58

effect on crustaceans, 329
effect on feeding response, 30
effect on fish, 330
effect on growth, 37
effect on molluscs, 330
effect on mortality, 337
effect on productivity and respiration, 139–141
effect on scallops, 330
effect on seaweed, 327
genetic effects, 49
heavy crudes, 321
light fraction, 321
sedimentation of, 320
short-term contamination, 331
spills, 311
  agents used to treat, 313, 360
  amounts and sites, 318, 360
  and water depth, 219
  assessment of damage, 330, 360
  ecological impact, 194, 313, 359
  effect of wind, 323
  effect of plaice, 352
  effect on reproduction and growth, 38, 333, 349
  evaporation, 323
  factor affecting damage from, 313, 318
  kinetics of decontamination, 343
  medium and long-term consequences, 338
  pathological effect, 345
  recovery from, 48
  short-term consequences, 326
  time of year, 323
  type of oil, 360
toxic volatile fractions, 320
viscosity, 322
Oligotrophic, and acidity, 136
Opportunistic species, 49, 87, 94
Organic
  content of sediment, 261
  detritus and pH, 115
  enrichment, 56, 67
  waste, 66
Organochlorines, 47, 140
Organometallic compounds, analytic procedures, 11
Organophosphorous pesticides, 30, 31, 45, 143
Osmoregulatory processes, 35
Osmotic–ionic balance, 34
Oxygen
  consumption, 35

Oxygen (*continued*)
  depletion, 293, 294
  dissolved levels in water, 199, 201, 204, 208, 224, 291, 293
  load, 200
  profiles, 292
  solubility of, 204
  stratification patterns, 291
Oxygen–nitrogen ratio, as an index, 35
Oysters
  effect of oil on, 330, 335, 339, 342, 349
  hydrocarbon content, 332, 344
  linear growth, and oil, 336
  oil decontamination, 343
  production of, 317
  purification kinetics, 341
Ozone, 132

Paraffins, 321
Parameters, for analysis of community structure, 53–54, 209, 256
PCBs, 9, 10, 39
Pentachlorophenate, 127
Percentage similarity indices, 75, 80, 88
Permafrost
  effect of, 274
  melting and, 301, 302
Persistence of ecosystems, 91, 105
Perturbation
  analysis of physical, 194
  and stability, 83
  assessing effects on communities, 54, 426
  ecosystem response to, 50, 144, 246, 274
  maximum size of, 83
  pollution-induced, 86, 144
  recovery form, 88
Pesticide (killer), 46
Pesticide–pollinator interactions, 397
Pesticides
  biological accumulation patterns, 389
  characterization of, 386
  effect on productivity and respiration, 139, 140
  effect on reproduction, 39
  effects of, 124–130
  nature of formulation, 386
  persistance of, 389
  predator–prey, 45
  primary indicators, 392
  secondary mobility, 389
  sequestering patterns, 389
  transport, 11

  use of, 124, 380–382
pH
  and phytoplankton, 238, 250
  effect on pollutants, 42
  effect on species richness, 56, 62
  effect on taxonomic complexity, 70
  neutralizing of, 115
  of sediments, 235
  related coefficient of variation, 249
Phase
  portraits, ecosystem stability, 248
  space, use of, 145
Phases, of recovery, 94
Phenol, effect of, 45
Phosphorus
  and acidity, 117
  and phytoplankton, 286
  content in air, 389
  effect of addition, 244
  effect of sediment and light on, 306
  loading, 276
Photochemical oxidants, 66
Photodegradation of fenitrothion, 388, 391
Photolysis, 392
Photosynthesis
  and acid rain, 131
  and heavy metals, 136, 139, 241
  and oil, 140
  and organochlorines, 140
  and sulphur dioxide, 132, 133
  effect of pollutants on, 36, 66, 93, 130, 132
Photosynthetic
  activity, 34
  energy, 104
  process, 130
  14C fixation, 132–133
Physical
  changes, hydroelectric development, 280
  models, 245
  stabilization, rate of, 302
Physico-chemical data, variables input, 219–220
Physiological processes, toxic effects on, 27, 34–36
Phytoplankton
  and acidity, 56, 236, 238, 342
  and light, 286–287
  biomass, 45, 138, 139, 236, 237
  blooms, 45
  community, 250
  cyclic activity, 21
  effect of base additions on, 243

SUBJECT INDEX

effect of phosphorous addition on, 244
number of taxa, 56, 237
productivity of, 286
vertical distribution, 236
Pike, mercury levels, 298
Plaice
  effect of oil on, 352, 355
  histopathological anomalies, 352
  population structure, 336
  reproduction of, 336
Planktonic crustaceans, and water levels, 288
Plant
  community morphometry, changes in, 82
  growth, 131, 334–335
  productivity, and heavy metals, 136
  reproduction, and sulphur dioxide, 134
Plants
  impact of pollution, 47, 131
  tolerance to metals, 66, 369
Point spills, 8
Pollinators
  effect of fenitrothion on, 396
  unique function, 60–61
Pollutant
  effects
    assessment of, 25–28, 36, 48, 60–61, 72, 193, 268, 425, 427
    concentration-response model, 24
    indices of assessment, 74, 75
    on competitive interactions, 47
  fate of, 386
  gradient, 48, 64, 269, 270, 427
  input, 53, 197, 230–232
  perturbation, ecosystem response to, 50, 86, 144
Pollutant–disease interaction, 47
Pollutant–environmental interactions, 41
Pollutant–induced
  changes in age structure, 48
  discontinuities, 79
  genetic changes, 32
  interpretation of changes, 27, 28, 52, 77
Pollutant-sensitive
  factors, 131
  species, 49
Pollutants
  behavioural response, 27, 29
  bioaccumulation of, 26
  biochemical response, 27, 31
  combined effects, 42, 427
  ecosystem-level assessment, 101–103, 395, 425

effect on decomposition, 108, 116, 123, 127, 128, 130
exposure to, 29
partitioning of, 392
transfer of, 12
Pollution
  airborne, 47, 230
  distance from source, 64
  effect on dominance, 70
  effect on environment, 13
  effect on food supply, 45, 46, 143
  effect on life stages, 40
  effect on metabolisms, 35
  effect on photosynthesis, 36, 131–132
  evaluation, 81
  impact on plants, 131
  influences, relative effects, 224
  interpretation of initial response, 426
  subacute and chronic effects, 26–27, 392
  Thames estuary, 197
  thermal, 2
Population
  changes in, 48
  characteristics, 27, 425
  extinctions, impact of, 62
  fluctuation in density, 94
  gene pool, change in, 48
  grazing, 45, 46
  interactions, 44, 47, 245
  microbial growth, 58
  regulation, 94
  response extrapolation, 49
  variation, measurement of, 86
Populations, resistant, 49
Potential energy, of a river, 301
Power, definition of, 144, 145
Predation
  natural stress, 46
  selective, 142
Predator
  and parasite populations, 141
  populations, reduction of, 45
Predator–prey, changes in relation, 45, 49, 143
Predators, effect of insecticides on, 383
Predispose, 42
Preferred prey, 49
Presence–absence data, 79
Prey
  abundance and pollution, 45
  reduction in, 46

Primary
  production
    and heavy metals, 136, 139, 241
    and impoundment, 286, 292, 302, 303, 305
    effect of nutrient availability, 373
    ratio to respiration, 139
    shifts in, 241
    variation of rates, 276
  productivity, 36, 105, 130
    and sulphur dioxide, 133
    effect of pesticides and oil, 139–141
Principal Component Analysis, 71, 81, 217, 220, 224
Productivity
  to biomass ratio, 92
  and stress, 130
  ecosystem, 49, 107, 130, 138
  terrestrial, 131
Protective standards, 89
Pulp mill effluents, benthic fauna, 56
Pyrrhotite, 419–420

Quantitative studies, 82

Radiation
  damage in plants, 82
  gamma, 88
Radioactive wastes, 10
Radioactivity, 9
Rainfall, chemical composition of, 232
Reciprocal averaging, 71, 81
Recolonization, 95
Recovery
  ability, 269
  factor in, 88, 89, 93
  from oil pollution, 140
  initial stages of, 91
  of ecosystems, 93, 95–99, 243
  of fish, 209, 251
  of Thames Estuary macrofauna, 207, 209
  phases of, 94
  prospects for, 243
  structural characteristics, 96, 98
  time, 84, 89
  to reference structure, 91, 269
Recycling, definition of, 145
Redundancy, 102
Regulation
  at physico–chemical level, 142
  disruption of, 143
  of energy, 143
  of nutrient cycles, 141–144

of pollutants, 26
of trophic functions, 103
Regulators, impact of, 141
Reproduction
  altered performance, 37–40, 43
  reduced juvenile settlement, 39
Reproductive
  behaviour, toxicology of, 38
  rate, 39
  success, 37, 47, 49
Resident communities, 52
Residual toxicity, 93
Resilience, definition of, 84, 246, 269
Resistance, to toxicity, 94
Resources
  exploitable marine, 315
  impact of oil on, 338
Respiration
  effect of pesticides and oil, 139–141
  effect of polluted air, 66
  physiological index, 35
  rates, and sulphur dioxide, 132, 133
Response
  analysis of frequency, 145
  threshold, 24
Restabilization, shoreline, 305
Restoration time, 88, 89
Retrogression, 92, 93, 109
Revegetation, programme, 408, 410
Richness
  indices, 72
  of species, measurement of, 213
River, diversion, 287–288, 303, 306
Root development, inhibition of, 137
Root-mycorrhiza, and acidity, 116
Rotenone, 45
Rotifer community, changes in, 239

Salinity, with cadmium, 41
Salmon, 205
Samples, required number, 19
Scallops
  and hydrocarbons, 330, 332, 357
  indicators of oil contamination, 337
Scope-for-growth index, 36
Sea trout, 205
Seasonal
  cycle, 211
  fluctuations, 55, 212, 223, 224
  influences, 48, 270
  variation of copper concentration, 261
Seaweed
  biomass, 329, 348
  brown and red, resource, 315

growth of, 316, 347
necrosis, 328
oil damage, 327
Sediment
and macrobenthos, 288
and metals, 235, 243, 258, 259
budgets, 284
chemistry, 233, 234
oil contamination of, 357
organic content, 261
pH of, 235
pollutant adsorption, 234, 256
suspended levels, 293, 305
Sedimentation
and erosion, 300, 302
of oil, 320
Sensitive species, 65
Sensitivity
key populations, 49
of shifts in composition and dominance, 70
property of stress index, 54, 55
to toxic chemicals, 43
Sensory organs, damage of, 33
Sensory perception, 30
Sequential comparison index, 77
Seral recovery, 99
Sewage
effluents, 198, 224
treatment, 200
Sexual maturation, 38
Sexually sterile, 38
Shannon diversity index, 72, 73, 77, 212, 256, 257, 264–266
Shell abnormalities, 33
Shoreline
erosion, 281, 292, 304
restabilization, 306
Similarity indices, 75, 79, 81, 99, 270
Simpson's indices, 74–76
Size, distribution, 56
Skeletal deformities, 33
Slumping, 301
Smelting, 365
Soil, composition, 404
Solar radiation, 130
Sole
effect of oil on population, 354, 355
growth of, 336
Spatial
changes, 79
heterogeneity, 82, 88, 245, 267–269
structure
aquatic, 82
terrestrial, 82
variability, 267
Spawning
activity, 209
and water level, 303
conditions of, 302
Specialists, 71, 94
Species
abundance, 61, 88, 224
assemblages, 65
characteristics, stress, 89
composition, effect of phosphorus addition on, 244
diversity
changes in, 77, 212, 245, 266, 305
indices of, 74, 224
extinction, 58, 59
identification of, 77
interactions, 88
intraspecific, variability in, 89
number of, 72, 139, 211, 264, 265
opportunistic, 87
richness
and diversity, 264–266
and pollution, 62, 64, 81, 209, 213, 224, 238, 239, 265, 266, 269
gamma radiation, 88
tolerant and sensitive, 65
Specific gravity, and oil, 322
Spiralling, nutrient, 145
Spruce budworm
function of, 377
lifecycle of, 378, 379
problem of, 378
Stability
and variability, 269
components affecting, 245
concept of, 83, 84, 86, 245
of ecosystems, 83
of populations, 269
relative, 92
sensitivity to, 91
short-term, 1
Stabilization, physical, 30
Stable points, 84
Stable state, 91
Stage, of community succession, 70
Stages, life, 43
Statistical
analysis, 15, 17–20
measurement of population variation, 86
methods, for comparing communities, 71

SUBJECT INDEX

Stream
  ecosystems, changes in, 270
  zone index, 67
Stress
  amplitude, 89
  assessment, 55, 61
  chemical, and population interactions, 44
  detecting severity of, 265
  effect on ecosystem indices, 426
  effects and tissue levels of pollution, 28
  response, 28, 103
  types of, 24
Structural
  changes, 52, 55
  characteristics, 51, 96, 103, 269, 279
    and copper, 260
    measure of, 54
    of macroinvertebrates, 263
  indices, 52, 54, 257, 270, 426
    assessing changes in, 255
    variability in, 87
  patterns, changes in, 256
  responses, to toxic pollutants, 52
  simplification, 64
Structure, analysis of community, 52, 209, 269
  spatial, 82
Sublethal response, 23
Succession, characteristics of, 92, 131
Successional, patterns, 93
Sulphur
  cycle, and acidity, 117
  dioxide, effect of, 66, 132–134
  gas, effects of, 21
Sulphate, deposition of, 231, 232
Sulphide ores, 403
Surface aeration, 203
Surface-active agents, 324–325
Surfactants, anionic, 324–325
Survival probability, effect of stress on, 49
Symbiotic interactions, 105
Symptoms, of metal toxicity, 137
Synergistic interactions, of pollutants, 42, 395
Synthetic detergents, 203

Tailings, mine, 405
Taxonomic
  composition, sensitivity to pollutants, 70
  indices, 55, 256, 265
  problems of identification, 77

Temefos, 30
Temperatures, low winter, 41
Temporal
  changes, 79
  heterogeneity, 267, 268, 269
  variability, 87, 267
Teratogenic effects, 33
Terrestrial
  acidification, 14
  ecosystem theory, 104, 107
  oil-pollutant effects on, 140
  plant communities, changes in, 80
  plants
    and stress, 131
    and sulphur dioxide, 133
  succession, 92
Thames Estuary, study on, 195–228
Thermal
  changes in structure, 291
  gradient, 286
  pollution, 204
  regimes, 284
  stratification, 29
  structure, 285–286, 291
Thermocline, 291
Threshold
  of metals, 26
  of response, 24
Tolerance
  limits, 89
  selection of, 48
  to low pH, 240
  to metals, 48, 237, 240
Tolerant species, 65, 92
Toxic
  materials, accumulation of, 144
  stress, 23, 28
  substances
    effects of, 23, 24, 66, 194
    transfer of, 46
  volatile fractions, of oil, 320
Toxicity
  characterization of pesticides, 386
  residual, 93
Toxicology of reproductive behaviour, 38
Transformation, processes of, 142
Transition zone, 93
Translocation, processes, 142
Trophic
  altered relationships, 304
  diversity, 54
  dominance, 71
  level, and biological changes, 44, 301, 303

SUBJECT INDEX 443

levels, energy transfer and acidification, 242
patterns, 49
regulation of functions, 103
structure, 70
Tumour incidence, 34
Turbidity, 294, 301

Ulcers, in fish, 47
Uncertainty analysis, 18

Variability
  cyclical and natural, 55
  intraspecific, 89
  of structural parameters, 269
  spatial and temporal, 21, 87, 267
Vegetation
  climax, 404
  program for revegetation, 408
Ventilation rates, 35
Vertical stratification, 88

Water
  balance, 36
  chemical composition of, 233
  column irradiance, 285, 286
  decline in level, 297
  depth and oil spills, 319
  exchange times, 290
  flow, and hydroelectric utility, 300
  improvement of quality, 209
  level, and fish production, 303
  levels, and plankton abundance, 302
  levels, and thermal stratification, 291
  quality, 196, 197, 217
Water-in-oil emulsions, 321, 323
Water–salt balance, 35
Wave energy, and erosion, 282
Whitefish
  abundance, 295, 296–297, 305
  changes caused by lake impoundment, 296
  mercury levels, 298
  stocks, 296
Whittaker's coefficient of community, 269
Woodlands, and metals, 365

Xenobiotics, biotransformation of, 31

Zooplankton
  abundance, and water levels, 280, 289, 302
  and acidity, 238
  and metals, 250
  biomass, 238, 249
  community interactions, 242
  diversion and impoundment, 288, 305
  effect of base additions on, 244
  natural fluctuation of, 21
  quantitative response, 302